Collins

ADVANCED MATHEMATICS

Statistics

Roger Fentem

The College of St Mark and St John, University of Exeter

SERIES EDITOR

John Berry

Centre for Teaching Mathematics, University of Plymouth

CONTRIBUTING AUTHORS

Steve Dobbs

Penny Howe

John White

PROJECT CONTRIBUTORS

Bob Francis, Ted Graham,

Howard Hampson, Rob Lincoln,

Sue de Pomerai, Claire Rowland,

Stuart Rowlands, Stewart Townend

Roger Fentem is retired and living in Cornwall. He was previously a Principal Examiner for the University of Oxford Delegacy of Local Examinations in Statistics at A Level and the Head of Maths at the College of St Mark and St John, a college of Exeter University.

Steve Dobbs teaches at Torquay Boys' Grammar School in Devon and is Deputy Headmaster of Curriculum there.

Penny Howe teaches Mathematics and Statistics at David Game College in central London. She is an Examiner for the Oxford Delegacy for Statistics and for Edexcel.

John White is Honorary Secretary of the Association of Teachers of Mathematics (ATM) and previously a secondary mathematics consultant for Swindon.

John Berry is Professor of Mathematics Education at the University of Plymouth and has acted as the Director of their Centre for Teaching Mathematics. As well as being research-active, John is the Mathematics Professor in Residence at Wells Cathedral School in Somerset. John is a consultant to the National Academy of Gifted and Talented Youth delivering workshops to members of The Academy and to primary pupils as part of the SWGate programme. He delivers CPD courses for teachers nationally and internationally and visits schools to run workshops on the teaching and learning of Mathematics. In addition, John also leads a team of teachers delivering a programme of GCSE and A level revision student workshops for TEACHERS FIRST.

William Collins' dream of knowledge for all began with the publication of his first book in 1819. A self-educated mill worker, he not only enriched millions of lives, but also founded a flourishing publishing house. Today, staying true to this spirit, Collins books are packed with inspiration, innovation and practical expertise. They place you at the centre of a world of possibility and give you exactly what you need to explore it.

Collins. Do more

Published by Collins
An imprint of HarperCollins*Publishers*
77-85 Fulham Palace Road
Hammersmith
London
W6 8JB

Browse the complete Collins catalogue at www.collinseducation.com

---

First published in 1996
This edition published in 2011

ISBN 978-0-00-742904-2

10 9 8 7 6 5 4 3 2 1

British Library Cataloguing in Publication Data
A Catalogue record for this publication is available from the British Library

Designed and illustrated by Ken Vail Graphic Design, Cambridge, UK
Project editor Joan Miller
Cover design by Julie Martin
Cover photograph iStock © Evgeny Kuklev
Additional illustration Tom Cross
Printed and bound by L.E.G.O. S.p.A. - Italy

STATISTICS

Contents

Acknowledgements

We are grateful to the following Examination Groups for permission to reproduce questions from their past examination papers and from specimen papers. Full details are given with each question. The Examination Groups accept no responsibility whatsoever for the accuracy or method of working in the answers given, which are solely the responsibility of the author and publishers.

Associated Examining Board (*AEB*)
Edexcel Foundation, London Examinations (formerly *ULEAC*)
Northern Examinations and Assessment Board (*NEAB*); also School Mathematics Project, 16–19 (*SMP 16–19*)
Oxford and Cambridge Schools Examination Board (*OCSEB*); also Mathematics in Education and Industry (*MEI*)
Scottish Examination Board (*SEB*)
University of Cambridge Local Examinations Syndicate (*UCLES*)
University of Oxford Delegacy of Local Examinations (*Oxford*); also Nuffield Advanced Mathematics (*Nuffield*)
Welsh Joint Examinations Council (*WJEC*)

Every effort has been made to contact all copyright holders. If any have been inadvertently overlooked the publisher would be pleased to make full acknowledgement at the first opportunity.

Preface

discovering advanced mathematics

Mathematics is not just an important subject in its own right, but also a tool for solving problems. Mathematics at A-level is changing to reflect this: in the course of your A-level work, you must study at least one area of the *application* of mathematics. This is what we mean by 'mathematical modelling.' Of course, mathematicians have been applying mathematics to problems in mechanics and statistics for many years. But now, the process has been formally included across A-level.

A second innovation is the inclusion of the mathematics of uncertainty – looking at data and probability – in all maths A-levels. In this book, we build on the work done at GCSE and show how important this topic is in modelling. The third innovation is the recognition that numerical methods play an important part in mathematics. This is much easier to study now that we have programmable calculators and computers.

Technology is advancing as well. Hand-held calculators that can produce graphs and even do simple algebra are revolutionising the subject. The Common Core for A-level expects you to know how to use appropriate technology in mathematics and be aware that this technology does have its limits.

We have written *discovering advanced mathematics* to meet the needs of the new A- and AS-level syllabuses and the Common Core for mathematics. The books provide opportunities to study advanced mathematics while learning about modelling and problem-solving. We show you how to make best use of new technology, including graphics calculators (but you don't need more than a good statistical calculator to work through the book).

In most chapters in this book, you will find:
- an introduction that explains a new idea or technique in a helpful context;
- plenty of worked examples to show you how the techniques are used;
- exercises in two sets, A and B (the B exercises 'mirror' the A set so that you can practise the same work with different questions);
- consolidation exercises that test you in the same way as real exam questions;
- questions from the Exam Boards.

We hope that you will enjoy advanced mathematics by working through this book.

Roger Fentem,
John Berry
September 1996

[illegible]

[illegible] described as quantitative. Such numerical information may be [illegible] into **continuous data** and **discrete data**.

STATISTICS

1

Data

Statistics is about gathering, communicating, analysing and interpreting data. In this chapter we begin the story by:

- *introducing various types of data*
- *presenting data in the form of bar charts, pie charts, frequency line diagrams, stem-and-leaf displays, histograms, frequency density polygons and curves.*

INFORMATION AND DATA

Statistics is all about gathering, communicating, analysing and interpreting **information**. The information is obtained from various sources, and in its basic form it is called **data**. There are two distinct types of data, which are **qualitative** and **quantitative**.

Qualitative data

Qualitative information, as the term suggests, is descriptive information. It may be expressed as a property, or in terms of categories, or as a quality for example: colour, gender, nationality, style, ...

Where the categories are well defined and distinct, the information may be described as **discrete**. For instance, the make of a calculator falls into exactly one of the following categories: Canon, Casio, Sharp, Texas, Hewlett Packard, Other. There is no inherent reason why Casio should always precede – or indeed should never follow – Texas. There is no natural hierarchy, no inbuilt order within these categories.

Quantitative data

Information that has a natural hierarchy, such as data in numerical form, is described as quantitative. Such numerical information may be separated into **continuous data** and **discrete data**.

When measurements are made, the data gathered are generally continuous since all values within a particular span are possible. An example of this is the length of a person's stride, which might be any length between, say, 0.3 m and 1.3 m.

Discrete data are usually obtained by counting. For example, the number of letters delivered daily to a house might be 0, 1, 2, 3, 4, 5, ... or any other **integer** but it cannot be any other sort of number. Discrete data may also arise when there is an order in the information, such as the sizes available in some clothes – large, medium or small.

Many brands of T-shirts are available in five sizes: XL, L, M, S, and XS; these sizes are discrete data. British shoe sizes are numbered as:

$$\ldots 4, 4\tfrac{1}{2}, 5, 5\tfrac{1}{2}, 6, 6\tfrac{1}{2}, 7, 7\tfrac{1}{2}, \ldots$$

A size 8 shoe is designed to fit a foot of length l inches where:

$$10\tfrac{1}{3} \le l < 10\tfrac{2}{3}.$$

Think about the nature of the shoe size and foot length data.

Exploration 1.1

Types of data

Categorise the following information into:

a) qualitative data,
b) discrete numerical data,
c) continuous numerical data.

- height of students
- area of noticeboards
- type of footwear
- age of teachers
- colour of flowers
- make of car
- number of open windows in a room
- number of items of jewellery worn

Limits

Age

When asked her age, a student said, 'I am 17 years old.' Does she mean that she is precisely 17? Or, is she 17 to the nearest whole year? Or, does she mean something else?

The way in which ages are recorded means that, if this student's age is x, then $17 \le x < 18$.

The student will have given her age as 17 from the day of her 17th birthday (i.e. the day precisely 17 complete years since she was born) up to the day before her 18th birthday.

Nearest

Measurements are frequently recorded 'to the nearest cm', or 'to the nearest kg' or 'to the nearest minute'. The length of a pencil that is actually 12.6 cm long would be recorded as 13 cm to the nearest cm, and one of actual length 12.49 cm would be recorded as 12 cm to the nearest cm. Pencils for which the lengths are recorded as 12 cm could have a length, p cm, where $11.5 \le p < 12.5$.

Frequency

When collecting information, for instance the colour of cars in a car park, there will be repeated examples of particular colours. There may be four yellow cars, 13 red cars, eight blue cars and 20 cars of other colours. The information is qualitative. The number of cars of each colour is the **frequency of observation** of that item of information.

Inevitably in gathering any information there will be a collection of frequencies associated with the items of data. The frequencies are not the data, they tell us something about the **distribution** of the data.

Presentation of data

Qualitative data may be presented in a **frequency table** such as the one below.

Colour of car	Number of cars observed
Yellow	4
Red	13
Blue	8
Others	20

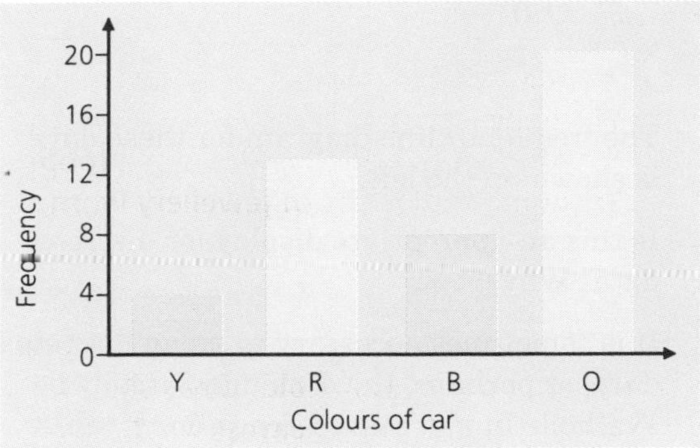

A pictorial presentation may make the data easier to interpret. **A block graph**, or **bar chart** is shown on the left. The height of a column is proportional to the frequency of the separate quality it represents.

Y	yellow
R	red
B	blue
O	other colour

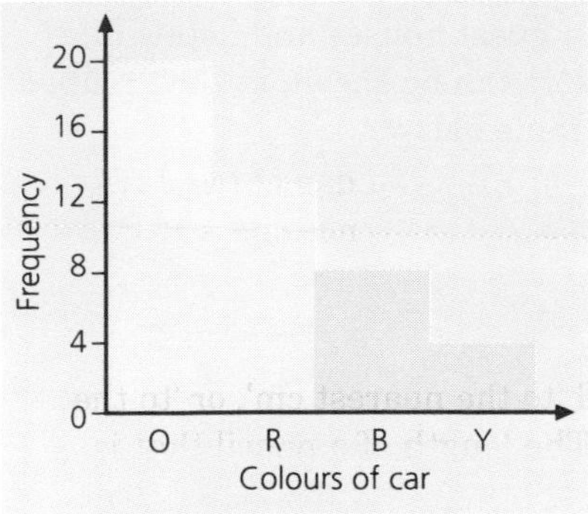

Each column should have the same width, the axes should be labelled but there is no need to present the data in the order given. It would be equally acceptable to present these data as shown on the left, here.

O	other colour
R	red
B	blue
Y	yellow

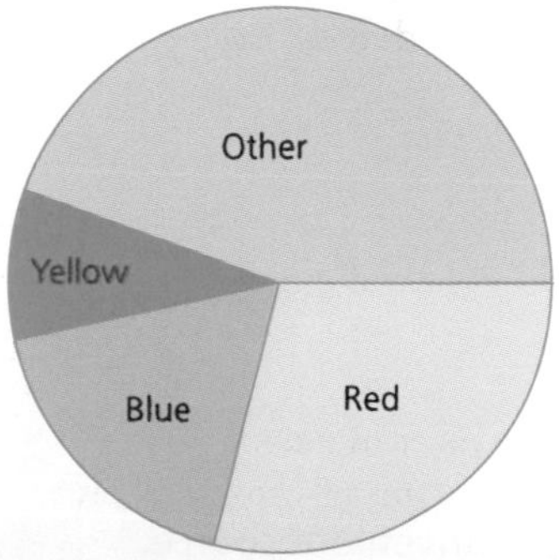

A **pie chart** is another effective way of displaying data of this nature.

A circle is divided into sectors, the angles of which are proportional to the frequency of the quality represented. In this case the angles are as shown.

other colour: $\frac{20}{45} \times 360° = 160°$

red: $\frac{13}{45} \times 360° = 104°$

blue: 64°

yellow: 32°

Note that someone looking at the pie chart is likely to think that the **area** of the sector represents the frequency of observation of the quality concerned, rather than the angle. Is this a fair representation? Why?

Discrete data may also be displayed in a **frequency line diagram**. Consider the following data where the number of people in a household was recorded for all the 93 houses in an estate.

Number in household	Frequency
1	18
2	33
3	21
4	15
5	4
6	1
7	0
8	1

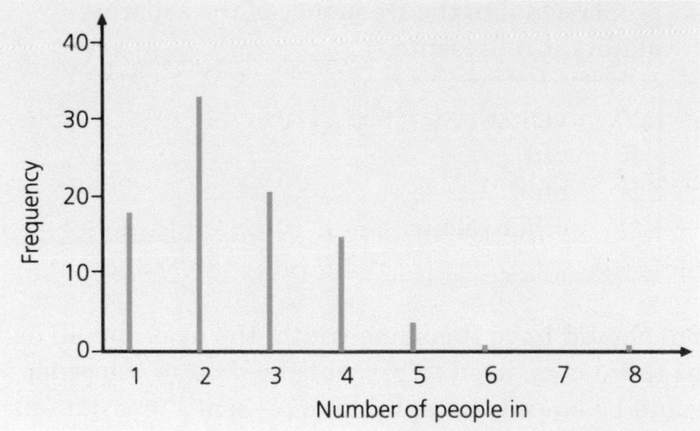

The frequency line diagram for these data is shown on the left.

Is this an appropriate display for discrete data? Why?

It is sometimes necessary to group discrete data, or perhaps, the data may already be available in a **grouped frequency table**. Data concerning the number of bedrooms available in guest houses and hotels in a seaside resort can be shown in the grouped frequency table below.

Number of bedrooms	Frequency
1–5	42
6–10	21
11–15	12
16–20	8
21–25	4
26–30	3
31–35	1
36–40	1
> 40	0

Exploration 1.2 *Collecting different types of data*

Identify and collect examples of:

- qualitative
- discrete
- continuous data from a car park or other convenient source.

EXERCISES

1.1A

1 Collect data on a range of topics, such as number of subjects, age of students, weight of students, colour of hair, make of calculator and number of desks. Categorise the data into:

a) qualitative data,
b) discrete numerical data,
c) continuous numerical data.

2 For each data set in question 1, construct a table, bar chart, line diagram or pie chart, choosing whichever of these is most appropriate.

Justify your choice of presentation.

3 A hotel manager used a questionnaire to ask diners how they rate the atmosphere, service, food quantity, food quality and prices in the hotel's restaurant. Each characteristic is rated on the scale Outstanding (O), Very Good (V), Good (G), Average (A), Poor (P).

The following are the ratings for food quality awarded by a random sample of 60 diners.

G O V V O V O V G G V O P O P V V O G G V A G O

V O G V O O O G A V V A V V G V G V G A A G A P A

A G A O V V G G A V V

Summarise these data in an appropriate table. What conclusions can you draw about the apparent quality of the food in the restaurant?

Illustrate the data in a diagram.

4 A census of the households in a village results in the following family sizes being recorded.

3 4 5 3 4 2 2 4 3 4 3 6 5 3 4 4 5 2 1 3 2 1 4 4 3 1 2 1 3 2 4 1

3 5 4 3 4 5 6 2 1 5 6 3 4 2 5 1 2 1 3 1 4 4 5 3 4 2 1 3 4 3 1 4

4 5 4 3 2 2 3 4 2 4 2 2 1 3 4 3 5 6 1 4 2 3 2 3 2 1 1 2 3 5 2

Construct an appropriate tabulation of these data and describe what they reveal.

Ilustrate the data in an appropriate diagram.

5 Mrs Smith decided to check the amount of post she received per day. She recorded the number of envelopes that were on her doormat each day for 120 days and the results were as follows.

Number of envelopes	0	1	2	3	4	5
Number of days	15	31	20	16	30	8

Summarise these data in an appropriate diagram.

6 In a survey into viewing habits, 50 TV owners were asked to name the channel which they had watched for the majority of time on the previous evening. The results are shown in this table.

Channel	Frequency
BBC1	17
BBC2	6
ITV	19
Channel 4	3
Satellite	5

a) Draw a diagram to illustrate this information.

This survey had also been done previously with the same 50 viewers two years ago, when the results were as shown in this table.

Channel	Frequency
BBC1	21
BBC2	5
ITV	17
Channel 4	6
Satellite	1

b) Draw a diagram to represent this data. Compare and contrast the two sets of results.

EXERCISES

1.1B

1 State whether the following data are discrete or continuous.

a) the daily rainfall at a particular location
b) the monthly numbers of telephone calls
c) the numbers of defective fuses produced
d) the duration of long-distance telephone calls
e) the ages of students in a class

2 Classify the following as qualitative or quantitative, and discrete or continuous.

a) gender (M/F)
b) height (cm)
c) GCSE grades in Statistics (A to G)
d) examination mark in Statistics (%)
e) blouse size (8 to 18)
f) car owner (0 = No, 1 = Yes)
h) weekly self-study time (% of 30 hour norm)

3 The following cars were seen in a car park. (B = British, F = French, G = German, I = Italian)

F B B B G I I B F F B F G F F I
B B G B I B B G I B G B G B F B

Construct a frequency table for these data and draw a bar chart and a pie chart.

4 Mandy was asked by her maths teacher to do a statistics project. She represented her school at hockey, so she decided to record the number of goals scored per match by her team over a season of 35 matches. She recorded her results in this table.

Number of goals	0	1	2	3	4
Number of matches	12	13	7	2	1

Display the data in an appropriate diagram.

Does this tell you anything about how successful Mandy's team was?

5 Discuss the kind of data for which a bar chart might be a clearer representation than a pie chart, and the kind for which the reverse might be true.

6 Thirty-eight people were asked the colour of their eyes and the responses were as follows.

Blue	14
Green	11
Hazel	9
Brown	3
Grey	1

a) Draw a diagram to show these data.

The person doing the survey had previously recorded what she thought the colour would be for each person, with the following results.

Blue	12
Green	11
Hazel	7
Brown	6
Grey	3

b) Illustrate these data with a similar diagram and make any appropriate observations about the two sets of results.

STEM-AND-LEAF DISPLAYS

A method of presenting data, using the items of data themselves, which provides a swift and convenient way of collecting, storing and displaying numerical data is known as a **stem-and-leaf display**. Suppose the items of data are:

274, 286, 345, 402, 193, 237, 318.

Each item has a **most significant part** and a **next most significant part**. For instance, the most significant part of 274 may be the 2 since it represents 200; then the next most significant part is the 7. The **stem** of the stem-and-leaf display is constructed from these significant parts. The stem acts as a vertical scale, so for the above data, the stem would be shown like this.

1	
2	
3	stem
4	

The **next most significant** part of the item of data becomes a **leaf** along an appropriate branch of the display.

1		
2	7	leaf from the item 274
3		
4		

along the branch defined by the 2 in the stem.

The collection of seven items of data could be represented by the stem-and-leaf display.

1	9		
2	7	8	3
3	4	1	
4	0		

Only the most significant and next-most-significant parts of the data are stored in the display. No attempt is made to round the data, the presentation is done by **truncating** the items, the 4 of 274 is lost. Thus, the 2-branch would contain any item in the range 200 to 299. Truncation is a feature of this method of presenting data.

Unit

At this stage, there is no indication of the magnitude of the displayed data. To remedy this it is necessary to indicate the unit of the display.

Unit is 10

1	9		
2	7	8	3
3	4	1	
4	0		

In this case the unit is 10 because a **factor of 10** is needed to scale a displayed item such as 27 to its true magnitude.

Grouped frequency tables

A stem-and-leaf display can be used to build up a **frequency table**. The number of leaves on any branch represents the frequency of occurrence of the items of data.

The following grouped frequency table can be constructed from the display above.

Class	Frequency
$100 \le x < 200$	1
$200 \le x < 300$	3
$300 \le x < 400$	2
$400 \le x < 500$	1

Exploration 1.3

Stem and leaf displays

Display the following collections of data in appropriate stem-and-leaf displays indicating the unit in each case.

a) 4818, 1212, 3425, 4478, 5891, 3257, 2176, 4665, 2217, 4119

b) 0.320, 0.044, 0.278, 0.413, 0.456, 0.163, 0.389, 0.418, 0.058

c) 8.2, 3.2, 3.0, 2.8, 1.5, 6.6, 5.3, 1.5, 8.9, 7.7, 4.0, 2.8, 5.6, 6.3, 5.7, 7.4

d) 894, 947, 904, 933, 906, 920, 916, 941, 899, 925, 935, 924, 947, 937, 916, 909

e) 0.013, 0.026, 0.014, 0.019, 0.023, 0.008, 0.013, 0.018, 0.007, 0.028, 0.034, 0.038, 0.006, 0.041, 0.028

Draw up frequency tables corresponding to the stem-and-leaf displays you have constructed.

Developing the stem-and-leaf display

The data in (d) of the above exploration may be displayed using a unit of 10, like this.

Unit is 10

8	9	9												
9	4	0	3	0	2	1	4	2	3	2	4	3	1	0

However, it is also possible to consider the most significant part of the data item 894 to be 89, then the leaf part would be 4.

Doing this to all the data leads to the following display (overleaf).

Unit is 1

89	4	9	
90	4	6	9
91	6	6	
92	0	5	4
93	3	5	7
94	7	1	7

This creates a somewhat different impression of the distribution of the data. Which is the more appropriate?

Exploration 1.4

Appropriate stems

Choose appropriate stems and display each of the following collections of data.

a) 0.98, 1.13, 1.31, 0.95, 1.13, 1.21, 0.96, 1.11, 0.88, 1.04, 1.17, 1.09, 1.16
b) 895, 1080, 888, 1152, 1035, 1041, 1162, 1082, 948, 920, 987
c) 10.02, 10.01, 10.00, 10.01, 10.02, 9.98, 10.03, 9.99, 9.98
d) 348, 342, 354, 340, 385, 327, 385, 358, 351, 330, 346, 358, 344, 351, 340, 353, 382

Two-part and five-part stems

Two-part stem

The sort of technique explored above is not always the most appropriate. Consider the data shown in the following stem-and-leaf display.

Unit is 1

0	4	3	7	0	6	7	3	1	0	8	2	3	7	3	3	1
1	1	5	1	8	3											
2	2															

The data span the range 0 to 22 and choosing a different 'most significant part' is not realistic. However, it is possible to split each stem into two equal parts – **lower** and **upper** – denoted by '–' and '+' respectively. Then the two-part stem display of the data looks like this.

0–	4	3	0	3	1	0	2	3	3	3	1
0+	7	6	7	8	7						
1–	1	1	3								
1+	5	8									
2–	2										

Note that the lower part contains the 0, 1, 2, 3, 4 leaves and the upper part contains the 5, 6, 7, 8, 9 leaves.

The same decay is still seen in the distribution of the data.

Using a two-part stem is rather like taking a magnifying glass to the data.

Five-part stem

On occasions we may want to tighten up the width of values of the data as shown by a simple stem-and-leaf display. Consider the following data.

0.9, 1.7, 2.7, 3.1, 3.4, 4.2, 4.8, 5.7, 5.6, 6.3, 6.2,
6.9, 7.4, 7.3, 8.4, 8.7, 9.8, 10.9, 11.5, 13.2, 15.4

Here is one possible stem-and-leaf display.

Unit is 0.1

0	9		
1	7		
2	7		
3	1	4	
4	2	8	
5	7	6	
6	3	2	9
7	4	3	
8	4	7	
9	8		
10	9		
11	5		
12			
13	2		
14			
15	4		

It is rather sparse, perhaps hinting at a degree of symmetry in the data or perhaps suggesting that the data are evenly distributed. Telescoping the display may help in gaining an appropriate view of the data. This is achieved by combining pairs of stems thus halving the number of 'branches'. For instance, the data: 2.7, 3.1, 3.4; are displayed on a **T**-stem (T being common to **T**wo and **T**hree). This would appear as:

T	2	3	3

Note that the data has had to be further truncated. Similarly, 4.2, 4.8, 5.7, 5.6 are displayed as:

F	4	4	5	5

(F being common to **F**our and **F**ive). The full five-part stem display is:

Unit is 1

O	0	1			
T	2	3	3		
F	4	4	5	5	
S	6	6	6	7	7
E	8	8	9		
1O	0	1			
1T	3				
1F	5				

(O representing Zero and One; S representing Six and Seven, E representing Eight and Nine). Then the underlying symmetry of the data becomes apparent.

Exploration 1.5

Truncated stem-and-leaf diagrams

■ Use a five-part stem to display these sets of data.

a) 4.9, 0.1, 1.2, 5.0, 0.2, 3.2, 0.3, 4.7, 1.9, 1.3, 1.5, 2.9, 2.6, 0.9, 7.2, 6.9, 9.9, 1.8, 1.8, 4.0, 14.1, 0.9, 5.7, 3.9, 3.3, 6.1, 1.7, 8.2, 7.5, 4.1, 2.6, 3.5, 2.9, 1.1, 3.0, 1.2, 3.8, 0.6, 1.7, 2.3, 5.1, 2.6, 9.2

b) 14, 13, 12, 12, 10, 22, 19, 11, 13, 11, 10, 10, 25, 11, 11, 16, 15, 20, 10, 14, 12, 14, 21, 13, 11, 10, 21, 23, 20, 23, 11, 15, 15, 10, 10, 12

■ Use a two-part stem to display these sets of data.

c) 26, 39, 42, 19, 29, 10, 5, 58, 25, 37, 53, 18, 14, 58, 26, 14, 36, 21, 22, 22, 8, 24, 19, 5, 25, 24, 44, 39, 1, 21, 33, 53, 41, 1, 3, 10, 9, 43, 40, 23, 15, 8, 27

d) 2.4, 4.4, 3.9, 0.1, 2.1, 3.3, 5.3, 4.1, 6.7, 0.1, 0.3, 1.0, 0.9, 4.3, 4.0, 2.3, 1.5, 0.8, 2.7, 2.6, 0.8, 2.2, 2.2, 5.0, 2.0, 3.1, 0.1, 2.7, 3.9, 3.0, 0.1, 3.4, 2.8, 0.6, 3.1, 1.7, 4.0

Histograms

The data shown in the display records the hand-spans, in centimetres, of a group of 55 children.

Unit is 0.1

12	0
13	
14	5
15	2 6
16	0 1 2 3 4 7
17	0 2 5 9
18	1 2 4 5 6 6 7 8
19	0 2 2 4 5 5 6 8 8 8 9
20	0 2 4 5 6 7 8 9
21	2 3 6 6 7
22	0 1 2 8
23	
24	2
25	
26	2 5 9

It is clear from this display that there was no child whose hand-span fell between 13 cm and 14 cm, but there is no reason to believe that such children do not exist. The distribution of these data appears reasonably symmetric with a peak in the 16 cm to 22 cm range.

A **histogram** is a means of displaying continuous data graphically, conveying the general characteristics of the data.

Consider the following representation of the data.

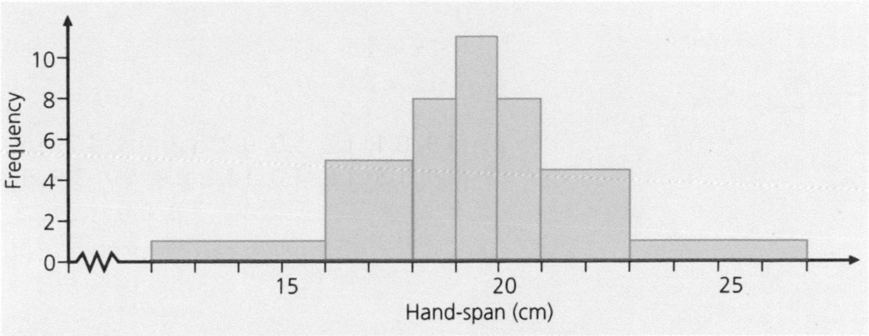

Along the 12 cm, 13 cm, 14 cm, 15 cm branches there are four items of data altogether. When these branches are combined and treated as one, the result will be one unit high and 4 cm wide. If the 16 cm and 17 cm branches are combined, the result is five units high and 2 cm wide containing ten items of data.

How many units high is the newly-created 21 cm and 22 cm branch? In a histogram, the **frequency** associated with any class of measures is represented by the **area** of the block in the display.

$$\begin{aligned}\text{Frequency} &\propto \text{area}\\ &= \text{width of interval} \times \text{height of block}\\ &= \text{class width} \times \text{frequency density}\end{aligned}$$

The terms **class width** and **frequency density** are commonly used when working with histograms.

A more formalised approach to constructing a histogram is illustrated in the following example.

Example 1.1

Construct a histogram to display these data using the classes given.

Class	Frequency
$0 \le x < 0.5$	12
$0.5 \le x < 1.5$	32
$1.5 \le x < 2.5$	20
$2.5 \le x < 4.5$	20
$4.5 \le x < 6.5$	6
$6.5 \le x < 10.5$	2

Solution

The first step is to calculate the widths of each of the classes. The first class is of width 0.5, the next is of width 1.0 and so on.

The next step involves the definition of frequency density inherent in the following relation.

Frequency $\propto$ class width $\times$ frequency density

The results of these two steps are recorded in the expanded table below.

Class	Frequency	Class width	Frequency density
$0 \le x < 0.5$	12	0.5	$12 \div 0.5 = 24$
$0.5 \le x < 1.5$	32	1.0	32
$1.5 \le x < 2.5$	20	1.0	20
$2.5 \le x < 4.5$	20	2.0	10
$4.5 \le x < 6.5$	6	2.0	3
$6.5 \le x < 10.5$	2	4.0	0.5

Finally, the histogram can be drawn using a linear scale on the x-axis.

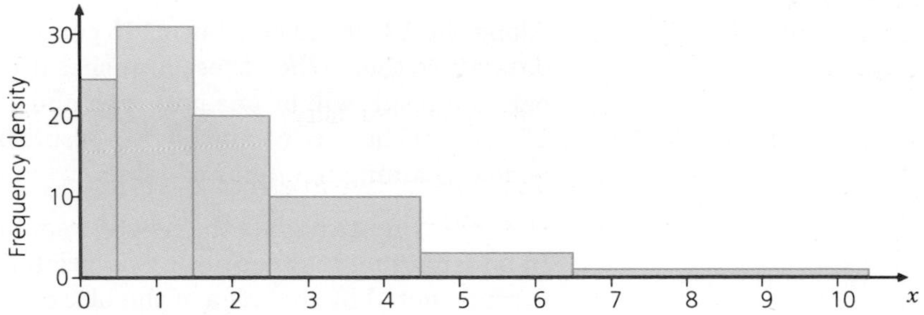

Example 1.2

The mass of each of 125 pizzas made for a frozen food company was recorded in grams, to the nearest 10 g. The results were tabulated as given in the table below.

Mass of pizza (g)	Number of pizzas
200–220	15
230–240	32
250	34
260	19
270–280	14
290–300	7
310–340	4

a) *What is the range of masses in the 200–220 class?*
b) *Identify the true class limits for each of the classes in the table.*
c) *Calculate the class widths for each class.*
d) *Construct a histogram of the data.*

Solution

a) *The mass, M, of a pizza is recorded to the nearest 10 g. Hence, if $195 \le M < 205$, the recorded mass would be 200. This means the class 200–220 would contain all pizzas in the range $195 \le$ mass < 225.*

b) and c)

Class	True class limits	Class width
200–220	$195 \le M < 225$	30
230–240	$225 \le M < 245$	20
250	$245 \le M < 255$	10
260	$255 \le M < 265$	10
270–280	$265 \le M < 285$	20
290–300	$285 \le M < 305$	20
310–340	$305 \le M < 345$	40

d) *To construct a histogram where the class widths vary, we need to calculate appropriate frequency densities. Since every class width is a multiple of 10, the following frequency densities are quite acceptable.*

Frequency	15	32	34	19	14	7	4
Frequency density	5	16	34	19	7	3.5	1

The classes have in effect, been treated as having widths of 3, 2, 1, 1, 2, 2 and 4.

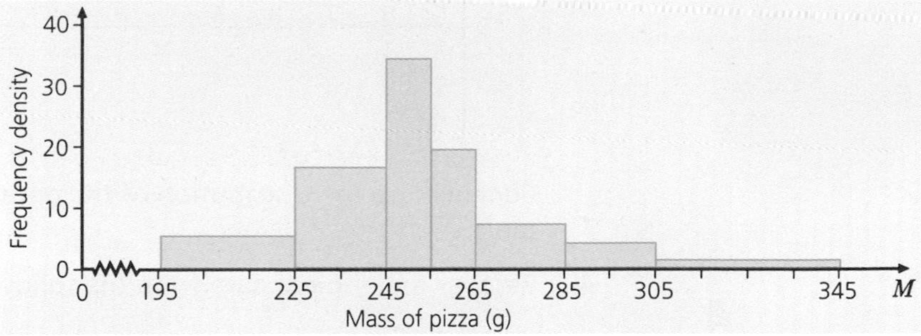

EXERCISES

1.2A

1 Construct a frequency table with appropriate classes for each of the following stem-and-leaf displays.

a) *Unit is 10*

3	5
4	7 7 8
5	0 3 3 3 5 6 8 8 9 9
6	1 3 4 8
7	0 1

b) *Unit is 100*

0	8 9 9 9
1	2 2 4 5 5 7 8 8
2	2 5 5 6 8 8 9
3	0

c) *Unit is 0.01*

6	7 7
7	2 3 5 5 9
8	5
9	0 1

Suggest, in each case, what the data might represent, giving suitable units of measurement.

2 Display the following data sets on appropriate stem-and-leaf displays.

a) 15, 19, 17, 17, 18, 18, 16, 16, 20, 21, 22, 19, 17, 16
b) 2100, 2350, 1800, 1850, 2500, 2000, 1900
c) 0.003, 0.008, 0.009, 0.010, 0.008, 0.011, 0.011, 0.008
Suggest, in each case, what the data might represent.

3 Recover the original data from the following two-part stem display.

Unit is 1

2–	3 0 4 0 3 2 2 1 0 1
2+	9 5 5 7 6 8 7 8 9 8 7 8
3–	0 2 1 1 1 3 2
3+	5 8 7 8 5
4–	3 1
4+	7 9 6
5–	0

Comment on the distribution of the data. Construct a frequency table.

4 Recover the original data from the following five-part stem display.

Unit is 1

F	4 5 5
S	6 7 7 7 7
E	8 8 9 9
1O	1 1
1T	2 2 2 3
1F	5
1S	

Construct a frequency table.

5 Use a two-part stem to display these data.

50.3, 51.2, 49.8, 49.2, 51.1, 53.7, 53.4, 52.5, 50.6, 52.1, 51.4, 53.0, 52.7, 51.7, 52.9, 53.1, 51.9, 53.2, 52.6, 50.1

6 Use a five-part stem to display these data.

5, 0, 12, 11, 1, 0, 2, 0, 0, 12, 3, 0, 10, 10, 1, 0, 1, 10, 1, 2, 12, 12, 9, 12, 1, 2, 1, 4, 8

7 The heights of 40 daffodils were recorded to the nearest centimetre and the results are given in this table.

Height class (cm)	10–14	15–19	20–24	25–29	30–39
Number of flowers	6	10	11	8	5

a) What are the:
i) minimum, **ii)** maximum,
heights of daffodils in the class 10–14?
b) Calculate the width of each class.
c) Calculate appropriate frequency densities and present the data in a histogram.

8 The teachers in a school were asked what age they were when they started teaching. The results are given in this table.

Age (years)	22–24	25–29	30–34	35–44	45–54
Number of teachers	60	20	12	5	1

a) What are the upper class limits for each of the age classes?
b) Write down the widths of each of the classes and calculate the corresponding frequency densities.
c) Draw a histogram of these data and comment on the distribution of ages at which these teachers commenced teaching.

9 The time, to the nearest minute, spent travelling to work was established for 100 people.

Time (minutes)	0–9	10–19	20–29	30–59	60–119
Frequency	19	38	20	18	5

a) Identify the true class limits and the widths of the classes.
b) Calculate the corresponding frequency densities and draw a histogram of these data.

EXERCISES

1.2 B

1 Rewrite the data in their original form from the stem-and-leaf diagrams below.

a) *Unit is 10*

0	1	1	2	7	
1	4	5	9	9	9
2	0	2	3		
3	1	4			
4	9				

b) *Unit is 0.1*

0–	4	2	0	1	
0+	8	6	7	7	9
1–	4	1	0	0	0
1+	7	9			
2–	0	2	2		
2+	6				

c) *Unit is 0.01*

O	0	1				
T	2	2	2	3		
F	4	4	4	4	5	5
S	6	6	7			
E	8	8	8	8	9	
1O	0	1	1	1		
1T	2	2	3			
1F	4					

2 Twenty nine-year-old children are each timed to tie their shoe laces with the following results, in seconds.

64, 75, 37, 48, 59, 50, 68, 65, 43, 52, 60, 79, 76, 83, 56, 88, 72, 65, 63, 79

Represent these data by a stem-and-leaf display.

3 Some data given in a stem-and-leaf diagram looked like this.

Unit is 1

78	4	9		
79	1	3	6	
80	2	2	4	7
81	1	4	9	9
82	3	3		
83	0	2	3	
84	1	4	5	

Rewrite this as a two-part stem-and-leaf diagram with unit 10.

4 The data below are the results of an experiment in which each of a group of children had to complete independently a particular puzzle. The values quoted are the completion times, in minutes.

2.4 2.0 1.6 1.4 1.8 1.1 1.6 1.7 1.6 2.1 2.0 1.4 2.2 2.3 2.5 2.3 1.7
1.7 1.8 1.9 1.4 1.4 1.9 1.8 1.6 1.7 2.1 1.8 1.7 1.9

Represent these data by a stem-and-leaf display using five-part stems.

5 A company manufactures a metal ring for industrial engines that usually weighs about 150 grams. A collection of 50 of these metal rings produced the following weights, in grams.

151 146 141 147 162 153 155 152 152 139 156 141 169 153 144 150 144
153 146 155 144 152 157 136 143 147 156 151 158 152 153 150 154 151
143 153 157 163 138 142 142 144 142 149 157 157 146 147 150 149

a) Illustrate these data in an appropriate stem-and-leaf display.
b) Construct a grouped frequency distribution for these data.

6 A speed camera recorded the speeds of 50 vehicles passing it in mph. The results, to the nearest mph, are listed below.

40 49 50 32 16 35 16 58 16 12 25 28 16 33 36 61 44 46 15 43
32 36 32 37 34 29 37 19 19 25 4 26 32 38 39 56 28 49 50 26
42 40 38 17 29 25 31 35 47 40

Tabulate these data into appropriate intervals ensuring that there are at least six intervals.

7 The data below represent the time intervals, in minutes, between successive vehicles on a quiet country road.

10.2 8.5 6.2 5.8 4.5 7.6 3.2 9.6 6.0 7.2 2.4 5.6 5.9 4.3 4.4 6.3
6.4 5.3 5.2 7.1 7.5 8.7 8.2 9.3 11.8 6.5 3.6 4.7 7.4 8.1 9.4 10.4
4.8 5.8 8.9 5.4 5.6 4.4 5.9 6.8

Construct a stem-and-leaf display of these data.

8 The following table shows the results of a survey of the people who lived in one road.

Age last birthday	15–24	25–34	35–44	45–64	65–70	Total
Number	52	41	43	29	3	168

a) Calculate the width of each class.
b) Plot a histogram to illustrate these data.

9 The following were results of a test of the numerical aptitude of a group of schoolchildren.

Score	10–29	30–39	40–49	50–59	60–69	70–79	80–100
Number of schoolchildren	80	140	20	10	5	0	45

a) Identify the true class limits and widths of these classes.
b) Draw a histogram for these data and comment on the distribution of scores.

FREQUENCY DENSITY POLYGONS AND CURVES

Polygons

It can take a long time to construct a histogram accurately. There is an alternative technique which conveys the same information rather more rapidly. This involves replacing the blocks of the histogram with points located at the midpoint of each block with the same frequency density. Joining the points with line segments produces a **frequency density polygon**. The hand-span histogram is reproduced below with the points in place.

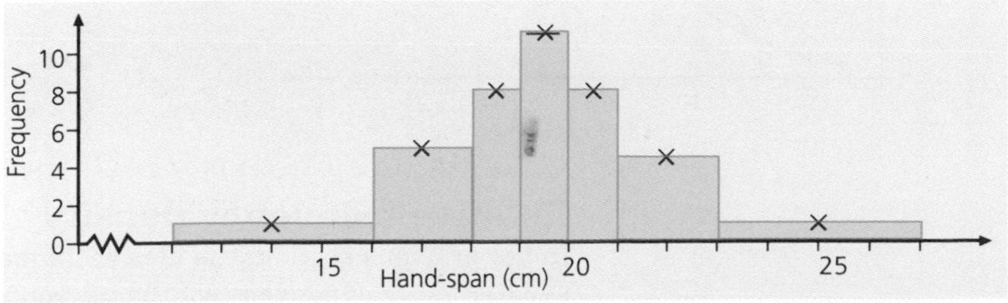

The frequency polygon for these data would look like this.

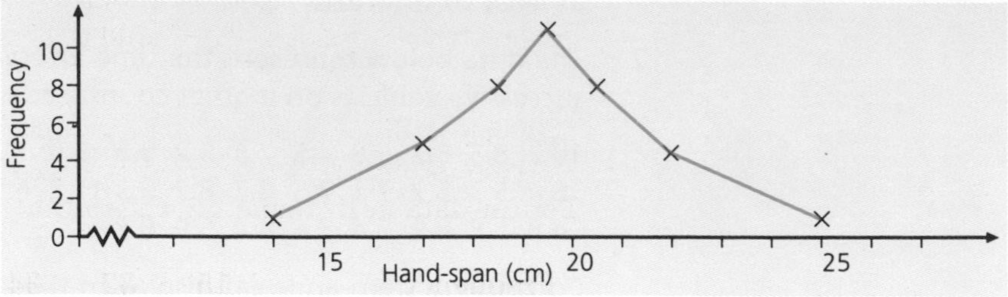

Curves

Class	Frequency	Class width density	Frequency	Midpoint
0–5	5	5	1	2.5
5–7.5	8	2.5	3.2	6.25
7.5–10	13	2.5	5.2	8.75
10–11	7	1	7	10.5
–12	9	1	9	11.5
–13	12	1	12	12.5
–14	16	1	16	13.5
–15	23	1	23	14.5
–16	28	1	28	15.5
–17	28	1	28	16.5
–18	21	1	21	17.5
–19	15	1	15	18.5
–20	13	1	13	19.5
–22.5	24	2.5	9.8	21.25
–25	13	2.5	5.2	23.75
–30	11	5	2.2	27.5

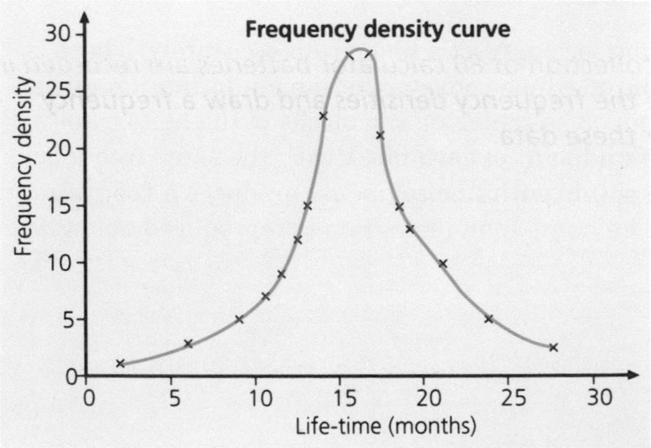

Consider the situation in the table above, where the class widths are relatively narrow, there are many classes, and there is a large total number of items of data. Under these circumstances the polygon will resemble a curve as shown here.

Relative frequency density

In the previous section, the idea of frequency density was introduced to cater for varying class widths in grouped frequency tables. This does have a drawback in that frequency densities are not unique; the class widths used in the solution to Example 1.2 were 3, 2, 1, ..., rather than 30, 20, 10, ... which produced frequency densities 5, 16, 34, ... as opposed to 0.5, 1.6, 3.4, ... Would the histogram have been essentially different using class widths of 30, 20, 10?

One approach to unifying this is to express the frequencies as proportions of the total frequency, giving rise to relative frequencies. For the data in Example 1.2, the results are as follows.

Frequency	15	32	34	19	14	7	4
Relative frequency	$\frac{15}{125}$	$\frac{32}{125}$	$\frac{34}{125}$	$\frac{19}{125}$	$\frac{14}{125}$	$\frac{7}{125}$	$\frac{4}{125}$
or	0.12	0.256	0.272	0.152	0.112	0.056	0.032

What is the sum of the relative frequencies? Will this always be the case?

If these relative frequencies are now used in conjunction with true class widths to produce **relative frequency densities**, and a histogram is then drawn, the total area of the blocks in the histogram will always be one square unit.

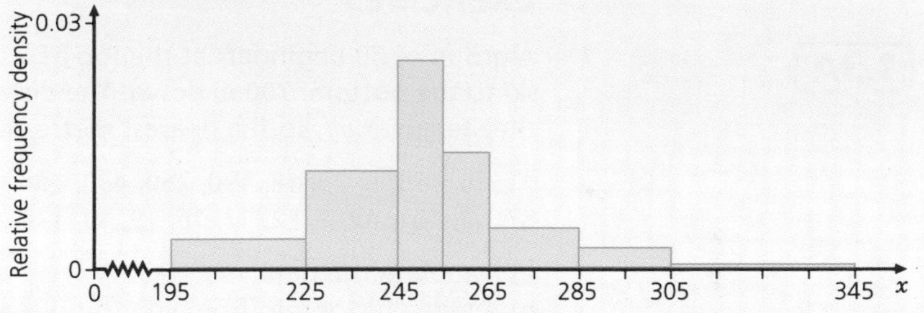

Example 1.3

The life-times of a collection of 80 calculator batteries are recorded in the table. Calculate the frequency densities and draw a frequency density polygon for these data.

Lifetime (months)	0–2	2–4	4–10	10–20	20–40
Number of batteries	32	17	18	9	4

Solution

Frequency densities require the class widths, so this ought to be included.

Life	0–2	2–4	4–10	10–20	20–40
Frequency	32	17	18	9	4
Class width	2	2	6	10	20
Frequency density	16	8.5	3	0.9	0.2

The frequency densities represent the vertical or y-coordinates for the frequency polygon. The horizontal coordinates are the midpoints of the corresponding classes. These are shown below.

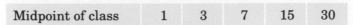

Midpoint of class	1	3	7	15	30

Hence the polygon will look like this.

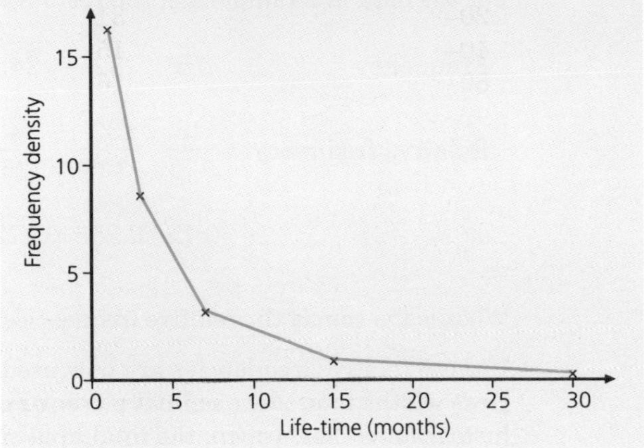

EXERCISES

1.3A

1 A group of 30 beginners at the top of a ski slope were asked to try to ski to the bottom, 700 m down. The distances they reached, before they fell over, m, to the nearest metre, were recorded.

41, 36, 150, 6, 29, 18, 190, 250, 430, 104, 12, 32, 84, 16, 62, 62, 151, 63, 87, 62, 101, 42, 2, 22, 12, 16, 19, 13, 175, 9

a) Tabulate the data.
b) Draw a histogram from your tabulated data.
c) Draw a frequency polygon based on the histogram.

2 Fifty smokers were asked the age at which they started smoking and the results are given in the table below.

Age	10–12	13–15	16–18	19–21	22–27
Frequency	8	19	15	4	4

a) Calculate the relative frequencies for these classes.
b) What are the **i)** maximum **ii)** minimum ages in the class 13–15?
c) Calculate the width of each class.
d) Calculate the appropriate relative frequency densities and present the data in a histogram.

3 The behaviour of children watching television has been studied a great deal. Now children are being studied as they play with toys. The following table summarises the play-period lengths, in seconds, for a particular five-year-old child.

Play-period length	Frequency
(s)	
0–	54
5–	44
10–	28
15–	21
20–	31
40–	15
60–	16
90–	5
120–180	8

a) Represent these data by a histogram and by a frequency density polygon.

b) **i)** What percentage of play-periods lasted at least 20 seconds?

ii) Estimate the proportion of periods which lasted between 25 and 75 seconds.

4 The volume, v cm³, of each pebble in a random sample of 50 taken from a beach, was calculated using a displacement jar.

Volume v (cm³)	Frequency
$0 \le v < 2$	4
$2 \le v < 5$	12
$5 \le v < 10$	10
$10 \le v < 15$	7
$15 \le v < 20$	4
$20 \le v < 30$	13

a) Calculate the frequency density for each class and scale these values so that the total area is one square unit.

b) Draw the resulting relative frequency polygon and use it to estimate the proportion of pebbles with volumes between 7 and 17cm² (inclusive).

EXERCISES

1.3 B

1 Forty students each measured the distance they travelled from home to their college to the nearest mile. The results were collated and they were then asked to produce a frequency polygon.

Distance from home to college (miles)	2–4	5–7	8–10	11–13	14–16
Frequency	5	8	13	10	4
Centre	3	6	9	12	15

a) Amend this table to give a relative frequency polygon.

b) Draw the relative frequency polygon.

c) Use your relative frequency polygon to estimate the proportion of students who live **i)** between two and five and **ii)** between seven and 14 miles from home.

2 The table below shows the weekly take-home pay of a group of 1000 people.

Pay x (£)	Frequency
$0 \le x < 160$	41
$160 \le x < 185$	43
$185 \le x < 190$	51
$190 \le x < 200$	142
$200 \le x < 210$	231
$210 \le x < 220$	212
$220 \le x < 240$	175
$240 \le x < 260$	105

a) Illustrate these data on a relative frequency density histogram where the total area of the histogram is one square unit.
b) Use this diagram to give an estimate of the proportion of workers with take-home pay between £155 and £213.

3 A large batch of metal tubes having specification limits for the diameter of 30.5 ± 0.2 mm was received from a supplier of machine parts. A measurement of the diameters of a random sample of 60 tubes resulted in the following data.

30.33 30.47 30.34 30.63 30.42 30.53 30.56 30.61 30.36 30.67 30.43
30.37 30.43 30.36 30.62 30.43 30.63 30.38 30.58 30.37 30.58 30.44
30.41 30.63 30.39 30.62 30.37 30.57 30.37 30.61 30.37 30.63 30.57
30.64 30.62 30.38 30.56 30.61 30.36 30.66 30.59 30.36 30.67 30.51
30.69 30.36 30.62 30.41 30.67 30.41 30.58 30.37 30.58 30.44 30.41
30.63 30.37 30.63 30.57 30.64

Construct a grouped relative frequency distribution.

4 The table below summarises the distance travelled, in miles, by delegates to a national conference.

Distance travelled (miles)	Number of delegates
1–50	5
51–100	26
101–150	49
151–200	63
201–300	85
301–500	92
501–700	38
> 700	12

Represent these data by a relative frequency histogram.

CONSOLIDATION EXERCISES FOR CHAPTER 1

1 State whether the following data are discrete or continuous.

a) waiting times for arrival of a bus
b) occupancy rate of hotel rooms
c) fish caught by an angler in a competition by weight
d) fish caught by an angler in a competition by number
e) FT share index
f) shop's daily takings

2 State whether each of the following variables are qualitative or quantitative, and discrete or continuous.

a) hair colour
b) journey time to college from home
c) postal code
d) attendance at cinema
e) house type
f) house size

3 A sample of 700 bus drivers, employed by public corporations, was selected and the numbers of traffic accidents in which each was involved during a four-year period were recorded as follows.

No. of accidents	0	1	2	3	4	5	6	7	8	9
No. of drivers	117	161	153	115	78	44	21	7	3	1

Construct a line diagram for these data.

4 A hospital's Accident and Emergency department records the following information on sports injuries.

Category of injury	Frequency
Sprain	63
Contusion	44
Fracture	35
Strain	31
Laceration	23
Dislocation	19
Concussion	15
Dental	13
Chronic	7
Total	250

Represent the percentage frequencies for these data using a bar chart. Comment on your results.

5 The table below shows the numbers of students on various types of course at a particular college.

Course	Number of students
Computing	232
Business Studies	296
Catering	161
Sociology	80
Economics	120
Humanities	161
Total	1050

Represent these data in an appropriate way.

6 The following table summarises part of the results of a survey undertaken by a supermarket which is planning to open on Sundays.

Likely form of Sunday transport	Percentage of customers
Car	60
Bus	15
Foot	10
Other	15

Represent these data by a pie chart.

7 An usherette at a cinema recorded at the end of the showing of a film the number of each type of coin she had in her tray. The results are show below.

Coin	1p	2p	5p	10p	20p	50p	£1
Frequency	17	36	12	28	10	9	25

Display this data in a suitable diagram.

8 At the end of the day, the owner of a small corner shop cashed up. He made a record of how many of each type of coins and notes he had in the till. The results were as follows.

Type of note/coin	1p	2p	5p	10p	20p	50p	£1	£5	£10	£20	£50
No. of that type	86	112	84	50	70	43	28	12	4	5	2

Display this data in an appropriate diagram. Are there any difficulties with this?

9 The accompanying data represent the daily numbers of customers visiting a shop over an 84-day period.

28 42 52 50 29 71 15 34 45 48 38 28 83 13 49 32 37 41 43 66 19 34 49
35 54 29 75 22 35 43 56 45 38 77 25 35 54 38 66 56 89 25 45 53 36 51
55 79 24 43 55 44 67 58 91 21 36 48 56 42 67 92 23 43 56 76 59 61 98
26 45 51 62 54 73 126 34 56 67 78 75 82 153 31

a) Present these data using a stem-and-leaf display.
b) Construct a grouped frequency distribution, not using constant class widths, from your stem-and-leaf display.

10 A student made a record of 50 observations of an experiment over a short period of time.

x	0	1	2	3	4	5	6	7	8
f	1	3	5	10	12	9	3	5	2

Plot the above sample of 50 observations as a histogram.

11 On a school sports day there were three javelin competitions: one for juniors, one for intermediates and one for seniors. The distances to the nearest 0.1 m thrown are given below (0 means there was a no throw).

Juniors	9.6	13.1	12.7	19.4	16.3	7.4	0	0
	21.3	8.4	18.6	14.2	9.6	0		
Intermediates	18.7	11.2	24.3	15.7	0	16.3	27.1	15.7
	11.4	16.2	19.6	31.2	15.1	22.4	25.7	
Seniors	0	0	25.3	26.7	19.4	11.1	38.3	39.7
	26.2	0	27.4	40.7	19.7	20.2	28.3	

a) Tabulate this data together in a grouped frequency table.
b) Is there any way in which grouping all the data together is not helpful?

12 The maximum height that 42 plants of the same species grew to was recorded. The results were recorded to the nearest cm and they are reproduced below.

Height (cm)	10–14	15–19	20–24	25–29	30–34
Frequency	6	12	11	8	5

a) What are the **i)** maximum **ii)** minimum heights in the class 20–24?
b) Draw a histogram of the given data.

13 Twenty-five competitors in a shot-putt competition had the lengths of three putts each measured to the nearest metre. This gave 75 results in all. The results are tabulated below.

Length (m)	0–9	10–12	13–15	16–18	19–22
Frequency	30	18	15	7	5

a) What are the **i)** maximum **ii)** minimum lengths in the class 13–15?
b) Calculate the width of each class.
c) Calculate appropriate frequency densities and present the data in a histogram.

14 The following data represents the distance from the centre of the board at which each of 12 darts lands (to the nearest cm).
14 cm, 12 cm, 14 cm, 17 cm, 5 cm, 26 cm, 16 cm, 8 cm, 17 cm, 22 cm, 19 cm, 31 cm, 20 cm

Show this in a two-part stem-and-leaf display indicating the unit.

15 The table below shows the hourly numbers of prescriptions dispensed by a large pharmacy over a period of time.

Number dispensed per hour	Number of hours
0–9	7
10–19	22
20–29	33
30–39	45
40–49	29
50–75	17
over 75	6
total	160

a) **i)** What type of data are these, discrete or continuous?
ii) Why does this cause a problem in constructing a histogram to represent the information?
b) **i)** What other problem has also to be overcome before the histogram can be constructed?
ii) Draw a histogram to represent these data.
c) Superimpose the frequency polygon on your histogram.

Summary

- In this chapter we have distinguished between information which is qualitative or quantitative in nature and identified **discrete** data and **continuous** data.
- Diagrammatic presentations of data include **line graphs**, **bar diagrams** and **pie charts**.
- The data may be used to present themselves in **stem-and-leaf displays**.

One-part

Unit is 1

4	2	5	3		
5	8	1	0	4	6
6	2	1	5	4	
7	4	3	6		

Two-part

Unit is 1

4–	2	3		
4+	5	7	6	
5–	0	4	3	2
5+	6	9	8	
6–	3	1		

Five-part

Unit is 10

F	4	5	5		
S	7	7	6	7	6
E	9	8	8	8	
1O	0	0	1		
1T	2	3			
1F	4				

- Histograms use area to represent frequency.

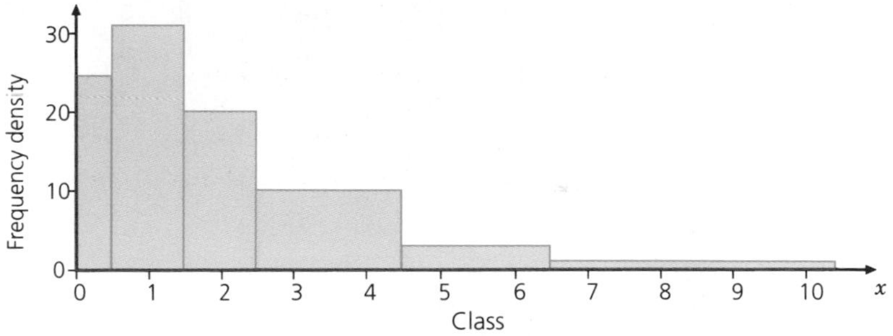

frequency × class width × frequency density

STATISTICS

2

Summary statistics 1

In this chapter:

- *we introduce, as measures of location or average for a collection of data, the median, the mean, the weighted mean and the geometric mean,*
- *we extend the methods of presenting data to include frequency tables and cumulative frequency polygons,*
- *we analyse symmetry in data and define skewness.*

Exploration 2.1

What do we need to know?

Consider the data presented in the stem-and-leaf diagram below.

Unit is 1

0	9	9	9	9					
1	3	7	3	7	2	5	8		
2	5	8	4	7	3	5	1	5	9
3	7	4	3	2	0	8			
4	5	6	2	5	3				
5	8	6	9						
6	4	1	8						
7	7	4							
8	3								
9									
10	4	7							
11	8								

This is a display of the time, in seconds, taken by 43 cars to pass through a continental motorway toll section. What would an experimenter need to take into account when collecting this information?

What we need to know

Among the many considerations, it would be important to have some idea of what a typical time was likely to be, and the likely range of times. This information is necessary so that an appropriate timing device can be selected. There would be little point in trying to use a watch's minute hand or an electronic timer capable of recording to $\frac{1}{1000}$ of a second. It is also unlikely to be necessary to record times as long as an hour, but the timer needs to be capable of recording more than a few seconds.

There are two measures representative of data which may be useful in cases like this – one representing the 'typical value' and another which indicates the 'sort of range of values' likely to be found. In statistical terms, these are measures of **location** or **average** and **dispersion** or **spread**.

Life would be simple if there were just one of each, or even an ideal measure of each quality. Unfortunately, this is not the case. There are many types of average and several kinds of spread. In this chapter we shall concentrate on averages.

Exploration 2.2

Appropriate values

In the light of the above discussion, what would be an appropriate value which represents:

a) the location of the data,
b) the dispersion of the data?

MEASURES OF LOCATION

Mode

The **mode** is the most commonly occurring value or item of data, or, in other words, the one that appears most frequently. In the context of the data under consideration, the most commonly occurring value is 9 seconds. Is it reasonable, though, to consider 9 seconds as being typical of the time taken to pass through this continental motorway toll? Almost certainly not! It may be more appropriate to consider the **modal class**. Referring back to the diagram, the 2-branch has the greatest frequency. Hence, it would be reasonable to say that the mode is a time between 20 and 30 seconds. This is the **longest branch in the stem-and-leaf** diagram.

The modal class may be the class with the highest frequency when the data are presented in a frequency table; but it may not!

Time t (s)	Frequency
$0 \le t < 10$	4
$10 \le t < 20$	7
$20 \le t < 30$	9
$30 \le t < 40$	6
$40 \le t < 50$	5
$50 \le t < 60$	3
$60 \le t < 120$	9

We need to proceed with great caution if the classes into which the data are grouped vary in width. Under these circumstances frequency density is a more accurate indicator of the modal class, as a histogram readily shows.

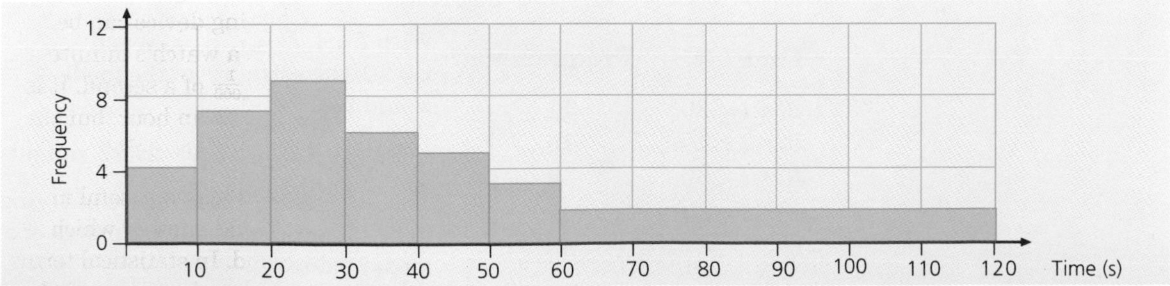

The prominence of the 20–30 seconds class is apparent. The 60–120 seconds class, which has the same frequency, is now seen more clearly!

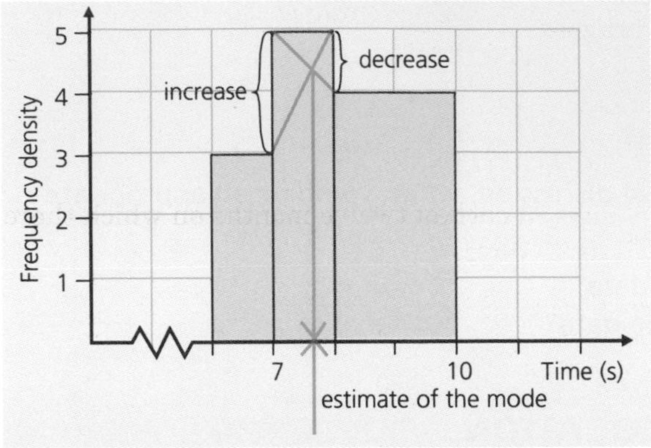

Having used the histogram to identify the modal class, it is possible to find an estimate of the mode. The diagram offers a simple graphical method of finding this estimate.

The method assumes that the mode divides the modal class in the **same ratio** as the *increase* in frequency density to the *decrease* in frequency density. In the diagram, this ratio is $(5-3):(5-4)$ which is equivalent to 2 : 1. Hence the mode divides the modal class in the ratio 2 : 1, and an estimate of the mode is $7\frac{2}{3}$. We need to ask ourselves, how valid is this process?

Example 2.1

Refer back to the motorway toll times frequency table. Determine an estimate of the modal time.

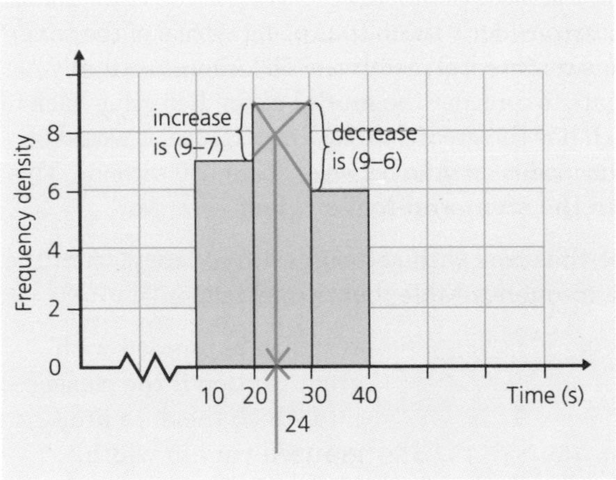

Solution

Here, the mode divides the modal class in the ratio 2 : 3. Hence an estimate of the mode is 24.

Is it possible to derive a formula for this estimate of the mode? Suppose the increase in frequency density is I, and the decrease is D. Then the estimate of the mode is given by:

$$\text{estimate} = x + \frac{I}{I+D} \times W$$

where W is the width of the modal class, and x is its lower bound.

In Example 2.1 $I = 2$, $D = 3$, $W = 10$ cm and $x = 20$ seconds.

Hence estimate of the mode = $20 + \frac{2}{5} \times 10 = 24$ seconds.

Caution

Suppose the same data were grouped differently, as in the following table.

Time t (s)	Frequency
$0 \le t < 20$	11
$20 \le t < 40$	15
$40 \le t < 60$	8
$60 \le t < 120$	9
	43

With $I = 4$, $D = 7$, $W = 20$ and $x = 20$, the estimate of the mode would be:

$$20 + \tfrac{4}{11} \times 20 = 27\tfrac{3}{11} = 27.27 \text{ seconds}$$

This illustrates the dangers in trying to be too precise about estimates of a modal value.

Discrete data

The simple definition of the mode as the most frequent item of data is directly applicable to discrete data, and techniques such as the one discussed above are entirely inappropriate.

The mode of the data collection:

12, 12, 12, 13, 14, 17, 17, 17, 17, 18, 20, 26

recording the number of days in each of twelve months on which there was a recordable rainfall, is 17 days.

The mode of the frequency distribution of discrete data:

x	0	1	2	3	4	5	> 5
f	17	12	15	4	1	1	1

where x records the number of children in a household is zero.

EXERCISES

2.1A

1 An Italian pizza restaurant records the number of people in each party for which a table was reserved, with the following results.

2 3 4 4 3 2 4 4 3 3 5 4 3 2 3 4 6 3 4 3 2 2 4 3 5 6 3 4 2 4 3 4 4 3 4 4

a) Construct a frequency table of these data.
b) Estimate the most frequent party size for reserved tables.

2 The numbers of visits to the doctor made by each of 20 children in one year are recorded below. What is the modal number of visits?

a) 0 6 0 3 2 0 0 1 1 0 0 0 1 2 1 2 0 5 0 0

Identify the mode of these data.

b) 156 159 156 151 156 159 157 157 159 159 155

3 Study these data.

Class (s)	Frequency
$0 \le x < 1$	2
$1 \le x < 2$	17
$2 \le x < 3$	31
$3 \le x < 4$	23
$4 \le x < 5$	6
$5 \le x < 7$	2

a) Write down the modal class.
b) Estimate the modal time, in seconds.

4 Study these data.

Class (cm)	Frequency
6–10	8
10–12	9
12–14	13
14–15	9
15–16	7
16–18	12
18–30	30

a) Write down the modal class.
b) Calculate an estimate of the modal length.

5 The marks gained by 20 students in a Statistics test were recorded in two ways, as shown below. Identify the modal class, and calculate an estimate of the mode, in each case.

a)

Class (marks)	Frequency
0–9	2
10–19	5
20–29	6
30–39	4
40–49	3

b)

Class (marks)	Frequency
0–10	2
11–20	5
21–30	6
31–40	4
41–50	3

EXERCISES

2.1B

1 The table below shows the number of rooms occupied in a particular hotel, for a random sample of 100 nights.

Number of rooms	Number of nights
Fewer than 10	5
10	8
11	12
12	23
13	16
14	17
15	12
More than 15	7

a) What is the modal number of rooms occupied?
b) The hotel has 20 bedrooms, Explain why it is not possible from the above data to determine the modal number of rooms unoccupied.

2 Below is a record of the numbers of rides at the funfair enjoyed by each of a class of 25 children. Find the mode(s) and comment on your findings.

2 5 0 2 0 0 0 5 0 5 4 0 3 3 5 4 0 5 5 5 4 4 0 0 4 5 4 5

3 Study these data.

Class (s)	Frequency
0	16
1	11
2	9
3	5
4	3
5	2
6	1

a) Write down the modal class.
b) Calculate an estimate of the modal time, in seconds.

4 Study these data.

Class (kg)	Frequency
0–6	16
7	17
8.5	16
10	12
12	8
14	5
20	6

a) Write down the modal class.
b) Estimate the modal weight, in kilograms.

5 The table below shows the distribution of the waiting times, in minutes, of 110 people at a station's taxi rank.

Waiting times	Frequency
0–	8
1–	14
2–	23
3–	24
5–	24
10–	12
15–	5

a) State the modal class.
b) Estimate the modal waiting time.

MEDIAN

Step-by-step

The centre or middle item of the data is known as the median. One approach to identifying the median is to:

- place the data in order
- locate the middle item
- hence, identify the median.

Suppose we need to identify the median of the following collection of data.

8, 15, 7, 10, 4, 3, 8, 6, 5, 7, 8

Placing the data in order yields:

3, 4, 5, 6, 7, 7, 8, 8, 8, 10, 15

The middle item is the one which is equidistant from the extreme values. Since there are eleven items of data, the middle is the sixth from either end.

sixth ↓

3 4 5 6 7 7 8 8 8 10 15

↑ median

Hence, the median of this collection of data is 7.

Even total

For the collection above, the total number of items is odd, which led to the median being one of the actual recorded items of data. In the following case, the total is an even number, which means the centre value of the data is midway between two of the recorded items.

4, 5, 0, 3, 9, 4, 8, 9, 9, 1

Ordering the data gives:

0, 1, 3, 4, 4, 5, 8, 9, 9

and locating the middle value yields:

middle ↓

0 1 3 4 4 5 8 9 9 9

Hence, the median, which is midway between 4 and 5, is 4.5.
Now try identifying the median of the motorway toll times.
You probably found that the median is 32 seconds, but you are likely to agree that the process is rather tedious when there are many items of data. Several techniques have been developed, which depend on the format in which the data are available.

Stem-and-leaf

When the data are presented in a stem-and-leaf display, the first step towards identifying the median is to **order** the leaves. This involves placing the leaves on every branch in order of magnitude. The **ordered stem-and-leaf display** for the motorway toll times (page 30) looks like this.

0	9	9	9	9					
1	2	3	3	5	7	7	8		
2	1	3	4	5	5	5	7	8	9
3	0	2	3	4	7	8			
4	2	3	5	5	6				
5	6	8	9						
6	1	4	8						
7	4	7							
8	3								
9									
10	4	7							
11	8								

The next step is to locate the middle of the data – the item which is the same distance from either end of the data. In this instance, there are 43 items of data, hence the middle item is the 22nd item since it has 21 items on either side of it.

← 21 items 22nd 21 items →
←———— 43 items ————→

In the most basic terms, what we need to do is to count from each extreme until the count reaches 21; the median is then the next item.

One method of recording this counting process is to keep a running total as we count along each branch, until the branch containing the median is encountered. (Why is it not realistic to record the count for this branch?)

Depth										
4	0	9	9	9	9					
11	1	2	3	3	5	7	7	8		
20	2	1	3	4	5	5	5	7	8	9
(6)	3	0	2	3	4	7	8			
17	4	2	3	5	5	6				
12	5	6	8	9						
9	6	1	4	8						
6	7	4	7							
4	8	3								
3	9									
3	10	4	7							
1	11	8								

This count of the position of the median *relative to the nearer extreme* is known as the **depth**. In this case the median has a unique depth of 22. Note that the branch containing the median does not have a depth recorded, instead it has (in brackets) the number of leaves on the branch.

Note that where the total number of items of data is even, the median lies midway between the middle two items of data.

If there are n items of data then the depth of the median is $\frac{n+1}{2}$.

Exploration 2.3

Using depths

Order the following stem-and-leaf display.

Unit is 10

0	7	8	8							
1	8	1	3	8	7					
2	9	8	8	4	8	8	4	5	4	0
3	3	3	3	3	3	3	5			
4	0	3	0	7	2	8	9	5	8	
5	8	3	8	3						
6	6	4								
7	4	4	9	0						
8	3	3	0							
9	1									

Use depths to identify the median of the data collection.

Grouped frequency table

If the data are presented in a frequency table, then it is only possible to obtain an **estimate** of the median. This is done either graphically or arithmetically.

Cumulative frequency

One graphical approach uses a process similar to calculating the depths of a stem-and-leaf display, except that only the depth from the *lower* extreme is calculated. This is known as the **cumulative frequency**. Then the cumulative frequency of the upper extreme is equal to the total number of items of data.

The cumulative frequencies for the earlier frequency table of the motorway toll times are given below.

Time t (s)	Frequency	Cumulative frequency
$0 \le t < 10$	4	4
$10 \le t < 20$	7	11
$20 \le t < 30$	9	20
$30 \le t < 40$	6	26
$40 \le t < 50$	5	31
$50 \le t < 60$	3	34
$60 \le t < 120$	9	43

What information can be gained from the cumulative frequency?

Consider the cumulative frequency of 20. This tells us that there are 20 items of data with values less than 30 seconds. Similarly, the cumulative frequency of 34 indicates that there are 34 times which are less than 60 seconds. There is a natural link between any given cumulative frequency and the upper bound of the corresponding class. Hence, when it comes to constructing a cumulative frequency graph, the points to be plotted come from the following series of data.

Upper class value (s)	**Cumulative frequency**
10	4
20	11
30	20
40	26
50	31
60	34
120	43

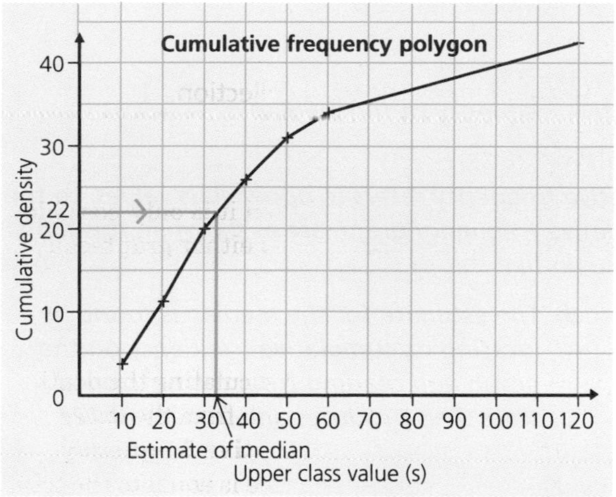

The resulting graph looks like this. It can be used to find the median.

Note that this does not necessarily produce the same value for the median as is found using the stem-and-leaf display. Why should this be the case?

The cumulative frequency of the median is $\frac{n+1}{2}$, where n is the total of the frequencies of the classes.

Sigma notation

There are many instances in studying Statistics which involve a process of 'adding up'. Because it occurs in so many cases, there is a standard notation which we use. This symbol Σ, called **sigma**, is a letter from the Greek alphabet. It is the equivalent of capital S. We use it to represent 'the sum of …'.

An example of its use is in identifying the cumulative frequency of the median of a grouped frequency distribution. The total of the frequencies is found by adding up the frequencies of all the classes. We write this as:

$$\sum f_i$$

This may be written in any of the following ways.

$$\sum_{\text{all data}} f_i \quad \text{or} \quad \sum_i f_i \quad \text{or} \quad \sum f$$

where f_i represents the frequency of the ith class.

Example 2.2

Study these data.

Time (s)	100	110	120	130	150–200
Frequency	3	8	5	4	3

a) *Identify the frequencies for each of the five classes.*
b) *Find the cumulative frequency of the median.*
c) *Construct a cumulative frequency graph.*
d) *Estimate the median.*

Solution

a) *The frequencies are:* $f_1 = 3, f_2 = 8, f_3 = 5, f_4 = 4, f_5 = 3$.

b) *To find the cumulative frequency of the median, we need to find the total, n, of the frequencies.*

$$n = \sum_{i=1}^{5} f_i \text{ (i.e. add up the five frequencies)}$$

$$= (3 + 8 + 5 + 4 + 3) = 23$$

Hence the cumulative frequency of the median is $\frac{23+1}{2}$ *which is 12.*

c) *Recall that cumulative frequency refers to upper class values, so the data for the cumulative frequency graph are as shown in this table. Hence, the graph looks like this.*

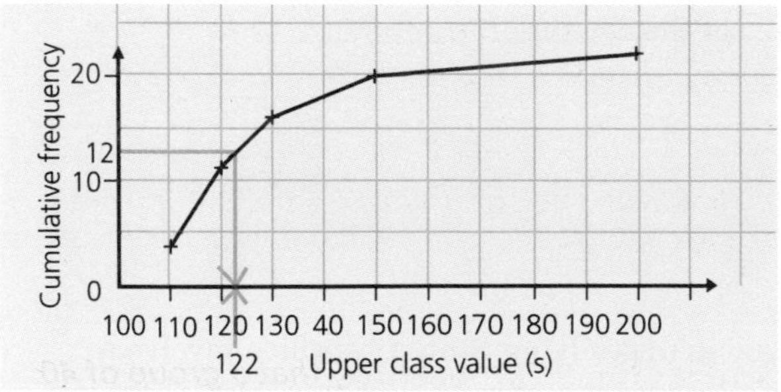

d) *The estimate for the median is found by locating its cumulative frequency on the graph and reading from the horizontal scale. This produces an estimate of 122 s.*

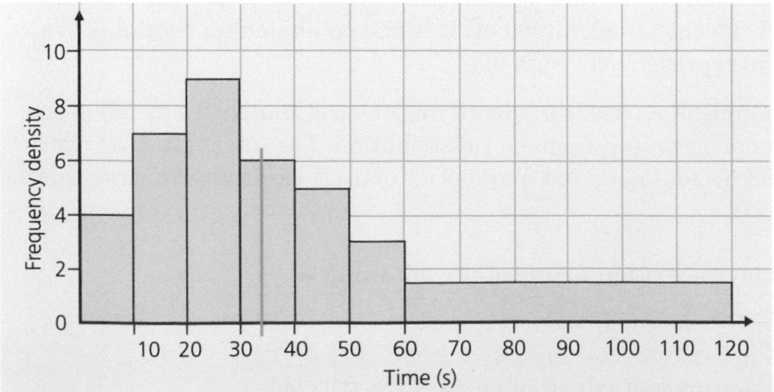

Histogram

A less common graphical approach makes use of a histogram and its property that area represents frequency. We need to locate the class containing the median, using perhaps a cumulative frequency table.

By definition, the median is in the middle of the data, so what we need to do is to find where a vertical line drawn in the histogram divides it so that the area of the blocks are the same on either side (see the diagram). The area of the first three blocks can be calculated.

$$(4 \times 10 + 7 \times 10 + 9 \times 10) = 200$$

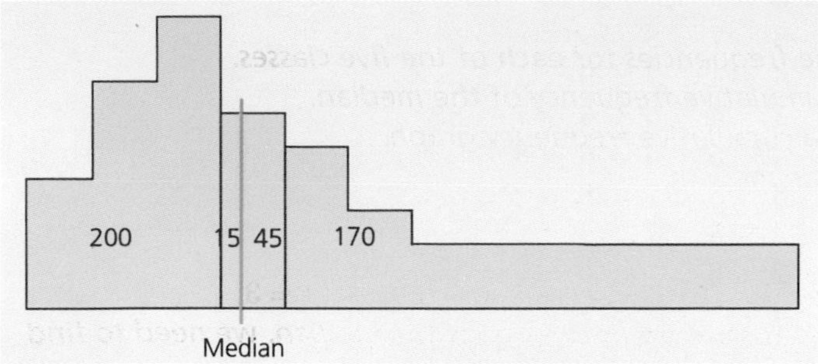

The area of the block containing the median is 60 and the remaining area is 170. Thus the total area can be considered to be 430.

The median must divide its class in the ratio 15 : 45 and so, this method of estimating the median from grouped data produces the estimate:

$$30 + \frac{15}{60} \times 10 = 32.5$$

Linear interpolation

The arithmetic approach starts from the cumulative frequency table and involves linear interpolation. Consider the section of the cumulative frequency graph where the median is located.

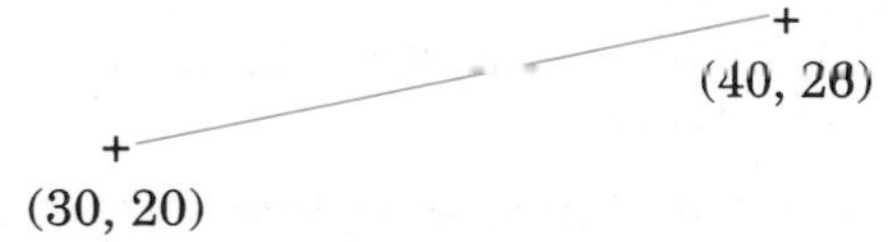

The cumulative frequency of the median is 22. Since 22 divides the cumulative frequencies in the ratio 2 : 4, the median will divide the class units 30 to 40 in the same ratio. Hence, this method of estimating the median produces this estimate.

$$30 + \frac{2}{6} \times (40 - 30) = 33\tfrac{1}{3}$$

This estimate should agree with the one given by the cumulative frequency graphical approach. Why?

Example 2.3

The data represent the distances, in centimetres, that a group of 40 children were able to achieve in 'standing' jumps. Use an ordered stem-and-leaf display to find the median standing jump.

Unit is 1

4	7						
5	4	2					
6	3	1	8	7			
7	9	7	4	1	4		
8	8	7	0	6	5	4	1
9	6	5	5	4	2	8	
10	8	7	3	1	5		
11	4	5	3	7			
12	8	1	2				
13	4	1					
14	2						

Solution

The ordered stem-and-leaf display and the associated depths are:

Depth
Unit is 1

1	4	7
3	5	2 4
7	6	1 3 7 8
12	7	1 4 4 7 9
19	8	0 1 4 5 6 7 8
(6)	9	2 4 5 5 6 8
15	10	1 3 5 7 8
10	11	3 4 5 7
6	12	1 2 8
3	13	1 4
1	14	2

Since there are 40 items of data, the median has a depth of $20\frac{1}{2}$. This means that the median lies halfway between two items of data. The 9-branch contains the median – as indicated by the absence of a depth for that branch. The leaf 2 on that branch has a depth of 20, as does the leaf 4 from the other extreme. The median lies halfway between these, and thus corresponds to a leaf value 3. This represents a standing jump of 93 cm.

Example 2.4

The 'standing jump' data are presented below in a grouped frequency table. Use this to construct a histogram and calculate an estimate of the median from this.

Jump (cm)	**Frequency**	**Class width**	**Frequency density**
40–60	3	2	1.5
–80	9	2	4.5
–90	7	1	7
–100	6	1	6
–110	5	1	5
–130	7	2	3.5
–150	3	2	1.5

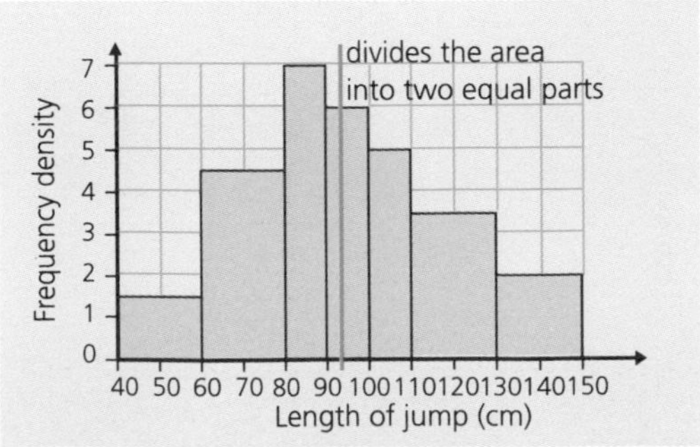

Solution

The histogram would be based on the frequency densities shown above.

The totals of the areas of the 'blocks' before and after the 'median block' can be considered to be 190 and 150; the median block would then be represented by an area of 60. The median is required to divide the area evenly into two sections of 200 units. Hence, the median divides its class in the ratio 10 : 50. Thus the estimate of the median based on the histogram is 91.7 cm.

Discrete data

The very nature of quantitative discrete data will often mean that the median of a collection of such data is one of the items of data. Two examples below illustrate ways of:

a) identifying the median in the case of an ungrouped frequency table of data,
b) estimating it where the data are presented in grouped frequency form.

a) Thirty-nine people were asked to state how many pairs of shoes they possessed.

Number of pairs of shoes	Frequency	Cumulative frequency
6	2	2
7	4	6
8	7	13
9	12	25
10	8	33
11	4	37
12	2	39

In the third column, the cumulative frequencies have been added to the frequency table. Because the data are discrete, the cumulative frequencies relate directly to the data values. In other words, the cumulative frequency 13 is associated with eight pairs of shoes; the cumulative frequency graph would include the point with coordinates (8, 13).

Why is it not appropriate to join the points of this cumulative frequency graph by line segments?

What might be an appropriate way of linking the points?

One approach is to link the points in a series of steps as shown below.

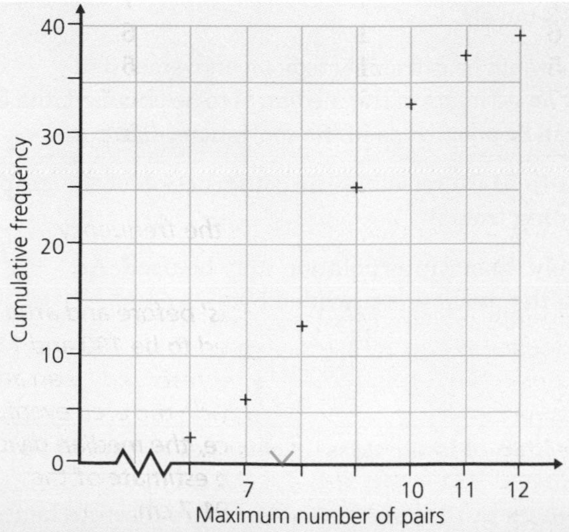

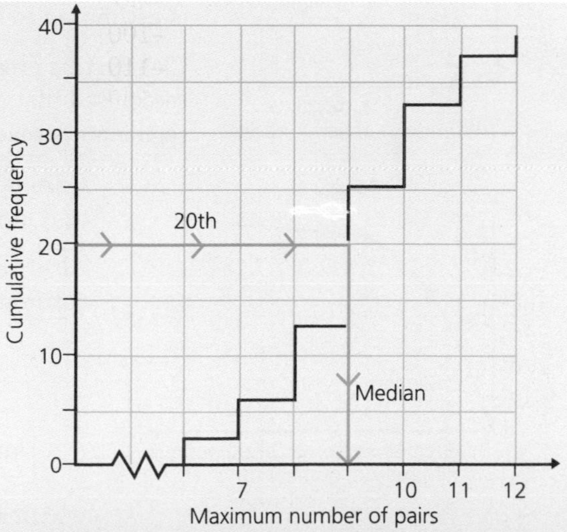

Why is this appropriate?

Since there are 39 items of data, the median item is the 20th which, from the graph, can be seen to be 9 pairs of shoes.

Alternatively, examining the table, the 13th item is 8, the 14th, 15th, up to the 25th are each 9, hence the median number of pairs of shoes is 9.

b) Sixty-seven children were asked how many novels they had read in the previous twelve months.

Number of books read	Frequency	Cumulative frequency
0–9	27	27
10–19	18	45
20–29	11	56
30–39	5	61
40–59	3	64
60–79	2	66
80–99	1	67

In this table, the data have been grouped. Because the data are discrete, the true class limits for a class such as 20–29 are 20 and 29 (not 29.5, 29.9, ... etc.) hence the cumulative frequency of 56 relates to the upper class value 29.

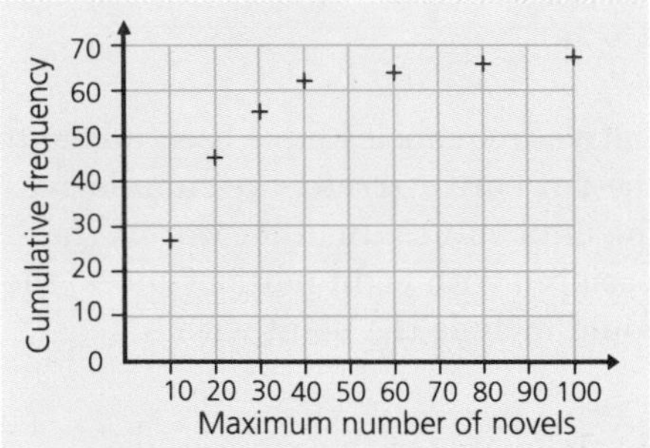

Constructing a step graph to link the points plotted for the cumulative frequency graph is not appropriate.

Why not?

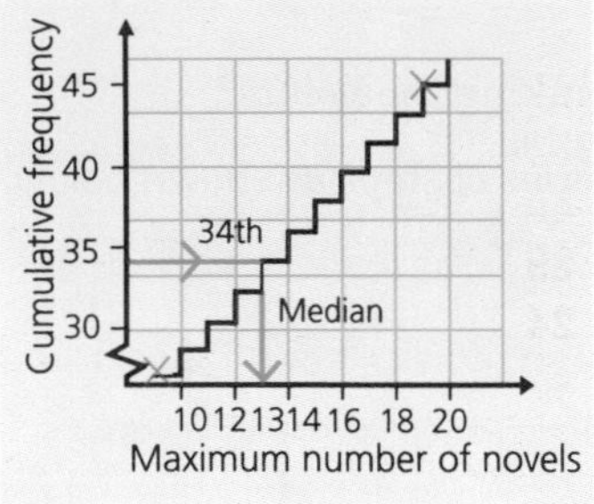

It could be argued that making the assumption that the number of novels read in any class is evenly distributed across the class would lead to linking consecutive points with a series of steps, as here.

The median is the 34th item of data. Hence, using the step graph, it is estimated to be 13 novels.

Clearly, this process would be extremely tedious and, since it is recognised that only an estimate of the median is to be obtained, the line segment approach can be adopted as in the continuous data case.

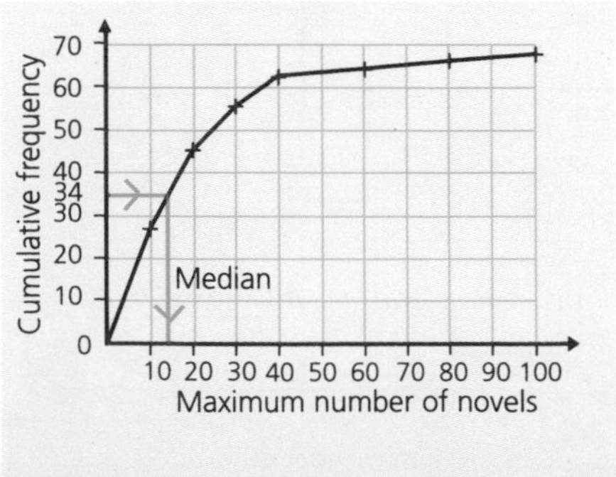

Care needs to be exercised in the construction of the graph, and in reading from it.

Alternatively, linear interpolation may be used. An estimate of the median is provided by:

$$9 + \frac{34 - 27}{18} \times 10$$

upper limit of lower class (pointing to 9) | class has a span of 10 (pointing to 10)

which amounts to 12.9. But as the data are discrete hence the median must be 13 novels.

Exploration 2.4

Comparing approaches

Discover how the three approaches:

- stepped cumulative frequency graph
- line segment cumulative frequency graph
- linear interpolation

are related.

EXERCISES

2.2A

1 This is a frequency table for the data in question 4 of Exercise Set 2.1A. What is the median number of people in a household?

Number in family	Number of families
1	15
2	20
3	22
4	23
5	11
6	4

2 The number of flowerheads on the foxgloves found in a small woodland were counted. The results are presented below.

Number of flowerheads	Number of foxgloves
0–19	6
20–29	22
30–39	33
40–49	45
50–69	29
70–99	24

Obtain an estimate of the median number of flowerheads:

a) using a cumulative frequency graph,
b) using linear interpolation.

3 Study the collections of data in question 1 of Exercise Set 1.2A.

a) Tabulate the cumulative frequencies.
b) Draw a cumulative frequency graph.
c) Use your graph to estimate the median.
d) Obtain an estimate of the median using linear interpolation.

4 Consider this grouped frequency distribution.

Length (cm)	**Frequency**
17.5–17.9	15
18.0–18.4	27
18.5–18.9	18
19.0–19.9	12
20.0–24.9	15
25.0–29.9	4

a) Identify the upper class values, if the measurements were recorded to the nearest millimetre.
b) Construct a cumulative frequency table and hence a cumulative frequency graph.
c) Use your graph to estimate the median length.
d) Obtain an estimate, in centimetres, correct to two decimal places, of the median length, using linear interpolation.

5 The populations in some Norfolk villages are given below.

933	410	956	1285	351	499	1414	2095	161	1488	1085	597
749	2444	1043	1303	3563	1124	4427	288	1099	1773	714	1314
623	893	234	532	2099	1243	709	521	472	655	308	613
483	442	189	198	541	351	940	705	439	399	732	101
341	452	324	3587								

a) Choose appropriate class intervals and present these data in a grouped frequency table.
b) Obtain an estimate of the median population using a cumulative frequency graph.
c) Compare the estimate in **b)** with the actual median of the data.

6 Obtain an estimate of the median for each of the stem-and-leaf displays of **a)** and **b)** of Exploration 1.5.

7 The table records the time, in seconds, taken by each member of a group of 165 girls, to complete a simple crossword.

Time (s)	**Number of girls**
480–539	22
540–569	29
570–599	54
600–629	33
630–659	17
660–719	10

a) Draw a histogram of these data.
b) Use your histogram to estimate the median time.
c) Compare your answer in **b)** to the estimate obtained using linear interpolation.

EXERCISES

2.2B

1 A lecturer has to pass through twelve sets of traffic lights on her route into university. She records the number of sets which she has to stop on each journey for 100 journeys.

These records are shown in this table.

Number of times stopped	Number of journeys
≤ 1	0
2	1
3	3
4	8
5	15
6	21
7	22
8	17
9	9
10	3
11	1
12	0

a) Construct a cumulative frequency table for these data.
b) Draw a cumulative frequency step graph.
c) Locate and write down the median number of occasions the lecturer is stopped at traffic lights on a journey.

2 The number of components produced each day by a production line is recorded in this table.

Number of components	Number of days
500–509	16
510–519	44
520–529	47
530–539	31
540–549	18

Obtain an estimate of the median number of components:

a) using a cumulative frequency graph,
b) using linear interpolation.

3 Study the collection of data recorded in question 5 of Exercise Set 2.1B.

a) Tabulate the cumulative frequencies.
b) Draw a cumulative frequency graph.
c) Use your graph to estimate the median.

4 The lengths of off-cuts of timber are summarised in the following grouped frequency table.

Length (cm)	**Frequency**
0–	17
25–	29
50–	38
75–	45
100–	23
150–	17
200	11

Obtain estimates of the median:

a) using linear interpolation, **b)** using a cumulative frequency graph.

5 A student records his journey times, in minutes, each morning for a period of six weeks. These times are as shown here.

15.2	16.7	20.5	17.8	21.2	14.7	16.8	17.3	15.6	15.3
16.8	18.7	19.3	19.6	18.2	15.6	18.8	17.4	16.7	18.2
15.6	16.9	17.3	18.7	20.2	20.4	18.4	16.7	17.9	18.4

a) Present these data in an ordered stem-and-leaf display.

b) Use your stem-and-leaf display to obtain an estimate of the median time.

6 Lecturers are asked to limit their lectures to 50 minutes. A Statistics student records the durations, in minutes, of the lectures she attends and presents them in the following display.

Unit is 0.1

45	2	3	5								
46	7	6	2	1	3						
47	5	8	7	1	2	1					
48	9	6	5	0	1	3	6				
49	2	7	6	5	3	2	8				
50	1	2	3	8	6	7	5	9	0	0	5
51	3	7	4	6	7	2	0	1	5		
52	2	7	0	3							
53	5	6	2								

a) Present the data in an ordered stem-and-leaf display, adding the depths of each branch.

b) Use your ordered display to identify the median time.

c) Construct a histogram of these data and use this to obtain an estimate of the median.

d) Explain any difference in the two values you have found for the median of these data.

7 Obtain estimates of the medians for the data in **c)** and **d)** of Exploration 1.5.

MEAN

The **mean** is the measure of location which is generally associated with the term 'average'. In simple terms, it is obtained by adding up the items of data and dividing by the number of items. Adding up all the motorway toll times of Exploration 2.1 produces a total of 1722 seconds and there are 43 such times, hence the mean time is:

$$\text{mean} = \frac{\sum t}{n} = \frac{1722}{43} = 40.05 \text{ seconds}$$

CALCULATOR ACTIVITY

Does your calculator have a statistical mode?

Many calculators have a statistics mode which will allow you to input data and to display the mean of the data. You need to put the calculator into appropriate mode. In many cases this is the SD mode. Find out if your calculator has a statistical mode by trying to persuade it to accept the following simple collection of data.

6, 8, 9, 10, 12

The specific key sequence for data input will depend on your calculator. You may have one which allows you to input data to a list. You will need to use an arrow key to move your cursor down the list after inputting each data item.

Having input the data, you need to extract the mean. Before doing this, it is useful to check that your calculator has been working on only the five items of data. Your calculator should be able to tell you the number of items of data. This may be found by keying 'n' or 'Σ' etc. If you have a calculator which **lists** the data, simply display the appropriate list.

When the data have been correctly stored in the calculator, find out how to display the mean.

Note: A few calculators with a graphical display may also be able to display the **median** of a data collection. Find out if your calculator will display the median of a collection of data.

Grouped frequency

Consider again the grouped frequency version of the motorway times from Exploration 2.1.

Class (s)	Frequency
$0 \le x < 10$	4
$10 \le x < 20$	7
$20 \le x < 30$	9
$30 \le x < 40$	6
$40 \le x < 50$	5
$50 \le x < 60$	3
$60 \le x < 120$	9

When the data are presented in this way, we need to take several decisions, in order to calculate an estimate of the mean of the data. Note the use of the word 'estimate'. Why is it used?

Firstly, how do we interpret, say, the frequency 6 and the associated class '30–40'? There are six items of data in the range 30–40 s. The best that can be done without any further information is to interpret this as six items with an average value 35 s. In other words, applying this principle to the grouped frequency table yields the following table.

Representative time (s)	Frequency
5	4
15	7
25	9
35	6
45	5
55	3
90	9

(Why is the representative value of the last class 90?)

In effect, the grouped frequency data has been replaced by:
four occurrences of 5 seconds,
seven occurrences of 15 seconds,
nine occurrences of 25 seconds,

and so on. Perhaps the easiest way of totalling the time represented in these data is to use a combination of simple multiplication followed by addition. The table below summarises this process.

Time	Frequency	Time × frequency
5	4	20
15	7	105
25	9	225
35	6	210
45	5	225
55	3	165
90	9	810
Total	43	1770

Hence, the estimate of the mean obtained by this method is found as follows.

$$\text{Estimate of mean} = \frac{\sum xf}{\sum f} = \frac{1770}{43} = 412 \text{ seconds}$$

where x is the representative value for a class, and f is the corresponding frequency.

Note: The mean of this distribution of data is often written as $\bar{x}$, which is x with a line above it. This symbol is often referred to as 'x-bar'.

$$\bar{x} = \frac{\sum xf}{\sum f}$$

Why is this not the same as the mean found earlier? Which is likely to be the more accurate?

CALCULATOR ACTIVITY

Inputting frequency data

Statistical calculators usually allow frequency data to be input. A typical key sequence may be like this.

[5] [;] [4] [DT]

or

[5] [×] [4] [DT]

Note: The order 'data ; frequency' may be appropriate to indicate to the calculator that the data item $x = 5$ has frequency $f = 4$.

You may find that you can store the items of data in one list and their corresponding frequencies in another list. Perhaps the same data may be input directly into the lists. Experiment with your own calculator.

Try to input the following frequency distribution.

x (cm)	6	8	9	10	12
f	2	5	12	7	4

Check that your calculator produces these values.

$n = 30$
$\Sigma x = 278$ cm
$\bar{x} = 9.2667$ cm

Remember to clear or erase any data before inputting new data.

Note: Graphical calculators may be used with frequency distributions of discrete data (but usually not grouped frequency data or continuous data) to find an estimate of the median. The key to the median may well be labelled like this.

[Med]

Check your calculator has this facility; the median of the above collection is 9.

Try inputting the following discrete data distribution and check that the calculator produces the correct median value of 19.5 (lying between the 11th and 12th item).

x	16	18	19	20	22
f	2	5	4	7	4

Exploration 2.5

Coding

Consider the following data which are the air pressure recordings in mb for one week.

1012, 1008, 1002, 996, 990, 1002, 1004

What is their mean?

Now consider the same values with 1000 mb subtracted.

12, 8, 2, –4, –10, 2, 4

Their total is readily obtained as 14, hence their mean is 2. To find the mean of the original recordings, you need only add 1000.

This is an example of transforming or **coding** data to simplify manual calculations of the mean. In this case, the transformation was a translation where:
coded value = air pressure – 1000

Hence mean (coded) values = mean (air pressure – 1000)
= mean (air pressure) – 1000

Inverting this leads to:
mean (air pressure) = mean (coded values) + 1000

Other transformations are also possible; among these are:

a) $\dfrac{\text{air pressure} - 1000}{2}$ **b)** $\dfrac{\text{air pressure}}{100}$

c) $\dfrac{\text{air pressure}}{2} - 500$ **d)** $\dfrac{\text{air pressure} - 990}{2}$

Applying **a)** to the data yields:
coded data: 6, 4, 1, -2, -5, 1, 2

and so the mean (coded data) = $\frac{7}{7}$ = 1. The inversion of the transformation mean (air pressure) = 2 × mean (coded data) + 1000
= 2 × 1 + 1000
producing the same result as before.

Have a go at applying the other transformations.

Example 2.5

The heights of 32 children in a class are recorded, to the nearest centimetre, in this table.

Height (cm)	140–149	150–159	160–169	170–179	180–189
Number of children	4	10	7	9	2

Calculate an estimate of the mean using the midpoint of the class 160–169 to translate the data and a scale factor of $\frac{1}{10}$.

Solution

Since these data are grouped, representative values are needed for each group. The data are recorded to the nearest centimetre, which indicates, for instance, that a height in the range 148.5–149.5 would be recorded as 149. So the true class limits are as shown below.

Height (cm)	**Frequency**	**Representative value**	**Coded value, x**
$139.5 \le h < 149.5$	4	144.5	–2
$149.5 \le h < 159.5$	10	154.5	–1
$159.5 \le h < 169.5$	7	164.5	0
$169.5 \le h < 179.5$	9	174.5	1
$179.5 \le h < 189.5$	2	184.5	2

The mean of the coded values is:

$$\frac{\sum xf}{\sum f} = -\frac{5}{32}$$

Hence an estimate of the mean height is:

$$10 \times \left(-\frac{5}{32}\right) + 164.5 = 162.9 \text{ cm}$$

EXERCISES

2.3A

1 The values below show the contents, x, of each of 15 match-boxes.

40, 41, 42, 42, 40, 39, 40, 40, 42, 41, 40, 41, 40, 40, 42

a) Calculate the mean content.
b) Use the coding; $y = x - 40$, to find y, and hence $\bar{x}$.

2 The table below records the number of students absent from 80 tutorials during a year.

Number absent	**Number of tutorials**
0	5
1	21
2	27
3	12
4	8
5	5
6	2

a) Calculate the mean number of absences.
b) Use your calculator to obtain the mean number of absences.
c) If your calculator has the facility, use it to estimate the median number of absences.

3 The masses, x grams, of the fruit contents in 200 examples of canned fruit are recorded in the table.

Mass of fruit (g)	300	325	350	375	400	425	450
Number of cans	10	24	59	51	31	18	7

a) Use the coding $y = (x - 350)/25$, to calculate $\bar{y}$, hence obtain an estimate of the mean mass of fruit per tin.
b) Use your calculator to obtain an estimate of the mean mass of contents directly from the table.

4 Obtain an estimate of the mean time for the data in question 3 of Exercise Set 2.1A.

5 **a)** What are the representative values for the length classes in question 4 of Exercise Set 2.1A?
b) Use your calculator to obtain an estimate of the mean length for these data.

6 Forty-five students were asked how long it took them to travel to school. The results are shown below.

Time take to travel to school (minutes)	**Frequency**
0–9	22
10–19	0
20–29	0
30–39	7
40–49	9
50–59	7

a) Estimate the mean time taken to travel to school.
b) Comment on the 0–9 (minutes) class.

7 Estimate the mean number of runs scored by cricketer X.

Number of runs	**Frequency**
0–5	5
6–10	0
11–20	0
21–30	0
31–40	7
41–60	4
61–80	2
81–100	1
101–150	1

Comment on the distribution of the data.

EXERCISES

2.3 B

1 Recorded below are the marks obtained by 20 students in a Statistics examination.

56, 62, 40, 26, 46, 41, 67, 52, 70, 56,
48, 79, 50, 78, 46, 81, 47, 0, 54, 59

a) Calculate the mean mark.
You are now told that the 'zero' was due to absence rather than lack of knowledge.
b) What is a more appropriate mean mark?

2 One Saturday there were 90 hockey matches played in the Midlands. The number of goals scored in each game was recorded.

Number of goals	Number of matches
0	17
1	25
2	18
3	14
4	10
5	3
6	1
7	2

a) Calculate an estimate of the mean number of goals per game.
b) Use your calculator to obtain an estimate of the mean.
c) If your calculator has the facility, use it to obtain an estimate of the median number of goals.

3 The heights, x, in centimetres, were measured of the intake of male Science students at a university. The results are given in the table.

Height (cm)	135	145	155	165	175	185	195
Number of students	14	56	342	401	143	77	23

a) Use the coding $y = (x - 155)/10$, to calculate $\bar{y}$, hence, obtain an estimate of the mean height.
b) Use your calculator to obtain an estimate of the mean height directly from the table.

4 Obtain an estimate of the mean time for the data in question 3 of Exercise Set 2.1B.

5 Use your calculator to obtain an estimate of the mean mass for the data in question 4 of Exercise set 2.2A.

6 Recorded below are the lengths of hospital stay for patients undergoing a certain operation. Estimate the mean length of stay.

Length of stay in hospital (days)	**Frequency**
1–2	7
3–4	16
5–8	1
9–16	0
17–80	1

7 Estimate the mean number of runs scored by cricketer Y.

Number of runs	**Frequency**
0–5	1
6–10	0
11–20	0
21–30	0
31–40	7
41–60	4
61–80	2
81–100	1
101–150	5

Compare his performance with that of cricketer X in question 7 of Exercises Set 2.3A.

SYMMETRY IN DATA

Exploration 2.6

Comparing averages

So far in this chapter we have concentrated on finding a measure of location, an average, for the motorway toll times. From the grouped frequency presentation of the data estimates were found to be:

mode	24 seconds
median	33.3 seconds
mean	41.2 seconds

Choose the average you feel to be most representative of the data and justify your choice.

Symmetry

These diagrams illustrate data which are perfectly symmetrically distributed.

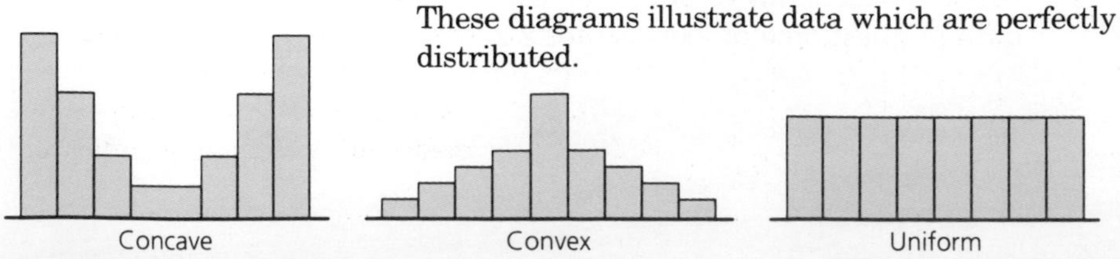

Whilst the natures of the distributions are different, they have one feature in common which is that there is no difference between their mean and median.

This difference '**mean – median**' is a simple measure of **symmetry**, or lack of it. If data are perfectly symmetric then the difference is zero. Data which are reasonably symmetrically distributed will have a difference close to zero. The **sign** of the difference provides an indication of the nature of the **lack of symmetry** or **skewness**.

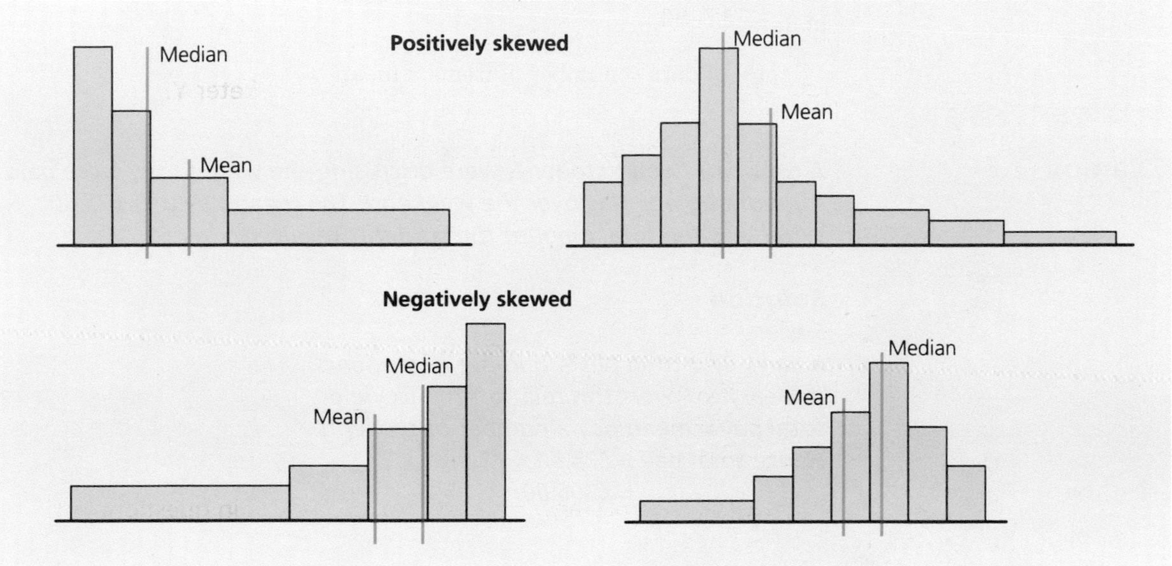

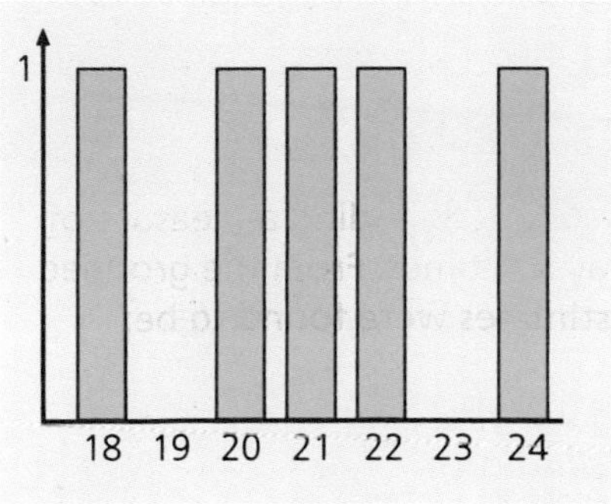

Why is the mean larger than the median for positively skewed data? Which measure of location is better? Consider the following data.

{18, 20, 21, 22, 24}

These could be displayed quite simply as shown here.

The mean is 21 and the median is also 21. The data are symmetrically distributed.

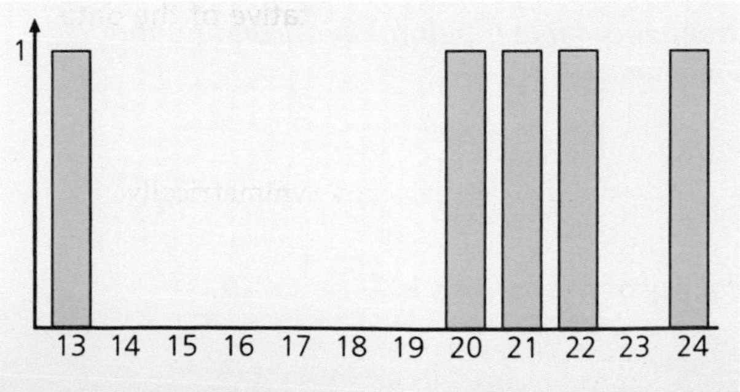

Now suppose that the 18 should have been 13, then this display would look like this.

Clearly the data is negatively skewed. What has happened to the two measures of location?

The median is still 21, but the mean has been reduced to 20. Thus the effect of skewing the data in this way has been to reduce their mean but it has had no effect on the median.

This property of the **median**, of not being affected by skew in data, leads to it being classified as a **robust** measure of location. But this does not make the median a better measure. The median's robustness is because it fails to take into account anything but the middle of the distribution. The mean, on the other hand, does take into account all the data. As a result of this it is affected by unusually large or unusually small values or by inaccurately recorded values.

One useful feature of the mean is that we can calculate the total of all the data from our known value of the mean and of the number of items of data.

total of data = number of items × mean

Example 2.6

A group of twelve students were discussing the wages they were paid for working one day over the weekend. The mean pay was £25.50. What was the total amount the students received?

Solution

Since:
mean pay = total pay ÷ number of students
it is easy to invert this relationship to yield:
total pay = mean pay × number of students
Hence total pay = £25.50 × 12
= £306.00

Example 2.7

The table records the weights, in kilograms, of the baggage of 79 passengers on a train.

Class (kg)	0–	1–	2–	5–	8–12
Frequency	11	21	33	6	8

a) *Using linear interpolation, calculate an estimate of median baggage weight.*
b) *Calculate an estimate of the mean weight.*
c) *Describe the symmetry of the distribution.*

Solution

a) *The cumulative frequency table is helpful in calculating the estimate for the median.*

Upper class value (kg)	1	2	5	8	12
Cumulative frequency	11	32	65	71	79

Cumulative frequency of the median is $\frac{79+1}{2} = 40$

Thus the median lies in the 2–5 kg class

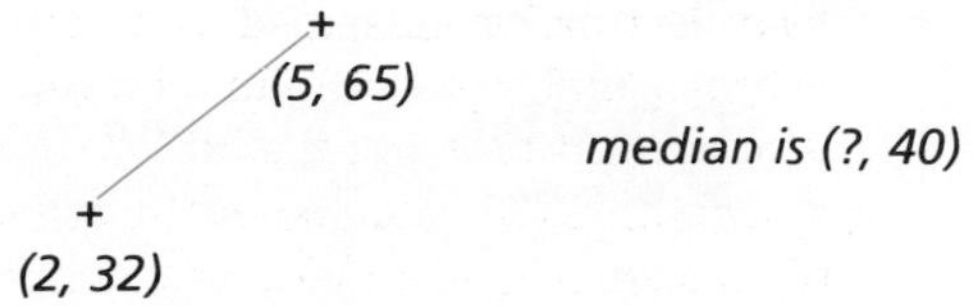

and it divides the class in the ratio 8 : 25.
Thus the estimate is:

$$2 + \frac{8}{33} \times 3 = 2.7 \text{ kg}$$

b) *The appropriate representative values for the classes, when calculating an estimate for the mean, are the mid-class values.*

Mid-class values x	0.5	1.5	3.5	6.5	10	
Frequency f	11	21	33	6	8	$\sum f = 79$
$x \times f$	5.5	31.5	115.5	39	80	$\sum xf = 271.5$

Hence

$$\bar{x} = \frac{271.5}{79} = 3.4 \text{ kg}$$

c) *Since the estimate for the mean is greater than that for the median, i.e.*

mean – median > 0

the data are positively skewed.

EXERCISES

2.4A

1 *Unit is 1*

0	6										
1	2	7	9	3	8	6					
2	5	6	9	4	6	8	8	1	6		
3	9	7	0	4	1	5	6				
4	0	3	7	8	1	0	3	9	9	6	5
5	1	4	7	0	3	6	0	2	5	9	
6	8	5	4	1	0	0	8	5	2	1	0
7	1	5	7	2	2	6	6				
8	9	3	2	4	7	2					
9	9	7									
10	6	5	9	4							
11	4	8	2	3							
12	7										

a) Use an ordered stem-and-leaf display to find the median of the data given above.

b) Write down the frequencies of the classes for the following collection of classes.

Collection of classes	Frequencies of classes
0–	
20–	
40–	
60–	
80–	
100–130	

and use these to calculate an estimate of the mean.

c) Comment on the distribution of these data.

2 Refer to the data in question 2 of Exercise Set 2.3A. Display these data in a frequency line graph. Use your display and estimates of the mean and median to comment on the distribution.

3 **a)** Present histogram of the data in question 3 of Exercise Set 2.3A.

b) Use a cumulative frequency graph, or linear interpolation, to obtain an estimate of the median mass of fruit per tin.

c) Comment on the symmetry of the data.

4 The table below records the weight loss, in kilograms, of each of 150 people attending a dietary clinic.

Weight loss x (kg)	Number of people
$-2 \le x < -1$	3
$-1 \le x < 0$	7
$0 \le x < 1$	12
$1 \le x < 2$	19
$2 \le x < 3$	38
$3 \le x < 5$	46
$5 \le x < 10$	25

a) What do you think the negative values indicate?

b) Obtain estimates of the values of:

i) the mean, **ii)** the median.

c) Present the data in a histogram.

d) Comment on the symmetry of these data.

5 A survey of the duration of telephone calls made from a Humanities department yielded the following results.

Duration (minutes)	0–	1–	3–	10–	15–	30–	60–
Number of calls	15	24	45	23	9	4	0

Comment on the symmetry of these data, justifying your comment.

EXERCISES

2.4 B

1 The reaction times, in seconds, of 38 drivers were measured with the results as shown in the table.

Reaction time (s)	< 0.3	0.3–	0.4–	0.5–	0.6–	0.8–1.0
Number of drivers	4	9	6	6	9	4

a) Present these data in a histogram and obtain an estimate of the mean reaction time.
b) Construct a cumulative frequency graph for these data and obtain an estimate of the median reaction time.
c) Comment on the symmetry of the data.

2 Refer to the data of question 2 of Exercise Set 2.3B. Display these data in a frequency line graph. Use your display and the estimates of the mean and median to comment on the distribution of these data.

3 **a)** Present the data in question 3 of Exercise Set 2.3B in a histogram.
b) Obtain an estimate of the median height.
c) Comment on the symmetry of these data.

4 Study these data.

Height of plant (nearest cm)	Number of plants
0–5	19
6–10	14
11–20	16
21–30	5
31–50	4

a) Draw a histogram of these data.
b) Obtain estimates of:
i) the mean, **ii)** the median,
height of plant.
c) Comment on the distribution of heights.

5 The table records the time, in minutes, that 50 people dining at a well-known fish restaurant had travelled to get to the restaurant.

Time (minutes)	0–	5–	10–	20–	40–	60–90
Number	11	15	8	7	6	3

a) Obtain estimates of:
i) the median, **ii)** the mean,
journey time.
b) Present a histogram of the data.
c) Comment on the distribution of journey times.

WEIGHTED MEANS AND INDEX NUMBERS

Exploration 2.7

Index numbers

Index numbers are frequently in the news. They are the subject of comment in newspapers, and on the radio, in television reports. The Retail Price Index is commonly quoted. Find out about the RPI. Find out about other index numbers.

Weighted means

Imagine you are running a café and that you offer your customers, when they order coffee, a choice from:

a) High Mountain,
b) Kenyan,
c) Mocha,
d) Columbian.

You charge the same amount irrespective of the coffee chosen.

Your task is to ascertain how much it is likely to cost you to provide this service to your customers. Once you know this you will be able to decide how much you should charge a customer for coffee.

Your coffee suppliers volunteer the information that the four coffees sell in the ratio 6 : 10 : 3 : 1 and cost 58p, 49p, 66p, 40p per 100 g respectively.

This is sufficient information to establish the likely average cost of a cup of coffee (i.e. the cost to *you* of the coffee). The data values are the prices per 100 g that you have to pay, the ratios can be considered to be 'relative' frequencies or, in this context, 'weights'. Hence, to estimate the average cost to you of a cup of coffee, you need to calculate the mean of the following distribution.

Cost per 100 g x (p)	**Weight** w
58	6
49	10
66	3
40	1

The weighted mean is thus:

$$\frac{\sum xw}{\sum w} = \frac{58 \times 6 + 49 \times 10 + 66 \times 3 + 40 \times 1}{6 + 10 + 3 + 1} = \frac{1076}{20} = 53.8$$

The interpretation of this result is that it is going to cost you, on average, 53.8 p to purchase 100 g of the customer's choice of coffee.

Note: The 'weights' used in this example could have been expressed in a variety of ways, all of which would lead to the same weighted mean. For instance, they could have been expressed:
as percentages: 30% : 50% : 15% : 5%

or as common fractions: $\frac{3}{10} : \frac{1}{2} : \frac{3}{20} : \frac{1}{20}$

Using the definition:

$$\text{weighted mean} = \frac{\sum x_i w_i}{\sum w_i}$$

leads to 53.8 in all cases.

Why? What advantage may there be in expressing weights in this way?

Index numbers

Suppose the coffee scenario was set two years ago. Today, the prices of the four coffees have risen to 65p, 55p, 75p, 45p.

The new weighted mean price is 60.5p. The **index** of coffee price today compared to two years ago is found as the ratio of the two prices expressed as a percentage.

$$\text{coffee price index} = \frac{60.5}{53.8} \times 100 = 112.5$$

Note that this indicates that the cost of coffee has risen by $12\frac{1}{2}\%$, on average, over the two years.

In simple terms, an index number for a value in year Y, based on year X, is:

$$\frac{\text{value in year } Y}{\text{value in year } X} \times 100$$

Example 2.8

The cost of salad vegetables in consecutive years is given in the table.

Vegetable	Cost in year 1 (p)	Weight	Cost in year 2 (p)
Tomato	45	0.4	54
Lettuce	27	0.3	28
Spring onion	15	0.1	20
Peppers	89	0.1	160
Sweetcorn	25	0.1	20

Calculate an index for the cost of salad vegetables in Year 2 based on their cost in Year 1 and interpret the index.

Solution

The weighted mean costs in years 1 and 2 are:

Year 1: $45 \times 0.4 + 27 \times 0.3 + 15 \times 0.1 + 89 \times 0.1 + 25 \times 0.1 = 39\text{p}$

Year 2: 50p

Hence the index number is:

$$\frac{50}{39} \times 100 = 128.2$$

This indicates that the cost of a salad has risen by 28.2% over the period.

Mean of combined groups of data

Two students independently collect leaves from a privet hedge and measure their lengths. One student, Christine, reports that she measured 80 leaves and her leaves had a mean length of 15.8 mm. David measured 50 leaves and his mean was 17.5 mm.

What is the mean length of a leaf?

In our first approach to this, we can use the mean to recover the total of the data. Christine's leaves totalled:

$(15.8 \times 80) = 1264\,\text{mm}$

and David's totalled:

$(17.2 \times 50) = 860\,\text{mm}$

Thus, the total length of all 130 leaves is 2124 mm. Hence, the mean leaf length is:

$2124 \div 130 = 16.3\,\text{mm}$

Another approach is to use a weighted mean.

Christine's data had a mean of 15.8 and represented 80 out of 130 lengths. David's data had a mean of 172 and represented 50 out of 130 lengths. Hence, the overall mean length is the weighted mean:

$$\frac{80}{130} \times 15.8 + \frac{50}{130} \times 17.2 = 16.3$$

Why are these approaches equivalent?

Geometric mean

A traditional method of estimating the age of a tree is to count its 'rings'. This method has a number of drawbacks: the tree has to be felled to expose the rings; the rings are not always readily distinguished.

In homogeneous woodlands (i.e. where the trees are all the same species), an alternative method can be used.

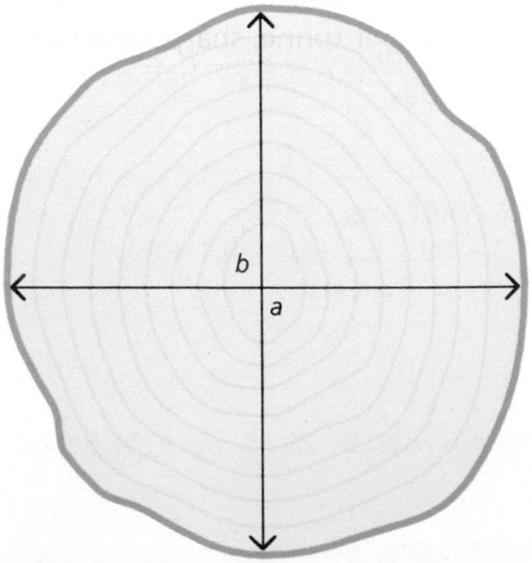

This method involves estimating the diameter of the trunk at the fell height. As no trunk is circular in cross section, two 'diameters', a and b, measured at right angles are recorded.

The square root of their product, $\sqrt{ab}$, is then taken to be the measure of the diameter of the trunk. This value is the **geometric mean** of the two 'diameters'.

The geometric mean of two positive numbers, x and y, is $\sqrt{xy}$.

This concept can be extended to three, four, five, ... positive numbers. For instance:

Geometric mean of x, y and z is $\sqrt[3]{xyz}$.

Geometric mean of n numbers: x_1, x_2, x_3, ... xn is the nth root of their product $\sqrt[n]{x_1 x_2 x_3 \ldots x_n}$.

The geometric mean is often used in quoting the average of several index numbers. Consider the following situation.

	Base year	Year 1	Year 2	Year 3
Index	100	110.0	121.0	133.1

The average index over the three years is:

$$\sqrt[3]{110.0 \times 121.0 \times 133.1} = 121$$

EXERCISES

2.5A

1 The costs, in pence, of certain food items in two consecutive years are recorded in the table. The relative importance of the items is indicated by the weights.

Item	Potatoes	Eggs	Cheese	Meat
Cost in Year 1	21	69	189	285
Cost in Year 2	24	89	209	315
Weight	2	5	4	9

a) Calculate weighted means for each year.
b) Calculate an index of food prices in Year 2 based on Year 1. Interpret the value of the index.

2 For each of the following sets of data, calculate both the arithmetic mean and the geometric mean.

a) 2, 4, 4, 8
b) 12.6, 15.8, 17.8
c) 125, 134, 96, 153
What do you notice about your values for each data set. Is this generally true? Can you prove it?

3 A type of concrete is made by mixing gravel, sharp sand and cement in the ratio 3 : 2 : 1. If the gravel costs £4 per tonne, sharp-sand £7 per tonne and cement £49 per tonne, calculate the cost of half a tonne of concrete.

4 Four groups of students sit the same Statistics examination, with the following results.

Group	Mean mark	Number of students
A	56	22
B	62	26
C	47	23
D	51	29

Calculate the mean mark for the 100 students.

5 The table below shows the mean cost of a householder's newspapers and magazines in consecutive years.

	Weight	Cost	Cost
		Year 1	Year 2
Morning papers	6	£0.44	£0.48
Evening papers	6	£0.25	£0.26
Sunday papers	1	£1.49	£1.64
Weekly paper	1	£0.28	£0.30
Magazine 1	1	£0.55	£0.65
Magazine 2	0.25	£2.60	£2.75

Calculate the percentage increase in this household's weekly expenditure on newspapers and magazines over the two years.

EXERCISES

2.5 B

1 The fat content of the four options for school lunch was as follows; the meals were chosen in the proportion 30% : 35% : 25% : 10%.

Meal	Fat (g)	Weight
A	63	6
B	24	7
C	45	5
D	32	2

a) Explain how the weights were obtained.
b) Calculate a weighted mean and explain your result.

2 The costs, in pence, of the ingredients of a sandwich are given in the table. Calculate an index for the cost of sandwich ingredients in Year 2 based on Year 1 and interpret the index.

Item	Bread	Butter	Cheese	Tomato
Cost in year 1	47	56	148	17
Cost in year 2	48	58	132	14
Weight	5	1	3	1

3 Two children baked biscuits. Jane's 22 biscuits had a mean mass of 12 g. Colin's 28 biscuits had a mean mass of 14.5 g. Calculate the mean mass of all the 50 biscuits produced.

4 Sample A has a mean of 10 whereas sample B has a mean of 30. The samples are combined. Calculate the resulting mean in each of these cases. Explain your answers.

a) $n_A = n_B = 50$
b) $n_A = 90 \quad n_B = 10$
c) $n_A = 20 \quad n_B = 80$

5 The table below shows the annual profit made by a small manufacturing company over four years.

Year	1	2	3	4
Profit (£)	26 310	32 425	29 985	36 250

a) Express the profits for each of years 2, 3 and 4 as a percentage of that for the previous year.
b) Use the geometric mean to find the average of your three percentages in **a)**. What is the company's mean rate of increase of profits for the four-year period?

CONSOLIDATION EXERCISES FOR CHAPTER 2

1 The following are the numbers of persons on a sightseeing coach tour on each of 100 days in the holiday season.

Class (number of persons)	**Frequency**
under 10	17
10–14	5
15–19	3
20–29	7
30–34	11
35–39	37
40–44	20

a) Identify the modal class.
b) Calculate an estimate of the mode using a histogram of the data.
c) Comment.

2 Calculate estimates of:

a) the mean,
b) the median,
for the following data.

Number of fish dinners eaten in three months	**Frequency**
0–4	11
5–9	15
10–19	8
20–39	7
40–59	6
60–89	3

3 Calculate estimates of:

a) the mean,
b) the median,

for the following data which shows the distribution of lengths of fish sampled from a trout farm.

Length (cm)	**Frequency**
0–4	12
–6	25
–8	17
–10	38
–12	12
–14	9
–16	5
–18	4
–20	3
–22	18
–24	26
–26	9
–28	2

4 The following are the ages of 25 people living in an old people's home.

65 72 73 81 60 69 75 78 86 82 87 64 73
72 66 80 69 80 76 74 81 93 65 72 72

a) Use a suitable stem-and-leaf diagram to find the median of these data.
b) Construct a suitable frequency table for these data and use it to calculate an estimate of the mean age.
c) Comment on the distribution of these data.

5 Find the mean and the median for each of these two data sets and construct a simple illustration for each.

a) 2 5 7 9 12
b) 1 2 6 7 10 11 12

Construct a data set of five values:

c) with mean 7 and median 4,
d) with mean 4 and median 7,
and illustrate each.
e) Comment on the symmetry or the lack of symmetry in each of the four data sets.

6 A weekly newspaper offers a free book to new subscribers. The five books on offer cost £7.50, £12, £6.99, £4.99 and £13.25 respectively and are chosen in the proportion:

18% : 15% : 31% : 28% : 8%.

Calculate the weighted mean cost to the newspaper of offering this incentive.

7 A company uses four raw materials in its production process. The cost per kg of each is given in this table:

Material	Cost per kg (£) in year 1	Weight	Cost per kg (£) in year 2
A	7.30	6	7.50
B	8.40	3	8.60
C	2.15	9	2.70
C	10.45	2	9.95

Calculate an index for the cost of production in Year 2 based on production costs in Year 1. Interpret the index.

8 There are three possible routes from Philip's home to his college. Over a term, he drove by route A 20 times with a mean journey time of 23.4 minutes. He also travelled by route B 20 times with a mean journey time of 25.7 minutes and by route C 15 times with a mean journey time of 26.0 minutes. Calculate his overall mean journey time.

9 Consider the situation as recorded in this table.

	Base year	Year 1	Year 2	Year 3	Year 4	Year 5
Index	100	105	113	103	94	87

Use the geometric mean to calculate, to the nearest whole number, the average index over the five years.

10 A packaging process is supposed to fill small boxes with approximately 50 raisins, so that each box will weigh the same. A count of the contents of each of a selection of 120 boxes resulted in the data below.

57	51	53	52	50	59	51	51	52	52	44	53
45	57	49	53	58	47	51	48	49	49	44	54
46	52	55	54	47	53	49	52	49	54	57	52
52	53	49	47	51	48	55	43	55	47	53	43
48	46	54	46	51	48	53	56	48	47	49	57
55	53	50	47	57	49	43	58	52	44	46	59
57	47	42	59	49	53	41	48	59	53	45	45
56	40	46	49	50	57	47	52	48	50	45	56
47	42	48	46	44	53	47	55	48	51	52	54
49	51	53	48	50	45	55	49	48	53	56	52

a) Use a two-part stem-and-leaf diagram to display these data.
b) Calculate the median number of raisins in the box.
c) Estimate the mean number of raisins per box.

11 The numbers of people per household in a road were recorded as shown in this table.

Number of people	1	2	3	4	5	6	7	8
Number of households	4	8	12	20	13	5	2	1

a) Use a step cumulative frequency diagram to find the median number of people per household.
b) Calculate the mean number of people.
c) Identify the modal number of people.
d) Which is the most appropriate measure of location for these data?

Summary

■ Measures of location

Mode: the item with greatest frequency

Median: the middle of the data
its cumulative frequency is $\frac{n+1}{2}$

Mean: $\bar{x} = \frac{\sum xf}{n}$

Weighted mean: $\frac{\sum xw}{\sum w}$

Geometric mean: $\sqrt{ab}$ for two items of data, a and b

$\sqrt[n]{x_1 x_2 x_3 \ldots x_n}$ for n items, $x_1, x_2, \ldots x_n$

■ Symmetry

Perfectly symmetric: mean – median = 0
Positively skewed: mean – median > 0
Negatively skewed: mean – median < 0

STATISTICS

3

Summary statistics 2

In this chapter:

- *we introduce various measures to describe the spread or dispersion of a range of data, including the range, variance and standard deviation*
- *we use box and whisker diagrams as visual displays of symmetry and spread of data.*

In Chapter 2, *Summary statistics 1*, we discussed some of the issues an experimenter working on the motorway tollgate data needed to consider, most notably typical times. In this chapter we shall consider the 'spread' of times.

MEASURES OF DISPERSION

Range

Perhaps the most simple measure of spread is the difference between the largest and smallest items of data i.e. the difference between the extremes. This is the **range**. In the case of the motorway toll times (page 30) the longest time recorded was 118 seconds and the shortest time was 9 seconds, hence the range of these data is given by:

range = 118 – 9 = 109 seconds

This measure of spread does not take into account anything about the distribution of the data other than the extremes. Neither is it very reliable or typical. Why?

Quartile spread

A more trustworthy measure is the range of the **middle half** of the data. To identify this range we need to find the items of data which are positioned halfway between the extremes and the median. Take the case of the data collection on page 36.

3	4	5	6	7	7	8	8	8	10	15
		↑			↑			↑		
					median					

By writing them in order, we can identify the middle 'half' of the data. We find that it lies between the values of 5 and 8.

It is now possible to identify the values between which the middle half of the motorway toll times lie. Take a moment to find them.

You should have found that the middle half of these times lies between 18 seconds and 58 seconds.

In general, the items of data lying midway between the median and the extremes are known as the **quartiles**. It is more usual to refer to them as the first or **lower** quartile and the third or **upper** quartile. The difference between them is called the **interquartile range** (IQR) or **quartile spread** (QS).

In this case of the motorway toll times, the quartile spread is 40 seconds.

What do you think the second quartile is called? Think about it for a moment.

Stem-and-leaf

As for the median, there are corresponding techniques for finding the quartiles. With the 43 motorway toll times, the depth of the median is 22, hence the depth of the quartiles is 11 (recall depth is a count of distance from the nearer extreme).

	4	0	9	9	9	9							
depth of quartile	11	1	2	3	3	5	6	6	8			*11th item from extreme*
	20	2	1	3	4	5	5	5	7	8	9	
	(6)	3	0	2	3	4	7	8				
	17	4	2	3	5	5	6					
depth of quartile in here	12	5	6	8	9							*11th item from extreme*
	9	6	1	4	8							
	6	7	4	7								
	4	8	3									
	3	9										
	3	10	4	7								
	1	11	8									

If there are n items of data, the quartiles have a depth of $\frac{n+1}{4}$.

Five-number summary

The values of the extremes, quartiles, and median of a collection of data provide a simple summary of the distribution.

	Lower		Upper
M		32	
Q	18		58
E	9		118

A graphical version of this five-number summary table, known as a **box-and-whisker diagram**, provides a visual presentation enabling us to focus on each of the four quarters of the data.

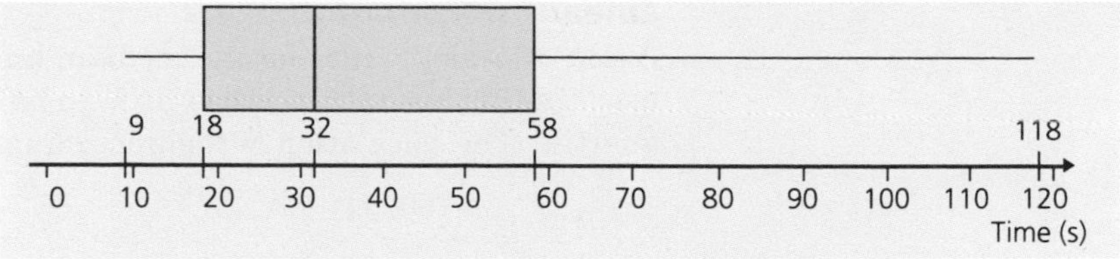

Vertical lines mark the quartiles and the median. These are joined to make a box containing the middle half of the data. From the quartiles, horizontal lines are drawn to the extremes – the lines are the 'whiskers'. What features of the distribution are apparent from a box-and-whisker diagram?

Grouped frequency table

The quartiles may also be found using cumulative frequencies. Unlike using depths, the cumulative frequencies of the first and third quartiles are not identical but are at $\frac{1}{4}(n + 1)$ and $\frac{3}{4}(n + 1)$ when there are n items of data.

The cumulative frequency table is reproduced below.

	Upper class value(s)	Cumulative frequency
	10	4
First quartile →	20	11
	30	20
	40	26
	50	31
Third quartile →		←
	60	34
	120	43

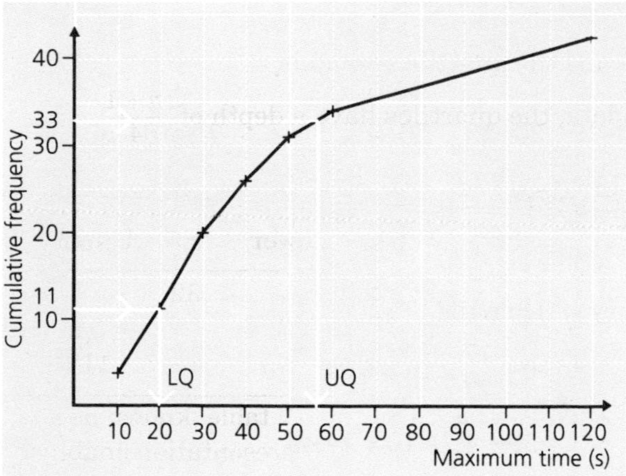

Graphical approach

For the motorway times, the quartiles are the 11th and 33rd items. The cumulative frequency graph can be used as illustrated where the values are:

lower quartile: 20 seconds
upper quartile: 56.7 seconds

Hence the quartile spread is:
56.7 – 20 = 36.7 seconds

Note: These values differ from those obtained using the stem-and-leaf display.

Linear interpolation

Again, as was the case for the median, linear interpolation may be used. In this instance, there is little to be done to calculate the lower quartile. The calculation of the upper quartile is:

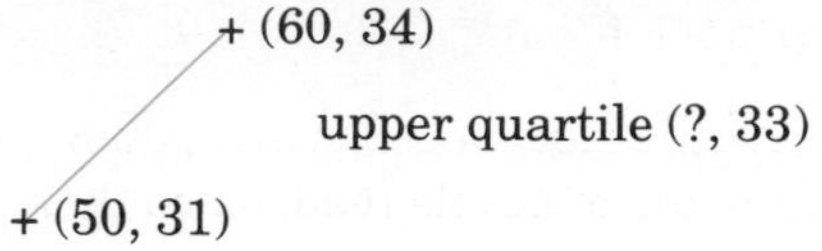

Upper quartile: $50 + \frac{2}{3} \times 10 = 56\frac{2}{3}$

Lower quartile: 20

Hence, linear interpolation yields the same result. Can you explain why?

Discrete data

If the data are discrete then their quartiles may be found as in the following illustrations.

Pairs of shoes (see page 43)	6	7	8	9	10	11	12
Frequency	2	4	7	12	8	4	2
CF	2	6	13	25	33	37	39

Since n = 39, the cumulative frequencies of the quartiles are:
cumulative frequency

Q1: $\frac{39+1}{4} = 10$ Q3: $\frac{3}{4}(39+1) = 30$

Interpreting the cumulative frequency table: there are six people with seven or fewer pairs of shoes and 13 people with eight or fewer. Hence the tenth item of data is eight pairs of shoes. Similarly, the 30th item of data is ten pairs of shoes.

$\Rightarrow$ Q1 = 8 and Q3 = 10 and the quartile spread is 2.

Number of novels read (see page 44)	0–9	10–19	20–29	30–49	50–69	70–79	80–99
Frequency	27	18	11	5	3	2	1
CF	27	45	56	61	64	66	67

	Cumulative frequency	Quartile
Q1	$\frac{1}{4}(67+1) = 17$	$0 + \frac{17}{27} \times 10 = 6.30$
Q3	$\frac{3}{4}(67+1) = 51$	$19 + \frac{6}{11} \times 10 = 24.45$

Recall that the data are discrete, hence the estimates of the quartiles are:

Q1 = 7 Q3 = 25

and the quartile spread is 18 novels.

CALCULATOR ACTIVITY

Calculating quartiles

A few graphical calculators will present estimates of the quartiles (but, as already mentioned, only for discrete data).

The quartiles for the shoe sizes, above, may be calculated but not for the numbers of novels read, since these are grouped.

Note: Not all calculators which present estimates of the median also present estimates of the quartiles.

You may find that your calculator will display data in a box-and-whisker diagram. However, since this presentation often uses the same quartiles, its relevance is restricted to discrete data.

Example 3.1

The heights of 49 earthenware jars found on an archaeological site are given in the table.

a) *Use linear interpolation to calculate estimates of the quartiles.*
b) *Provide an estimate of the interquartile range.*
c) *Draw a box-and-whisker diagram of the data.*

Class (cm)	**Frequency**
16–	6
18–	9
20–	8
22.5–	6
25–	7
27.5–	9
30–32	4

Solution

a) *The first step is to calculate the cumulative frequency distribution.*

Upper class value (cm)	18	20	22.5	25	27.5	30	32
Cumulative frequency	6	15	23	29	36	45	49

There are 49 items of data, hence the cumulative frequencies of the quartiles are:

Lower quartile CF: $\frac{1}{4}(49+1) = 12.5$

Upper quartile CF: $\frac{3}{4}(49+1) = 37.5$

Lower quartile: this lies in the 18–20 class

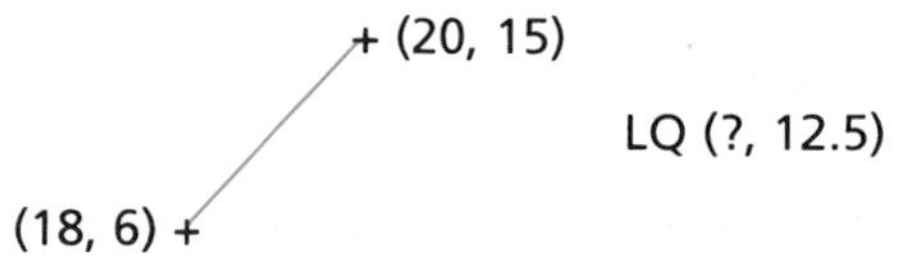

$\text{LQ} = 18 + \frac{6.5}{9} \times 2 = 19.4 \text{ cm}$

Upper quartile: is in the 27.5–30 class

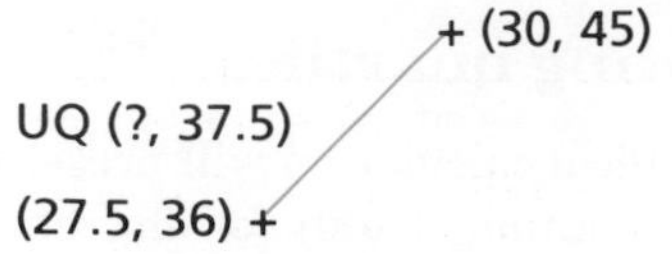

$UQ = 27.5 + \frac{1.5}{9} \times 2.5 = 27.9$ cm

b) *An estimate of the IQR is 27.9 – 19.4 i.e. 8.5 cm.*

c) *The median is needed for a box-and-whisker display of the data and is:*

$22.5 + \frac{2}{6} \times 2.5 = 23.3$ cm

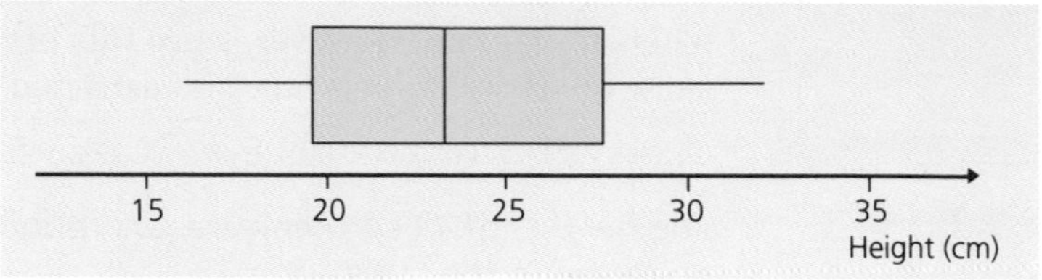

Note: With grouped frequency, the lower and upper extremes are estimates of the lower and upper limits of the first and last classes.

Exploration 3.1 *Comparing distributions*

Three package holiday companies were asked to provide prices for 220 different 14-day all-inclusive holidays. The data returned by the companies are given in this table.

Company A		Company B		Company C	
Cost (£)	**Number of holidays**	**Cost (£)**	**Number of holidays**	**Cost (£)**	**Number of holidays**
100–	62	0–	5	150–	10
200–	20	100–	30	250–	45
250–	30	200–	40	300–	65
300–	73	300–	50	350–	50
500–	30	400–	45	400–	30
550–600	5	500–	35	450–550	20
		600–700	15		

Estimates for the median and quartiles of the costs for the holidays offered by company A are:

median $250 + \frac{28.5}{30} \times 50 = £297.50$

lower quartile $100 + \frac{55.25}{62} \times 100 = £189.11$

upper quartile $300 + \frac{53.75}{73} \times 100 = £373.63$

Hence, a five-number summary for these data is:

	Lower		Upper
Median		297.50	
Quartiles	189.11		373.63
Extremes	100		600

Verify that a five-number summary for the costs of the holidays offered by Company B is:

	Lower		Upper
Median		371.00	
Quartiles	250.63		490.56
Extremes	0		700

Obtain a five-number summary for the data provided by Company C.

Exploration 3.2

Value for money

A convenient visual method of comparing the three collections of data is to present a 'box and whisker' of each on one schematic diagram with a single scale.

Schematic diagram

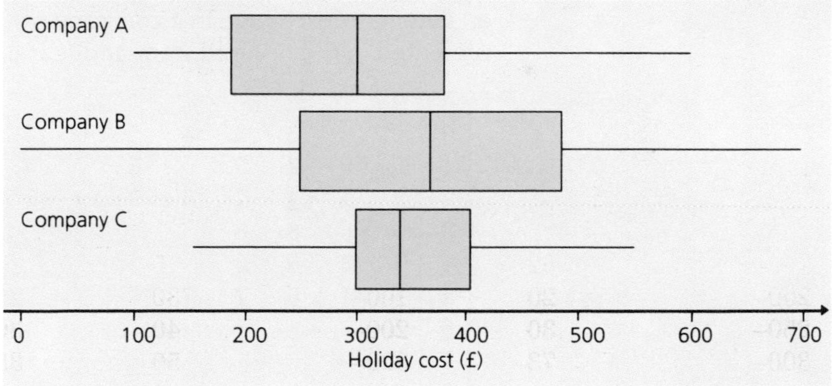

Which company would you choose for your fortnight's holiday? Discuss your answer.

EXERCISES

3.1A

1 Calculate the range and quartile spreads for each of the following data collections.

a) 6, 6, 8, 9, 9, 10, 11

b) 0.1, 0.2, 0.2, 0.3, 0.4, 0.4, 0.4, 0.5, 0.5, 0.6, 0.7

c) 312, 313, 315, 316, 320, 328, 330, 333, 358

d) 4.1, 4.2, 4.3, 4.5, 4.5, 4.6, 4.6, 4.6, 4.6, 4.7, 4.8, 4.9

2 Calculate the median for each data collection in question 1 and draw box-and-whisker diagrams for each collection.

3 **a)** Using a cumulative frequency graph, calculate estimates of the median and quartiles for the data recorded in the table.

Distance (cm)	0–10	–20	–30	–50	–100	>100
Frequency	25	35	20	15	5	0

b) Provide an estimate of the quartile spread for these data.
c) Present the data in a box-and-whisker diagram.

4 The lengths, to the nearest millimetre, of 99 screws were measured and recorded in the table.

Length (mm)	25–29	30–34	35–39	40–49	50–69
Number of screws	5	12	33	28	21

a) Use linear interpolation to estimate:
i) the median length,
ii) the upper and lower quartile lengths.
b) Display the data in a box-and-whisker diagram.
c) Comment on the distribution of these lengths.

5 The lengths of time, in minutes, of 49 telephone calls were logged by an automatic exchange in a company office. These times are presented in the ordered stem-and-leaf diagram below.

Unit is 0.1

0	0	2	3	3	4	5	6	7	8	9	9	9
1	0	0	2	3	5	6	7	7				
2	1	1	4	7	8	9	9					
3	3	6	8	8								
4	1	3	5	6								
5	0	1	6	7	7							
6	5											
7	7	9										
8	3											
9	1											
10	1											
11	3											
12	3	3										

For these data:
a) identify the median and quartile times,
b) present the data in a box-and-whisker plot.

6 The differences, in seconds, in time taken, by each of a group of 27 students, to write a sentence with their non-writing hand and then with the other hand are recorded below.

0.3, 0.4, 0.6, 0.7, 0.9, 1.4, 1.4, 1.9, 2.0, 2.1, 2.2, 2.3, 2.3, 2.7, 2.7, 2.7, 2.7, 2.8, 3.1, 3.7, 4.4, 4.5, 4.7, 5.3, 6.7, 7.1, 8.9

a) Present these data in an appropriate ordered stem-and-leaf display.
b) Produce a five-number summary for these data.
c) Draw a box-and-whisker diagram of the data.

7 The heights of 15 boys and 17 similarly-aged girls are recorded in the frequency table below.

Height (h cm)	Number of boys	Number of girls
$140 \le h < 145$	1	1
$145 \le h < 150$	2	0
$150 \le h < 155$	3	1
$155 \le h < 160$	5	4
$160 \le h < 165$	3	6
$165 \le h < 170$	1	3
$170 \le h < 175$	0	2

a) **i)** Obtain estimates of the median height for the boys' data.
ii) Estimate the quartile height for the boys.
iii) Record your estimates in a five-number summary.
b) Repeat **a)** for the girls' data.
c) Present box-and-whisker diagrams in a schematic diagram and compare the distributions of heights of the boys and girls.

8 In a survey, 27 guest houses and hotels were randomly selected and the price of one night's bed-and-breakfast was collected for four different times of the year.

	Frequency			
Cost of B&B (£x)	Spring	Winter	Autumn	Summer
$0 \le x < 10$	1	0	0	0
$10 \le x < 20$	3	3	7	2
$20 \le x < 25$	4	0	2	0
$25 \le x < 30$	2	4	4	3
$30 \le x < 35$	3	3	6	3
$35 \le x < 40$	6	3	3	2
$40 \le x < 50$	1	10	3	10
$50 \le x < 60$	6	2	1	3
$60 \le x < 70$	1	2	1	3
$70 \le x < 80$	0	0	0	0
$80 \le x < 100$	0	0	0	1

a) For each period of the year, calculate estimates of the median and quartile costs, and produce a five-number summary of the data.
b) Draw box-and-whisker diagrams for the data of each period in a single schematic diagram.
c) Discuss the similarities and differences in the distributions of prices.

EXERCISES

3.1B

1 Calculate the range and interquartile range for each of the following sets of data.

a) 2, 3, 4, 5, 5, 6, 7, 9, 11, 12, 12
b) 45, 46, 52, 55, 59, 62, 68, 73, 75, 84
c) 101.3 101.4 101.6 101.7 101.7 101.8 102.0 102.3 102.3 102.4 102.6
d) 506 503 507 504 510 511 526 513 517 508 515 513 508 509 516

2 Calculate the median for each data set in question 1 and hence draw the associated box-and-whisker diagrams.

3 Calculate, using a cumulative frequency graph, estimates of the three quartiles for the following data.

Time (s)	1–10	11–20	21–30	31–40	41–50	51–100
Frequency	2	12	28	21	11	6

Hence construct the associated box-and-whisker plot.

Do you think you would have obtained the same results from the original data? Explain your answer.

4 The fuel usage of 35 cars in a test drive is shown in this table.

Miles per gallon	10–19	20–24	25–29	30–34	35–39	40–49
Number of cars	2	4	5	13	8	3

a) Estimate the median and quartiles for the data.
b) Provide a five-number summary for the data.
c) Present the distribution in a box-and-whisker diagram.

5 The TV audience figures (in millions) for 59 editions of a new programme were estimated. The audience figures are presented in the ordered stem-and-leaf display.

Unit is 100 000

0	1 3 3 5 8
1	0 1 1 2 3 3 5 5 6 7 8 8 9
2	0 0 1 2 2 3 4 4 4 5 5 6 6 7 8 8 8 9
3	1 3 3 3 3 4 5 5 6 6 6 8
4	0 0 1 1 2 4 6 7
5	2 8
6	
7	
8	3

a) Identify the median and quartile audience figures for the data.
b) Present the data in a box-and-whisker plot.

6 The numbers of lunches served in a restaurant on each of 23 days are given below.

58, 80, 85, 90, 91, 96, 98, 99, 101, 102, 106, 111,115, 118, 118, 120, 121, 126, 129, 135, 146, 148, 158

a) Present the data in an ordered stem-and-leaf display.
b) Present the data in a five-number summary.
c) Present the data in a box-and-whisker plot.

7 The weights of 99 cod landed in each of two ports are recorded in the table.

Weight (lbs)	0–	10–	12–	14–	16–	18–	20–40
Number of fish Port A	33	22	21	10	7	3	3
Number of fish Port B	12	15	32	21	10	4	5

a) Estimate the median and quartile weights of fish landed in each port.
b) Provide a five-number summary for the data in each port.
c) Display the distribution of weights of these cod in a schematic diagram.

Comment on the distributions of fish weights in the two parts.

BOX-PLOTS AND OUTLIERS

Using a refinement of the box-and-whisker diagram, the same features of symmetry, location and spread can be demonstrated at a glance and unusually large or small items of data can be highlighted.
The **box-plot** is not suitable for data presented as a grouped frequency distribution since its intention is to give details about actual data.

Note:
QS quartile spread
LQ lower quartile
UQ upper quartile

Fences are values defined in relation to the quartiles of the distribution.

Inner fences: the lower quartile is calculated as – 1QS
the upper quartile is calculated as + 1QS
Outer fences: these are calculated as LQ – 2QS
UQ + 2QS

In many distributions, approximately 95–96 per cent of the items will be between the inner fences. So it would be an unusual item of data which would lie further away from the median than an inner fence. Such items of data are referred to as **outliers**.

In many distributions, less than 0.1 per cent of data lie outside the outer fences. These items are sometimes referred to as **extreme outliers** or they may be described as **far out**!

For the motorway toll times (page 30), the quartiles are 18 s and 58 s, and the quartile spread is 40. Hence the fences have the following values.

	Lower	Upper
Inner fences:	18 – 40 = –22	58 + 40 = 98
Outer fences:	18 – 80 = –62	58 + 80 = 138

Note that, in this case, the lower fences are negative and so there are no items of data outside these. Note also that the upper outer fence is larger than any of the items of data. However, there are items of data, 104, 107, 118 with values larger than the upper inner fence. It would be appropriate to describe these items as outliers.

The refined box-plot of these data looks like this.

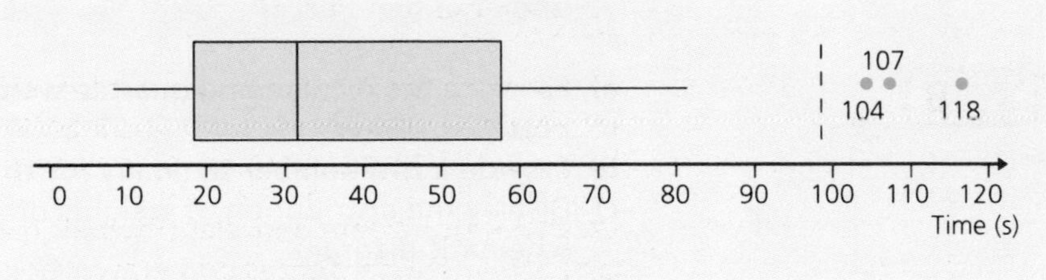

Notice that:

a) the 'whiskers' now only go out to the item of data just inside the inner fence(s),
b) the fences are shown as dotted vertical lines,
c) the outliers are individually marked.

EXERCISES

3.2A

1 The heights, in centimetres, of 29 foxglove plants are recorded below.
29.8, 32.1, 33.2, 36.4, 36.8, 36.9, 37.1, 37.4, 37.8, 38.1, 38.7, 39.2, 39.3, 39.5, 40.0, 41.3, 42.8, 42.9, 43.0, 43.01, 43.04, 43.05, 43.05, 43.06, 44.07, 44.08, 46.02, 47.07, 48.06, 53.03

a) Identify the median and quartile heights.
b) Calculate the values of the inner fences, and hence identify any outliers in the data.
c) Draw a box-plot of the data.

2 Determine the outliers, if any, in the data collections of question 1, Exercise Set 3.1A.

3 **a)** Calculate inner and outer fences for the data shown in question 5, Exercise Set 3.1A.
b) Identify the outliers in the data.

4 Identify the outliers in the data in question 6, Exercise Set 3.1A and draw a box-plot of the data.

5 a) Order these data and identify their median and quartiles.
b) Calculate inner fences and draw a box-plot of the data.
c) Identify outliers in the data.

4.53	2.18	7.93	1.98	6.40	3.73	3.59	0.04	0.94	3.04
1.18	2.52	9.94	7.63	8.81	1.27	6.36	2.70	3.01	0.13
6.25	5.27	6.28	8.60	5.56	2.84	8.59	5.14	8.86	0.18
7.71	0.77	2.36	2.46	7.94	8.92	8.26	8.39	1.88	2.63
2.67	1.66	5.06	4.87	4.35	4.15	8.03			

6 a) Calculate the interquartile range for the following data.
b) Identify the outliers in the data.
c) Draw a box-plot of the data and comment on their distribution.

5.41 2.89 4.28 7.12 3.37 6.50 6.23 8.62 4.99 3.79 7.26 1.77 5.62
8.11 3.45 3.80 6.82 9.85 6.63 9.35 5.43 4.79 6.55 1.88 0.04

EXERCISES

3.2B

1 The numbers of ferry passengers on 43 trips are shown below.
65 70 12 71 80 62 68 68 43 73 57 64 80 78 61 20 52 80 54 60 80
63 58 56 74 80 79 48 15 80 50 68 75 38 54 80 46 70 76 69 78 72 80

a) Order these data and identify their median and quartiles.
b) Calculate inner fences and draw a box-plot of the data.
c) Identify outliers in the data.

2 Determine the outliers, if any, in the data colletions of question 1, Exercise Set 3.1B.

3 a) Calculate inner and outer fences for the data shown in question 5, Exercise Set 3.1B.
b) Identify any outliers in the data, highlighting the extreme outliers.

4 Identify the outliers in the data in question 6, Exercise Set 3.1B and draw a box-plot of the data.

5 The stem-and-leaf display below summarises the times taken by 50 eleven-year-old girls to complete a circuit-training exercise in PE.

Unit = 0.1

5	2	7										
6	8	0	3	4								
7	6	2	7	5	4	3						
8	9	7	2	9	2	6	7	1				
9	0	3	2	6	5	8	9	0	4	6	1	
10	2	5	3	1	6	1	0	7				
11	4	3	9	6	2							
12	2	1	8									
13	8											
14												
15	6	9										

a) Copy this diagram, ordering the values to find the median and quartiles.
b) Construct a box-plot, identifying outliers as appropriate.

6 The heights of a random sample of 23 daffodil leaves were measured with the following ordered results, in centimetres.

20.0 26.5 26.5 27.0 27.5 28.0 28.0 28.0 28.5 28.5 29.0 30.1
30.5 30.5 31.5 32.0 32.0 33.5 34.0 34.0 34.5 38.0 44.5

a) Identify the median and the upper and lower quartiles.
b) Hence determine the inner and outer fences.
c) Construct a box-plot of the data.
d) Identify any outliers, noting which if any are far out.

STANDARD DEVIATION

The interquartile range measures the spread of the middle 'half' of the data and is closely linked to the median. We can define a measure of dispersion, taking into account all the data, which is linked instead to the mean. There are several possibilities:

a) the average deviation from the mean,
b) the average magnitude of the deviation from the mean,
c) the average squared deviation from the mean.

Consider the following collections of data.
I 18, 20, 21, 22, 24
II 15, 19, 21, 23, 27
III 3, 5, 6, 7, 9
IV 13, 20, 21, 22, 24

a) Deviation from the mean

The mean of collection I is 21, hence the deviations from the mean are:

–3, –1, 0, 1, 3

and the average (or mean) of these deviations is their sum divided by 5. This comes to zero, i.e.

$$\frac{1}{n}\sum\left(x-\bar{x}\right)=0$$

The mean of collection II is also 21 and the deviations from the mean are:

–6, –2, 0, 2, 6

which are different from those of collection I. The mean of these deviations is, however, also zero.

As an exercise, calculate the mean deviation from the mean for each of collections III and IV.

You may well have correctly reached the conclusion that the mean deviation from the mean is always zero. So, this would appear to be unhelpful as a measure of spread. The reason for this is that the mean is rather like a balance point for the data. This is clearly seen in the case of collections I to III which are symmetrical, but in collection IV some diagrammatic explanation might help.

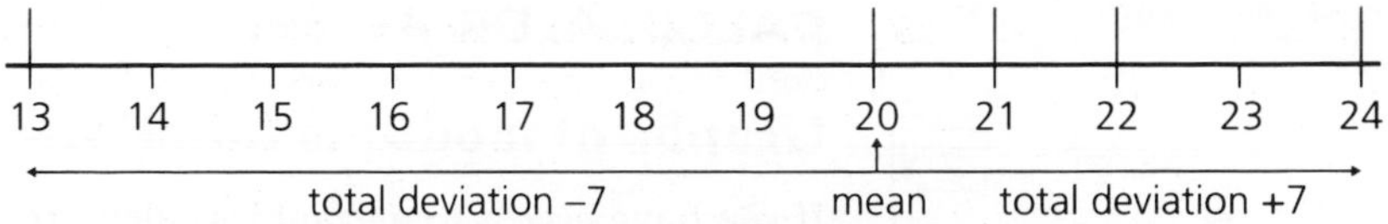

The total deviation of the collection of data is the same above the mean as below it, hence the mean deviation is zero.

b) Magnitude of deviation from the mean

One method of overcoming the problem above is to ignore the sign of the deviation i.e. to take the **modulus** or **magnitude** of the deviation. For collection I the magnitudes of the deviations from the mean are:

3, 1, 0, 1, 3

Hence, their mean is $\frac{8}{5}$ i.e. 1.6. For the other collections the mean magnitudes are:

II 3.2
III 1.6
IV 2.8

There seems to be little to quarrel with in these values. Collections I and III appear equally spread and the spread of II seems to be twice that of I and III.

Unfortunately, the modulus function is notoriously difficult to deal with; sketch graphs of some modulus functions are shown below.

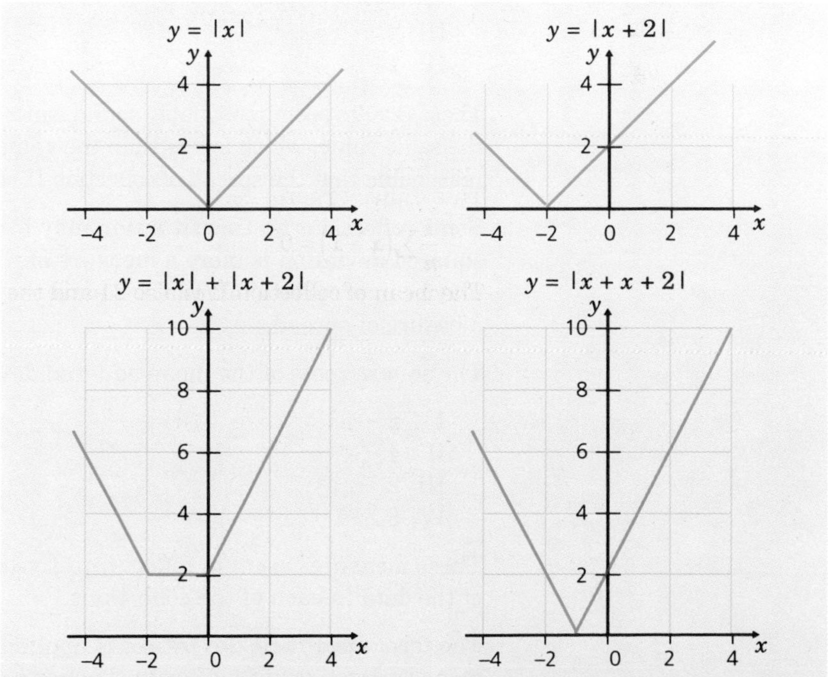

Sum of the modulus functions

Modulus of the sum of the functions

CALCULATOR ACTIVITY

Graphs of modulus functions

If you have access to a graphics calculator or a computer graphics package, create the above graphs using:

$X_{min} = -5$
$X_{max} = 5$
$Y_{min} = 0$
$Y_{max} = 10$

You may like to draw some other modulus functions.

Note: ABS(x) is the same as MOD(x) or $|x|$.

c) Squared deviation from the mean

Ignoring the sign of the deviation seems a sound principle, since the deviation of 18 from 21 is as much as the deviation of 24 from 21. A similar effect can be achieved by **squaring** the deviations from the mean.

For collection I this results in the following squared deviations:

9, 1, 0, 1, 9

and the mean squared deviation is $\frac{20}{5}$ i.e. 4.

Verify for yourself that the **mean squared deviations** for the other three collections of data are:

II 16
III 4
IV 14

These results seem reasonably sound, since collections I and III produce the same value, which is less than the value for II. However, is it reasonable that the spread of collection II is four times that of collection I?

Some reflections on this situation may lead to the conclusion that this squared deviation is more a measure of the square of the spread. So, it may be appropriate to find the square root of this, to use as the measure of spread.

The square roots of the mean squared deviations are:

I 2
II 4
III 2
IV 3.742

These measures seem to reflect intuitive impressions of the dispersion of the data in each of the collections.

The term **standard deviation** is applied to this square root of the mean squared deviation about the mean.

Standard deviation = $s = \sqrt{\frac{1}{n}\sum\left(x-\bar{x}\right)^2}$

The **mean squared deviation** from the mean from which the standard deviation is obtained is known as the **variance**.

Variance $s^2 = \frac{1}{n}\sum\left(x - \bar{x}\right)^2$

Exploration 3.3

Mean, variance and standard deviation

Find the mean, the variance, and the standard deviation for each of the following collections of data.

- 4, 6, 8, 10, 12
- 2, 3, 3, 4, 4, 4, 5, 5, 6
- 12, 13, 13, 14, 14, 14, 15, 15, 16, 20

Alternative method of calculation

Consider the following data collection.

3, 4, 6, 9, 8, 7, 2, 3, 7, 8, 8, 5, 6

The mean is $\frac{76}{13} = 5\tfrac{11}{13} = 5.846\,15...$ and variance is $4\tfrac{126}{169} = 4.745\,56....$

Evaluating $\left(x - \bar{x}\right)$ for each item of data involves rather a lot of tedious arithmetic; squaring these deviations makes matters worse. There is a method which arrives at the same value for the variance but only involves the mean of the data once rather than n times.

This method involves finding the mean of the squares of the data and subtracting from this the square of the mean of the data.

In this case the mean of the squares = $\frac{\sum x^2}{n} = \frac{506}{13} = 38\tfrac{12}{13} = 38.923\,08...$

and the square of the mean = $(\bar{x})^2 = \left(\frac{76}{13}\right)^2 = 34\tfrac{30}{169} = 34.177\,51.$

Hence, the variance = $\frac{\sum x^2}{n} - (\bar{x})^2 = 38.923\,076\,9 - 34.177\,514\,8 = 4.745\,562\,1$

which agrees with the previous value.

So it seems that it may be considerably easier to evaluate the variance of a collection of data using:

variance = mean of squares minus square of mean

Justification

Consider the sum of the squared deviations for n items of data, x_i:

$$\sum_{i=1}^{n}\left(x_i - \bar{x}\right)^2$$

and expanding yields:

$$\sum_{i=1}^{n}\left(x_i^{\,2} - 2x_i\,\bar{x} + \bar{x}^2\right)$$

Distributing the summation over the three parts of this expansion produces:

$$\sum_{i=1}^{n} x_i^2 - \sum_{i=1}^{n} 2x_i\bar{x} + \sum_{i=1}^{n} \bar{x}^2$$

Considering each term:

$\sum_{i=1}^{n} x_i^2$ is the sum of the squares of the collection of data

$\sum_{i=1}^{n} 2x_i\bar{x}$ since 2 and $\bar{x}$ are fixed for any collection of data, this term is the same as $2\bar{x}$ times the sum $\sum_{i=1}^{n} x_i$ of the data

$\sum_{i=1}^{n} \bar{x}^2$ means 'add up n lots of $\bar{x}^2$ and is equal to $n\bar{x}^2$.

Hence:

$$\sum_{i=1}^{n}\left(x_i - \bar{x}\right)^2 \equiv \sum_{i=1}^{n} x_i^2 - 2\bar{x}\sum_{i=1}^{n} x_i + n\bar{x}^2$$

and the variance, which is the mean of this expression, is:

$$\frac{\sum_{i=1}^{n}\left(x_i - \bar{x}\right)^2}{n} \equiv \frac{\sum_{i=1}^{n} x_i^2 - 2\bar{x}\sum_{i=1}^{n} x_i + n\bar{x}^2}{n}$$

$$= \frac{1}{n}\sum_{i=1}^{n} x_i^2 - 2\bar{x}^2 + \bar{x}^2 = \frac{1}{n}\sum_{i=1}^{n} x_i^2 - \bar{x}^2$$

$$\text{Variance} = s^2 = \frac{\sum x^2}{n} - \bar{x}^2$$

Example 3.2

Calculate the standard deviation of the following collection of data.

4, 4, 5, 5, 5, 5, 6, 6, 6, 6, 6, 6, 6, 8, 9, 10, 11

Solution

The sum is 108 and the sum of the squares is 750. Hence, the variance is

mean of squares – square of means = $\frac{750}{17} - \left(\frac{108}{17}\right)^2 = 3.757\,87\ldots$

The standard deviation is $\sqrt{3.7578} = 1.94$ *correct to two decimal places.*

Exploration 3.4

The alternative method of calculating variance

Use the alternative method described above to calculate the variance for each of the following data collections.

- 2, 3, 4, 4, 5, 6, 8
- 0, 0, 0, 1, 1, 1, 2, 2, 3, 4, 6
- 13, 15, 18, 19, 19, 20, 20, 20, 20, 21, 21, 22, 24

Frequency distribution

In the same way that the mean can be evaluated from data presented as a frequency distribution, so can the variance. Consider the following data.

Length x (cm)	0.2	0.3	0.4	0.5	0.6	0.7
f	2	4	7	9	5	2

The mean of this distribution is found as $\frac{\sum xf}{\sum f}$.

The mean of the squares will similarly be found from $\frac{\sum x^2f}{\sum f}$.

A convenient way of laying this out in a table is as follows.

x (cm)	0.2	0.3	0.4	0.5	0.6	0.7	Σ
f	2	4	7	9	5	2	29
xf	0.4	1.2	2.8	4.5	3.0	1.4	13.3
x^2f	0.08	0.36	1.12	2.25	1.8	0.98	6.59

The mean of the data is $\frac{13.3}{29}$ and the mean of the squares is $\frac{6.59}{29}$.
Hence the variance is

$$\frac{6.59}{29} - \left(\frac{13.3}{29}\right)^2 = 0.0169...$$

Now consider these data.

Mass m (g)	**Frequency**
0–	4
10–	9
20–	3
30–	3
40–60	1

A representative value is needed for each class, and this is the midpoint of the class. Hence the grouped frequency distribution is represented in this table.

Mass m (g)	5	15	25	35	50
f	5	9	3	3	1

Then the calculation can proceed as in the previous problem yielding:

$$\text{mean mass} = \frac{\sum xf}{\sum f} = \frac{390}{21}$$

$$\text{mean of squares} = \frac{\sum x^2f}{\sum f} = \frac{10\,200}{21}$$

Hence, the variance is $\dfrac{10\,200}{21} - \left(\dfrac{390}{21}\right)^2 = 140.8163$ and the estimate of the standard deviation is $\sqrt{140.8163...} = 11.83$.

Exploration 3.5 *Coding*

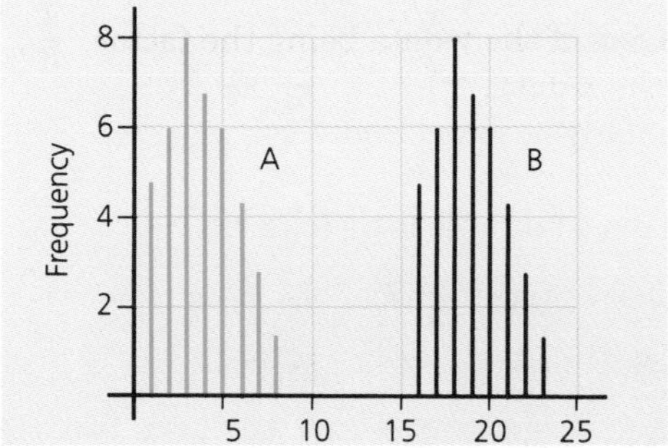

In Chapter 2, *Summary statistics 1*, we saw the effect that coding, by translating or by scaling, had on the mean of a data distribution. Consider the two distributions illustrated here.

Are there any differences between collections A and B?

They are in different places on the number line – hence their locations are different. But they are spread out in the same way, and so their dispersions (or standard deviations) should be the same.

Check that they are the same, given that their frequency distributions are as in the table.

a	1	2	3	4	5	6	7	8
b	16	17	18	19	20	21	22	23
Frequency	5	6	8	7	6	4	3	1

(For distribution A, s is 1.8735.)

Now consider the two distributions below.

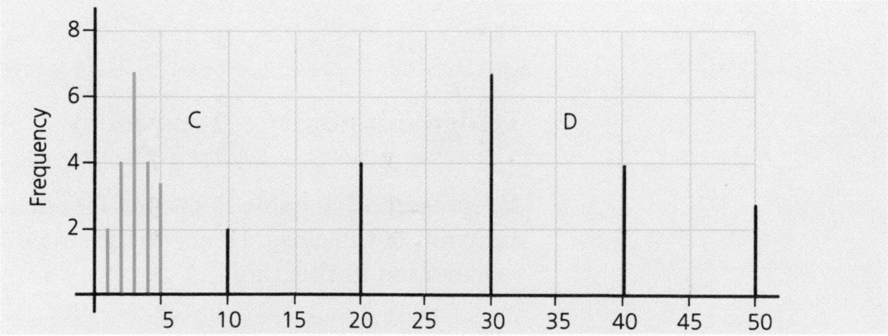

They are differently located, and differently dispersed. Collection C ranges from 1 to 5 and collection D ranges from 10 to 50 in a similar manner. Show that the standard deviation of D is ten times that of C where:

a	1	2	3	4	5
b	10	20	30	40	50
Frequency	2	4	7	4	3

In this exploration, you have seen that coding data using a translation has no effect on their spread i.e. does not change their variance or standard deviation. But, using scaling in coding alters the standard deviation by the same scale factor. What effect does it have on the variance?

A combination of translating and scaling:

$$y = \frac{(x \pm \alpha)}{\beta} = \frac{x}{\beta} \pm \delta$$

will result in the standard deviation of the y-data being the factor, $\frac{1}{\beta}$, times the standard deviation of the x-data.

Example 3.3

Use the coding:

$$y = \frac{x - 0.55}{0.1}$$

to facilitate the calculation of:

a) *the mean,*

b) *the variance,*

c) *the standard deviation,*

of the distribution of lengths of off-cuts of timber recorded in the table.

Length (x m)	**Frequency**
$0.2 \le x < 0.3$	6
$0.3 \le x < 0.4$	5
$0.4 \le x < 0.5$	8
$0.5 \le x < 0.6$	10
$0.6 \le x < 0.7$	8
$0.7 \le x < 0.8$	5
$0.8 \le x < 0.9$	8

Solution

Coded midpoint y	**Frequency** f	yf	y^2f
−3	6	−18	54
−2	5	−10	20
−1	8	−8	8
0	10	0	0
1	8	8	8
2	5	10	20
3	8	24	72
Σ	50	6	182

The distribution of the coded data is shown in the table above, along with the calculations needed to find the variance of y.

Variance of $y = s_y^2 = \frac{182}{50} - \left(\frac{6}{50}\right)^2 = 3.6256$

and hence the standard deviation of y is 1.904 10.

Since the translation has no effect on spread, the standard deviation of x is

$s_x = 0.1 \times s_y = 0.190$ correct to two decimal places

Thus the answers are:

a) *Mean of* x: $0.1 \times \frac{6}{50} + 0.55 = 0.562$ *correct to two decimal places*
b) *Variance of* x:$(0.1)^2 \times 3.6256 = 0.036\,256 = 0.0363$ *correct to four decimal places*
c) *Standard deviation of* x: $0.190\,41 = 0.19$ *correct to two decimal places*

EXERCISES

3.3A

1 Calculate the mean of the following collection of data. Hence, calculate the variance using:
a) the squared deviations from the mean,
b) the mean of the squares of the data.

1, 2, 2, 3, 3, 3, 3, 4, 4, 4, 5, 5, 6, 7, 8

2 Calculate the standard deviation of the following distribution of shoe sizes.

Size	4	5	6	7	8	9	10	11	12
f	5	8	12	10	10	15	10	8	2

a) Use the squared deviation about the mean.
b) Use the mean of the squares of the data.

3 Use the coding:

$$y = \frac{x-5}{10}$$

to estimate the mean, variance and standard deviation of the following distribution of weights.

Weights (g)	0–	10–	20–	30–	40–50
Representative value, x	5	15	25	35	45
Frequency	8	22	45	15	10

4 The speeds of cars travelling on a dual carriageway are recorded in the table.

Speed (mph)	20–30	–40	–50	–60	–70	70–100
Number of cars	2	12	18	32	28	28

Calculate an estimate of the standard deviation of the speeds.

5 Records of barometric pressure were made over a period of 65 days and are shown in the table below. Calculate estimates of:

a) the mean pressure,
b) the standard deviation of the pressure.

Pressure (mb)	< 960	960–	980–	990–	1000–	1010–1020
Number of days	0	2	17	13	18	15

EXERCISES

3.3B

1 The following data collection gives the age last birthday of a group of ten sixth-formers.
17 17 16 19 16 18 17 18 18 17

Calculate the mean of the data. Hence calculate the variance using:
a) the squared deviations from the mean,
b) the mean of the squares of the data.
c) Comment on the accuracy of the mean.

2 Calculate the standard deviation of the following distribution using:

a) the squared deviations from the mean,
b) the mean of the squares of the data.

Number of pets owned	0	1	2	3	4	5 or more
Number of families	6	7	5	1	1	0

3 Study these data.

Length x (cm)	0–	1–	2–	3–	4–	5–	6–7
Frequency	5	12	31	18	14	6	4

a) Use the coding $y = x - 2.5$ to calculate estimates of the mean, variance and standard deviation of y.
b) Hence, write down estimates of these measures for the length, x.

4 As a simple statistical activity, a primary school teacher has his class of 32 children measure the lengths of all the pencils in the classroom. Their results are summarised in the grouped frequency table on the right.

Calculate an estimate of the standard deviation of the length.

Length x (cm)	Number of pencils
$3 \le x < 5$	3
$5 \le x < 7$	8
$7 \le x < 9$	14
$9 \le x < 10$	14
$10 \le x < 11$	19
$11 \le x < 12$	24
$12 \le x < 13$	17
$13 \le x < 14$	11
$14 \le x < 15$	6
$15 \le x < 17.5$	4

5 A record of the amount spent at the buffet car by passengers on a particular train journey is summarised below.

Amount spent x (£)	Number of passengers
0.50– 0.99	46
1.00–1.99	85
2.00–2.99	93
3.00–4.99	52
5.00–7.49	15
7.50–9.99	10

Calculate estimates of the mean and standard deviation of the amount spent per passenger.

CALCULATOR ACTIVITY

Standard deviation

The statistical facility on most calculators will provide estimates of the standard deviation using the same initial key sequences as are used to find the mean of a data distribution.

Clear the statistical registers of your calculator and input the following data.

0, 7, 4, 9, 3, 9, 2, 8, 4, 9

Now check that you have input the data correctly. There are ten items of data and their mean is 5.5. Use the key sequence appropriate to your calculator to reveal the standard deviation σ.

Data presented in a frequency table may be input in the same way as to find the mean. Clear your statistical registers and input the following data, carefully choosing appropriate representative values for the following classes.

Time (s)	0–	1–	2–	3–	4–	5–	6–	7–	10–15
f	1	5	7	7	5	7	5	8	5

Check that you have correctly input the data.

$n = 50$ and $\bar{x} = 5.48$

Now, display the estimate of the standard deviation (3.2341 ...).

To obtain the value of the variance from the calculator you need only square the standard deviation.

You may have noticed that your calculator has a choice of keys for standard deviation. The key explored so far has been the one labelled σ_n or s_n or xs_n – the significant part of the label is the n. The alternative is a similar key where n is replaced by $n - 1$, such as σ_{n-1} or s_{n-1} or xs_{n-1}.

The keys output different values because they do not represent the same quantity. The difference is in their definitions. They are the standard deviations associated with the following relationships.

Key	Related to	Defined as
σ_n or s_n	Variance	$s_n{}^2 = \dfrac{\sum(x-\bar{x})^2}{n}$
σ_{n-1} or s_{n-1}	Unbiased estimate of population variance	$s_{n-1}{}^2 = \dfrac{\sum(x-\bar{x})^2}{n-1}$

What is the relationship between s_n and s_{n-1}?

From the definitions, you may see that:

$$n \times s_n{}^2 = (n-1) \times s_{n-1}{}^2$$

You can check this relationship from the data stored in your calculator (remember the calculator displays the standard deviations not variances, and so you have to use the square function to recover the estimates of variance).

Hence, the relationship is:

$$s_{n-1} = s_n \times \sqrt{\frac{n}{n-1}}$$

Exploration 3.6 Combining collections of data

In Chapter 2, *Summary statistics 1*, we saw that it is possible to calculate the mean of a combined collection of data, if we know the individual means of the individual collections. Knowing the mean and variance of individual collections also enables us to calculate the variance of the combined collection.

Suppose a collection of six items of data has a mean of $\frac{22}{6}$ and a variance of $\frac{35}{9}$.

The total of their squares, Σx^2, can be obtained using the 'mean of squares etc.' definition of the variance.

$$\text{Variance} = \frac{\sum x^2}{n} - \bar{x}^2$$

therefore:

$$\frac{35}{9} = \frac{\sum x^2}{6} - \left(\frac{22}{6}\right)^2$$

and so,

$$\sum x^2 = 6\left(\frac{35}{9} + \frac{484}{36}\right)$$

This results in

$$\sum x^2 = 104$$

Suppose another collection of data, this time ten items, has a mean of 4 and variance of 1.4. What is the total, Σy^2, of the squares of this collection of data?

The same process as before yields:

$$\sum y^2 = 10(1.4 + 16) = 174$$

Imagine, now, that the two collections of data are combined to form one collection. What are the mean and variance of the combined collection?

In Chapter 2, *Summary statistics 1*, we discovered how to calculate the mean of the combined collection.

$$\text{Mean of combined} = \frac{\text{total value of data}}{\text{number of items of data}} = \frac{\sum x + \sum y}{6 + 10} = \frac{22 + 40}{16} = 3.875$$

Similarly, the variance can be found, using the mean of squares minus squares of mean approach.

$$\text{Variance of combined} = \frac{\text{total of squares}}{16} - 3.875^2$$

$$= \frac{\sum x^2 + \sum y^2}{16} - 3.875^2 = \frac{104 + 174}{16} - 3.875^2$$

$$= 2.359\ 375 = 2.359 \text{ correct to three decimal places}$$

Example 3.4

Two collections of data are summarised in the table below.

Collection	Number of items	Mean	Variance
x	32	3	$2\frac{3}{16}$
y	28	6	$3\frac{1}{14}$

The data are combined into a single collection. Calculate:

a) *the mean,*

b) *the variance,*

c) *the standard deviation,*

of the collection.

Solution

a) *The total of the individual collections are:*

$\Sigma x = 32 \times 3 = 96$

$\Sigma y = 28 \times 6 = 168$

Hence, the mean of the combined data $= \dfrac{96 + 168}{32 + 28} = 4.4$

b) *The totals of the squares are:*

$$\sum x^2 = 32(2.1875 + 3^2) = 358$$

$$\sum y^2 = 28(3\tfrac{1}{14} + 6^2) = 1094$$

Then the variance of the combined collection $= \dfrac{358 + 1094}{60} - 4.4^2 = 4.84$

c) *The standard deviation is merely the square root of the variance i.e. 2.2.*

Outliers and standard deviation

Earlier in this chapter (see page 81), the concept of an outlier was introduced. The definition used there is **robust**; it is not affected by the presence or otherwise of outliers in the data collection.

Another definition is sometimes used: items of data which are **more than two standard deviations** away from the mean are described as outliers.

The mean and standard deviation for the motorway toll times are:

$\bar{x} = 40.0 \quad s = 27.5$ to one decimal place.

Using the two standard deviations definition for outliers leads to items of data greater than or less than $40.00 \pm 2 \times 27.5$ being described as outliers. Hence, items of data which are:

less than −15
or more than 95

would be described as outliers.

Which motorway toll times (page 30) are outliers using the two standard deviations definition?

Exploration 3.7

Outliers

Suppose the longest recorded time in the motorway toll example (page 30) had been 5 minutes (i.e. 300 seconds) instead of 118 seconds.

What would the mean and standard deviation of the times be? Which times would now be described as outliers using the two standard deviation definition?

Which items does the definition based on quartiles identify as outliers in this case?

Discuss the validity of the two definitions. Use the $\bar{x} \pm 2s$ definition to identify outliers in the data collections of:

- Question 1 Exercise Set 3.2A
- Question 5 Exercise Set 3.2A
- Question 1 Exercise Set 3.2B
- Question 5 Exercise Set 3.2B

Standardised values

There are occasions when we need to compare items from different data collections. A student studying Engineering gained a mark of 52 in her Soil Mechanics examination and a mark of 57 in the examination on Mathematical Modelling. Superficially, it appears that she did better in Modelling than in Mechanics. However, if the mean marks in the two examinations were:

Examination	Mean mark
Soil Mechanics	44
Modelling	60

then, her score of 52 which is above the exam average is relatively better than the mark of 57 which is below the exam average.

Suppose this student also sat examinations in Structures and in Dynamics with the following results.

Examination	Mark	Mean mark
Structures	58	50
Dynamics	63	57

It is now more difficult to determine in which exam she performed better, since her mark in each is higher than the mean. It could be argued that she did better in Structures, where her mark is 8 above the mean, than in Dynamics, where she was only 6 marks above the mean. However, suppose that the spread of marks in the two examinations is as follows.

Examination	Standard deviation
Structures	6.8
Dynamics	2.6

Now her mark in Dynamics can be seen to be particularly noteworthy, since it is more than two standard deviations above the mean, whereas the mark in Structures is not.

Relating items of data to the mean and expressing the difference from the mean in terms of the standard deviation of the data is a convenient way of **standardising data**. In the case of this Engineering student, her standardised examination scores are obtained.

Examination	Mark	Mean	Standard deviation	Standardised mark
Structures	58	50	6.8	$\frac{58-50}{6.8} = 1.23$
Dynamics	63	57	2.6	$\frac{63-57}{2.6} = 2.31$
Soil Mechanics	52	44	3.1	
Modelling	57	60	5.0	

What are her standardised scores in the other two examinations? In which exam did she do best?

Re-expression

Standardised scores such as 1.23, 2.31, etc. from the above do not seem like marks. For the sake of convenience, these may be re-expressed to give the appearance of marks, whilst retaining the comparability that standardisation offers.

Suppose all the examination marks are to be presented as if each examination produced marks with a mean of 55 and a standard deviation of 15. Then the standardised score is multiplied by the new standard deviation (15) and to this is added the assumed mean (55).

The student's score in Structures would be presented as:

$1.23 \times 15 + 55 = 73.45$

What would her other scores be?

Exploration 3.8 *Re-expression*

In one school, Russian is offered as an option to Year 12 students. The marks that the students studying Russian obtained were:

28, 33, 43, 46, 54, 57, 64, 70, 83, 95

- Calculate the mean and standard deviation of these marks.
- Find the standardised score for each of the marks, giving it correct to two decimal places.
- Re-express the scores to have a mean 50 and standard deviation 12.
- The student who obtained 46 in the Russian examination also studies Latin. He scored 57 in his Latin examination where the mean mark was 66 and the standard deviation 10.3. In which exam did he do better?

CONSOLIDATION EXERCISES FOR CHAPTER 3

1 A group of 55 boys were practising high-jumping. In one week, their individual best jumps using the 'straddle' were recorded. The following week, their individual best jumps using the 'flop' were recorded. The data represents the difference between the 'straddle' and the 'flop' results, measured in centimetres.

Depth
Unit is 0.1

1	−7	2
1	−6	
1	−5	
1	−4	
2	−3	9
3	−2	4
11	−1	9 8 6 5 4 2 1 0
27	−0	9 8 7 7 6 4 3 3 3 2 2 1 1 0 0 0
(19)	0	0 0 1 1 1 2 3 4 4 4 5 5 6 6 6 7 7 8 8
9	1	1 2 2 3 6
4	2	1 2 3
1	3	
1	4	0

a) Identify the median and quartiles.
b) Write down the interquartile range.
c) Present the data in a box-and-whisker diagram.

2 **a)** Calculate the interquartile range for these data.

0.05 0.36 0.49 0.67 0.42 0.42 1.43 0.98 0.68 0.14 2.28 0.24 0.82
0.74 1.28 1.25 1.12 0.95 0.03 0.61 1.30 1.51 1.42 1.26 0.19

b) Identify the outliers in the data.
c) Draw a box-plot of the data and comment on their distribution.

3 **a)** Explain how the median and quartiles of a distribution can be used when describing the shape of the distribution.
Summarised below is the distribution of masses of new potatoes, in grams to the nearest gram.

Mass (g)	**Frequency**
19 or less	2
20 – 29	14
30 – 39	21
40 – 49	34
50 – 59	39
60 – 69	42
70 – 79	13
80 – 89	4
90 or more	2

b) Use linear interpolation to estimate the median and quartiles of this distribution. Hence describe its skewness.
c) Draw a box-and-whisker plot to illustrate these data.

(ULEAC, Question 7, Specimen paper T1, 1994)

4 The following table refers to all marriages that ended in divorce in

Age of wife (years)	16 – 20	21 – 24	25 – 29	30 and over
Frequency	4966	2364	706	524

Scotland during 1977. It shows the age of the wife at marriage.

a) Draw a cumulative frequency curve for these data.
b) Estimate the median and the interquartile range.

The corresponding data for 1990 revealed a median of 21.2 years and an interquartile range of 6.2 years.

c) Compare these values with those you obtained for 1977. Give a reason for using the median and interquartile range, rather than the mean and standard deviation for making this comparison.

The box-and-whisker plots that follow (on page 101) refer to Scotland and show the age of the wife at marriage. One is for all marriages in 1990 and the other is for all marriages that ended in divorce in 1990 (the small number of marriages in which the wife was aged over 50 have been ignored).

d) Compare and comment on the two distributions.

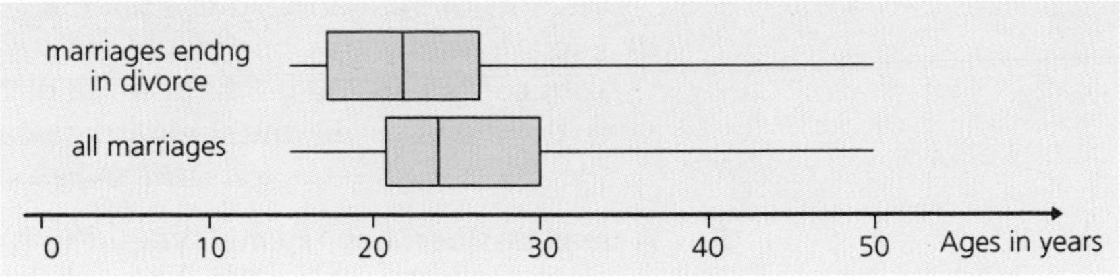

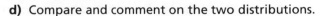

(AEB, Question 5, Specimen paper 2, 1996)

5 The distribution of speeds of a sample of 250 vehicles on an inner city road is summarised in the following table. The road is subject to a 30 mph speed limit.

Speed v (mph)	Number of vehicles
$10 < v \leq 20$	18
$20 < v \leq 25$	26
$25 < v \leq 30$	90
$30 < v \leq 35$	58
$35 < v \leq 40$	32
$40 < v \leq 50$	18
$50 < v \leq 70$	8

a) Represent these data by a cumulative frequency diagram.
b) Estimate the median, lower quartile and upper quartile speeds of the vehicles. Hence draw an approximate box plot.

Comment briefly on the extent to which vehicles break the speed limit based on the evidence in the diagram.

(NEAB, Question 13, Specimen paper 1, 1996)

6 A class of pupils sat end-of-year examinations in English and Mathematics. The class marks for English were distributed with mean 56 and standard deviation 8. The Mathematics marks had mean 64 and standard deviation 15.

Brian scored 71 marks for English and 82 marks for Mathematics. Transform each of his marks into a standardised score, mean 50 and standard deviation 10. Hence compare Brian's performance in the two subjects.

(NEAB, Question 3, Specimen paper 8, 1996)

7 Tubs of margarine are claimed on the label to contain 250 grams. The actual weights of margarine in a random sample of 15 tubs were as follows. The weights are given correct to the nearest gram.

261	252	242	240	254
257	252	250	265	249
253	248	259	246	252

a) Construct a stem-and-leaf diagram to represent these data, using a two-part stem.
b) Comment on the shape of the distribution of weights, with particular reference to the claim that tubs contain 250 grams.

c) Use your calculator to find the mean and standard deviation of the weights of margarine in this sample.

d) Explain briefly how one would ensure that a greater proportion of tubs contained 250 grams or more or margarine by adjusting:

i) the mean, ii) the standard deviation.

(MEI, Question 1, Paper S1, June 1992 (modified))

8 A small business has 12 employees. Their weekly wages, £x, are summarised by:

$\sum x = 2501$ $\sum x^2 = 525\,266.8$

a) Calculate the mean and standard deviation of the employees' weekly wages.

A second business has 17 employees. Their weekly wage, £y, has a mean of £273.20 and a standard deviation of £23.16.

b) Find $\sum y$ and show that $\sum y^2 = 1\,277\,969$.

c) Now consider all 29 employees as a single group. Find the mean and standard deviation of their weekly wages.

(MEI, Question 1, Paper S1, January 1993)

9 A frequency diagram for a set of data is shown here.

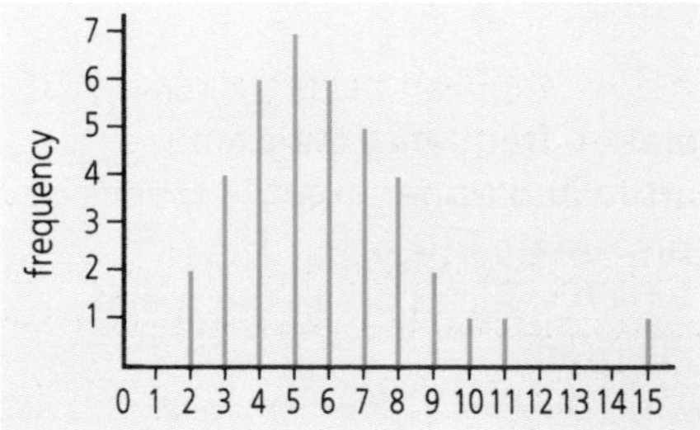

a) Find the median and the mode of the data.

b) Given that the mean is 5.95 and the standard deviation is 2.58, explain why the value 15 may be regarded as an outlier.

c) Explain how you would treat the outlier if the diagram represents:

i) the ages (in completed years) of children at a party,

ii) the sums of the scores obtained when throwing a pair of dice.

d) Find the median and the mode of the data after the outlier is removed.

e) Without doing any calculations state what effect, if any, removing the outlier would have on the mean and on the standard deviation.

f) Does the diagram exhibit positive skewness, negative skewness or no skewness? How is the skewness affected by removing the outlier?

(MEI, Question 1, Paper S1, June 1993)

10 The table gives an analysis of a random sample of 200 sales of unleaded petrol at a petrol station.

Volume of petrol (litres)	Number of sales
5 or less	6
10 or less	20
15 or less	85
20 or less	148
25 or less	172
30 or less	184
35 or less	194
40 or less	200

a) Use linear interpolation to estimate:
  i) the median volume of unleaded petrol sales,
  ii) the upper and lower quartiles.
  Hence write down the interquartile range.

b) Construct, on graph paper, a box-plot of the distribution of sales of unleaded petrol.

(Oxford, Question 8, Specimen paper 3, 1996)

11 An angler made a record of the weights (in lb) of the 200 fish he caught during one year. These are summarised in the table.

Weight of fish (mid-class value) (lb)	0.5	1.25	1.75	2.25	2.75	3.5	4.5	5.5	7.0	10.5
Number of fish in the class	21	32	33	24	18	21	16	12	11	12
Class width	1	0.5	0.5	0.5	0.5	1	1	1	2	5

Using the information supplied in the table:

a) calculate suitable frequency densities and, on graph paper, construct a histogram of the data,

b) calculate estimates of the mean and standard deviation of the weights of the fish.

How are the estimates of the mean and standard deviation likely to differ from the true values? Give brief reasons for your answers.

(Oxford, Question 6, Specimen paper 53, 1996)

12 On 1 September 1992 the grouped frequency distribution of the ages (in completed years) of 1000 pupils aged under 16 in a comprehensive school was as given in the following table.

Age (in completed years)	11	12	13	14	15
Frequency	165	184	216	231	204

a) Calculate, to three significant figures, estimates for the mean and the standard deviation of the ages of these pupils on 1 September 1992.

b) Draw a cumulative frequency polygon and estimate, to three significant figures, the median age of the pupils on 1 September 1992.

c) Given in addition that there were 222 pupils aged 16 or over, estimate, to three significant figures, the median age of all the pupils in the school on 1 September 1992.

(NEAB, Question 8, Paper 2, June 1993)

Summary

- **Range:** the difference between the largest and smallest values in a collection of data
- **Quartile spread:** the 'middle half' of the data between the lower and upper quartiles
- **Variance:** $s^2 = \frac{1}{n}\sum\left(x - \bar{x}\right)^2 = \frac{1}{n}\sum x - \bar{x}^2$

  $s^2 = \frac{\sum x^2 f}{\sum f} - \left(\frac{\sum xf}{\sum f}\right)^2$ for grouped data.
- **Standard deviation:** $s = \sqrt{\text{variance}}$

  variance = mean of squares
  – square of the mean
- The measure of location and spread can be displayed in a box-and-whisker diagram of the data.

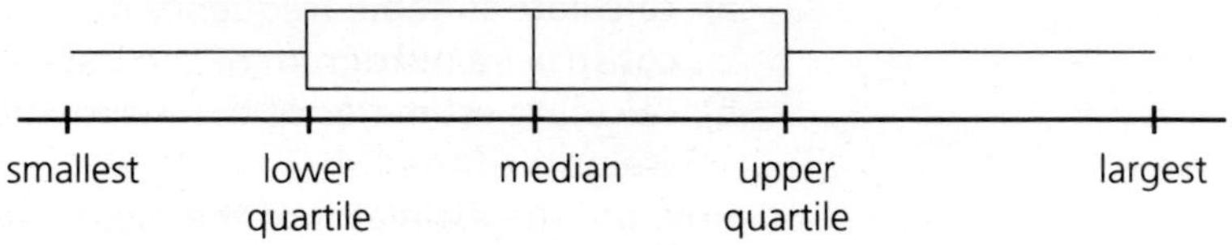

STATISTICS

4

Probability

In this chapter we look at the mathematical structure of likelihood, chance and probability, processes which underpin all data analysis. After working through it you should be able to:

- *set up models to study data,*
- *draw inferences from data,*
- *make sensible judgements based on your conclusions,*
- *use the techniques to generate conclusions and predictions, drawn from various data sources.*

THE LANGUAGE OF CHANCE

In everyday speech we use the words 'possible' and 'probable' to indicate degrees of likelihood. A typical weather forecast may sound like this: '*... there is a possibility of showers in the east, but, in the west, rain is probable before dawn.*' What do you think the forecaster means?

In the language of mathematics and statistics, both 'rain in the west' and 'showers in the east' are **possibilities**, or **events**, or **situations that could occur**. The forecaster was trying to indicate that there was a chance that either would happen but, that in her opinion, there was a greater likelihood of 'rain in the west'.

The forecaster had in mind a 'likelihood scale' such as:

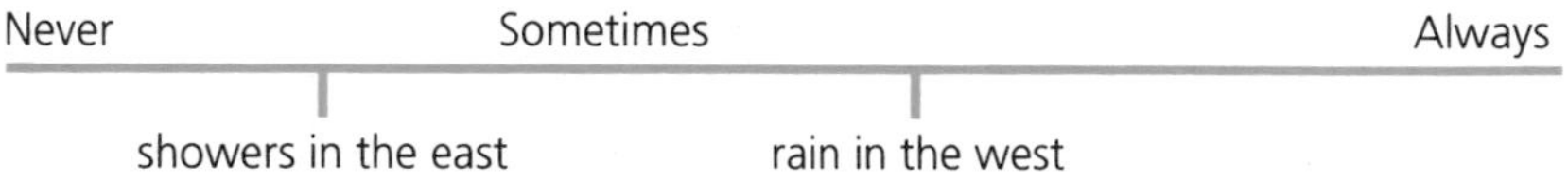

where she had mentally located the two events mentioned in the forecast.

The probability scale

Where do you think the following events should be placed on the scale?

A The United Kingdom will win more gold medals than any other country at the next Olympics.
B The next baby born in Scotland will be a girl.
C It will rain tomorrow.
D The 4:15 train will leave on time.
E There will be change of government at the next general election.
F A coin will land 'heads'.

One possible result would be:

Never	Unlikely	Sometimes Evens	Highly likely	Always
↑	↑	↑	↑	↑
A	*C*	*B*, *F*	*E*	*D*

where the 'sometimes' has been expanded. Events such as the landing of a coin where arguably there are two equally plausible and equally likely outcomes may have the label 'evens', or 'fifty-fifty', or '50%' attached. In this context, 50 per cent indicates the **proportion** of occasions that coins land heads.

What proportions do you think could be attached to events A, B, C, D and E?

Probable results

There seems to be little likelihood of event A occurring, and so five per cent or less would be reasonable. It could be argued that a girl is just as likely as a boy and so 50 per cent would be appropriate for event B. As far as event C is concerned, there are many factors which could have an influence on this. However if the diagram above is a guide, then 30 per cent to 40 per cent is plausible. Those who run the railways may argue that 85 per cent of trains leave on time. General elections do not happen very often and arguments over politics consume much of the intervening time. The diagram indicates that a response of about 75 per cent is anticipated.

Proportions, or percentages, such as 50%, 75%, 30% could be written as $\frac{1}{2}$, $\frac{3}{4}$, 0.3 or in any appropriate form. These numbers are the **probabilities** associated with the various events.

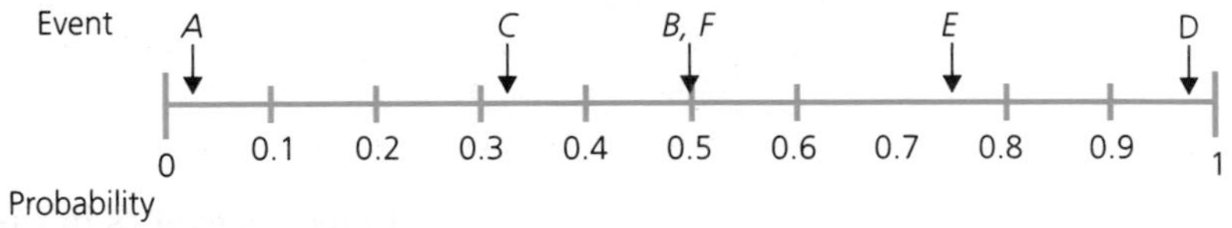

Probability is a number, lying between zero and one, which measures the likelihood of occurrence of the event to which it refers.

Notation

A common notation for the probability of an event is P(event).

Thus, for the events described: $P(B) = \frac{1}{2}$

Write down the probability of the events A, C, D, E and F using this notation.

Exploration 4.1

Events and liklihoods

Define ten events and discuss the likelihood of each happening. Display the events on a likelihood scale and assign probabilities to each one.

Symmetry and equal likelihood

The probability that a coin lands 'heads' is $\frac{1}{2}$. Why? Two assumptions have been made:

- there are only two possible ways in which a coin can land,
- these ways are equally likely.

An ordinary cubical die has six faces which show the numbers 1, 2, 3, 4, 5 and 6, respectively. What is the probability that the die will land, having been rolled, so that the face with 5 on it is uppermost?

If we assume that the die is a cube and is perfectly symmetrical, then the likelihood of any face landing uppermost is the same. There is only one way for a 5 to appear uppermost, but there are six *possible* outcomes. Thus the probability of a 5 appearing is $\frac{1}{6}$.

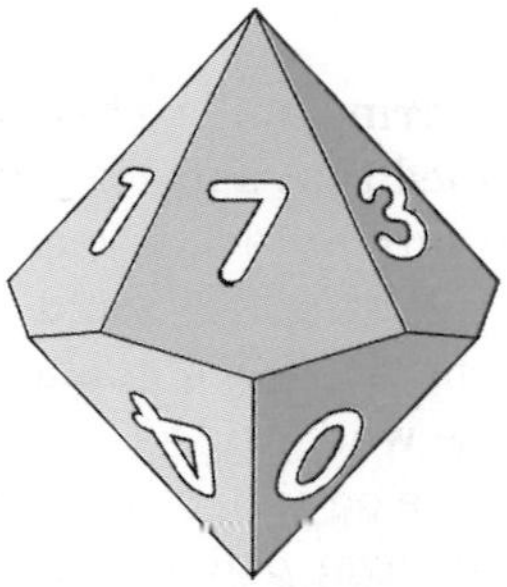

Consider a ten-faced die where each face is equally likely to appear uppermost. The die has the numbers 0, 1, 2, .., 8, 9 on its faces. What is the probability of a multiple of 3 appearing?

There are ten possibilities. Of these, the multiples of three are 3, 6, 9. Thus the probability is $\frac{3}{10}$.

This leads to the definition:

$$\text{probability of an outcome} = \frac{\text{number of events in the outcome}}{\text{total number of possible events}}$$

where it is assumed that the events are equally likely, and that the outcome consists of a number of equally likely events. The previous example had 'die resulting in a multiple of three' as the outcome. The equally likely events which contributed to this outcome are the 3, 6 or 9. There were ten possible events. Hence, the probability must be $\frac{3}{10}$.

The term 'outcome space' is sometimes used in place of outcome, and 'possibility space' or 'sample space' may be used instead of 'all possibilities'. So, you may come across the following definition.

$$\text{P(outcome)} = \frac{\text{number of events in the outcome space}}{\text{number of events in the possibility space}}$$

Exploration 4.2

Determining probability

Determine the probability of each of the following outcomes.

- a prime number on a cubical die
- a prime number on a ten-faced die
- at least one head when two coins are flipped
- an ordinary pack of cards being cut at a heart

In each case make sure you identify the equally likely events in the sample space and the ones in the outcome space.

Relative frequency and empirical probability

Is it always raining where you live? It almost certainly isn't, but what is the chance that it will rain tomorrow? One possible way of finding out is to consult meteorological records and use relative frequencies. Suppose the records show that, over a period of seven days, there were three rainy days. In these circumstances a reasonable estimate for the chance of a rainy day is $\frac{3}{7}$. The following table represents an extract from the records.

Week	Number of rainy days
1	3
2	1
3	3
4	0
5	1
.	.
.	.
.	.

During the second week recorded there was only one rainy day. The relative frequency, or proportion, of wet days during that particular week was $\frac{1}{7}$. However, over the first two weeks, the proportion of wet days was $\frac{4}{14}$. Which of these two proportions do you think is the better to use as an estimate for the probability of a rainy day?

Perhaps you preferred the individual weekly relative frequencies. For the five weeks shown, these are:

$\frac{3}{7}, \frac{1}{7}, \frac{3}{7}, \frac{0}{7}, \frac{1}{7}$

Written as decimals, to two places, they are:

0.43, 0.14, 0.43, 0.00, 0.14

There is a good deal of variation in these, which is not surprising. However, it is difficult to decide which, if any, of these should be used to tackle the question about rain tomorrow.

The cumulative relative frequencies for this period are:
$\frac{3}{7}, \frac{4}{14}, \frac{7}{21}, \frac{7}{28}, \frac{8}{35}$ or, as decimals:

0.43, 0.29, 0.33, 0.25, 0.23.

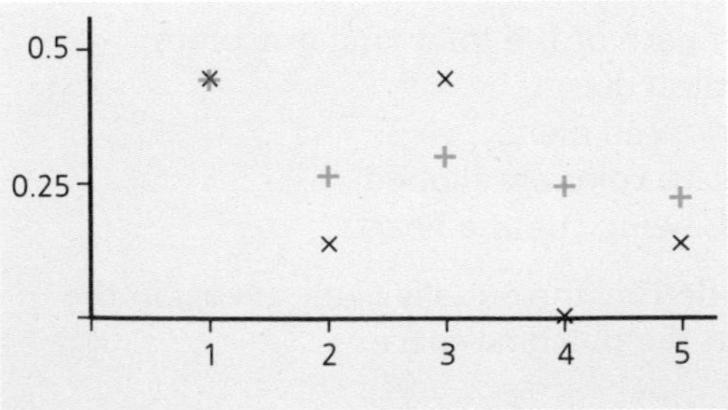

These cumulative relative frequencies still vary, although much less than the individual weekly proportions. Accumulating the information in this way using **cumulative relative frequencies** provides an estimate of the longterm **experimental** or **empirical probability**.

We can present this analysis graphically, as in this diagram, where × and + represent individual and cumulative frequencies respectively.

Exploration 4.3 *Deciding whether the weather will be wet*

- An extract from the meteorological records over a period of six months for a different region yields the following data.

Week	1	2	3	4	5	6	7	8	9	10	11	12	13	14	15	16	17	18	19	20	21	22	23	24	25	26
Number of wet days	3	0	1	0	4	1	5	6	2	5	2	1	1	0	1	2	1	3	0	1	0	1	0	5	3	7

Calculate:
a) individual weekly relative frequencies,
b) successive cumulative relative frequencies,
for the proportion of wet days, and plot a graph of the results. Discuss the outcome of your work.

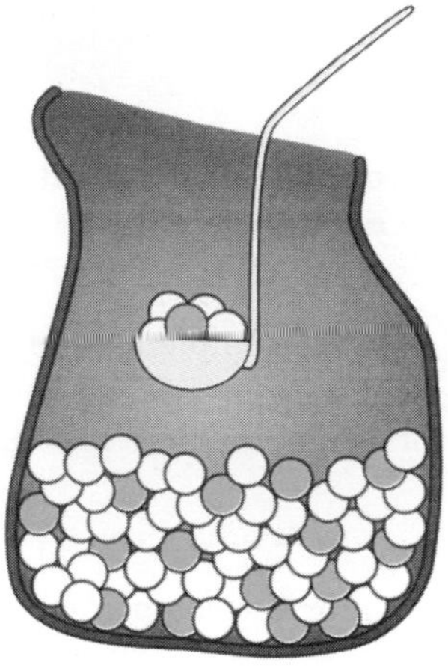

- Conduct an experiment to see how appropriate longterm relative frequency is to measure a known proportion. You need a large collection (say 50, 100, 200) of identically-shaped objects (tiddly winks, beads or marbles). A known proportion ($\frac{1}{5}, \frac{1}{4}, \frac{1}{3}$) of the objects need to be distinguishable from the others in appearance (for example a different colour, say red).
  Place the beads in a container (bag, box).
  - Randomly distribute them within the container (shake the bag). Obtain, randomly, a sample of beads (a spoon is a useful device, or the lid of a small box). Record the number of beads in the sample and the number of red beads included in it. Return the sample to the bag. Now repeat the process.
  - When you have conducted enough repeats, you are in a position to examine the effectiveness of cumulative relative frequency as a measure of proportion (or probability). Calculate individual sample relative frequencies and the cumulative relative frequencies. Plot a graph of the results. Interpret the results of your experiment and compare with the known proportion red beads.
  - The diagram overleaf shows the results of an experiment recorded using a spreadsheet. What proportion of the beads are red?

Sample	No. of Red	No. in Spoon	Cumul. Reds	Cumul. total	Relative Freq.	Cumul. RF	Sample	No. of Red	No. in Spoon	Cumul. Reds	Cumul. total	Relative Freq.	Cumul. RF
1	2	7	2	7	2/7	2/7	11	2	9	30	130	2/9	0.231
2	0	8	2	15	0	2/15	12	5	15	35	145	1/3	0.241
3	3	16	5	31	3/16	5/31	13	5	17	40	162	5/17	0.247
4	3	11	8	42	3/11	0.190	14	2	13	42	175	2/13	0.240
5	4	13	12	55	4/13	0.218	15	1	12	43	187	1/12	0.230
6	4	13	16	68	4/13	0.235	16	6	15	49	202	2/5	0.243
7	2	13	18	81	2/13	0.222	17	3	12	52	214	1/4	0.243
8	0	10	18	91	0	0.198	18	3	15	55	229	1/5	0.240
9	5	15	23	106	1/3	0.217	19	3	13	58	242	3/13	0.240
10	5	15	28	121	1/3	0.213	20	2	12	60	254	1/6	0.236

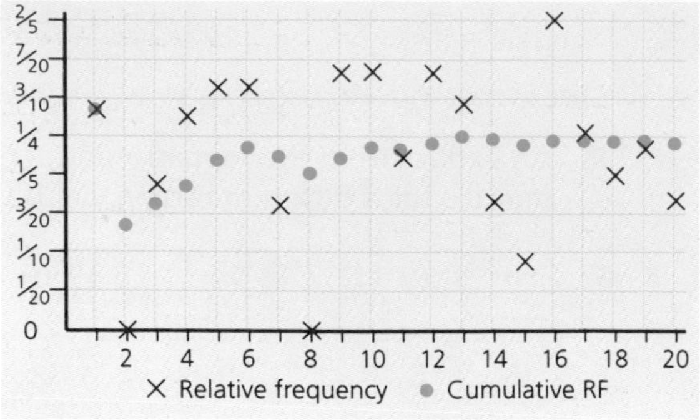

Some classical results

There are several important results concerning probability which always hold true. We have already encountered the first of them: the probability of an event is never more than 1, nor is it less than zero.

Complementary events

The second result is very simple but extremely valuable, as we shall see as the chapter progresses: the probability of an event and its complement add up to 1.

In the previous section we considered rainy days. The complement of a rainy day is a dry day, or alternatively, a day that is not rainy. What this result indicates is that, if the probability of a rainy day is $\frac{3}{7}$, then, the probability of a dry day is $1-\frac{3}{7}$ or $\frac{4}{7}$.

Notation

If E represents an event, then a common notation for the complement of E is $\bar{E}$. Sometimes the alternative notation of E' is used. In this textbook, the former notation will be used. Hence:

$$P(E) + P(\bar{E}) = 1$$

or:

$$P(E) = 1 - P(\bar{E})$$

The third and fourth results concern combinations of events. Consider the following weather records for twelve randomly chosen days.

Day	Weather		
a	dry	hot	still
b	wet	hot	windy
c	wet	cold	still
d	wet	hot	still
e	dry	cold	still
f	dry	hot	windy
g	wet	hot	windy
h	dry	hot	still
i	wet	cold	windy
j	wet	cold	still
k	dry	hot	windy
l	dry	hot	still

What is the probability that one of the randomly chosen days was hot?

Altogether, there are twelve days, and eight of these are described as hot, hence:

$P(\text{hot}) = \frac{8}{12}$

What is the probability that one of the days is both hot and wet?

There are three days that are both hot and wet.

$P(\text{hot} \cap \text{wet}) = \frac{3}{12}$

Now let's focus our attention solely on those days described as hot. These are the days **a**, **b**, **d**, **f**, **g**, **h**, **k** and **l**. What is the probability that the weather is wet given that it was hot?

There are eight hot days; of these, three are described as wet. Hence: P(wet given that the day was hot) = $\frac{3}{8}$

This is known as the **conditional probability** of a day being wet knowing that it is hot.

Notation

A common notation for this is:

$P(\text{wet} \mid \text{hot}) = \frac{3}{8}$

The vertical line indicates that the event of interest is a day being wet given that it is hot. What is $P(\text{hot} \mid \text{wet})$?

To find out, we need to focus only on those days that were wet. There are six wet days. Of these, three are hot (we know this already – how?). Hence, the probability of a hot day given that it is wet is $\frac{3}{6}$. This is not the same as $P(\text{wet} \mid \text{hot})$.

Note that: $P(\text{wet} \cap \text{hot}) = P(\text{hot}) \times P(\text{wet} \mid \text{hot}) = P(\text{wet}) \times P(\text{hot} \mid \text{wet})$

Exploration 4.4

Examining conditional probability

- Return to the twelve randomly chosen days, above. Define your own two events, say events *A*, *B*.
  - **a)** Explore the relationship between the three probabilities: $P(A \cap B)$, $P(B)$, $P(A \mid B)$
  - **b)** Explore the relationship between: $P(A \cap B)$, $P(A)$, $P(B \mid A)$
- Now repeat the first part for another pair of events.

The multiplication law

The third result which concerns conditional probabilities is:

$P(A \cap B) = P(A) \times P(B \mid A)$

or:

$P(A \cap B) = P(B) \times P(A \mid B)$

This result is always true, for any events *A*, *B*. The special case is where the conditional probability of event *A* occurring is the same as the unconditional probability. In other words, when:

$P(A|B) = P(A)$

Under these circumstances A and B are said to be **independent events**. A direct consequence of this is:

$P(A \cap B) = P(A) \times P(B)$

when A and B are independent. This result, known as the **multiplication law** for independent events, is very important. It can be extended for any number of events which are independent.

$P(A \cap B \cap C) = P(A) \times P(B) \times P(C)$ etc.

Exploration 4.5

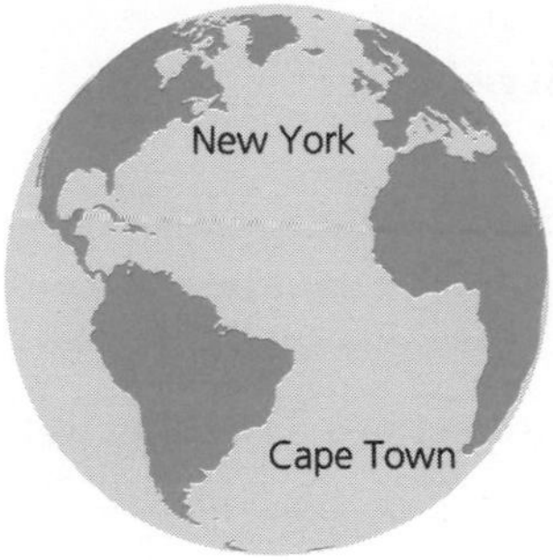

Independent events

- Define an event. Now define another event which you feel is independent of your first event. (For example: if the first event is 'It will rain in Cape town, South Africa, tomorrow', the second event could be 'It will be windy in New York, USA, tomorrow'.)
- Define five other pairs of independent events.

Inclusive events

Returning to the weather data on page 110, what is the probability that a day is *either* wet *or* hot? There are eleven of these days (every day apart from day **e)**, so the probability is $\frac{11}{12}$.

The common notation for this event is:

$P(\text{wet} \cup \text{hot}) = \frac{11}{12}$

Note that in saying 'wet or hot', we include all those events where the day was both wet *and* hot. Previously, we found that:

$P(\text{wet}) = \frac{6}{12} \quad P(\text{hot}) = \frac{8}{12} \quad P(\text{wet} \cap \text{hot}) = \frac{3}{12}$

Can we find a relationship between the probabilities of the four events:

wet hot wet *and* hot wet *or* hot?

We find that:

$P(\text{wet} \cup \text{hot}) + P(\text{wet} \cap \text{hot}) = P(\text{wet}) + P(\text{hot})$

Exploration 4.6

Inclusive events

- Return to the twelve days and define your own two inclusive events, say A and B. Explore the relationship between the probabilities of the events:
  $A, B, A \cup B, A \cap B$
- Repeat this for another pair of events.

The addition law

The fourth result concerning the probability of inclusive events is:

$P(A \cap B) + P(A \cup B) = P(A) + P(B)$

This result is always true regardless of the definition of events A and B.

The special case applies when the events in question do not occur together. Under these circumstances the events are called **mutually exclusive**. Since mutually exclusive events do not occur together:

$$P(A \cap B) = 0$$

which, in turn, implies that:

$$P(A \cup B) = P(A) + P(B)$$

This is a result know as the **addition law** for exclusive events.

Example 4.1

Three ordinary coins are flipped. What is the probability that at least one head appears?

$$\textit{Probability of outcome} = \frac{\textit{number of events in outcome space}}{\textit{number of events in sample space}}$$

Solution

One approach to solving this is to identify all the ways where at least one head occurs.
These are:
{HTT, THT, TTH, HHT, HTH, THH, HHH}
This set of outcomes is the outcome space. We now need to identify the sample space i.e. all possible ways that three coins could fall, which is:
{TTT, HTT, THT, TTH, HHT, HTH, THH, HHH}
We are now in a position to use the definition:

$$\textit{P(outcome)} = \frac{\text{number of events in the outcome space}}{\text{number of events in the possibility space}}$$

Thus, P(at least one head) = $\frac{7}{8}$

Alternative solution

It is equally valid and, perhaps, quicker to use some of the probability results. The question concerns the outcome 'at least one head'. It is often worth considering the complementary outcome; the complement to 'at least one' is 'less than one'. In this case, 'less than one head' is 'no head'. Then:
P(at least one head) + P(no head) = 1

Hence, the very useful result:
P(at least one head) = 1 – P(no head)

The outcome corresponding to 'no head' is quite simply 'three tails'. The probability of three coins landing tails can be found using the result for independent events:
$P(A \cap B \cap C) = P(A) \times P(B) \times P(C)$
In this case, we require:
P(tail and tail and tail) = P(tail) × P(tail) × P(tail) = $\frac{1}{2} \times \frac{1}{2} \times \frac{1}{2} = \frac{1}{8}$
hence:
P(at least one head) = 1 – P(no head) = 1 – $(\frac{1}{8}) = \frac{7}{8}$

Example 4.2

A cubical die with faces numbered 1, 2, ... 6, is rolled. Two coins are independently flipped. What is the probability that the coins land showing at least one head or the die shows a multiple of three?

Solution

This is a question about inclusive combined events. Its solution could be obtained by using:

$P(A \text{ or } B) + P(A \text{ and } B) = P(A) + P(B)$

In this case, event A might be 'a multiple of three' and event B might be 'at least one head'. For the die, two of the six equally likely outcomes are multiples of three, hence:

$P(A) = P(\text{multiple of three}) = \frac{2}{6}$

For the two coins:

$P(B) = P(\text{at least one head}) = 1 - P(\text{no head}) = 1 - (\frac{1}{2})^2 = \frac{3}{4}$

It is reasonable to assume that the outcomes of rolling the die and flipping the coins are independent, hence:

$P(A \text{ and } B) = P(A) \times P(B) = \frac{2}{6} \times \frac{3}{4} = \frac{1}{4}$

We are now in a position to answer the question.

$P(A \text{ or } B) = P(A) + P(B) - P(A \text{ and } B) = \frac{1}{3} + \frac{3}{4} - \frac{1}{4} = \frac{5}{6}$

Alternative solution

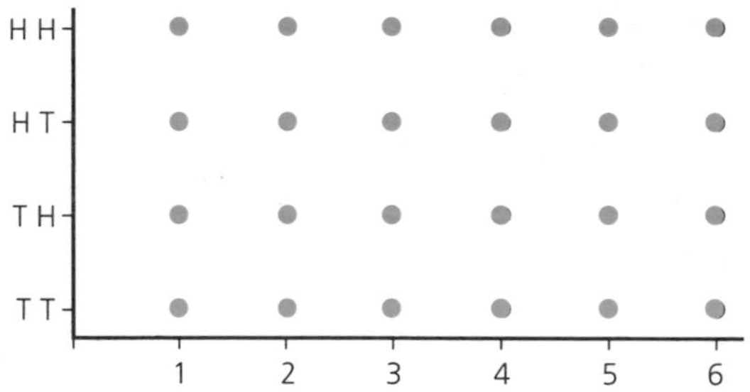

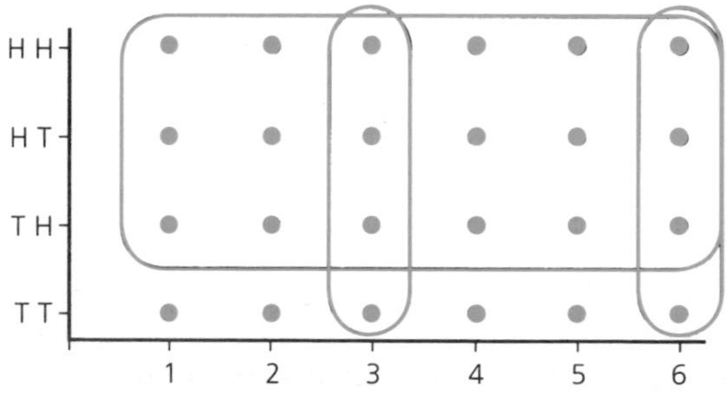

Again there is another way to approach the problem. The method of identifying equally likely elements in sample and outcome spaces can be used. Whilst it might be considered to be rather lengthy, the approach outlined here introduces a graphical way of identifying sample spaces.

The method requires us to identify the sample spaces for the die and the coins separately. These are:

Die: {1, 2, 3, 4, 5, 6}
Coins: {HH, HT, TH, TT}

Now we display the possible contributions as points on a grid.

Each of the 24 points represents an equally likely event in the sample space. The elements of the outcome space can be identified and counted.

There are 20. Hence:

$P(\text{at least one head or a multiple of three}) = \frac{20}{24} = \frac{5}{6}$

Example 4.3

A market gardener grows some vegetables organically and others conventionally, using fertilisers. She uses fertilisers for $\frac{4}{5}$ of her produce. Carrots make up 15 per cent of her conventional produce and 30 per cent of her organic produce.

a) *What proportion of the gardener's produce are carrots?*

b) *A thief breaks into the market garden and is caught stealing carrots. What is the probability that they were organically grown?*

Solution

There is quite a lot of information to take in. It may help to display this information diagrammatically. One useful approach is known as a ***tree diagram****. The first bit of information we have is about organic or conventional methods of growing. The second, about the type of vegetable, is dependent on the growing method.*

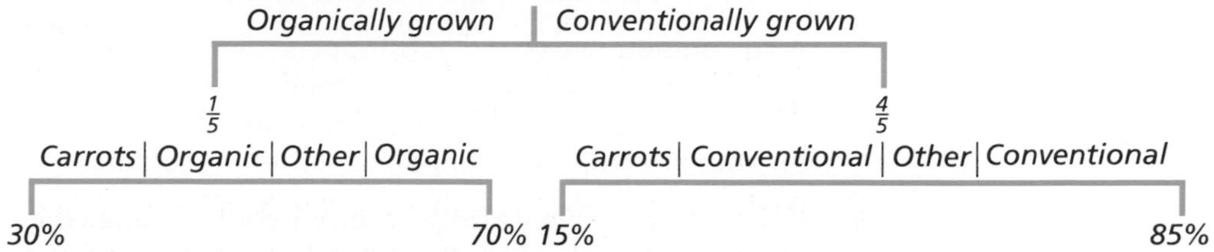

Notice the conditional events.

a) *30 per cent of organically grown vegetables are carrots and* $\frac{1}{5}$ *of the vegetables are grown organically. Hence:*

P(organically grown and a carrot) = $\frac{1}{5} \times 30\% = 6\%$

Similarly:

P(conventionally grown and a carrot) = $\frac{4}{5} \times 15\% = 12\%$

Hence:

P(carrot) = P(C and organic) + P(C and conventional) = 6% + 12% = 18%

b) *The question seeks the probability :* P(organic | carrot).

Since we know that:

P(carrot and organic) = P(C) × P(organic | C)

This can be rearranged to yield:

P(organic | C) = P(C and organic) ÷ P(C) = 6% ÷ 18% = $\frac{1}{3}$

EXERCISES

4.1A

1 Two cubical dice are rolled. What is the probability of at least one 5 appearing?

2 A tetrahedral die with faces numbered 1, 2, 3, 4 is rolled and two coins are flipped.

a) What is the probability of at least one tail and a multiple of 2 appearing?

b) What is the probability of at least one tail or one multiple of 2 appearing?

c) What assumptions have you made in arriving at your answers?

3 From an ordinary pack of 52 cards, one card is selected.

a) What is the probability that the card is:

i) a club, **ii)** the king of clubs, **iii)** a black king?

b) Consider the outcomes of selecting a club, a king, a red king. Which of the three possible pairs of outcomes are:

i) independent, **ii)** mutually exclusive, **iii)** neither?

4 A tetrahedral die with faces numbered 1, 2, 3, 4 is rolled. A cubical die with faces numbered 1, 2, 3, 4, 5, 6 is rolled.
a) Display the sample space graphically.
b) Use your graphical display to determine the number of elements in the following outcome spaces:
i) the sum of the outcomes on the two dice is 5,
ii) the sum of the outcomes on the two dice is at least 5,
iii) the outcome on the cubical die is a multiple of 3, or the outcome on the tetrahedral die is prime,
iv) the product of the outcome on the two dice is at least 16.

5 A jar contains twelve red gobstoppers and eight white gobstoppers. Two gobstoppers are taken out randomly (without replacement). Using a tree diagram, find the probability that:
a) both are red, **b)** they are of different colours.

6 Box X contains two white discs, box Y contains one white and one black disc. Two coins are flipped, if the outcome is two tails then a disc is selected at random from box X. Otherwise, a disc is randomly selected from box Y.
a) What is the probability that a white disc is selected?
b) Given that a white disc is selected what is the probability it came from box Y?

EXERCISES

4.1B

1 Two octahedral dice with faces numbered 1, 2, 3, 4, 5, 6, 7, 8 are rolled. What is the probability of at least one 6 appearing? What is the probability that the outcome on one die is a least 6 and the outcome on the other is less than 6?

2 Three coins are flipped. A cubical die is rolled.
a) What is the probability of a least one head *and* an even number appearing?
b) What is the probability of at least one head or an even number appearing?

3 **a)** For two events A, B, explain why $\mathrm{P}(A) = \mathrm{P}(A \text{ and } B) + \mathrm{P}(A \text{ and } \bar{B})$
b) John swims about once a week. Gill swims on three-quarters of the days that John does, but only a third of the days that John does not. What is the probability that Gill has a swim?

4 Two cards are selected randomly and without replacement from an ordinary pack of 52 cards. The ace in each suit is assumed to be numbered 1. The court cards (jacks, queens, kings) are assumed to have the values 11, 12, 13 respectively.
a) A is the outcome that the first card is prime, B is the outcome that the second card is even. Calculate:
i) $\mathrm{P}(A)$ **ii)** $\mathrm{P}(B)$ **iii)** $\mathrm{P}(A \mid B)$ **iv)** $\mathrm{P}(B \mid A)$.
b) X is the outcome that the first card is less than 11, Y is the outcome that the second card is 10 or more. Calculate:
i) $\mathrm{P}(X)$ **ii)** $\mathrm{P}(Y)$ **iii)** $\mathrm{P}(X \mid Y)$ **iv)** $\mathrm{P}(Y \mid X)$.

5 Two octahedral dice with faces numbered 0, 1, 2, 3, 4, 5, 6, 7 are rolled.
a) Display the possibility space in a diagram.
b) Calculate the probabilities of the following events:
i) the sum of the outcomes of the two dice is 5,
ii) the sum of the outcomes of the two dice is at least 5,
iii) there is a multiple of 3 on one die and a prime number on the other die,
iv) there is a multiple of 3 on one die or a prime on the other,
v) the product of the outcomes on the two dice is at least 25.

6 A box contains three red and three blue cubes. Another box contains four red and three blue cubes. A cube is removed randomly from the first box and placed in the second box. A cube is now randomly selected from the second box. Use a tree diagram to calculate the probability that:
a) the cube finally selected is red,
b) if the selected cube is red then the cube removed from the first box was also red.

SOME WAYS OF COUNTING ELEMENTS IN A SAMPLE SPACE

Factorial notation

Have you ever overheard this sort of thing?

'It's been a dreadful week. On Monday it was foggy all day, Tuesday it blew a gale, and Wednesday it didn't stop raining ... '

Sometimes weather conditions like these do occur, even if not always in that particular order. A summary of the order of events might be:

	Monday	Tuesday	Wednesday
	fog	gale	rain
or, more briefly:	F	G	R

Suppose that these events *did* occur on three consecutive days, in what other sequences might they occur? There are six distinct sequences. FGR FRG GFR GRF RFG RGF

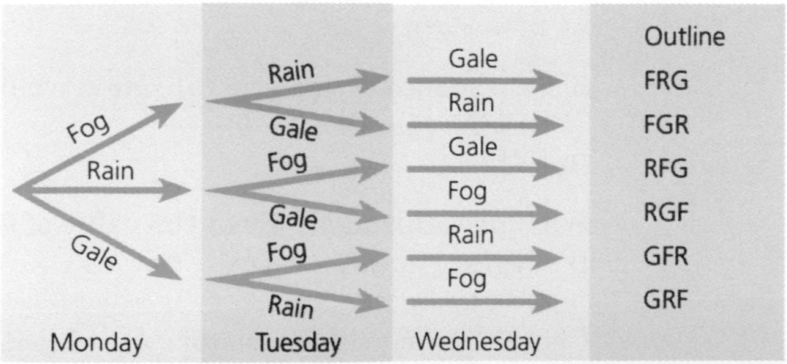

This approach is not necessary if all we want to do is to determine the number of different arrangements. Notice that for Monday, there are three possible outcomes (F, G, R). Then, there are only two possible outcomes for Tuesday. This leaves one for Wednesday. Hence the total number of possible arrangements of F, G, R is: $3 \times 2 \times 1 = 6$.

Suppose it is known that on four consecutive days the weather was fog, rain, gale, and cold. How many different arrangements of these events are possible?

There are 24, since there is a choice of four weather types for the first day, then a choice of three for the next (hence a total of 4×3 possibilities so far), and a choice of two for the third. Finally, only one possibility is left for the final day. Hence the number of possible arrangements of these four weather types is: $4 \times 3 \times 2 \times 1 = 24$.

This sort of number pattern: $3 \times 2 \times 1 \quad 4 \times 3 \times 2 \times 1$
occurs frequently in the study of Statistics. Consequently, it is given its own notation: $\mathbf{3 \times 2 \times 1}$ is written as **3**! and this is called **three factorial** or, equally correctly, **factorial three**.

CALCULATOR ACTIVITY

Evaluating factorials

Your calculator will probably evaluate factorials directly. If it does, check that 3! is 6 and that 4! is 24.

Evaluate: **a)** 5! **b)** 6! **c)** 10! **d)** 1!

Factorial zero

Notice that:
4! is 4 times 3! and
5! is 5 times 4! and
6! is 6 times 5!.
Why is this?

Using the number patterns, we can see that:
$3! = 3 \times 2 \times 1$ and
$4! = 4 \times 3 \times 2 \times 1 = 4 \times 3!$
and similarly for the other results.

A generalisation of this is:
$n! = n \times (n-1)!$

A consequence of this generalisation would seem to be that:
$1! = 1 \times (1-1)! = 1 \times 0!$
But: $1! = 1$

What does this imply about the value of 0!?
It suggests that:
$0! = 1$
Check this on your calculator.

Example 4.4

Statistically speaking, what do these two situations have in common?

a) *Five children: John, Andrew, Philippa, Kate, and Colette, are drawing lots to be last to have an injection. What is the probability that they will be drawn in alphabetical order?*

b) *A restaurant offers its customers a meal consisting of three courses. For the first course there is a choice of one of the following dishes:*
pâté, melon, prawn cocktail, avocado.
For the main course, the diners have a choice of:
venison, turbot, steak, stir-fry vegetables, pulse pie.
For the final course, there is:
ice cream, cheese.
How many different meals are possible?

Solution

a) *Clearly, there is only one arrangement which results in the children being in alphabetical order. This is:*
Andrew Colette John Kate Philippa
The total number of possible orders is 5! since there is a choice of five for the first in line to receive the injection, the second can then be anyone of four, and so on. This leads to the number of distinct orders being $5 \times 4 \times 3 \times 2 \times 1$ or 5!.
Hence, the probability that they will be in alphabetical order is $\frac{1}{120}$.

b) *In this problem we can readily arrive at the following response:*

	First course	Main course	Final course
Choice of	4	5	2

Hence, there is a total of $4 \times 5 \times 2 = 40$ possible meals.

Distinct sequences

Let's return to the conversation about the weather that we overheard on page 117.
'Then the gales returned on Thursday and Friday ... '
So the weather pattern for these five weekdays was FGRGG.
How many distinct sequences of these five are there?

If all five days had different weather, then there would be 5! distinct sequences. But that is not the case here. Consider one possible sequence: FRGGG.

If the GGG had all been different they could have been arranged in 3! distinct ways. Instead, there is *only one* arrangement. Consider another, possible sequence: GFGGR.

Again, there is only one distinct arrangement of the three Gs in place of the 6 (i.e. 3!) possible ways were they all different. Thus there is only $\frac{1}{6}$ of the number of sequences possible if all days were different. Hence, the number of distinct **arrangements of five** events where **three are identical** is $5! \div 3!$

Suppose that the week in question culminated in: '... As for the weekend, well that was a total washout. It rained the whole time ... '

So, what we've now got is the following sequence: FGRGGRR

How many distinct sequences of these seven events are possible?

The same sort of arrangement which led to $\frac{5!}{3!}$ for five events, three of which were identical can be applied here. In this case, there are seven events, three alike of one sort, and three alike of another sort. Hence, the total possible arrangements are $\frac{7!}{3!3!}$, or 140.

This result can be generalised.

The number of distinct arrangements of n objects where p are allied of one type, q alike of another, r alike of a third type, etc. is $\frac{n!}{p!q!r!}$

Example 4.5

a) How many distinct arrangements of the five letters of the following words are there?
i) RADAR ii) ERROR iii) HELEN

b) If the letters of RADAR are randomly arranged, what is the probability that the resulting sequence will be symmetric?

Solution

a) i) In RADAR, two letters are As and two are Rs, hence the number of distinct arrangements is $\frac{5!}{2!2!}$. This comes to 30.

ii) In ERROR, there are three Rs. Hence, the number of distinct arrangements is $\frac{5!}{3!} = 20$.

iii) In HELEN, there are two Es. Hence the number of distinct arrangements is $\frac{5!}{2!} = 60$.

b) The symmetric arrangements of the letters of RADAR are: RADAR and ARDRA. Hence, the probability required is $\frac{2}{30}$ or $\frac{1}{15}$.

Two types of object

There is a particular case of the previous result which will recur in the study of statistics. This is where there are n objects but only two different types. For example, the seven days we have heard about in this section might have been described as:

wet dry wet dry dry wet wet

where only two different descriptions are used. The total number of different sequences of these description is $\frac{7!}{4!3!} = 35$

Where there are n objects of which r are of one type and the rest, $(n - r)$, are of another type, the number of distinct arrangements is $\frac{n!}{r!(n-r)!}$

Example 4.6

a) How many distinct ways are there of arranging the letters ABBA? List all of them.

b) *A bicycle rack has spaces for ten cycles. Three identical bicycles are to be placed in the rack. In how many ways may this be done?*

c) *In how many ways can a group of eight people be divided into a group of three and a group of five?*

Solution

a) *In ABBA there are four letters consisting of two As and two Bs. Hence, the number of distinct arrangements is:* $\frac{4!}{2!2!} = 6$.
The six arrangements are AABB, ABAB, ABBA, BABA, BAAB, BBAA.

b) *This problem can be modelled by the sort of arrangements considered above. The way to do this is to consider the rack as consisting of 'occupied spaces' and 'not-occupied spaces'. An occupied space could be written as O, and not-occupied as N. Thus, this becomes a problem about having ten letters in a line, of which three are Os and seven Ns. There are* $\frac{10!}{7!3!}$ *distinct ways of arranging these letters. Hence there are 120 different ways of placing the three cycles in the rack.*

c) *This can be treated in the same way as in part* **b)**. *The people chosen for the smaller group could be labelled S, and those chosen for the larger group L. Then this problem is equivalent to finding how many ways eight letters consisting of three Ss and five Ls can be arranged. The answer is* $\frac{8!}{3!5!} = 56$.

EXERCISES

4.2A

1 The lunch-time menu of a seaside café offers the following.
First course: soup or fruit juice
Second course: plaice and chips or ham salad or cheesy baked potato
Third course: ice cream or apple pie with cream

a) How many different three-course lunches are possible?

b) What is the probability of a customer having a meal which includes ham salad?

c) State the assumptions you have made in answering **b)**.

2 How many distinct five-digit numbers can be formed using the following sets of numbers?

a) 9, 8, 7, 6, 5 **b)** 4, 3, 2, 1, 0

3 In the staffroom of a sixth-form college there are 50 pigeon-holes for staff mail. The principal of the college has been given six free tickets to the theatre. She decides to place them randomly in the 50 pigeon-holes. In how many ways may this be done?

4 Contestants in a lottery are required to choose four of the ten digits: 0, 1, 2, 3, 4, 5, 6, 7, 8, 9.
A computer is used to generate four of the digits randomly. What is the probability of a contestant winning the top prize if:

a) it is given for choosing the same digits in the same order as the computer,

b) it is given for choosing the same digits as the computer?

5 A squad of 17 players is named for a football team to play in a final of a competition. How many different teams of eleven players can be chosen from the squad if:
a) all players can play in all positions,
b) the players are specialists and:
two only play in goal (one is needed in the team)
five play only in defence (three are needed in the team)
six play only in mid-field (four are needed in the team)
four play only in attack (three are needed in the team)?

EXERCISES

4.2B

1 A man has two jackets, eight pairs of shorts, three ties and five pairs of trousers. How many distinct outfits are possible?

2 **a)** How many distinct arrangements of the work PLYMOUTH are possible using all the letters?
b) How many of the arrangements in **a)** begin with two vowels?

3 A man who works a five-day week likes to vary the means of transport he uses. He has a car, a motorbike, a bicycle, a bus, or his feet available. In how many different ways can he arrange a week's travelling if:
a) he uses different means of transport on each day of the week,
b) there are no other restrictions placed on the means of transport each day?

4 There are 32 children in a class. Six disco tickets are given randomly to the children.
a) How many different ways are there of doing this?
b) There are 20 girls and twelve boys in the class. In how many ways will:
i) three tickets go to boys and three go to girls,
ii) all six go to girls,
iii) four go to girls and two to boys?

5 A squad of twelve players is named for a seven-a-side competition. How many different teams of seven players can be chosen if:
a) all players can play in any position,
b) the players are specialists and:
five only play in the forwards (three are needed)
five only play in the backs (three are needed)
two only play at full back (one is needed)?

APPLICATION OF ARRANGEMENTS

One Monday my friend asked, 'Why does it always rain more at the weekend?' I couldn't answer the question, but I wondered what had prompted it. My friend said that the rainfall for the previous seven days had been:

	Monday	Tuesday	Wednesday	Thursday	Friday	Saturday	Sunday
Rain (mm)	1.2	0.4	4.8	0.7	0.5	4.6	1.1

This is how I approached the answer. I distinguished between the weekend and the rest of the week, and coded the days as E for weekend and D for week day. Then the above data becomes:

Coded day	D	D	D	D	D	E	E
Rain (mm)	1.2	0.4	4.8	0.7	0.5	4.6	1.1

I then thought about the way the question was worded, and in particular the use of the word 'more'. It occurred to me that it was not the *actual amount* of rain that mattered, but the **relative amount**. A simple way of dealing with this was to **rank** the rainfall. I decided to give the wettest day a rank of 1, down to the driest with a rank of 7. This resulted in:

Coded day	D	D	D	D	D	E	E
Ranked rainfall	3	7	1	5	6	2	4

Putting them in order:

Coded day	D	E	D	E	D	D	D
Ranked rainfall	1	2	3	4	5	6	7

This is just one arrangement of five Ds and two Es out of the $\frac{7!}{5!2!} = 21$ different possible arrangements. In this arrangement the ranks associated with the two Es are 2 and 4. These ranks seem low and low ranks are consistent with more rain. Perhaps, the evidence of this week indicates that it does rain more at the weekend. I thought a little further. This arrangement is not the 'wettest'. Arrangements of which associated ranks are consistent with wetter weekends are:

	Ranks	1	2	3	4	5	6	7
a	arrangement	E	E	D	D	D	D	D
b	arrangement	E	D	E	D	D	D	D
c	arrangement	E	D	D	E	D	D	D
d	arrangement	D	E	E	D	D	D	D

These showed 'wetter' weekends because of the lower totals for the ranks of the two Es. In arrangement **a**, the sum of the ranks is 3 which is the smallest possible sum of two ranks. Arrangement **b** has the next smallest possible sum of two ranks, 4.

What can we say about arrangements **c** and **d**?

The sum of the ranks of the two Es in each of these arrangements amounts to 5. It would be reasonable to say that they correspond to equally wet weekends.

Are there any other arrangements which could be said to be as wet as the weekend in question, i.e. DEDEDDD?

In this case, the sum of the ranks of the two Es is 6. One other arrangement, EDDDEDD, has the same rank sum.

Summarising the findings so far:

Wettest possible	Ranks of the two Es
less wet ↓	1, 2
	1, 3
	1, 4 or 2, 3
	1, 5 or 2, 4

I wanted to know the chance of getting a weekend at least as wet as the one in question. Assuming that there is no tendency for one day to be wetter than any other, it is reasonable to argue that each of the 21 possible arrangements of two Es and five Ds is equally likely. Of these 21 arrangements, there are six which correspond to a weekend 'at least as wet' as the one in question. The chance is, therefore, $\frac{6}{21}$ or about 29 per cent. I concluded that this then wasn't particularly unusual and told my questioner that the evidence did not support the assertion that it rains more at the weekend.

This application will be explored more extensively in Chapter 17, *Non-parametric tests*, where it is encountered as 'Wilcoxon's Rank Sum Test'.

Example 4.7

a) *Assume the seven days considered above had the following rainfall figures.*

	Monday	Tuesday	Wednesday	Thursday	Friday	Saturday	Sunday
Rain (mm)	0.7	0.8	3.2	0.4	0.6	1.1	3.6

Does this evidence support the contention that it rains more at the weekends?

b) *The heights, in cm, of nine students studying A-level mathematics are recorded in this table.*

Boy	165, 183, 175, 172
Girl	173, 164, 168, 155, 163

Do these data support the view that the boys who study A-level mathematics are taller than the girls?

Solution

a) *Code the days as before and rank the rainfall similarly (low ranks corresponding to rainy days).*

D	D	D	D	D	E	E
5	4	2	7	6	3	1

Start by considering the smallest possible sum of ranks and build up a table.

Rank sum	Ranks of two Es	Number of arrangements with this rank sum
3	1, 2	1
4	1, 3	1

So, there are two arrangements of two Es, five Ds which correspond to weekends as wet as this one. There are 21 possible arrangements, and assuming that each of these is equally likely leads to the conclusion that the probability of an arrangement corresponding to a rank sum of 4 or fewer is $\frac{2}{21} \approx 9\frac{1}{2}\%$*. In this context, the assumption that each arrangement of two Es and five Ds is equally likely is equivalent to saying that no day is more likely to be wet than any other day. Consequently, the probability of getting a weekend as relatively wet at this one is less than ten per cent.*
Perhaps, it would be reasonable to conclude that the evidence does support the view that weekends are wetter.

Note: *The evidence does not* ***prove*** *that weekends are wetter. In fact, what we have done is to indicate that we shall get this sort of result one week in ten when there is no tendency for one day to be any wetter than any other. By drawing the conclusion that weekends are wetter,we are, thus, likely to be wrong about ten per cent of the time.*

b) *In this case, the students can be coded B, G and their heights ranked (with 155 cm being given the rank 1, say). This results in:*

Code:	G	G	G	B	G	B	G	B	B
Rank:	1	2	3	4	5	6	7	8	9

Thus we have five Gs and four Bs, of which there are: $\frac{9!}{5!4!} = 126$ *distinct arrangements.*
Because of the way we have chosen the ranks, high ranks correspond to the taller students. The largest rank sums are summarised in the table over the page.

Rank sum	Ranks of the four Bs	Number of arrangements in the arrangement
30	9, 8, 7, 6	1
29	9, 8, 7, 5	1
28	9, 8, 7, 4, or 9, 8, 6, 5	2
27	9, 8, 7, 3, or 9, 8, 6, 4 or 9, 7, 6, 5.	3

That data in this problem correspond to a rank sum of 27. Assume that each of the 126 possible arrangements is equally likely. The probability of an arrangement having a rank sum of at least 27 for the four Bs is $\frac{(3+2+1+1)}{126} \approx 5\frac{1}{2}\%$*. So there is a small chance of getting a rank sum of 27 or more when there is no tendency for boys to be taller than girls. It is reasonable to conclude that boys are taller than girls. (**Note:** The chance of being wrong is between five and six per cent).*

EXERCISES

4.3A

1 A group of eight children with learning difficulties are taught writing skills. Four of the children have access to a word processor. Their scores on creative writing are given in the table.

Children with access to WP	80.3	84.2	89.6	75.3
Children without access to WP	83.2	79.1	75.4	60.8

a) Rank the scores giving the score 89.6 a rank of 1.
b) Code the ranked scores using A to represent 'access', and B to represent 'without access'.
c) Explain why there are 70 distinct arrangements possible of four As and four Bs.
d) In how many of the arrangements does the sum of the ranks of the four As amount to:
i) 10 **ii)** 11 **iii)** 12 **iv)** 13?
e) Explain whether you feel there is sufficient evidence in these data to suggest that children with access to a word processor achieve higher creative writing scores.

2 The concentration of a chemical pollutant, in parts per million, is measured at a total of nine places in two rivers R and T. The results are shown below.

River R	25.2	25.7	25.8			
River T	24.3	25.9	26.1	28.2	30.4	31.7

a) Rank the concentration levels giving 24.3 the rank of 1.
b) Code the ranked levels R, T appropriately.
c) How many distinct arrangements of three Rs and six Ts are possible?
d) Is there sufficient evidence in these data to support the view that river R has lower levels of pollutant than river T?

3 The reaction times of a group of smokers and a group of non-smokers were measured in milliseconds. The results are shown below.

Smokers	17	25	26	38	46
Non-smokers	13	15	16	24	

Is there sufficient evidence that the reaction time of non-smokers is less than that of smokers?

4 A measure of social adjustment was obtained for a group of inner-city children and a group of rural children. The results are shown below.

Inner-city	155	157	162	158	134	163
Rural	142	153	148	154		

Is there sufficient evidence to suggest that rural children attain lower measures of social adjustment?

5 The breaking strains of two types of fibre are measured. The results are displayed in the table. Is there sufficient evidence to support the view that the fibre with the new additive has a higher breaking strain?

Old fibre	97.0	97.3	97.4	97.5	97.8	97.9
New fibre	96.5	96.7	96.8	97.1	97.2	97.6

EXERCISES

4.3B

1 A group of seven children have their computer aptitude measured. Four of the children are known to have learning difficulties and three are classified as not having learning difficulties. The aptitude scores are given below.

Learning difficulties	276	268	266	271
No learning difficulties	274	275	279	

a) Rank the aptitudes giving 279 the rank 1.
b) Code the rank using L to represent 'learning difficulty' and N to represent 'no learning difficulty'.
c) Explain why there are 35 distinct arrangements of four Ls and three Ns possible.
d) In how many of the arrangements does the sum the ranks of the three Ns amount to:
i) 6 **ii)** 7 **iii)** 8?

e) Explain, with reasons, whether you feel there is sufficient evidence in these data to support the view that children without learning difficulties have higher computer aptitude scores.

2 A group of three cocker spaniels and six beagles were assessed for obedience. The results are recorded in this table.

Cocker spaniels	85.4	88.3	84.0			
Beagles	82.7	85.3	85.1	86.3	81.3	80.3

a) Rank the obedience assessments giving 88.3 the rank 1.
b) Code the ranked assessments C, B appropriately.
c) How many distinct arrangements of three Cs and six Bs are possible?
d) Is there sufficient evidence in these data to support the view that cocker spaniels are more obedient than beagles?

3 The concentration, in parts per million, of an alcohol derivative in the blood of rodents found in urban and rural habitats is recorded in this table.

Rural rodent	4.3	4.6	3.7	2.8	4.9	5.0
Urban rodent	4.8	5.6	7.4	4.7		

Is there sufficient evidence to support the view that urban dwelling rodents have higher concentrations of this derivative?

4 The coefficients of viscosity of syrup and treacle are recorded in this table.

Treacle	38.2	38.5	34.8	37.6	35.1
Syrup	35.2	37.5	32.4	31.3	30.8

Is there evidence to support the view that syrup is less viscous than treacle?

5 The amounts of carbon monoxide, in parts per million, produced by car exhaust systems are recorded in this table.

Standard system	3.8	2.6	3.4	2.3	4.5	4.6
Catalytic system	1.5	2.7	1.8	2.4		

Does the available evidence indicate that catalytic systems produce less carbon monoxide?

PERMUTATIONS AND COMBINATIONS

The following is a list of descriptions that could be applied to the weather on a day: misty, cold, dry, freezing, windy, raining, sunny, hot.

Each of a group of people was asked to list, in reverse order of preference, the three types of day they least like. How many different ways of doing this are there?

The least favourite could be any one of the eight types, then the next least favourite could be chosen from seven, the final selection is from the remaining six. This gives: $8 \times 7 \times 6$ different selections where the order of selection matters. A name often applied to this is a **permutation**.

A permutation is an **ordered arrangement of *r* objects from *n* objects**.

In this particular case, the group of people were asked to list a permutation of three descriptions from eight. The total number of permutations of three objects from eight unlike objects is:

$$8 \times 7 \times 6$$

and this can be written in factorial form as:

$$\frac{8 \times 7 \times 6 \times 5 \times 4 \times 3 \times 2 \times 1}{5 \times 4 \times 3 \times 2 \times 1} = \frac{8!}{5!} = \frac{8!}{(8-3)!}$$

The notation, ${}_8P_3$ or ${}^8P_3$, is used for this.

In general:

$${}_nP_r = \frac{n!}{(n-r)!}$$

Suppose the group of people had instead, been asked to list their three least favourite types of day, but not necessarily to indicate the order of preference? How does this affect the number of arrangements?

Consider one permutation, say 'freezing, windy, misty'.

As a permutation, this indicates that 'freezing' is the least desirable, 'windy' the next least followed by 'misty'. Again as a permutation this is different from 'windy, freezing, misty'.

There are 3! permutations of windy, freezing and misty in total. Hence, the number of selections of three descriptions from the eight is:

$$8 \times 7 \times 6 \div 3! \text{ or } \frac{8!}{(5!3!)}$$

The term **combination** is often used for the process of **selecting *r* objects from *n* objects** when the **order** of selection **does not matter**. The notation used is:

$${}_nC_r = \frac{n!}{(n-r)!r!} \quad \text{or sometimes, } {}^nC_r\text{, or even } \binom{n}{r}$$

Take care! There is potential for confusion with other areas of mathematics with this last notation.

Example 4.8

a) Evaluate the following.

i) ${}_5P_2$ *ii)* ${}_5C_2$ *iii)* ${}_9P_4$ *iv)* ${}_{10}C_7$

b) i) How many different hands of five cards is it possible to deal from an ordinary pack of 52 cards?

ii) How many five card hands will have no high cards? (A high card is a king, queen, jack or ace.)

iii) What is the probability that a five-card hand has no high card?

Solution

a) i) $_5P_2 = \dfrac{5!}{(5-2)!} = \dfrac{5!}{3!} = \dfrac{5 \times 4 \times 3 \times 2 \times 1}{3 \times 2 \times 1} = 5 \times 4 = 20$

ii) $_5C_2 = \dfrac{5!}{3!2!} = \dfrac{5!}{3!} = \dfrac{5 \times 4}{2 \times 1} = 10$

iii) $_9P_4 = \dfrac{9!}{(9-4)!} = \dfrac{9!}{5!} = 9 \times 8 \times 7 \times 6 = 3024$

iv) $_{10}C_7 = \dfrac{10!}{3!7!} = \dfrac{10 \times 9 \times 8}{3 \times 2 \times 1} = 120$

b) i) *Dealing five cards from a pack of 52 is equivalent to selecting the five when the order does not matter. Hence, the number of possible five-card hands is:*

$$_{52}C_5 = \frac{52!}{5!47!} = \frac{52 \times 51 \times 50 \times 49 \times 48}{5 \times 4 \times 3 \times 2 \times 1} = 2\,598\,960$$

(i.e. nearly 2.6 million).

ii) *The number of five-card hands with no high cards can be found by considering the selection to be a pack of 36 'low cards'. Hence the number is:*

$$_{36}C_5 = \frac{36!}{31!5!} = \frac{36 \times 35 \times 34 \times 33 \times 32}{5 \times 4 \times 3 \times 2 \times 1} = 376\,992$$

(i.e. nearly 380 thousand).

iii) *The probability required is:*

$$\frac{376\,992}{2\,598\,960} = 0.145\,05$$

CALCULATOR ACTIVITY

Permutations and combinations

Many calculators have nPr and nCr programmed into the available functions. Explore your calculator for either or both of these facilities and try to evaluate the permutations and combinations in Example 4.8.

CONSOLIDATION EXERCISES FOR CHAPTER 4

1 Two cubical dice are rolled. The sum of the outcomes is noted. Use an appropriate diagram to identify the possibility space. Use the possibility space to find:

a) $P(A)$, where A is the set of all events in which the sum is even,

b) $P(B)$, where B is the event that the sum is a multiple of 4,

c) $P(C)$, where C is the event that the sum is prime.

2 A tetrahedral die with faces numbered 1, 2, 3, 4 is rolled three times. Use a tree diagram to find the probability of the following events:

a) there are at least two even outcomes,

b) there are at least two consecutive even outcomes,

c) the outcomes are EOE or OEO, where E represents even outcome, O represents an odd outcome.

3 A coin is flipped five times. What is the probability that:
a) the sequence of outcomes is HTHTH or THTHT,
b) the first flip results in H and the fifth results in T,
c) the first and last flips result in the same outcome?

4 A bag contains four milk chocolates and two plain chocolates. Two chocolates are randomly removed and eaten. What is the probability that they are both the same type?

5 A cubical die with faces numbered 1, 2, 3, 4, 5, 6 and an octahedral die with faces numbered 0, 0, 1, 1, 2, 2, 3, 3 are rolled.
a) Display the sample space in a suitable diagram.
b) Calculate the probability that:
i) exactly one die shows a score of 1,
ii) the total of the outcomes on the two dice is 5,
iii) the product of the outcomes on the two dice is 0,
iv) the outcome on the octahedral die is 2 less than the outcome on the cubical die.

6 A standard pack of 52 cards is shuffled. The top card is examined. The card is returned and the pack is shuffled again. The top card is again examined. Calculate the probability that:
a) both cards are black,
b) the two cards are of different colours,
c) both cards are court cards (J, Q, K or ace),
d) at least one card is a court card,
e) the first card is a king and the second is red.

7 A bag contains s sweets and t toffees. Two are removed and eaten. Calculate the probability, in terms of s and t, that:
a) both are toffees,
b) neither is a toffee,
c) there is one sweet and one toffee.

8 A woman has two pairs of shoes, four skirts and three blouses. How many different outfits consisting of a blouse, a pair of shoes and a skirt can be made out of these?

9 a) How many six-digit numbers can be formed using all the digits 3, 4, 5, 6, 7, 8?
b) How many of these are odd?
c) How many are greater than 600 000?
d) How many are odd numbers greater than 600 000?

10 a) How many distinct arrangements are there using all the letters of the word LETTER?
b) How many of these arrangements have:
i) the L and the R together,
ii) the L and the R separated?

11 Fifteen teachers, six men and nine women, volunteer to sit on the curriculum committee. The head teacher decides that the committee is to have only six members.

a) How many different ways can the six members of the committee be chosen from the 15 volunteers?

b) How many of these ways consist of:

- **i)** exactly four men and two women,
- **ii)** at least one man?

12 Each of eight people has a coin. They each flip their coin.

a) In how many different ways can exactly five heads and three tails result?

b) Calculate the probability that, when eight coins are flipped, exactly five heads occur.

13 Six black marbles, four red marbles and two white marbles are placed in a straight line.

a) In how many distinct ways can this be done?

b) In how many ways will there be a white marble at each end?

c) In how many ways will there be the same colour marble at each end?

14 A pack of 52 cards is shuffled. A hand of five cards is dealt.

a) How many distinct hands are possible?

b) How many hands consist of five diamonds?

c) What is the probability that a hand of five cards consists of five cards of the same suit?

15 A tetrahedral die has its faces numbered 1, 1, 1, 4.

a) Calculate the probability of not getting a 4 when:

- **i)** the die is rolled once,
- **ii)** the die is rolled twice,
- **iii)** the die is rolled five times,
- **iv)** the die is rolled N times?

b) What is the smallest value of N, if the probability in **a) iv)** is less than 0.0001?

16 My car is rather unreliable. The battery functions properly on only 90 per cent of the occasions I try to use it. When the battery does function properly, the engine starts with probability $\frac{2}{3}$. What is the probability that the engine starts?

17 Two cubical die – one red and one blue – are rolled together.

a) Illustrate the possible outcomes in a suitable diagram.

b) If it is known that the sum of the outcomes on the two dice is 6:

- **i)** what is the probability that the outcome on the blue die is 2 more than the outcome on the red die,
- **ii)** what is the probability that the difference between the outcomes on the two dice is 2,
- **iii)** what is the probability that the difference between the outcomes on the two dice is 3?

c) If it is known that the difference between the outcomes on the two dice is 4, what is the probability that the sum of the outcomes is 8?

18 In the town square, there are two large flowerbeds. One has 40 flowers of which 15 are *salvia*. The other flowerbed has 30 flowers of which twelve are *salvia*. A flowerbed is selected and a flower is chosen at random from that bed. The bed with the larger number of flowers is twice as likely to be selected.

a) Write down the probability that the bed of 40 flowers is chosen.
b) Draw a labelled tree diagram to model the scenario described.
c) Calculate the probability that a *salvia* is chosen from the larger bed.
d) Calculate the probability that a *salvia* is chosen.
e) Given that a *salvia* is chosen, calculate the probability that it came from the first bed.

19 In a certain town, 20 per cent of the adult population had at some time been innoculated against 'flu. It is known that there is a chance of 1 in 10 of an adult catching 'flu when innoculated. The chance of catching flu rises to 75 per cent when not innoculated.

a) Calculate the probability of an adult catching 'flu.
b) Given that an adult has caught 'flu, what is the probability the person had been innoculated?

20 There are two different routes A, B into town from college. The chance of encountering a delay on the two routes is:

A: $\frac{2}{5}$ B: $\frac{3}{5}$

Three lecturers X, Y, Z, independently choose route A with probabilities:

X: $\frac{4}{7}$ Y: $\frac{3}{5}$ Z: $\frac{8}{35}$

Find the probability that:

a) X **b)** Y **c)** Z

is delayed on their way to town.

21 At a large community college, the probability that a blue-eyed student is left-handed is 0.15. The probability that a left-handed student is blue-eyed is 0.3. The probability that a student is either blue-eyed or left-handed is 0.2.

Let l, b, x represent the probability that a student is left-handed, blue-eyed, left handed *and* blue-eyed respectively.

a) Demonstrate that $b = 2l$.
b) Express the probability that a student is either blue-eyed or left-handed in terms of l, b, x, hence, in terms of l.
c) Calculate the probability that a student is left-handed and blue-eyed.

22 X and Y are events. $P(X) = \frac{8}{15}$, $P(Y) = p$, $P(X$ and $Y) = \frac{1}{3}$, and $P(X$ given $Y) = \frac{4}{7}$.

a) Evaluate p.
b) Calculate: **i)** $P(Y$ given $X)$ **ii)** $P(Y$ given $\overline{X})$.
c) Explain whether X and Y are:
i) independent, **ii)** exclusive, **iii)** neither.

23 **a)** How many distinct arrangements are there of four As and five Bs?
b) The first-year undergraduates at four new universities and five old universities are rated for their economic activity. The university where the students are least active is given a rank of 1, and the one where they are most active is given rank 9. The result is indicated in the diagram.

Rank	1	2	3	4	5	6	7	8	9
Type of university	A	A	B	A	B	B	B	A	B

(The code A refers to a new university.)
Use a rank sum method to determine if there is sufficient evidence in these data to support the view that students at old universities are more economically active than the students at new universities.

24 Three samples of wire A and seven samples of wire B are tested to destruction. The forces at which the samples broke are given below.

Sample	A	A	A	B	B	B	B	B	B	B
Force (N)	2602	2524	2636	2688	2646	2508	2788	2674	2724	2670

a) How many arrangements of three As and seven Bs are possible?
b) Is there sufficient evidence in these data to indicate that wire A breaks with a lesser applied force than wire B?

25 A lottery consists of selecting five different letters of the 26 letters in the alphabet.
a) How many ways may this be done assuming that the order of selection does not matter?
b) The first prize goes to contestants who select the same five letters as those chosen randomly by a computer. What is the probability of winning the first prize?
c) Only those selections which include at least three of the computer's selected numbers win prizes in the lottery. What is the probability that a selection fails to win a prize?

26 A national lottery requires contestants to select six different numbers from the numbers 1, 2, 3, ..., 48, 49.
a) How many different ways may this be done (the order of selection does not matter)?
b) What is the probability that a contestant's selection matches the selection made by the lottery computer?
c) Any selection which includes at least three of the lottery computer's selection wins a prize. Calculate the probability that a contestant's selection:
i) contains no number also selected by the computer,
ii) contains exactly one number also selected by the computer,
iii) contains exactly two numbers also selected by the computer,
iv) wins a prize.

27 The staff employed by a college are classified as academic, administrative or support. The following table shows the numbers employed in these categories and their sex.

	Male	Female
Academic	42	28
Administrative	7	13
Support	26	9

A member of staff is selected at random.
A is the event that the person selected is female.
B is the event that the person selected is academic staff.
C is the event that the person selected is administrative staff.
($\bar{A}$ is the event not A, $\bar{B}$ is the event not B, $\bar{C}$ is the event not C).

a) Write down the values of:
i) $P(A)$ **ii)** $P(A \cap B)$ **iii)** $P(A \cup \bar{C})$ **iv)** $P(\bar{A} | C)$

b) Write down one of the events which is:
i) not independent of A,
ii) independent of A,
iii) mutually exclusive of A.
In each case, justify your answer.

c) You are given that 90 per cent of academic staff own cars, as do 80 per cent of administrative staff and 30 per cent of support staff.
i) What is the probability that a staff member selected at random owns a car?
ii) A staff member is selected at random and is found to own a car. What is the probability that this person is a member of the support staff?

(AEB Question 2, Paper 9, November 1991)

28 In a large group of people it is known that ten per cent have a hot breakfast, 20 per cent have a hot lunch and 25 per cent have a hot breakfast or a hot lunch. Find the probability that a person chosen at random from this group:
a) has a hot breakfast and a hot lunch,
b) has a hot lunch, given that the person chosen had a hot breakfast.

(ULEAC Question 2, Paper S1, June 1992)

29 In a group of six students, four are female and two are male. Determine how many committees of three members can be formed containing one male and two females.

(ULEAC Question 1, Paper S1, January 1993)

30 A and B are two independent events such that $P(A) = 0.2$ and $P(B) = 0.15$. Evaluate the following probabilities.

a) $P(A|B)$ **b)** $P(A \cap B)$ **c)** $P(A \cup B)$

(ULEAC Question 2, Paper S1, June 1993)

31 An urn contains three red, four white and five blue discs. Three discs are selected at random from the urn. Find the probability that:

a) all three discs are the same colour, if the selection is with replacement,

b) all three discs are of different colours, if the selection is without replacement.

(ULEAC Question 6, Paper S1, June 1993)

32 Show that for any two events E and F:

$$P(E \cup F) = P(E) + P(F) - P(E \cap F).$$

Express in words the meaning of $P(E|F)$.

(ULEAC Question 1, Paper S1, 1994)

33 Doctors estimate that three people in every thousand of the population are infected by a particular virus. A test has been devised which is not perfect, but gives a positive result for 95 per cent of those who have the virus. It also gives a positive result for two per cent of those who do not have the virus.

Suppose that someone selected at random takes the test and that it gives a positive result. Calculate the probability that this person really has the virus.

(Nuffield Question 5, Specimen paper 2, 1996)

34 My youngest child enjoys skiing on a dry-ski slope. Inevitably, this involves him in many falls. He is at the stage of development in skiing where he can execute two types of turn: a snowplough turn (S), a parallel turn (L). He is neither wholly nor equally successful in these. The chances that he falls as a result of attempting these turns are recorded in the table, as are the probabilities that he uses the turns.

	Probability	
Type of turn	**Falling**	**Using**
Snowploughs (S)	0.1	0.4
Parallel (L)	0.2	0.6

Let S, L be the events that he uses a snowplough, parallel turn respectively, and let F be the event that he falls. Calculate the following probabilities.

a) $P(S \cap F)$ **b)** $P(L \cap F)$ **c)** $P(F)$

(Oxford Question 9, Specimen, Module 3, 1996)

35 An insurance company offering comprehensive policies to car drivers classifies each applicant as being high-risk, average-risk or low-risk. The probability that a high-risk driver will submit a claim in any year is 0.4, the corresponding probabilities for an average-risk driver and a low-risk driver are 0.2 and 0.1 respectively. The proportions of the policy holders who have been classified as high-risk, average-risk and low-risk are 0.3, 0.6 and 0.1, respectively.

a) Calculate the probability that a randomly-chosen policy holder will submit a claim in a year.
b) Given that a randomly-chosen policy holder did make a claim in one year, calculate the conditional probability that the policy holder was a high-risk driver.
c) Two brothers have comprehensive policies with the company, one is classified as average-risk and the other as low-risk. Calculate the probability that exactly one of the two brothers will submit a claim in a year.

(WJEC Question 7, Specimen paper A3, 1994)

36 In Britain, one per cent of pregnancies results in twins. The probability that twins are identical is $\frac{1}{3}$, and in that case the twins will be of the same sex. Twins which are non-identical are, independently of one another, equally likely to be of either sex.
a) Calculate the probability of each of the following events.
(A) A pregnancy results in identical twins.
(B) A pair of non-identical twins are of the same sex.
(C) A pair of twins are of the same sex.
b) At a maternity hospital, two sets of twins are due on the same day. What is the probability that all four babies are of the same sex?
c) Find the probability that a pair of twins are identical given that they are of the same sex.

(MEI Question 2, Paper S1, January 1993)

37 To withdraw money from a cash machine with my plastic card, I need to type in my Personal Identification Number (PIN). The PIN is four digits long, and can take any value from 0000 to 9999 inclusive.
a) How many different PINs are there?
Unfortunately, the only thing I can remember about my PIN is that all four digits are different.
b) How many possibilities are there for my PIN?
On further thought, I can recall that the digits in my PIN are 1, 4, 5, 9, but I cannot remember the order.
c) How many possibilities are there now for my PIN?
The cash machine allows me up to three attempts to get my PIN right, but if all three attempts are incorrect then the card is confiscated. I choose randomly from the possible PINs (as in part **c)** above), but I take care not to try a wrong PIN twice. Find the probability of each of the following events.
d) I get the PIN right at the first attempt.
e) I get the PIN right at the third attempt.
f) The card is confiscated.

(MEI Question 3, Paper S1, January 1993)

38 A test of proficiency can be taken up to three times by each of a large group of people. No one who passes takes the test again. Sixty per cent of those attempting the test for the first time pass, as do 75 per cent of those attempting it for the second time, and 30 per cent of those attempting it for the third time.

a) Find the probability that a randomly-chosen person fails at all three attempts.
b) Find the probability that a randomly-chosen person fails at the first attempt but passes at either the second or third attempt.
c) Write down the probability that a randomly-chosen person passes the test.
d) Given that the person passed the test, find the probability that exactly two attempts were required.

(MEI Question 1, Paper S1, May 1994)

39 When one card is selected at random from a pack of playing cards,
A is the event 'the card is an ace',
B is the event 'the card is black',
C is the event 'the card is a club',
D is the event 'the card is a diamond'.
For each of the following pairs of events state whether the events are mutually exclusive, independent, neither or both.
a) *A* and *B*, **b)** *B* and *C*, **c)** *C* and *D*.

(NEAB Question 1, Specimen paper 8, 1994)

40 Vehicles approaching a crossroad must go in one of three directions – left, right or straight on. Observations by traffic engineers showed that of vehicles approaching from the north, 45 per cent turn left, 20 per cent turn right and 35 per cent go straight on. Assume that the driver of each vehicle chooses direction independently.
a) What is the probability that of the next three vehicles approaching from the north:
   i) all go straight on,
   ii) all go in the same direction,
   iii) two turn left and one turns right,
   iv) all go in different directions,
   v) exactly two turn left?
b) Given that three consecutive vehicles all go in the same direction, what is the probability that they all turn left?

(AEB Question 6, Specimen paper 2, 1994)

41 A child has a bag containing twelve sweets of which three are yellow, five are green and four are red. When the child wants to eat one of the sweets, a random selection is made from the bag and the chosen sweet is then eaten before the next random selection is made.
a) Find the probability that the child does not select a yellow sweet in the first two selections.
b) Find the probability that there is at least one yellow sweet in the first two selections.
c) Find the probability that the fourth sweet selected is yellow given that the first two sweets selected were red ones.

(ULEAC Question 6, Specimen paper T1, 1994)

Summary

- **Probability:**
  probability of an outcome
  $$= \frac{\text{number of events in the outcome}}{\text{total number of possible events}}$$

- **Notation:**
  probability of event $E \equiv P(E)$
  probability of complementary event $\equiv P(\bar{E}) = 1 - P(E)$
  conditional probability of A given that B has occurred $\equiv P(A \mid B)$
  probability of events A and $B \equiv P(A \cap B)$
  probability of events A or $B \equiv P(A \cup B)$

- The **multiplication law**:
  $P(A \cap B) = P(A) \times P(B \mid A) = P(B) \times P(A \mid B)$

- The **addition law**:
  $P(A \cup B) = P(A) + P(B) - P(A \cap B)$

- **Independent events**:
  $P(A \mid B) = P(A) \Rightarrow P(A \cap B) = P(A) \times P(B)$

- **Mutually exclusive events**:
  $P(A \cap B) = 0 \Rightarrow P(A \cup B) = P(A) + P(B)$

- A **permutation** is an ordered arrangement of r objects from n objects.
  $${}_nP_r = \frac{n!}{(n-r)!}$$

- A **combination** is the process of selecting r objects from n objects when order does not matter.
  $${}_nC_r = \frac{n!}{r!(n-r)!}$$

STATISTICS 5

Expectation

- *When trying to predict the possible financial outcome of an event, it is useful to consider the expected gain.*
- *In this chapter we shall explore the relationship between expected gain, fair games, mean and variance.*

FAIR GAMES

Exploration 5.1

Some you win ...

A sign at a Summer Scout Fete said: 'Win £1 for just a 10p stake!'

Anyone who wanted to have a try paid 10p and was handed four discs, each marked 'Head' on one side and 'Tail' on the other. They rolled the four discs separately and received £1 (and the 10p stake) if all four discs landed showing heads. Otherwise the 10p stake was forfeited. Obviously, this was a game of chance, or luck, but what is the chance of winning? How much might the Scouts expect to make out of the game? Is this a fair game? We shall explore all these aspects in this chapter.

To explore what might happen when this game is played, we could try:

- calculating theoretical probabilities,
- rolling four coins and recording the outcome repeatedly,
- simulating the game using a calculator, spreadsheet, or other means.

The theoretical approach will be discussed later in the chapter. You might like to explore coin rolling yourself. In this section we shall consider how we can simulate the game using random numbers.

Simulation

The outcome of rolling a disc can be one of two possibilities (head or tail). If we assume that these are equally likely, we can use a simple correspondence between a random digit and the outcome of the roll: an even digit means a head and an odd digit means a tail.

Then any group of four random digits represents one 'go' of the game, worth 10p each time.

Try a simulation of the game using random digits. Does your exploration provide any answers to the questions above?

1	3	2	9
T	T	H	T
2	7	3	2
H	T	T	H
4	1	5	8
H	T	T	H

Example 5.1

The output from a calculator producing three-digit random decimals is given below. Convert these decimals into groups of four digits and code them as heads and tails.

0.745	0.560	0.772	0.885	0.559	0.900
0.597	0.122	0.536	0.006	0.174	0.692
0.291	0.886	0.866	0.902	0.136	0.084
0.924	0.133	0.608	0.006	0.306	0.696
0.219	0.079	0.175	0.200	0.080	

Solution

We start by rewriting the decimal fractions from the table above, omitting the first zero and the decimal point in each one. Then we replace odd digits with 'T' and even digits (or zero) with 'H'.

7	4	5	5	6	0	7	7	2	8	8	5	5	5	9	9	0	0	5	9	7	1	2	2
T	H	T	T	H	H	T	T	H	H	H	T	T	T	T	T	H	H	T	T	T	T	H	H
5	3	6	0	0	6	1	7	4	6	9	2	2	9	1	8	8	6	8	6	6	9	0	2
T	T	H	H	H	H	T	T	H	H	T	H	H	T	T	H	H	II	II	H	H	T	H	H
1	3	6	0	8	4	9	2	4	1	3	3	6	0	8	0	0	6	3	0	6	6	9	6
T	T	H	H	H	H	T	H	H	T	T	T	H	H	H	H	H	H	T	H	H	H	T	H
2	1	9	0	7	9	1	7	5	2	0	0												
H	T	T	H	T	T	T	T	T	H	H	H												

This now represents 21 goes at the game. In these, 19 result in the Scouts gaining the contestant's 10p stake money, and two (boxed in the table above) result in the contestant winning £1 and regaining the stake.

Expected gain

The simulation shows that:

- the chance of winning £1 is only $\frac{2}{21}$,
- the Scouts make a loss of 10p having paid out 2 × £1 and taken 19 × 10p,
- the game appears to be reasonably fair, since neither the contestants nor the Scouts seem, on average, to have much of an advantage.

The actual net income to the Scouts is:

$$10\text{p} \times 19 + (-100\text{p}) \times 2$$

This was for 21 attempts at winning. It is very unlikely that the Scouts would know in advance exactly how many people were going to play their game. It might be more reasonable to consider the net income per game:

$$\frac{10\text{p} \times 19 + (-100\text{p}) \times 2}{21} = -\frac{2}{21}\text{p per game}$$

This can be written as:

$$10\text{p} \times \frac{19}{21} + (-100\text{p}) \times \frac{2}{21}$$

This gives the net income per game as:

the sum of (net gain × experimental probability of winning)

This simulation is just an example of what might happen when someone plays the game. Other simulations will produce different results. The Scouts might be more interested in the likely outcome in the long term. They want to know if, on average, they can expect to make a profit out of this game.

Looking at the theoretical probabilities helps to gain a longterm view. Assuming that the probability of a head is $\frac{1}{2}$, the chance that four discs independently show heads is:

$$\left(\frac{1}{2}\right)^4 \quad \textit{or} \quad \frac{1}{16}$$

On average, once in every 16 games the Scouts can expect to pay out £1

i.e. they expect to pay $\frac{1}{16} \times 100\text{p}$ *per go.*

On 15 out of 16 occasions, they can expect to gain 10p

i.e. they expect to gain $\frac{15}{16} \times 10\text{p}$ *per go.*

The **expected net gain** to the Scouts is:

$$\frac{15}{16} \times 10\text{p} - \frac{1}{16} \times 100\text{p} = 3\frac{1}{8}\text{ p } \textit{per go.}$$

There are two very important ideas in the work we have done so far.

Expected gain is the expected net gain per go to the Scouts, and can be thought of as:

probability of outcome — 4 heads

$$\text{Expected gain} = \frac{15}{16} \times 10\text{p} \quad + \quad \frac{1}{16} \times -100\text{p}$$

not 4 heads — net gain

$$= \sum \text{gain} \times \text{probability}$$

A **fair game** is one where the expected gain is the same for both contestant and organiser.

This means that the **expected gain is zero**.

The theoretical probabilities show that, in the long term, the organisers of the game should gain $3\frac{1}{8}$p for every 10p staked. Perhaps it is not a fair game in providing both organiser and contestant the same expected gain but it was all for charity!

Example 5.2

In a game, a coin is flipped until the first head appears. To take part in the game, each player pays the banker 20p. The banker pays the player an amount according to the number of times the coin has to be flipped to get a head.

No. of flips	Amount banker pays
1	20p
2	10p
3	5p
4	2p
5	1p
6 or more	zero

No player is allowed more than five flips.

Calculate the expected gain to the player. Is the game fair?

Solution

*At the end of a game, the gain to the player is the amount the banker has paid the player, less the 20p the player has paid to take part in the game. So the expected gain to the player is the expected amount the banker pays **minus** 20p.*

*We need to find the probabilities associated with each number of flips, to find this expected amount paid. To calculate these probabilities, we also need to make assumptions about **independence**, i.e. the result of flipping one coin does not depend on the result of flipping another. For instance, the probability that the*

first head appears on the third flip requires the flips to be: tail (T) then tail (T) then head (H).
And so we require:

$$P(TTH) = P(T) \times P(T) \times P(H)$$
$$= \frac{1}{2} \times \frac{1}{2} \times \frac{1}{2}$$

assuming that individual flips of the coins are independent.

Amount banker pays	Probability
20	$\frac{1}{2}$
10	$\frac{1}{4}$
5	$\frac{1}{8}$
2	$\frac{1}{16}$
1	$\frac{1}{32}$
0	$\frac{1}{32}$ (= 1 – sum of other probabilities)

Expected bank pay out $= 20 \times \frac{1}{2} + 10 \times \frac{1}{4} + \ldots + 0 \times \frac{1}{32}$
$\approx 13.3\text{p}$

So the expected gain to the player is 20p less than this, which gives –6.7p. Since this is not zero, the game is not fair.

EXERCISES

5.1A

1 At the School Fete, Year 7 decide to offer a prize of £10 to anyone who scores five heads when rolling five coins. Anyone who tries but fails has to pay 10p.

Using random digits, simulate 25 goes at this game. Answer the following questions:
i) based on your simulation, and
ii) based on theoretical probabilities.
a) What is the probability of winning the prize?
b) What can Year 7 expect to gain in offering this game?
c) Is the game fair?

2 The Parent-Teacher Association runs a stall offering a £10 prize to anyone who, for a 10p stake (not returnable), can roll three ordinary dice and get three 6s.
a) Is this a fair game?
b) What is the expected gain to the PTA?

3 A simple fruit machine has three windows.

The windows show apples, bananas or cherries. The probability of this varies according to the window and the fruit as shown in this table.

Prizes are given when the three windows all display the same fruit.

	Apple	Banana	Cherry
First window	0.3	0.2	0.5
Second window	0.3	0.1	0.6
Third window	0.2	0.1	0.7

Three bananas win £10, three apples win £2 and three cherries win 50p.

Work out a way of simulating 30 attempts at winning a prize using this fruit machine.

What does your simulation suggest the probability of winning the top prize is?

If it cost 20p a go, calculate the theoretical probability of:

a) three bananas,

b) winning a prize,

and evaluate the expected net gain to the owner of the fruit machine.

4 A businessman has £250 000 to invest. He is considering investing in a business enterprise where he estimates the chance of differing financial returns, in two years' time are as shown in this table.

Return	£400 000	£300 000	£150 000	£50 000
Probability	0.25	0.5	0.2	0.05

a) What is the expected return of this investment?

b) Is he likely to have a larger return by placing his money on deposit where there is a guaranteed annual interest rate of five per cent?

5 An insurance company estimates the probability of a disaster such as a fire or flood completely demolishing a house in any year is $\frac{1}{750}$. The company insures 50 000 houses against such a disaster and the average insured value is £80 000.

a) The company charges an annual premium of £100 to each householder. Is this likely to be sufficient to cover the expected payout?

b) How much should the premium be if the company wants to make a total annual profit of 1 million pounds from the premiums?

6 The faces of three tetrahedral dice give scores of 1, 2, 3 and 4. The dice are rolled, and the player's counter is moved a number of squares on the game board, according to the outcome of the roll of dice.

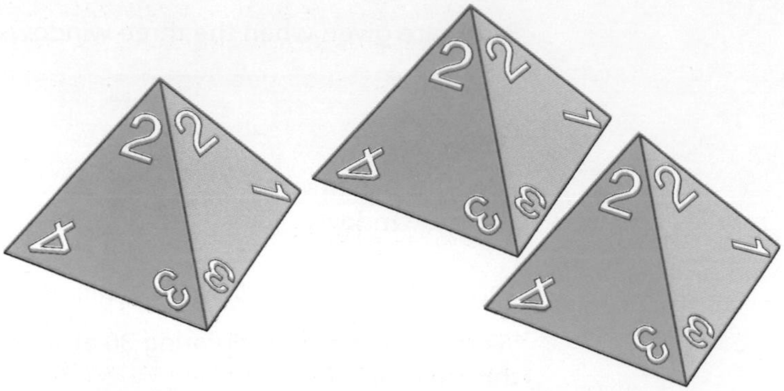

Outcome	Action
4,4,4	3 squares clockwise
one or two 4s	1 square clockwise
no 4	1 square anticlockwise

a) Calculate the probability of each of the outcomes.
b) Calculate the expected move.

EXERCISES

5.1B

1 A stallholder in a fairground offers £5 for a 25p stake to anyone who can roll three ordinary dice and get a multiple of three on each die. Use random numbers to simulate 30 attempts at winning this prize. Does your simulation indicate that this is a fair game? Calculate the theoretical expected gain to the stallholder. (Winners received their 25p stake money back.)

2 The following notice was seen at a church fete.

> Pick a card from each pack and win **£10**!

The rules were:
- there were three ordinary packs of cards,
- a player won the £10 only by picking an ace from each pack,
- each attempt cost 20p,
- the only consolation prize was £1 for picking three court cards,
- a person winning either prize got the stake money back.

a) What is the likely outcome for a thousand attempts?
b) Is this game fair?

3 A child's roulette wheel consists of a disc with twelve equal sectors marked 1 to 12, which are coloured green, red or black. The instruction booklet suggests the following method of play.

One child acts as banker, whilst others play as gamblers. The gamblers place a stake of their choice on a colour and then they receive a prize as follows, if their colour appears on the wheel.

Number	Colour	Prize
1, 6, 7, 12	green	zero
2, 3, 4, 9, 10, 11	red	double the stake
5, 8	black	four times the stake

In all cases the stake is forfeited.

a) Devise a method of simulating the game and work out the outcome for 36 spins of the wheel.

b) Is the game fair?

c) Calculate the theoretical expected gain for a gambler who always places a stake of 50p on the black.

4 It is estimated that four out of every five businesses fail within their first year. Three-quarters of those that succeed manage only to break even (i.e. regain the original investment) in the first year. On average, the rest produce a return amounting to five times the original investment. What is the likely return to an investor who risks £100 000 in a business?

5 A motor car insurance company estimates that the risk of its having to pay a claim for any motorist depends on the details as shown in the table. The overall average claim costs the company £1000.

Category	Young	Single and inexperienced	Married and experienced	Advanced
Risk	0.25	0.10	0.05	0.01

a) Assume the company has the same number of policy holders in each category.

  i) How much is it likely to have to charge as a premium in order to break even?

  ii) How much should the premium be to make a profit of £250 000, assuming it has 50 000 policy holders?

b) **i)** It is argued that it is fairer to motorists to make the premium reflect the risk. The company decides to charge a premium to each category in the ratio 5 : 2 : 2 : 1.
If the company is to break even, how much should it charge motorists in each category?

  ii) What should the premiums be to make a profit of £250 000, if the numbers of motorists in each category are 2000 : 13 000 : 30 000 : 5000?

6 Five ordinary dice are rolled in a game. If they all give a score of 6, a prize of £50 is given for a 50p stake. Any other outcome resulting in at least one 6 yields a prize of £1.

a) Calculate the probability of:
  i) winning the £50 prize,
  ii) winning the £1 prize.
b) What is the expected gain to the player of the game?
c) What stake would make this a fair game?

MEAN AND VARIANCE

Mean

In the last section we thought of expected gain as longterm average gain. Why is the expected gain the same as the average – or **mean** – gain? Once again, the Scout's game can be used to illustrate the reason for this.

Imagine that the game is played 16 times and that the contestant wins exactly once. The results of these 16 plays are as shown in this table.

Gain to Scouts	Number of times this gain occurs
10p	15
–100p	1

The mean gain is $\dfrac{10\text{p}\times 15+\left(-100\text{p}\right)\times 1}{16}$

which is the same as the expression for the expected gain per game.

So mean gain = expected gain
= sum of products of gain × probability of outcome resulting in the gain

$$\text{E(gain)} = \sum_{\text{all outcomes}} \text{gain} \times \text{probability of related outcome.}$$

More formally, if X is a random variable, the link between the mean and the expected value is defined as:

$$\text{mean value} = \text{E}(X)$$
$$= \sum_{\text{all } x} x \times \text{P}(X = x)$$

Example 5.3

The random variable, X, represents the sum of the outcomes on two tetrahedral dice with possible scores 1 to 4. X can take the values 2, 3, 4, 5, 6, 7 or 8 with probability $\frac{1}{16}, \frac{2}{16}, \frac{3}{16}, \frac{4}{16}, \frac{3}{16}, \frac{2}{16}, \frac{1}{16}$.

What is the expected value of X?

Solution

The expected value of X is:

$$\begin{aligned}E(X) &= \sum_{x=2}^{8} x \times P(X = x) \\ &= 2\times\frac{1}{16} + 3\times\frac{2}{16} + 4\times\frac{3}{16} + 5\times\frac{4}{16} + 6\times\frac{3}{16} + 7\times\frac{2}{16} + 8\times\frac{1}{16} \\ &= \frac{80}{16}\end{aligned}$$

Hence, the mean of X is 5.

Variance

The mean of squares was introduced with the work on variance. For a random variable this can be defined as the expected value of the square of the random variable:

$$E(X^2) = \sum x^2 \times P(X = x)$$

Then, the variance of X can be written as:

$$\text{variance of } X = E(X^2) - \{E(X)\}^2$$

using the 'mean of squares minus square of mean' definition.

Example 5.4

The mean of the squares for the tetrahedral dice in Example 5.3 can be found as:

$$\begin{aligned}E(X^2) &= \sum_{x=2}^{8} x^2 \times P(X = x) \\ &- 2^2 \times \frac{1}{16} + 3^2 \times \frac{2}{16} + \cdots + 8^2 \times \frac{1}{16} \\ &= \frac{440}{16}\end{aligned}$$

Find the variance.

Solution

Using the equation above, the variance is:

$$\begin{aligned}\text{Var}(X) &= E(X^2) - \{E(X)\}^2 \\ &= 27.5 - 5^2 = 2.5\end{aligned}$$

Example 5.5

A bag contains four orange discs and four green discs which are identical in all other aspects. Two players play a game in which they take turns to select a disc randomly from the bag. The disc is not returned after selection. The game ends when the first green disc is drawn. Calculate the mean and variance of the number of selections.

Solution

First, we need to determine the possible number of selections and the associated probability for each number. The outcomes are not independent, since the discs are not returned to the bag.
The possible number of selections is 1, 2, ..., 5. The associated probabilities may be established from a tree diagram.

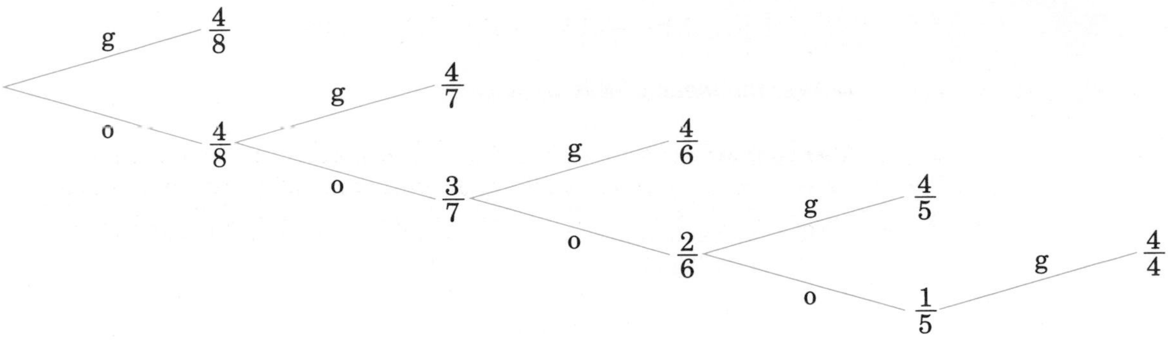

Multiplying along the branches produces the following probabilities.

No. of selections	Probability
1	$\frac{4}{8}=\frac{1}{2}$
2	$\frac{4}{8}\times\frac{4}{7}=\frac{2}{7}$
3	$\frac{4}{8}\times\frac{3}{7}\times\frac{4}{6}=\frac{1}{7}$
4	$\frac{4}{8}\times\frac{3}{7}\times\frac{2}{6}\times\frac{4}{5}=\frac{2}{35}$
5	$\frac{4}{8}\times\frac{3}{7}\times\frac{2}{6}\times\frac{1}{5}\times\frac{4}{4}=\frac{1}{70}$

Hence the expected number of selections is:

$$1\times\tfrac{1}{2}+2\times\tfrac{2}{7}+3\times\tfrac{1}{7}+4\times\tfrac{2}{35}+5\times\tfrac{1}{70}$$

which is equal to 1.8.

The variance may be established using

$$\mathrm{E}(X^2)-\text{mean}^2$$

giving

$$\left(1^2\times\tfrac{1}{2}+2^2\times\tfrac{2}{7}+3^2\times\tfrac{1}{7}+4^2\times\tfrac{2}{35}+5^2\times\tfrac{1}{70}\right)-1.8^2$$

which gives 0.96.

EXERCISES

5.2A

1 Use expectation to calculate the mean and variance of the outcome on a cubical die with faces numbered 1 to 6.

2 A tetrahedral die with faces numbered 1 to 4 is rolled. Calculate the expected value of the number on the face on which the die lands and the expected value of the square of the number on the face on which it lands. Hence find the variance of the outcome.

3 The distribution of the weekly number of breakdowns, X, of a machine which produces sticks of lead for pencils has the following probability function.

x	0	1	2	3	4
$P(X = x)$	0.10	0.15	0.45	0.25	0.05

Calculate the mean and variance of X.

4 Three biased coins for which $P(H) = \frac{1}{3}$ are spun and the number of heads which occur is X. Write down the probability distribution of X and hence find the mean number of heads and the variance of the number of heads.

5 Two dice are rolled, and the score is taken as the value of the larger number. If the value is the same on both dice, the score is taken as the value on either die. List the possible pairs of results from the throws of the dice, then give all the possible scores, and their probabilities. Calculate the mean and variance of the scores.

6 A motorist passes through two sets of traffic lights. The probability that the first set is at green is $\frac{1}{2}$; the probability that the second set is at green is $\frac{1}{3}$.

a) Draw a tree diagram and find the probabilities that the motorist has to wait: **i)** at neither set, **ii)** at just one set, **iii)** at both sets.
b) Find the expected number of waits and the variance of the number of waits.

7 Two cubical dice – one red and the other blue – are rolled. The outcome on the blue is subtracted from that on the red. What are the possible differences? What are the associated probabilities? Use expectation to calculate the mean difference and the variance of the difference.

EXERCISES

5.2B

1 Use expectation to calculate the mean and variance of the outcome on an octahedral die with faces numbered 1 to 8.

2 A game is played in which an ordinary pack of 52 cards is shuffled and then the top card is turned over. The player scores the value of this card where:
- an ace counts as 1,
- picture cards count as 10,
- all other cards are face value.

a) Calculate the expected value of the score.
b) Calculate the variance of the score.

3 The number of emergencies requiring the assistance of one particular fire-station in a day have probabilities as shown in this table.

x	0	1	2	3	4	5
$P(X = x)$	0.05	0.25	0.35	0.15	0.15	0.05

Calculate the mean and variance of X.

4 A game is played using two tetrahedral dice, one red and one blue. The score is worked out as the outcome on the red minus the outcome on the blue. List all the pairs of possible outcomes and hence find the possible scores and their probabilities. Calculate the mean and the variance of the score.

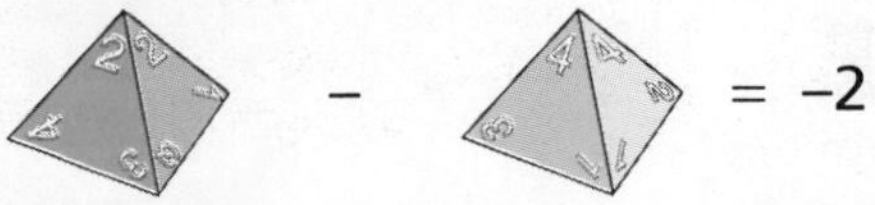

5 **a)** A multiple choice test consists of five questions, each with two options. A student guesses all the answers. Find the mean and the variance of the number of correct answers.
b) Repeat part **a)** for a test with three options for each answer.

6 A disc with faces labelled H and T and a cubical die with faces numbered 1 to 6 are thrown together. If the disc shows H, the score recorded is that shown on the die. If the disc shows T, the score recorded is twice that shown on the die.

Find the mean and variance of the score recorded.

7 Let X represent the outcome on a tetrahedral die with faces numbered 1 to 4. Let Y represent the outcome on a cubical die with faces numbered 1 to 6. What are the mean and variance of X and of Y?

A random variable, Z, is found by doubling the outcome on the tetrahedral die and adding it to the outcome on the cubical die, i.e.

$$Z = 2X + Y$$

Write down the possible values of Z and the associated probabilities. Find the expected value of Z and its variance.

CALCULATING VARIANCE: THE GENERAL CASE

The way in which variance has been calculated above included an example of finding the expected value of a function of a random variable. The function involved was the square of the random variable:

$$\text{Var}(X) = \text{E}(X^2) - \text{mean}^2$$

expected value of the square of X

Calculating $\text{E}(X^2)$ involved summing products of x^2 and $\text{P}(X = x)$.

Note: It is $\text{P}(X = x)$ i.e. the probability of the random variable having the value x rather than x^2.

This is an example of a more general extension to other functions of random variables where the expected value of $\text{f}(X)$ is defined as:

$$\text{E}(\text{f}(X)) = \sum_{all\ x} \text{f}(x)\text{P}(X = x)$$

We can apply this general principle, in the following example and exercises.

Example 5.6

Louise works for an estate agency. She receives a fixed monthly income, and commission based on the number of houses she sells per month. Her total income, in pounds, is $200X + 400$ where X is the number of houses she sells in the month. The probability of her selling various numbers of houses is shown in this table.

x	1	2	3	4	5	6	7	≥ 8
$\text{P}(X = x)$	0.05	0.1	0.25	0.3	0.15	0.1	0.05	0

What is her mean monthly income?

Solution

Louise's income, T, is a function of the random variable X. We need to find the expected value of this function:

$$\begin{aligned}E(T) &= E(200X + 400)\\ &= \sum_{x=1}^{7}(200x + 400)P(X = x)\\ &= 600 \times 0.05 + 800 \times 0.1 + 1000 \times 0.25 + 1200 \times 0.3 + 1400 \times 0.15\\ &\quad + 1600 \times 0.1 + 1800 \times 0.05\\ &= 1180\end{aligned}$$

So, her mean monthly income is £1180.

EXERCISES

5.3A

1 The random variable, X, has probability function $P(X = x)$ for $x = 1, 2, 3, 4$ as shown in this table.

x	1	2	3	4
$P(X = x)$	0.1	0.2	0.3	0.4

Calculate each of the following.

a) $E(X)$ **b)** $E(2X)$ **c)** $E(2X + 3)$ **d)** $E(5X - 3)$

e) $E(X^2)$ **f)** $E(2X^2)$ **g)** $2E(X) + 3$ **h)** $5E(X) - 3$

Comment on your answers to **c), d), g)** and **h)**.

2 A double-glazing salesman receives a commission for each sale he makes, together with a fixed daily income. The number of sales, X, he makes in a day lies between 0 and 5 with probability as shown in the table.

x	0	1	2	3	4	5
$P(X = x)$	0.4	0.2	0.3	0.05	0.03	0.02

His daily income, Y, is given by:

$$Y = 30X + 20$$

Calculate his expected weekly income assuming he works on five days per week. What is his mean weekly income if he works six days per week?

3 A number, X, can take values 0, 1, 2, 3, 4, 5 and 6 with probabilities as in the table.

x	0	1	2	3	4	5	6
$P(X = x)$	0.005	0.11	0.22	0.33	0.22	0.11	0.005

Calculate:
a) the mean of Y where $Y = 3X - 2$,
b) the variance of Y,
c) the variance of X.

What is the relationship between the variances of X and Y?

4 The number of days that it will take a contractor to complete a particular construction may be represented by a random variable X with the following probability distribution.

x	8	9	10	11	12
$P(X = x)$	0.2	0.4	0.2	0.1	0.1

Calculate the expected value of X.

If the contractor's profit is $Y = £1000(11 - X)$, calculate his expected profit from the construction. Find the probability that the contractor makes a profit of more than £1000 on this construction.

5 A man uses the cashcard machine X times a month, with probabilities as given in the table. He draws out £Y on each occasion where:

$$Y = \frac{300}{X}$$

Find the expected amount of money he draws out on each visit to the cashcard machine, and its variance.

x	3	4	5	6
$P(X = x)$	0.4	0.3	0.2	0.1

6 The number of students, N, in a German conversation class varies from 1 to 4, with probability as shown in this table.

x	1	2	3	4
$P(N = x)$	0.2	0.25	0.25	0.3

The price, £P per hour, that a student has to pay is given by the following relation.

$$P = 16 - \tfrac{1}{2}N^2$$

a) Find the expected cost per hour for a student willing to join a class of any size, and the variance of the cost.
b) Find the expected income per hour of the teacher, and its variance.

EXERCISES

5.3B

1 The probability function, $P(X=x)$, for the random variable X is set out in this table.

x	1	2	3
$P(X=x)$	0.2	0.5	0.3

Calculate each of the following.

a) $E(X)$ **b)** $E(1-2X)$ **c)** $E(2X)$
d) $1-2E(X)$ **e)** $E(X^2)$ **f)** $E(X^2+X)$

Comment on the relation you see:
i) between the results in **a)**, **b)**, **c)** and **d)**,
ii) between the results in **a)**, **e)** and **f)**.

2 The number, X, of insurance policies sold weekly by a saleswoman and the associated probabilities are given in this table.

x	0	1	2	3	4	5	6	≥ 7
$P(X=x)$	0.3	0.2	0.15	0.1	0.1	0.1	0.05	0

The saleswoman receives a fixed income of £50 per week plus £100 for every insurance policy she sells. Calculate her mean weekly income and its variance.

3 A number, X, can take values 1, 2, 3 and 4 with probabilities given in this table.

x	1	2	3	4
$P(X=x)$	0.2	0.3	0.4	0.1

Calculate:
a) the mean of Y where $Y=2X-1$,
b) the variance of Y,
c) the variance of X.

What is the relationship between the variances of X and Y?

4 The cost of van hire is £24 per day plus a fixed charge of £10. A man estimates the number of days he will need to hire a van and the associated probabilities as shown in this table.

x	2	3	4
$P(X=x)$	0.4	0.35	0.25

a) Express the cost of van hire, C, as a function of the number of days, X.
b) Find his expected cost and the variance.

5 The relationship between two variables P and V is known to be

$$P = \frac{6000}{V}$$

and V can take the values 1, 2, 3, 4, 5 and 6 with probabilities as in the table.

x	1	2	3	4	5	6
$\mathrm{P}(V = x)$	0.001	0.024	0.075	0.2	0.3	0.4

a) Calculate the mean and variance of V.
b) Calculate the mean and variance of P.

6 Experience has shown that vehicles emerging from a tunnel have speeds V with the following distributions.

Speed v (mph)	30	40	50	60	70
$\mathrm{P}(V = v)$	0.1	0.2	0.3	0.3	0.1

The braking distance at speed v is $v + \frac{v^2}{20}$ feet.
What is the mean braking distance for a vehicle coming out of the tunnel? What is the probability that a vehicle coming out will exceed the mean braking distance?

CALCULATOR ACTIVITY

Does your calculator accept probabilities?

Some statistical calculators will accept data with associated probabilities (as compared to frequencies). You could find out whether yours will, by trying to input the following data in the same way that you would input frequencies.

x	1	2	3	4
$\mathrm{P}(X = x)$	0.1	0.3	0.4	0.2

You should find that:

$n = 1$

$\sum x = 2.7$

$\sum x^2 = 8.$

if your calculator accepts probabilities.

Σx is really the expected value, $\Sigma x\mathrm{P}(X = x)$ and Σx^2 is the expected value of the square of X, $\Sigma x^2\mathrm{P}(X = x)$.

Use your calculator to find the variance of X. It should come out at 0.81.

How might you find the expected value of a function, using your calculator?

Suppose $Y = 15 - \frac{12}{x}$, then the values of Y and corresponding probabilities are as in this table.

y	3	9	11	12
$P(Y = y)$	0.1	0.3	0.4	0.2

Putting this distribution into the calculator produces:

mean of $Y = 9.8$
variance of $Y = 6.36$.

CONSOLIDATION EXERCISES FOR CHAPTER 5

1 A fairground stallholder invites contestants to shuffle a pack of cards and turn over the top card. The contestant wins a prize (and receives the stake money back) according to the poster. If the stake money is 10p, calculate the expected gain to the stallholder.

Turn the Ace of Spades into £5

2 Two bags each contain five identical discs. Each set of discs has the numbers 1 to 5 painted on them. A gambler randomly selects one disc from each bag and adds the two numbers. The gambler then receives a prize or pays a forfeit according to the sum obtained.

Sum	Prize	Forfeit
2	£1.00	
3	20p	
4		10p
5		10p
6		40p
7		10p
8		10p
9	20p	
10	£1.00	

Determine the expected gain to the gambler.

3 A boy and a girl play a game where they take turns to select (at random) one ball from a bag containing four blue and four red balls. They keep the ball selected on each occasion. The winner is the first one to select a red ball. Calculate the probability that the game ends with the first, second, third, fourth, fifth ball drawn. What is the mean number of balls drawn?

4 In the competition 'Brain of Britain', contestants are asked a series of questions until they fail to provide a correct answer or they provide five correct answers consecutively. They receive a point for every correct answer and a bonus point for five correct answers. What is the expected number of points a contestant is awarded if her chance of correctly answering a question is $\frac{1}{2}$?

5 Two ordinary dice are thrown. If the sum of the outcomes is 12 the player receives £1.00. The player receives 50p if the sum is 10 or 11. For any other sum, the player pays 10p. What is the expected win for the player?

6 By choosing a suitable example, or otherwise, establish that:

$$E(X^2) - \{E(X)\}^2 \neq 0$$

What is this difference known as? Under what circumstance is this difference zero?

7 A random variable, X, is known to take values with the probabilities as shown in this table.

x	1	2	3	4
$P(X = x)$	$7a$	$5a$	$3a$	a

a) Use the fact that the sum of the probabilities is 1 to find a.
b) Calculate the mean and variance of X.

8 A discrete random variable, X, has probability function given by:

$P(x) = Kx^2 \quad x = 1, 2, 3$
$P(x) = 0 \quad$ otherwise.

Find the value of:
a) K, **b)** $E(x)$, **c)** Var(x).

(ULSEB, Specimen S1, 1990)

9 A discrete variable X has probability distribution specified in this table.

x	-1	0	1	2
$P(X = x)$	0.25	0.10	0.45	0.20

a) Find $P(-1 \leq x < 1)$.
b) Find $E(2X + 2)$.

(ULSEB Question 2, S1/S10, June 1993)

10 A box contains eight discs of which five are red and the remainder green. Three discs are withdrawn at random without replacement. Calculate:

a) the probability that all three are green,
b) the probability that none of the three is green,
c) the expected number of green discs withdrawn.

11 A quiz team of three is to be selected from a group of four women and five men. If the members of the team are selected at random, calculate the expected number of women and the variance of the number of women.

12 A child's roulette wheel has 32 equal sectors numbered with the digits 1, 2, 3, 4, 5 and 6 in such a way that the probability that the wheel stops on a particular number is as in this table.

Number	1	2	3	4	5	6
Probability	$\frac{10}{32}$	$\frac{6}{32}$	$\frac{6}{32}$	$\frac{4}{32}$	$\frac{4}{32}$	$\frac{2}{32}$

Calculate the mean and the variance of the number on which the wheel stops.

13 A company uses computers in its business. The company replaces a certain number of computers each year, depending on the reliability of individual machines in the previous year. If the number of computers purchased is represented by the random variable X, the probabilities associated with the possible values of X are shown in this table.

x	0	1	2	3	≥ 4
$P(X = x)$	$\frac{1}{10}$	$\frac{3}{10}$	$\frac{2}{5}$	$\frac{1}{5}$	0

The cost of a new computer is £1200. A reduction of £$50x^2$ is offered by a supplier. What amount does the company expect to spend on new computers at the end of the year?

14 A wealthy businessman has his own private aeroplane. He wants to insure it for £250 000. An insurance company estimates that the likelihood of:

- a total loss is 0.0025,
- a 50 per cent loss is 0.01,
- a ten per cent loss is 0.1.

The company wants to make an expected annual profit of £2000 in providing this insurance.

What premium should it charge?

15 A blacksmith's records reveal that he is asked to produce very small numbers, X, of specific designs according to the following probability distribution.

x	1	2	3	4	5
$P(X = x)$	0.10	0.15	0.35	0.20	0.20

The cost, £Y, of individual items in such cases is calculated according to the function:

$$Y = a + \frac{b}{x}$$

where a and b are constants determined by the actual design. Calculate the expected value of Y if $a = 50$ and $b = 100$.

The blacksmith's charge, £Z, to his customers is given by $Z = 1.1(20 + Y)$. Calculate his expected profit.

16 A simplified game of *Pontoon* is played as follows.

A player has two packs of cards in front of him and he picks one card at random from each pack. The score for each card is the numerical face value, except that jacks, queens and kings are worth 10 and an ace is worth 11. The overall score for a game is the sum of the scores on the two cards. By considering all possible pairs, write down the distribution of the total score on the two cards and find its mean and variance. If the rules said a player had to beat 14 to win a game, would he be happy?

17 A newsagent stocks twelve copies of a specialist magazine every week. He has regular orders for nine copies but the number of additional copies required varies from week to week. He estimates the probability for each possible number of copies sold (including the nine regular orders) as follows.

Number of copies	9	10	11	12
Probability	0.20	0.35	0.30	0.15

a) Calculate an estimate of the mean number of copies he sells in a week. The newsagent buys the magazines at 65p each and sells them at £1.15 each. Any copies not sold are destroyed.

b) Find his total profit in a week in which he sells eleven copies.

c) Write out a probability distribution table for the newsagent's weekly profit from the sale of these magazines. Hence, or otherwise, calculate an estimate of his mean weekly profit.

(JMB Question 1, Paper II, June 1992)

18 A hotel caters for business clients who make short stays. Past records suggest that the probability of a randomly-chosen client staying X nights in succession is as follows.

r	1	2	3	4	5	6+
$P(X = r)$	0.42	0.33	0.18	0.05	0.02	0

a) Draw a sketch of this distribution.
b) Find the mean and the standard deviation of X.
c) Find the probability that a randomly chosen client who arrives on Monday evening will still be in the hotel on Wednesday night.
d) Find the probability that a client who has already stayed two nights will stay at least one more night.

(MEI Question 1, Paper S1, January 1994)

19 The random variables, X and Y, take the values 1, 2, 3, 4, 5 and 6 with probabilities as shown in this table.

Value	1	2	3	4	5	6
$P(X = \text{value})$	0.05	0.15	0.3	0.3	0.15	0.05
$P(Y = \text{value})$	0.05	0.2	0.15	0.45	0.1	0.05

a) For each random variable, calculate:
i) its mean, **ii)** its variance.
b) Sketch a probability line graph for each random variable. Comment on the similarities and differences between the variables.
c) A measure of the symmetry, or lack of symmetry (skewness), for a random variable may be found by calculating

$$E(X^3) - \mu(3\sigma^2 + \mu^2)$$

where μ and σ^2 are the mean and variance of the random variable. Calculate the measure of symmetry for each of the random variables. Comment on your result in the light of your answer to **b)**.

20 **a)** Calculate the mean and variance for each of the random variables, X and Y, with probabilities as shown in this table.

Value	1	2	3	4	5	6	7
$P(X = \text{value})$	0.04	0.11	0.2	0.3	0.2	0.11	0.04
$P(Y = \text{value})$	0.06	0.09	0.1	0.5	0.1	0.09	0.06

b) Draw probability line graphs for each random variable.
c) A measure of the peakedness (known as **kurtosis**) of these random variables can be found by calculating

$$E\left(\frac{X-4}{\sqrt{2}}\right)^4 \text{ and } E\left(\frac{Y-4}{\sqrt{2}}\right)^4.$$

Calculate these measures of kurtosis and comment in the light of the graphs you drew in **b)**.

Summary

- The expectation of a function is the same as the mean.
- The expected value of a function is the mean value of the function.
- In games of chance:

  expected gain = E(gain) = Σgain $\times$ probability
- For a fair game, E(gain) = 0.
- For probability distributions:

  $$\text{expected value} = \text{E}(X) = \sum_{all\ x} x \times \text{P}(X = x) = \text{mean}$$

  variance = $\text{E}(X^2) - (\text{mean})^2$
- Expected value of a function $= \text{E}(\text{g}(X)) = \sum_{all\ x} \text{g}(x) \times \text{P}(X = x)$

STATISTICS

6

Probability distribution

In this chapter:

- *we discover how the concepts of probability and the associated results may be used to model real-world situations,*
- *we meet a probability distribution as a model for a real random variable,*
- *we explore the process of mathematical modelling,*
- *we discover how a calculator may be used to explore properties of random variables.*

MOST LIKELY EVENT – MODE

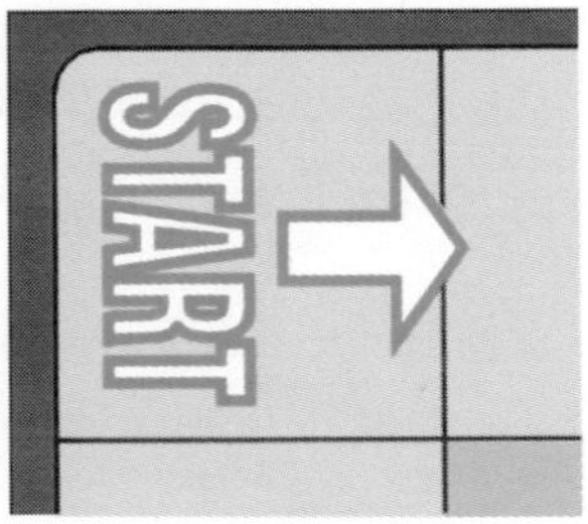

In many board games, players have to throw a six or roll a double to start. If the player fails, the forfeit is to wait for a turn before trying again. Imagine you are taking part in such a game. You have to roll a double (any double) using a pair of ordinary dice to start playing.

What is the probability that you will start playing on your first attempt?

The pair of dice can land in any one of 36 equally likely ways. Of these, exactly six result in a double. Hence:

$$P(\text{double}) = \frac{6}{36} = \frac{1}{6}$$

This is the chance that the dice will land showing a double on the first attempt. Not every player succeeds in getting started on their first attempt. What is the most likely number of attempts that a player will need to roll the dice to start the game? Is it twice, three times, six times, twelve times, ...? Before we provide a definitive answer, try this.

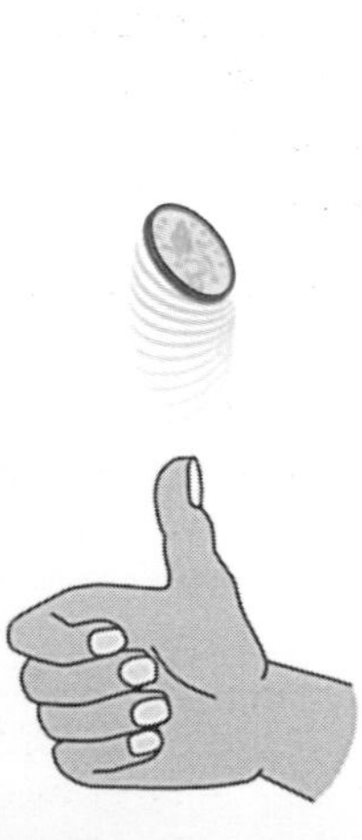

Suppose the rule to start the game is different. Instead of throwing a double with a pair of dice, you need to flip a head with a coin. Have a go at flipping a coin until you get a head. How many times did you need to flip the coin, was it once, twice, three times, ...?

What is the probability of obtaining a head on the first flip?
A coin may be assumed to be symmetric with only two possible outcomes, hence:
$P(H) = \frac{1}{2}$

What is the probability of not getting a head until the second flip? third flip? ... If we do not get a head until the second flip then the outcome must have been tail then head. The probability of this outcome is:

$P(TH) = \frac{1}{2} \times \frac{1}{2} = \frac{1}{4}$

Similarly, the probability of not getting a head until the third flip is:

$P(TTH) = \frac{1}{2} \times \frac{1}{2} \times \frac{1}{2} = \frac{1}{8}$ etc.

What is the most likely number of flips needed to get a head?

The answer is 1, because the probability of getting a head in one go is $\frac{1}{2}$ and the probability of taking any other number of attempts must be less than $\frac{1}{2}$.

Suppose, now, that the rules are again different. This time you must roll a tetrahedral die and get a 3 (the vertices of the die are numbered 1, 2, 3, 4).

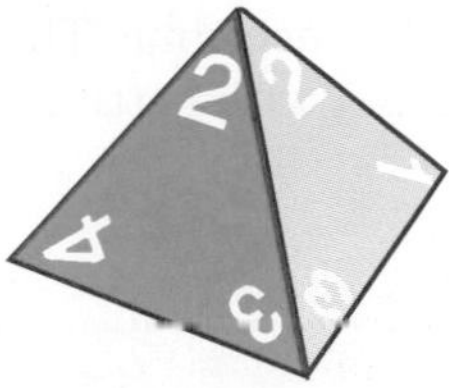

What is the probability of getting the first 3 on:

- the first roll.
- the second roll,
- the third roll?

It is reasonable to assume that the die is symmetrical and then:

$P(3) = \frac{1}{4}$

Hence, the probability of a 3 on the first attempt is $\frac{1}{4}$. If it takes two rolls to get the first 3 the outcome must have been 'not 3' then '3'. Hence:

$$\begin{aligned} P(\text{needing 2 rolls}) &= P(\text{'not 3' and then '3'}) \\ &= P(\text{not } 3) \times P(3) = \tfrac{3}{4} \times \tfrac{1}{4} \end{aligned}$$

which is, of course, less than $\frac{1}{4}$, the probability of a 3 on the first attempt. (Why 'of course'?)

$P(\text{needing three rolls}) = \frac{3}{4} \times \frac{3}{4} \times \frac{1}{4}$

which is less than P(needing two rolls).

What is the most likely number of rolls needed to get a 3? The answer is 1. 'One roll' has the largest probability $\frac{1}{4}$, since all subsequent probabilities are $\frac{3}{4}$ of the preceding probability. So, a player is more likely to start the game on their first attempt than any other attempt.

Is this also true when the rule is 'throw a double to start'?

The probability of throwing a double on the first attempt is $\frac{1}{6}$. If it takes two throws to get started, it means that the first throw was not a double and the second was a double. The probability of this combined outcome can be written:

$$P(\overline{D} \text{ and } D) = P(\overline{D}) \times P(D) = \tfrac{1}{6} \times \tfrac{5}{6}$$

which is $\frac{5}{6}$ of the probability of getting started on the first attempt. So a player is less likely to need two attempts than just one attempt.

What is the probability of taking exactly:

a) three, **b)** four, **c)** n,

attempts to get a double?

a) Taking three attempts means that the first two did not result in a double and that the third did:

$$P(\overline{D}\,\overline{D}\,D) = P(\overline{D}) \times P(\overline{D}) \times P(D) = \tfrac{5}{6} \times \tfrac{5}{6} \times \tfrac{1}{6} = \left(\tfrac{5}{6}\right)^2 \times \tfrac{1}{6}$$

which is $\frac{5}{6}$ of the probability of getting started on the second attempt.

b) $P(\overline{D}\,\overline{D}\,\overline{D}\,D) = P(\overline{D}) \times P(\overline{D}) \times P(\overline{D}) \times P(D) = \tfrac{5}{6} \times \left(\tfrac{5}{6} \times \tfrac{5}{6} \times \tfrac{1}{6}\right) = \left(\tfrac{5}{6}\right)^3 \times \tfrac{1}{6}$

which is $\frac{5}{6}$ of the previous probability.

c) P(exactly n attempts)

$$= \underbrace{P(\overline{D}) \times P(\overline{D}) \times \;\ldots\; \times P(\overline{D})}_{(n-1)} \times P(D) = \left(\tfrac{5}{6}\right)^{n-1} \times \tfrac{1}{6}$$

The probabilities are getting smaller by a factor of $\frac{5}{6}$ each time. The conclusion is that the largest probability is associated with getting started on the first attempt.

CALCULATOR ACTIVITY

Flipping two coins

In Chapter 5, *Probability*, we saw how a calculator could be used to simulate flipping a coin. How might it be used to simulate the outcome of flipping two coins? Suppose that the rules of a game require a player to get two 'heads' when two coins are flipped. Use a calculator to simulate 60 players successfully attempting to get started. Record the number of flips each player takes. What is the **modal** number of attempts? Compare the number of players who are successful on their first, second, third, ... attempts.

Probability distribution

So far, we have introduced a way of modelling the likelihood of any given number of attempts at getting started. A player might get started on the first attempt, or it might take two, three, four, ... attempts.

No. of attempts	Probability
1	$\frac{1}{6}$
2	$\frac{5}{6} \times \frac{1}{6}$
3	$\left(\frac{5}{6}\right)^2 \times \frac{1}{6}$
4	$\left(\frac{5}{6}\right)^3 \times \frac{1}{6}$
n	$\left(\frac{5}{6}\right)^{n-1} \times \frac{1}{6}$

The table shows the number of possible attempts together with their probabilities when the probability of a successful outcome is $\frac{1}{6}$. It represents a **probability distribution** which aims to model the occurrence of a particular **random variable**. The random variable, R, is the number of attempts required to get started. The probability distribution can be summarised with the following probability function:

$$P(R = n) = \left(\tfrac{5}{6}\right)^{n-1} \times \tfrac{1}{6} \qquad n = 1, 2, 3, \ldots$$

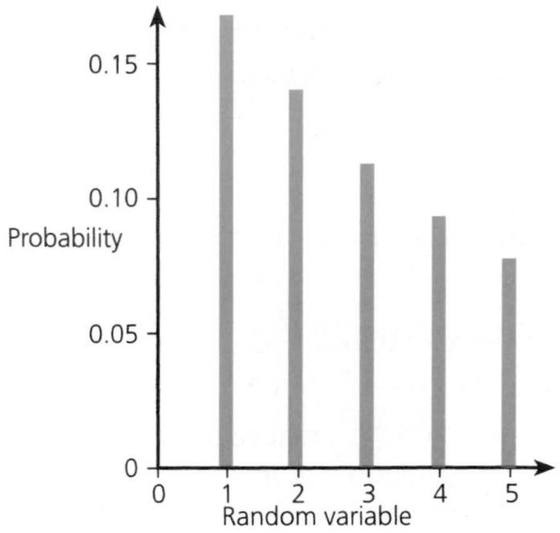

A graphical display of the distribution is shown in the diagram.

A probability distribution has the following features:

a) all possibilities must be identified,

b) all associated probabilities must lie between 0 and 1,

c) the total of all the probabilities must be 1.

Are these features met in the probability distribution here?

a) *All possibilities* – a player could get started on their first go, or it could take two attempts, or three, or ..., or thirty-three, ... It could take any positive whole number of attempts. So, the random variable, R, could be any positive whole number.

b) *Lie between 0 and 1* – the largest probability is $\frac{1}{6}$. The next largest is $\frac{5}{6} \times \frac{1}{6}$, then $\frac{5}{6}$ of this, etc. All probabilities are a positive fraction of $\frac{1}{6}$. So, the probabilities lie between 0 and 1.

c) *The total*, T, of the probabilities is:

$$T = \tfrac{1}{6} + \tfrac{5}{6} \times \tfrac{1}{6} + \left(\tfrac{5}{6}\right)^2 \times \tfrac{1}{6} + \left(\tfrac{5}{6}\right)^3 \times \tfrac{1}{6} + \ldots$$

It is far from obvious what this comes to. One simple approach to evaluating T is firstly to find $\frac{5}{6}T$.

$$\tfrac{5}{6}T = \tfrac{5}{6}\left\{\tfrac{1}{6} + \tfrac{5}{6} \times \tfrac{1}{6} + \left(\tfrac{5}{6}\right)^2 \times \tfrac{1}{6} + \left(\tfrac{5}{6}\right)^3 \times \tfrac{1}{6} + \ldots\right\} = \tfrac{5}{6} \times \tfrac{1}{6} + \left(\tfrac{5}{6}\right)^2 \times \tfrac{1}{6} + \left(\tfrac{5}{6}\right)^3 \times \tfrac{1}{6} + \ldots$$

Now find the difference between T and $\frac{5}{6}T$.

$$T - \tfrac{5}{6}T = \left\{\tfrac{1}{6} + \tfrac{5}{6} \times \tfrac{1}{6} + \left(\tfrac{5}{6}\right)^2 \times \tfrac{1}{6} + \left(\tfrac{5}{6}\right)^3 \times \tfrac{1}{6} + \ldots\right\} - \left(\tfrac{5}{6} \times \tfrac{1}{6} + \left(\tfrac{5}{6}\right)^2 \times \tfrac{1}{6} + \left(\tfrac{5}{6}\right)^3 \times \tfrac{1}{6} + \ldots\right) = \tfrac{1}{6}$$

Since the two expressions are almost identical, but, treated as an algebraic expression: $T - \frac{5}{6}T = \frac{1}{6}T$.

thus $\frac{1}{6}T$ and $\frac{1}{6}$ represent the same value: $\frac{1}{6}T = \frac{1}{6} \Rightarrow T = 1$.

This means that the total of all the probabilities is 1, as required for a probability distribution.

Exploration 6.1 *Two heads*

Develop a probability distribution to model the number of times you need to flip a pair of coins to get two heads. State your probability function, and show that the three features required are present in your model. Ensure you give a graphical presentation of your distribution.

Geometric distribution

The probability distributions considered in this chapter are examples of the **geometric probability distribution**. A simple notation often used for this distribution would be:

$\text{Geo}\left(\frac{1}{6}\right)$

in the case of the 'double' on the pair of dice. In general terms: $\text{Geo}(p)$ is used where p is the probability of succeeding at the first attempt.

A geometric sequence is one in which there is a constant ratio between successive terms. Examples of geometric sequences are:

3, 6, 12, 24, 48, 96, 192, 382, ... and

$\frac{1}{4}, \frac{3}{16}, \frac{9}{64}, \frac{27}{256}, \frac{81}{1024}, \frac{243}{4096}$

In the first sequence, the constant ratio is 2 and in the second sequence, it is $\frac{3}{4}$.

The general geometric sequence is often written:

$a, ar, ar^2, ar^3, ar^4, ar^5, \ldots$

The sum of all the terms in the sequence is $\frac{a}{1-r}$, where a is the first term, and r the ratio (provided the magnitude of r is less than 1).

Exploration 6.2

The probability distibution

- Consider again, the probability distribution, $\text{Geo}\left(\frac{1}{6}\right)$.
  The possible values of the random variable, x, associated with this distribution and the relevant probabilities can be written:

x	1	2	3	4	5	6	...
$\text{P}(x)$	$\frac{1}{6}$	$\frac{1}{6}\left(\frac{5}{6}\right)$	$\frac{1}{6}\left(\frac{5}{6}\right)^2$	$\frac{1}{6}\left(\frac{5}{6}\right)^3$	$\frac{1}{6}\left(\frac{5}{6}\right)^4$	$\frac{1}{6}\left(\frac{5}{6}\right)^5$	...

  Identify the first term and the constant ratio of the sequence of probabilities. Hence, find the sum of all the probabilities.
- Now repeat the process for:

  a) $\text{Geo}\left(\frac{3}{4}\right)$ **b)** $\text{Geo}\left(\frac{2}{3}\right)$ **c)** $\text{Geo}(0.8)$.

Example 6.1

Potatoes being prepared for packing are carried in a single file along a conveyor belt. The potatoes are visually inspected for blemishes. Those which are blemished are rejected. The inspector, having just rejected a potato, counts the number of potatoes he inspects until he next rejects another one.

If the proportion of blemished potatoes is five per cent calculate:
a) *the probability that the inspector counts exactly five potatoes,*
b) *the probability that the inspector counts at least five potatoes.*

Solution

a) *Let B represent a potato which is blemished, and S a potato which is sound. Then the combined outcome sought is:*

1st	2nd	3rd	4th	5th potato
S	S	S	S	B

If we assume that the quality of one potato is independent of the quality of any other potato, the probability of this combined event can be found from the multiplication law as:

$$\begin{aligned} P(SSSSB) &= P(S) \times P(S) \times P(S) \times P(S) \times P(B) \\ &= 0.95 \times 0.95 \times 0.95 \times 0.95 \times 0.05 \\ &= (0.95)^4 \times 0.05 \end{aligned}$$

$P(\text{exactly } 5) = 0.0407$

b) *There are at least three different ways in which we might approach the solution of this part.*

i) *A fairly direct approach can be found using the formula for the sum of all the terms of a geometric sequence. If the number of potatoes counted is at least 5, then it could be exactly 5, exactly 6, exactly 7, exactly ... . The probability could be calculated from:*

$$\begin{aligned} P(\text{at least } 5) &= P(5 \text{ or } 6 \text{ or } 7 \text{ or } \ldots) \\ &= P(5) + P(6) + P(7) + \ldots \\ &= 0.95^4 \times 0.05 + 0.95^5 \times 0.05 + 0.95^6 \times 0.05 + \ldots \end{aligned}$$

Notice, this is a geometric sequence with the first term equal to $0.95^4 \times 0.05$ and the common ratio equal to 0.95. Hence, using $\frac{a}{(1-r)}$, its sum is:

$$\frac{0.95^4 \times 0.05}{(1-0.95)} = \frac{0.95^4 \times 0.05}{0.05} = 0.95^4$$

$P(\textit{at least } 5) \approx 0.8145$

ii) *An alternative approach is to consider the complementary event to 'at least 5', which is 'fewer than 5', i.e.*

$$\begin{aligned} P(\textit{at least } 5) &= 1 - P(\textit{fewer than } 5) \\ &= 1 - \{P(1) + P(2) + P(3) + P(4)\} \\ &= 1 - \{0.05 + 0.95 \times 0.05 + 0.95^2 \times 0.05 + 0.95^3 \times 0.05\} \\ &= 1 - \{0.05 + 0.0475 + 0.045\,125 + 0.042\,868 \ldots\} \\ &= 1 - 0.185\,49 \ldots \end{aligned}$$

$P(\textit{at least } 5) \approx 0.8145$

iii) *A third, and perhaps the most direct approach, considers the implications of the count being at least five. This means that the first four potatoes were all sound.*

$$\begin{aligned} P(\textit{at least } 5) &= P(\textit{first four sound}) \\ &= P(SSSS) = P(S) \times P(S) \times P(S) \times P(S) = (0.95)^4 \end{aligned}$$

$P(\textit{at least } 5) \approx 0.8145$

Example 6.2

A tetrahedral die, with vertices numbered 1, 2, 3 and 4, is rolled repeatedly until a 3 is scored. What is the probability that the first 3 is scored in fewer than eight rolls?

Solution

Again, there are several different approaches which we could adopt to solve this problem. Let T denote the event that the outcome of rolling the die is 3, and N describe the event that the outcome is not a 3.

a) *For the first 3 to appear in fewer than eight rolls, one of the following must take place:*
T or NT or NNT or NNNT or NNNNT or NNNNNT or NNNNNNT
So, the probability required is:
P(*T or NT or NNT or ... or NNNNNNT*)
$= \mathrm{P}(T) + \mathrm{P}(NT) + \mathrm{P}(NNT) + \ldots + \mathrm{P}(NNNNNNT)$

$$= \tfrac{1}{4} + \tfrac{3}{4} \times \tfrac{1}{4} + \left(\tfrac{3}{4}\right)^2 \times \tfrac{1}{4} + \ldots + \left(\tfrac{3}{4}\right)^6 \times \tfrac{1}{4}$$

This can be found, perhaps using a calculator, to be approximately 0.8665.

b) *There is a formula for summing n terms of a geometric sequence:*

$$S_n = a + ar + ar^2 + \ldots + ar^{n-1} = \frac{a\left(1 - r^n\right)}{1 - r}$$

In this case, $n = 7$, $a = \tfrac{1}{4}$, $r = \tfrac{3}{4}$ hence:

$$\mathrm{P}(\textit{fewer than } 8) = S_7 = \frac{\tfrac{1}{4}\left(1 - \left(\tfrac{3}{4}\right)^7\right)}{1 - \tfrac{3}{4}} \approx 0.8665$$

c) *A fairly direct solution can be found by considering the complementary event. In this case the complement of 'fewer than 8' is '8 or more'. Thus:*
P(*fewer than 8*) = 1 − P(*at least 8*) = 1 − P(*none of the first 7*)

$$= 1 - \left(\tfrac{3}{4}\right)^7 \approx 0.8665$$

EXERCISES

6.1A

1 A game requires a player to roll a six with a cubical die before the player may start the game.
  a) What is the most likely number of rolls of the die the player needs to get started?
  b) What is the probability that a player will need exactly four rolls?
  c) What is the probability that a player will need at least four rolls?

2 A random variable may be modelled by the distribution Geo(0.6). Calculate the probability:
  a) P(variable = 1) **b)** P(variable = 2) **c)** P(variable = 3)
  d) P(variable = 4) **e)** P(variable > 4).
  Illustrate the occurrence of the variable using a probability line graph.

3 A pack of cards is shuffled and the top card is turned over. The process is repeated until a heart is turned up.

a) What is the probability that this occurs after exactly one shuffle?
b) What is the probability that it takes fewer than four shuffles?
c) What is the probability that it takes at least four shuffles?

4 A car salesman knows that 15 per cent of the cars he sells are red.

a) What is the probability of his next five sales not being red?
b) What is the probability of his selling at least one red car in the next five sales?
c) What is the probability of him having to make at least five sales before he sells a red car?

5 Calculate the probability of having to roll a cubical die fewer than ten times before the first six appears.

6 How many times does a cubical die need to be rolled for the probability of getting at least one 6 to be greater than 0.995?

EXERCISES

6.1B

1 In a game there is a stage where a player could land in 'jail'. The player must throw a double with an ordinary pair of cubical dice to get out of jail.

a) What is the probability that a player will get out of jail on the first attempt?
b) What is the probability that a player will take exactly three attempts to get out of jail?
c) What is the probability that a player will not get out of jail in three attempts?
d) What is the most likely number of attempts a player will make to get out of jail?

2 A random variable follows the geometric distribution Geo($\frac{1}{4}$).

a) Calculate the probability that the random variable:
   i) is exactly one,
   ii) is exactly two,
   iii) is exactly three,
   iv) is more than three.
b) Write down the modal value of the variable.
c) Illustrate the occurrence of the variable using a probability line graph.
d) Calculate the probability that the variable:
   i) is at least four, ii) is at least ten.

3 A student is conducting a survey of road traffic. One part of the survey involves the country of manufacture of road vehicles. A simple classification is used: home or foreign. The student has to record the country of manufacture of each vehicle passing her. The proportion of foreign vehicles is 65 per cent.

a) What is the probability that the first two vehicles are foreign?

b) What is the probability that there is no foreign vehicle in the first four that pass her?
c) What is the probability that she records at least six vehicles before the first foreign vehicle?
d) What is the probability she records more than ten vehicles before the first home-manufactured vehicle?

4 A tetrahedral die is rolled.
a) What is the probability of not getting a 4 in five rolls of the die?
b) What is the probability of getting at least one 4 in five rolls of the die?
c) What is the probability of having to roll the die at least five times to get the first 4?

5 What is the probability that a coin needs to be flipped fewer than six times before the first head occurs?

6 How many times does a coin need to be flipped if the probability of not getting a head is less than 0.0001?

MEAN

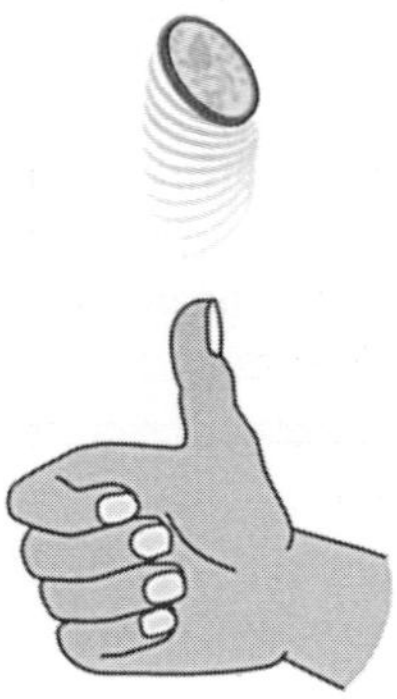

We have considered the geometric probability distribution, Geo(p), for a variety of values of the probability p. For instance, in the first section of this chapter, the value of p was $\frac{1}{6}$; in Exploration 6.1, p was $\frac{1}{4}$, etc. In every case, the modal value of the random variable was 1. Is this also true for the mean value of the random variable? Let's find out.

Consider the distribution used to model the number of flips of a coin to get a head. This is Geo($\frac{1}{2}$). Suppose we conduct an experiment where we flip a coin repeatedly and record the outcome: H, T; on each occasion.

The results are likely to be rather like those in the diagram.

How many flips did it take to get the first head? How many more flips were needed to get the next head? ... the next one? ... the next?

The diagram indicates that the first flip resulted in a head, so it took one flip. The next head appeared on the third flip after that, so four flips were needed. The next occurred one flip later. The next ... it is not possible to tell from the diagram.

However, suppose we had recorded the results of, say, 200 flips. Why is it reasonable to expect 100 of these to be heads?

The probability of a head is assumed to be $\frac{1}{2}$. We should expect half of the outcomes to be heads and half of 200 is 100! Let us now think about the situation where a half of the outcomes are heads – i.e. where, on average, every other outcome is a head. What is the average number of flips before the next head?

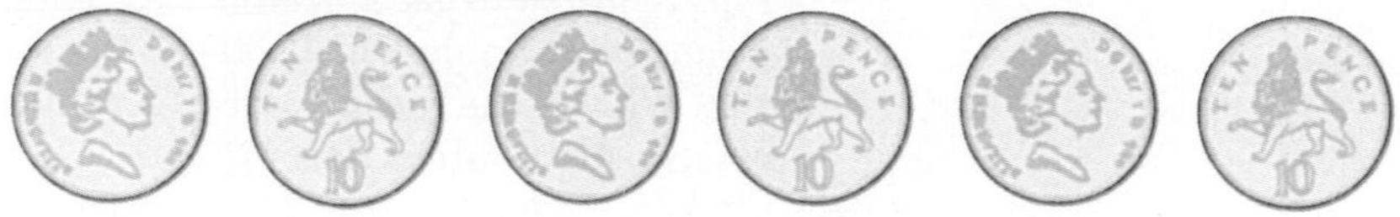

Since, on average, the outcomes alternate between tails and heads, the number of flips is two. The discussion so far leads to the following suggestion regarding the mean for Geo($\frac{1}{2}$):

mean of Geo($\frac{1}{2}$) = 2

CALCULATOR ACTIVITY

Using expected value

It should be possible to verify this result using the 'expected value' approach discovered in Chapter 5, *Probability*. In particular:
mean of X = E(X) = $\sum x$P($X = x$)
In Geo($\frac{1}{2}$), x can be any positive integer i.e. 1, 2, 3, 4, ... and the probability associated with the value is $(\frac{1}{2})^x$.
The following table presents part of the probability distribution.

x	P($X = x$)	x P($X = x$)
1	$\frac{1}{2}$	$1 \times \frac{1}{2}$
2	$(\frac{1}{2})2$	$2 \times (\frac{1}{2})^2$
3	$(\frac{1}{2})3$	$3 \times (\frac{1}{2})^3$
4	$(\frac{1}{2})4$	$4 \times (\frac{1}{2})^4$
⋮	⋮	⋮

Clearly, the number of possible flips of the coin needed to get a head could be extremely large and it could be quite tedious keying in the calculations needed to evaluate the mean, i.e:
$1 \times \frac{1}{2} + 2 \times (\frac{1}{2})^2 + 3 \times (\frac{1}{2})^3 + \ldots$

If you have a programmable calculator you could use a program based on the following flowchart where N represents the number of terms in the series $\sum_{r=1}^{N} r \times \left(\frac{1}{2}\right)^r$ and M is the mean.

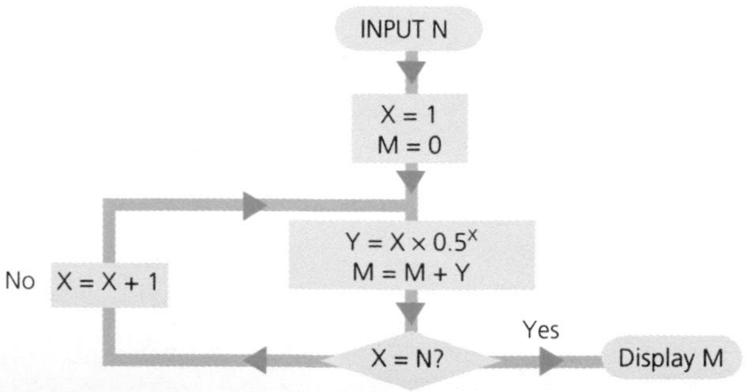
INPUT N
X = 1
M = 0
Y = X × 0.5^X
M = M + Y
No
X = X + 1
X = N?
Yes
Display M

Note: M will always be slightly less than the true expected value since N cannot be infinite.

Try for N = 10, 20, 30; what do you think the expected value is?

If your calculator does not have the facility to run programs, you could input the values of X and P($X = x$) in its statistics mode and find the mean in that way.

Interpreting the results

The work in this section has led us to conclude that the mean number of flips required to get a head is two. This suggests that the mean value of a random variable which follows the geometric distribution, Geo($\frac{1}{2}$), is two. What is the mean value of a random variable which follows the geometric distribution Geo($\frac{1}{6}$)?

Earlier, we decided that Geo($\frac{1}{6}$) could be used to model the number of rolls of two dice required to get started in a board game. Imagine the scene at the annual convention for players of such a board game. There are hundreds, perhaps a thousand or more, players. Each player has to roll a double with a pair of ordinary dice. Suppose at some stage in this convention there have been 600 rolls of the dice, how many of these rolls could we expect to have resulted in a double? Since a double occurs one roll in every six, on average, it would be reasonable to expect that 100 of the 600 would result in a double. This suggests that the average number of rolls to get a double would be six, i.e.
mean of Geo($\frac{1}{6}$) = 6

CALCULATOR ACTIVITY

Testing the assertion

It ought to be possible to test the assertion just made about the mean of Geo($\frac{1}{6}$). How might the program you have written for Geo($\frac{1}{2}$) be adapted for Geo($\frac{1}{6}$)? The following flowchart is one possible solution.

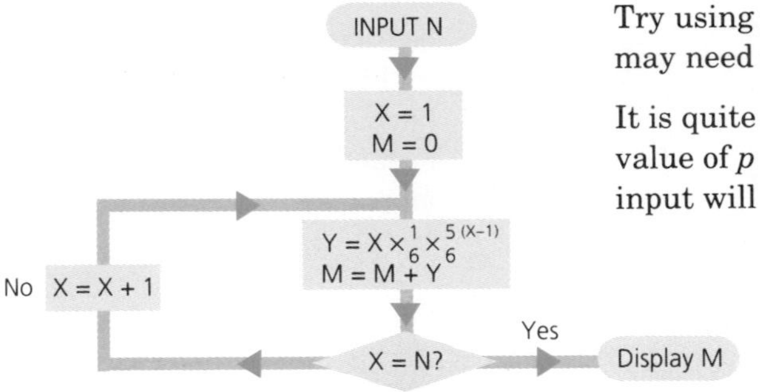

Try using your calculator to find the mean of Geo($\frac{1}{6}$); you may need to set N to 30, 40, 50, 60 or more.

It is quite possible to adapt the program to cope with any value of p and then to find the mean of Geo(p). An additional input will be required which is the value of p. Then the program will need to evaluate the complementary probability, $1 - p$, which is often labelled q. This flowchart presents one way of carrying this out.

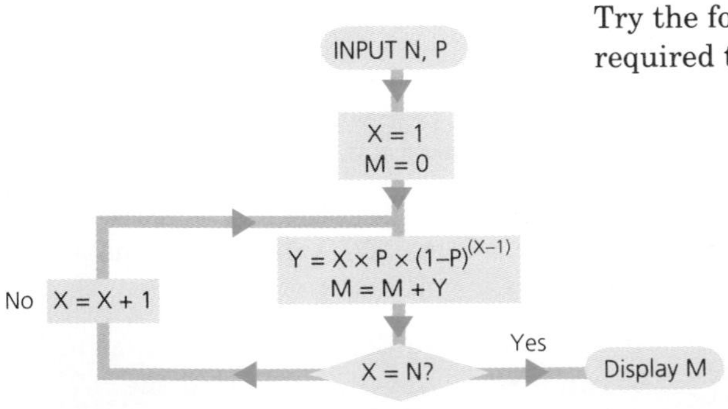

Try the following values for p and N (larger values of N are required the smaller p is).

p	N (at least)
$\frac{1}{2}$	20
$\frac{1}{6}$	75
0.8	10
0.75	15
0.6	20
$\frac{1}{3}$	30
$\frac{1}{4}$	40

The results you obtain could be displayed in a table like this.

Value of p	Mean of Geo(p)
$\frac{1}{2}$	2
$\frac{1}{6}$	6
0.8	1.25
0.75	$1\frac{1}{3}$
0.6	$1\frac{2}{3}$
$\frac{1}{3}$	3
$\frac{1}{4}$	4

What do you think the relationship between the value of the probability, p, and the mean of the geometric distribution is?

Mean of Geo(p) = $\frac{1}{p}$

Variance

In Chapter 5, *Probability*, we introduced the concept of the variance of the probability distribution involving expectation:

$$\text{Var}(X) = \text{E}(X^2) - \text{mean}^2$$

where $\text{E}(X^2) = \sum x^2 \times \text{P}(X = x)$

Exploration 6.3

Mean and variance

Use this relation to explore the relationship between the mean and the variance of the geometric probability distribution Geo(p).

As an example, consider Geo(0.8), with mean $\frac{1}{0.8} = 1\frac{1}{4}$.

x	$\text{P}(X = x)$
1	0.8
2	0.8 0.2
3	0.8 0.2^2
4	0.8 0.2^3
⋮	⋮

The expected value of the square, $\text{E}(X^2)$ is:

$$\text{E}(X^2) = \sum_{x=1}^{\infty} x^2 \times 0.8 \times 0.2 + 3^2 \times 0.8 \times 0.2^2 + 4^2 \times 0.8 \times 0.2^3 + \ldots \approx 1.875$$

(using a calculator to sum the first 15 terms).

Then the variance can be found as:

$$\text{Var}(X) = \text{E}(X^2) - (\text{mean})^2$$
$$= 1.875 - \left(1\tfrac{1}{4}\right)^2$$
$$= \tfrac{5}{16}$$

Then, for Geo($\frac{4}{5}$), the mean is $1\frac{1}{4}$ and the variance is $\frac{5}{16}$.

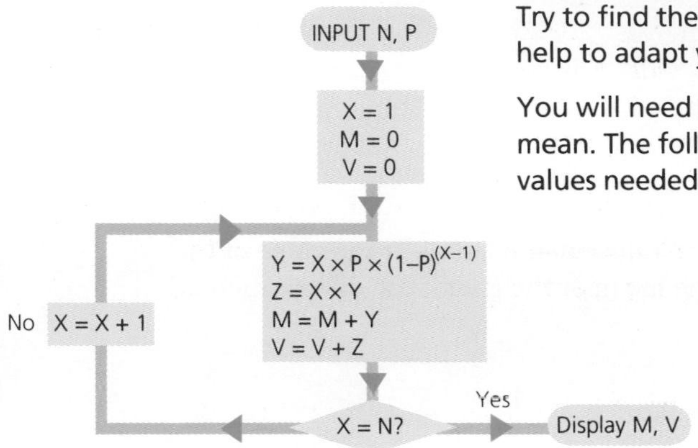

Try to find the variance for other geometric distributions. It might help to adapt your program as indicated in this flowchart.

You will need larger values of N than were needed in finding the mean. The following table provides an indication of the sort of values needed.

p	N
$\frac{1}{2}$	25
$\frac{1}{6}$	100
0.8	15
0.75	15
0.6	25
$\frac{1}{3}$	40
$\frac{1}{4}$	75

The results you may obtain are given in this table.

p	Mean	Variance
0.8	$1\frac{1}{4}$	$\frac{5}{16}$
0.75	$1\frac{1}{3}$	$\frac{4}{9}$
0.6	$1\frac{2}{3}$	$1\frac{1}{9}$
$\frac{1}{2}$	2	2
$\frac{1}{3}$	3	6
$\frac{1}{4}$	4	12
$\frac{1}{6}$	6	30

What do you think is the relationship between the variance and the mean of Geo(p)?

Variance = mean × (mean − 1)

Example 6.3

a) *If X is a random variable which is distributed* Geo(0.2), *find values for:*
i) E(X) ***ii)*** Var(X) ***iii)*** P(X < mean).

b) *The standard deviation of a geometric distribution is 6. What is its mean?*

Solution

a) **i)** *In general for* Geo(p), *mean* $= \frac{1}{p}$, *so for* Geo(0.2):

mean $= \frac{1}{0.2} = 5$

ii) *The variance is mean × (mean – 1), so for* Geo(0.2):

variance $= 5 \times 4 = 20$

iii) $P(X < \text{mean}) = P(X < 5) = P(X = 1) + P(X = 2) + P(X = 3) + P(X = 4)$

$= 0.2 + 0.2 \times 0.8 + 0.2 \times 0.8^2 + 0.2 \times 0.8^3 = 0.5904$

b) *If the standard deviation is 6, then the variance is* 6^2*, i.e. 36. So, using the result variance = mean × (mean – 1):*

$36 = \text{mean} \times (\text{mean} - 1) \Rightarrow \text{mean}^2 - \text{mean} - 36 = 0$

This is a quadratic equation with solution:

$$\text{mean} = \frac{1 \pm \sqrt{145}}{2} = 6.5208 \text{ or } -5.5208$$

But a random variable which is geometrically distributed cannot be negative, hence the mean cannot be negative.

mean $= 6.5208$

EXERCISES

6.2A

1 A random variable, X, is geometrically distributed with parameter $p = 0.7$.

a) Write down the mode of X.

b) State the mean of X.

c) Write down the variance of X.

2 Illustrate Geo(0.7) and Geo(0.3) on a single diagram using a probability line graph.

3 Write down the mean and standard deviation of Geo(0.9).

4 If X is known to be geometrically distributed with standard deviation $\frac{6}{5}$, what is its mean?

5 The mean of a geometric distribution is 1.21. Calculate its standard deviation.

EXERCISES

6.2B

1 A random variable is geometrically distributed Geo(0.2).

a) Calculate the probability of the first success on the third attempt.

b) Write down its mean.

c) Calculate the variance of the random variable.

2 Illustrate Geo(0.1) and Geo(0.9) on the same probability diagram.

3 A random variable is modelled by Geo(0.25). Write down its expected value and the expected value of its square.

4 The standard deviation of a random variable, Y, which is geometrically distributed is $\frac{3}{5}$. Calculate the probability: $P(Y = 1)$.

5 Calculate the mean of a random variable which is geometrically distributed with variance $\frac{15}{4}$.

MODELLING THE OCCURRENCE OF A RANDOM VARIABLE

We have used the term 'model' earlier, in connection with the probability distribution introduced in this chapter. The process of modelling a real-world situation can involve several different stages. These are shown as boxes in our modelling flowchart. The first of these is a description of that real situation or the real problem requiring a solution.

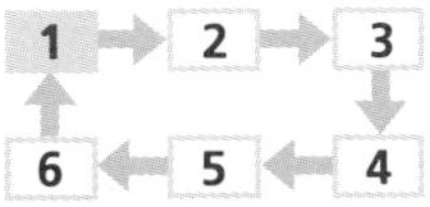

Specify the real problem

In the discussion at the start of this chapter the real-world situation involved the number of rolls of two dice required to get a double in order for a player in a game to start playing.

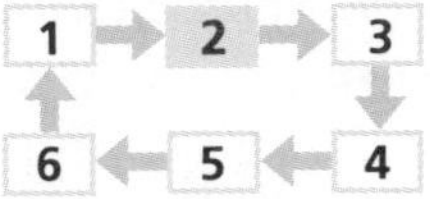

Set up a model

Once the real situation is grasped, the process of setting up an appropriate model can begin. Geo($\frac{1}{6}$) was proposed as a model for the number of rolls of two dice required to get a double.

What assumptions led to this model?

The process which leads to Geo($\frac{1}{6}$) might start like this.

1 A die can land with any one of six faces uppermost.

2 Each face is as likely as any other face.

3 For two dice, there are 36 possible equally likely outcomes.

4 Exactly six of these outcomes result in a double.

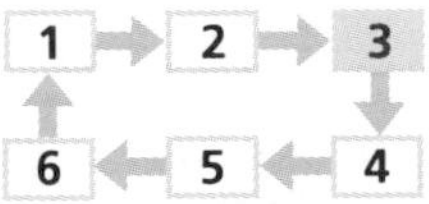

Formulate the mathematical problem

The next stage in the process is the formulation of the associated mathematical problem.

This leads us to assigning some numerical values to the likelihood of a 'double' being the outcome of rolling two dice.

5 The probability of a pair of dice landing as a double is $\frac{6}{36}$ i.e. $\frac{1}{6}$.

6 The probability that the dice do not land as a double is $1-\frac{1}{6}$, i.e. $\frac{5}{6}$.

7 The result of rolling the two dice is not affected by the result of any previous rolling, i.e. the outcomes are independent.

Having formulated the mathematical problem, the next step is to solve it.

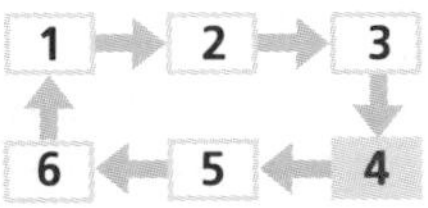

Solve the mathematical problem

8 The probability that a 'double' appears on the first roll of the dice is $\frac{1}{6}$.

9 To take exactly two rolls of the dice to get the first double means that the first roll was not a double but the second was a double. Since the outcomes are independent the probabilities can be multiplied to yield:
P(first double on second roll) $=\frac{5}{6}\times\frac{1}{6}$

10 The argument presented in (9) extends to:

P(first double on third roll) $=\frac{5}{6}\times\frac{5}{6}\times\frac{1}{6}=\left(\frac{5}{6}\right)^2\times\frac{1}{6}$

P(first double on fourth roll) $=\left(\frac{5}{6}\right)^3\times\frac{1}{6}$

...

P(first double on nth roll) $=\left(\frac{5}{6}\right)^{n-1}\times\frac{1}{6}$

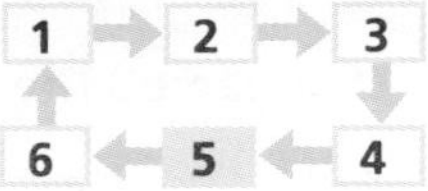

Interpret the solution

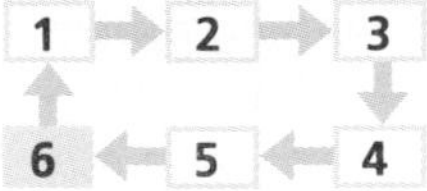

Compare with reality

We now need to interpret the solution of the mathematical problem. This suggests that in $\frac{1}{6}$, or about 16 per cent to 17 per cent, of the times a player has a go at rolling double he will be successful. It also indicates that he will need two attempts on between 13 per cent and 14 per cent $\left(\frac{5}{36}\right)$ of the occasions he has to roll a double to get started. Then, three attempts are required on about $11\frac{1}{2}$ per cent of the games. The model predicts that there is a reduction in the likely frequency as the number of attempts increases. The most likely number of attempts it predicts is 1. How these predictions, and interpretation compare with the reality is the next stage in the modelling process.

11 The model suggests that the modal number of attempts is 1. Is this borne out in reality?

12 The model suggests that the mean number of attempts is 6. Is this borne out in reality?

13 The model suggests that the variance of the number of attempts is 30. Is this borne out in reality?

14 The models suggests that the probability reduces at a constant rate. Again is this borne out in reality?

If the model seems to be appropriate then a report may be relevant. Otherwise it might be relevant to review the modelling process and to refine the proposed model. Thus the modolling process would be recommenced. Here, however, the model seems to be quite acceptable.

Exploration 6.4

Modelling

Look back at Example 6.1 and carry out the modelling process.

1 Specify the real world problem.
2 Set up the model.
3 Formulate the mathematical problem.
4 Solve the mathematical problem.
5 Interpret the implications of the solution.
6 Compare with reality.
7 Report.

Exploration 6.5

Identify the modelling stages

- Refer to Example 4.7(a); identify all the stages of the modelling process involved in arriving at an answer to the question posed.
- Repeat this for Example 4.7(b).

Applying the geometric distribution

Example 6.4

A rowing club ran a stall during regatta week. The club offered a prize of £5 for a 20p stake to anyone who succeeded in rolling a double 6 with an ordinary pair of dice. If a contestant succeeded in rolling just one 6 they got a free go. A contestant who rolls a double 6 also gets their stake back. Is this a fair game?

Solution

In Chapter 5, Probability, we met the concept of a 'fair game'. The expected gain must be zero, where:

$\text{E(gain)} = \sum \text{gain} \times \text{probability of the gain}$

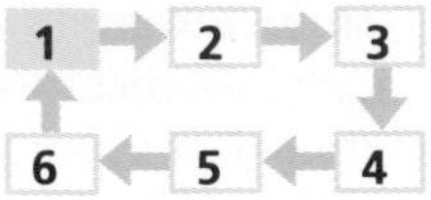

Specify the real problem

It is simple enough to identify the gain. A contestant who rolls a double 6 gains £5. Otherwise, the gain is –20p. But, what about the probability of these gains? We need to use the modelling process. The real-world situation has been specified. The model is based on symmetry and equal likelihood.

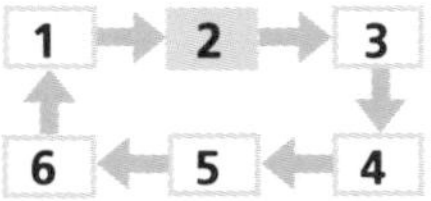

Set up a model

This leads to the familiar result which models the outcome space for a pair of dice as 36 equally likely events. Only one of these results in two 6s; ten result in exactly one 6; 25 have no 6. Hence, the model corresponds to:

$\text{P(double 6)} = \frac{1}{36}$

$\text{P(exactly one 6)} = \frac{10}{36}$

$\text{P(no 6)} = \frac{25}{36}$

The mathematical problem has already been formulated i.e. $\text{E(gain)} = 0$.

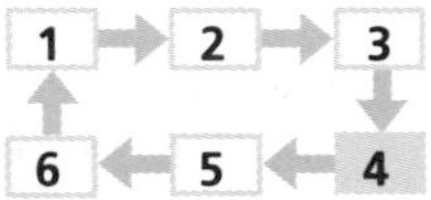

Solve the mathematical problem

The solution of the mathematical problem requires us to establish the probabilities associated with each of the possible gains. The gain, £5, comes about when a contestant gets a double 6. The probability that this happens on the first roll of the dice is $\frac{1}{36}$*. This is not the only opportunity that a contestant would conceivably have to get a double 6. How else could a contestant get a double 6?*

Imagine the contestant who rolls exactly one 6, then on the free go, gets a double 6. The probability that this happens is:

$\frac{10}{36} \times \frac{1}{36}$ *(What modelling has led to this result?)*

Then, there is the contestant who gets exactly one 6 on the free go and on the next free go gets a double 6. The probability for this is:

$\frac{10}{36} \times \frac{10}{36} \times \frac{1}{36}$

and so on. The possibilities can be illustrated in a tree diagram.

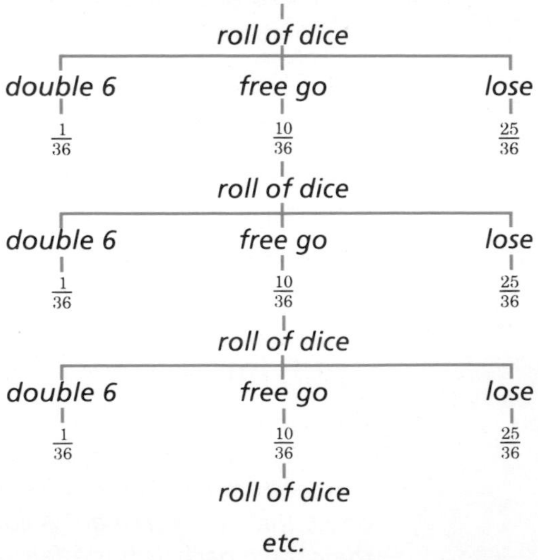

The probability of gaining £5 is modelled as:

$\frac{1}{36} + \frac{1}{36} \times \frac{10}{36} + \frac{1}{36} \times \left(\frac{10}{36}\right)^2 + \frac{1}{36} \times \left(\frac{10}{36}\right)^3 + \ldots$

which is a geometric series with first term equal to $\frac{1}{36}$ *and common ratio equal to* $\frac{10}{36}$. *The sum of the series is:*

$$\frac{\frac{1}{36}}{\left(1 - \frac{10}{36}\right)} = \frac{1}{26}$$

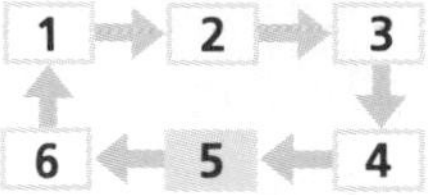

Interpret the solution

The probability of gaining –20p must be $\frac{25}{26}$.
(This can be checked since it should be the sum of

$\frac{25}{36} + \frac{25}{36} \times \frac{10}{36} + \frac{25}{36} \times \left(\frac{10}{36}\right)^2 + \frac{25}{36} \times \left(\frac{10}{36}\right)^3 + \ldots$ *)*

A summary of this is:

Gain	Probability
£5	$\frac{1}{26}$
–20p	$\frac{25}{26}$

Then, the expected gain for the game is: $= 500 \times \frac{1}{26} + -20 \times \frac{25}{26} = 0$
The interpretation which can be put on this solution is that the game is fair.

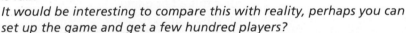

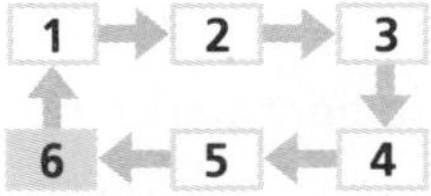

Compare with reality

It would be interesting to compare this with reality, perhaps you can set up the game and get a few hundred players?
A brief report: the modelling undertaken here suggests that, on average, one contestant in every 26 will win the £5 prize. Twenty-five contestants will forfeit their stake money. The game is fair but it is not likely to lead to the rowing club making any money.

CONSOLIDATION EXERCISES FOR CHAPTER 6

1 The organisers of a lottery claim that the chance of winning a prize with one ticket is 1 in 54. I decide to buy one ticket each week until I win a prize. Establish a probability model for the number of weeks. Use your model to:

a) determine the expected number of weeks I have to wait for a prize,
b) estimate the chance I will win a prize in less than four weeks,
c) estimate the chance I will have to wait for more than a year,
d) estimate the most likely number of weeks I have to wait.

2 A woman decides she will give birth to one child each year until she bears a daughter. Establish a model for the number of children she has.

a) What is the expected number of children?
b) What is the modal number of children?
c) Suggest a refinement you might make to your model so that it compares more favourably with reality.

3 An octahedral die with faces numbered 1, 2, ..., 8 is repeatedly rolled until a seven is obtained. Establish a model for the number of rolls required. Use your model to:

a) determine the most likely number of rolls,
b) estimate the mean number of rolls,
c) predict the variance of the number of rolls,
d) estimate the likelihood of requiring at least ten rolls.

4 A driving school claims that, on average, its pupils only need to take one driving test to qualify for a full driving licence. Establish a model for the number of tests a pupil needs assuming that the probability of a pupil passing the test is p.

a) If the claim of the school is true, which 'average' is being quoted?
b) Use your model to demonstrate that the claim is false if the school is using the mean number of tests.
c) Suggest a reason why your model might need to be refined.

5 A game in an amusement park is played in which a contestant pays 25p for the privilege of rolling two dodecahedral dice each numbered 1, 2, 3, ..., 12. A prize of £10 is offered for any contestant getting a double 12. Contestants who get exactly one 12 are given a free go. All other contestants forfeit their stake money.

a) Is this a fair game?
b) Who stands to gain?
c) What is the expected gain per go to the amusement park?

6 I estimate that whenever I play a game of darts and I try to hit the bull my chance of success is $\frac{1}{8}$. What is the probability that I need:

a) exactly eight attempts,
b) fewer than eight attempts,
c) more than eight attempts,

to hit the bull.

7 A tetrahedral die with its vertices numbered 1, 2, 3 and 4 is rolled respectively until a 2 is scored. What is the probability:

a) that five rolls are required,
b) that fewer than ten rolls are needed,
c) that at least four rolls are required?

8 How many times does an ordinary die need to be rolled for the probability of getting at least one number greater than 4 to exceed 0.95?

9 If five per cent of the vehicles in your town are registered as taxis and you are waiting for a taxi to come into view, calculate the probability that:

a) the first taxi is the sixth vehicle to come into view,
b) the first taxi is not among the first six vehicles which come into view.

10 A fairground stall owner has two airguns, A and B. The guns are identical in appearance, but the chance that the pellet fired from gun A hits its target is $\frac{2}{3}$. For gun B the chance is $\frac{1}{2}$.

a) I fire five pellets from gun A. What is the probability that:
   i) none of them hit the target,
   ii) the first four miss the target but the fifth is successful,
   iii) at least one of them hits the target?

b) If I use gun B, what is:
   i) the most likely number of shots I need to hit the target,
   ii) the probability of needing more than five shots to hit the target?

11 A computer is programmed to produce single digits 0, 1, 2, ..., 9 randomly.

a) What proportion of the digits produced do you anticipate being zeros?

b) What is the probability that none of the first 30 digits produced is a zero?

c) What is the probability that two consecutive digits produced are zeros?

d) If the computer is programmed to produce double digits 00, 01, 02, ..., 99 randomly, what is the probability that none of the first:
   i) 30 pairs contain a double zero,
   ii) 300 pairs contain a double zero?

12 A random variable, X, has the probability distribution defined by:
$P(X = n) = p^n \qquad \text{for } n = 1, 2, 3, \ldots$

a) Calculate p.

b) Find $P(X > 3)$.

13 The Beavers run a stall to raise funds. For a stake (non-returnable) of 25p, contestants are invited to shuffle a pack of cards and cut the pack. If a picture (king, queen or jack) is cut, the contestant is given £5. If an ace is cut, the contestant is given a free go. Otherwise the contestant loses.

a) Calculate the probability that a contestant loses.

b) Calculate the probability that a contestant wins the £5.

c) What is the expected win/loss to the Beavers per go?

14 In the game of *Monopoly*, players may find themselves in jail. There are a number of ways of getting out of jail: rolling a double with two ordinary dice; paying £50; or using a special 'Get out of jail' card. The rules of the game dictate that if a player elects to roll the dice and fails to get a double in the first three attempts then the player must use one of the other options.

a) Assume that, when I play *Monopoly*, I always elect to roll the dice if I find myself in jail. What is the probability:
   i) I get out of jail on my first attempt,
   ii) I need two attempts to roll a double
   iii) I fail to get out of jail using dice alone?

b) Suppose, when I find myself in situation **iii)**, I always pay £50. Calculate the expected cost of my stays in jail.

15 Two players, X and Y, play a game. The probability that X wins the first game is 0.5. The probability for Y is 0.3. The probability that the first game is drawn is 0.2. If it is drawn another game is played, with the same probabilities. Calculate:

a) the probability that X is the first to win,

b) the probability that Y is the first to win,

c) the expected number of games played.

16 The probability that my brother is at home on any evening at 7.00 pm is 0.7. On those evenings when he is at home, the probability that he is on the telephone at 7.00 pm is 0.2. I try to telephone him on successive evenings at this time. Find the probability that I am first able to speak to him on the fourth evening I try.

(ULEAC Question 4, Specimen paper T4, 1994)

17 Three players, Ann, Brian and Claire, play on a fruit machine in turn, independently in the order Ann, Brian, Claire, Ann, ..., etc. until one of them wins. The probability of winning at each turn is $\frac{1}{3}$.

a) Find the probability that Claire wins on her first turn.
The random variable X denotes the number of turns at the machine until one of them wins.

b) Find the probability that X equals n where $n = 1, 2, 3, \ldots$

c) Calculate the mean number of turns needed until one of the players wins.

(ULEAC Question 6, Specimen T4, 1994 (modified))

18 My computer has 16 memory chips in it and there is a problem which indicates that exactly two of these chips are faulty. To locate the faulty chips I have to remove them one by one for testing.

a) How many different ways are there of choosing two chips from 16, irrespective of order?

b) Write down the probability that:

- **i)** the first chip I test is faulty,
- **ii)** the first and second chips I test are both faulty.

c) Find the probability that the faulty chips are the first and eighth that I test.

d) Write down the probability that the faulty chips are the second and eighth that I test.

e) Obtain the probability that I need exactly eight tests to find the two faulty chips.

f) Find the probability that I need at most eight tests to find the two faulty chips.

(MEI Question 3, Paper 1, June 1993)

19 One plastic toy aeroplane is given away free in each packet of cornflakes. Equal numbers of red, yellow, green and blue aeroplanes are put into the packets.
Faye, a customer, has collected three colours of aeroplanes but still wants a yellow one. Find the probability that:

a) she gets a yellow aeroplane by opening just one packet,

b) she fails to get a yellow aeroplane in opening four packets,

c) she needs to open exactly five packets to get the yellow aeroplane she wants.

(MEI Question 3, Paper 1, January 1994)

20 What is the necessary condition for two events A and B to be described as statistically independent?
A group of students conducts an experiment in which each shuffles an ordinary pack of 52 playing cards and notes the suit of the card on the top of the pack after the shuffle. They stop the process when a heart appears. Then each records the number of shuffles it took to get the heart. The whole process is repeated a number of times.

a) Explain why the students may have been endeavouring to illustrate a geometric distribution.

b) A sixth-former is waiting for a bus to take him to town. He passes the time by counting the number of buses, up to and including the one he wants, that come along his side of the road. If 30 per cent of the buses travelling on that side of the road go to town, what is:

i) the most likely count he makes to the arrival of the one that will take him to town,

ii) the probability that he will count at most four buses?

(Oxford Question 3, Module S5, Specimen paper, 1994)

21 A darts player practises throwing a dart at a bulls-eye on a dart board. Independently for each throw, her probability of hitting the bulls-eye is 0.2.

a) Find the probability that she is successful for the first time on her third throw.

Let X be the number of throws she makes, up to and including the first success.

b) Write down the probability function of X, and give the name of its distribution.

(ULEAC Question 2, Paper S1, 1993)

22 The probability that a door-to-door salesman convinces a customer to buy is 0.7. Assuming sales are independent find the probability that the salesman makes a sale before reaching the fourth home.

(ULEAC Question 1, Paper 1, June 1992)

Summary

- A probability distribution models the occurrence of a particular random variable.
- For a geometric probability distribution, Geo(p), where p is the probability of succeeding at the first event, $P(R = n) = p(1 - p)^{n-1}$.
- The mean of Geo(p) = $\frac{1}{p}$
- The variance of Geo(p) = mean (mean – 1) = $\frac{1}{p}(\frac{1}{p} - 1)$

STATISTICS

7

The binomial probability model

In this chapter we shall:

- *discover how to conduct statistical tests (one-tailed and two-tailed tests) and interpret their results,*
- *build a versatile model for a commonly occurring random variable for independent events,*
- *develop modelling skills in a new context,*
- *use the null hypothesis and the alternative hypothesis,*
- *consolidate results discussed in earlier chapters.*

MOVING TOWARDS THE BINOMIAL MODEL

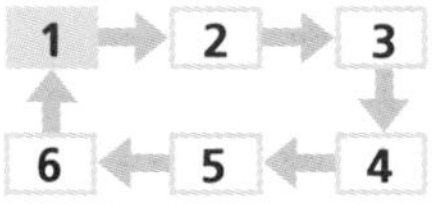

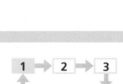

Specify the real problem

It has been estimated that 40 per cent of the world's population are of blood type A. What is the probability that, in a group of five people, exactly three will have type-A blood?

One approach to this problem is to work within a modelling framework. The real-world problem has been specified. A possible way of setting up the model is to let A represent a person with blood of type A and X represent a person who does not have a blood of type A.

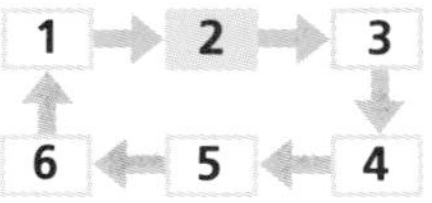

Set up a model

Then the sort of situation where exactly three people in a group of five have blood of type A can be represented by:

AAAXX or AXAXA or XAAXA or ...

There are ${}_5C_3 = \dfrac{5!}{2!3!}$ arrangements of five people, three of whom have blood type A and two of whom do not.

In formulating this as a mathematical problem we have had to make certain assumptions about the people in the group. It might be reasonable to assume that:

- the blood type of one person is independent of that of any other person,
- the probability that a person's blood group is type A is constant – the same as the estimate given, i.e. 40 per cent or 0.4.

These assumptions are equivalent to saying that:

$$P(A) = 0.4 \qquad P(X) = 0.6 \qquad P(AX) = P(A) \times P(X)$$

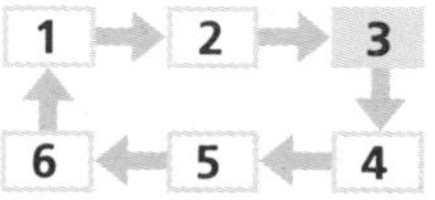

Formulate the mathematical problem

So, the mathematical problem could be presented as:

If five letters are randomly chosen from a very large collection of As and Xs, what is the probability that the selection contains exactly three As when 40 per cent of the collection are As?

We are in a position to solve the mathematical problem. Consider one selection where the letters chosen are, in order:

AAAXX

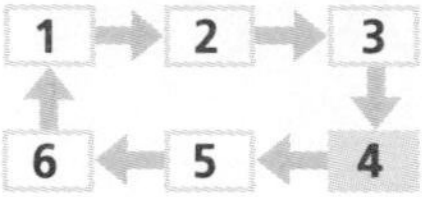

Solve the mathematical problem

The probability that this selection occurs is:

$$\begin{aligned}P(AAAXX) &= P(A) \times P(A) \times P(A) \times P(X) \times P(X)\\ &= 0.4 \times 0.4 \times 0.4 \times 0.6 \times 0.6 = 0.4^3 \times 0.6^2\end{aligned}$$

However, this is just one of $\frac{5!}{2!3!}$ possible selections which result in exactly three As. Hence:

$$P(\text{exactly three As}) = \frac{5!}{2!3!} \times 0.4^3 \times 0.6^2 = 0.2304$$

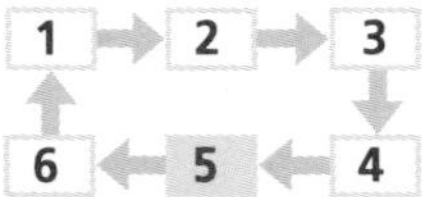

Interpret the solution

An interpretation of this is that there is a probability of about 0.23 (23 per cent) that in a randomly selected group of five people there will be exactly three whose blood type is A.

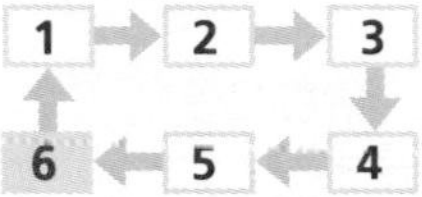

Compare with reality

Try comparing this result with reality. Your report should focus on how the proportion of groups of five people that include three whose blood type is A compares with the model's prediction of 23 per cent.

It is worth considering why there might be discrepancies. Possible reasons are:

- random variation,
- appropriateness of the assumption of independence,
- 40 per cent as the estimate of people with blood type A.

Example 7.1

Use the modelling framework to calculate the probability that, in a group of seven people, exactly two will have blood of type A.

Solution

The specification of the real problem is given above. The model can be set up as before, using A and X. Then the sort of situation sought with exactly two people having blood of type A in a group of seven can be represented by:

AAXXXXX or AXAXXXX or ...

There are $_7C_2 = \frac{7!}{(5!2!)} = 21$ *arrangements with two As and five Xs.*

The formulation assumes that the probability of A is constant and is 0.4. Assuming that one person's blood type is independent of the blood type of any other person allows us to say that:

$P(AAXXXXX) = (0.4)^2 \times (0.6)^5$

It is only one step to the solution which says that there are $_7C_2 = 21$ *possible outcomes where there are exactly two people of blood type A.*

$P(\text{exactly two As}) = 21 \times 0.4^2 \times 0.6^5 \approx 0.2613$

The answer is given to four decimal places for convenience.

This can be interpreted to indicate that in just over a quarter (26 per cent) of all randomly-selected groups of seven people there will be exactly two who have type-A blood.

Exploration 7.1 *Examining blood groups*

Use the modelling framework to answer the following questions.

- What is the probability that, in a group of five people, exactly four will be of blood type A?
- What is the probability that in a group of seven people:
  a) exactly three have type-A blood,
  b) exactly four have type-A blood,
  c) exactly five have type-A blood?

The binomial distribution

In a group of five people, all five might be of blood type A; or there could be exactly four, or exactly three, or exactly two, or exactly one, or none of blood type A. Following the same modelling process as in the previous section leads to the following distribution of probabilities.

For a group of five people No. people with blood type A	Probability
0	$(0.6)^5 = 0.077\,76$
1	$5(0.4)(0.6)^4 = 0.2592$
2	$10(0.4)^2(0.6)^3 = 0.3456$
3	$10(0.4)^3(0.6)^2 = 0.2304$
4	$5(0.4)^4(0.6) = 0.0768$
5	$(0.4)^5 = 0.010\,24$

Note that the sum of the probabilities is 1. Why should this be so?

What we have done here is to list all possibilities when there are five people and so the total of the associated probabilities must be one. The resulting probability distribution is an example of a **binomial probability distribution**.

'Binomial' refers to there being only two possible blood types for each person i.e. type A or not type A. In reality, there are many more blood types but it is still valid to regard blood as being either of one type or not of that type.

Another binomial probability distribution arises if we consider all the possibilities when there are seven people.

For a group of seven people No. people with blood type A	Probability
0	$(0.6)^7 = 0.027\,993\,6$
1	$7(0.4)(0.6)^6 = 0.130\,636\,8$
2	$21(0.4)^2(0.6)^5 = 0.261\,273\,6$
3	$35(0.4)^3(0.6)^4 = 0.290\,304$
4	$35(0.4)^4(0.6)^3 = 0.193\,536$
5	$21(0.4)^5(0.6)^2 = 0.077\,414\,4$
6	$7(0.4)^6(0.6) = 0.017\,203\,2$
7	$(0.4)^7 = 0.001\,638\,4$

The difference between the two distributions arises from the different numbers of people involved. In this case there were seven people, in the first there were five. The probability of an individual having blood type A is assumed to be constant.

Exploration 7.2

People with blood type O

It has been estimated that 45 per cent of the world's population have blood type O. Use an appropriate model to obtain the probability distribution for the number of people who have blood type O in a group of:

a) five people,

b) seven people,

c) four people.

Notation

The binomial distribution is usually written as $B(n, p)$, where n is the number of people in the group, and p is the probability.

The number of people in a group of five who have blood of type O may be modelled by a binomial distribution similar to the blood type A example already considered. The difference is in the blood type. For 'type O' the probability is 0.45. The appropriate binomial distribution may be represented by B(5, 0.45).

This indicates that if a random variable, R, follows a binomial distribution, B(5, 0.45) then the following are the probabilities:

$P(R = 0) = {}_5C_0(0.45)^0(0.55)^5 \approx 0.050\,33$

$P(R = 1) = {}_5C_1(0.45)^1(0.55)^4 \approx 0.205\,89$

$P(R = 2) = {}_5C_2(0.45)^2(0.55)^3 \approx 0.336\,91$

$P(R = 3) = {}_5C_3(0.45)^3(0.55)^2 \approx 0.275\,65$

$P(R = 4) = {}_5C_4(0.45)^4(0.55)^1 \approx 0.112\,77$

$P(R = 5) = {}_5C_5(0.45)^5(0.55)^0 \approx 0.018\,45$

***Note*:** The final probabilities have been given to five places of decimals.

Example 7.2

Identify the binomial distribution which is appropriate to model the following random variables:

a) *the number of heads which result when six coins are flipped,*

b) *the number of 3s which result when five tetrahedral dice are rolled,*

c) *the number of left-handed children in a class of 30 children if the proportion of left-handed people in the area is 15 per cent.*

Solution

a) *Here, the probability that a coin lands head is* $\frac{1}{2}$, *and there are six coins, hence the binomial distribution is* $B(6, \frac{1}{2})$.

b) *The probability that a tetrahedral die lands as a 3 is* $\frac{1}{4}$ *and there are five dice, hence,* $B(5, \frac{1}{4})$.

c) *The appropriate distribution is B(30, 0.15) given there are 30 children, and the probability that any individual child is left-handed is assumed to be 0.15.*

Probability function for the binomial distribution

Looking back at the probability distribution B(5, 0.45) we can see that the probability for each possibility can be written in a consistent way:

$$P(R = x) = {}_5C_x(0.45)^x(0.55)^{5-x}$$

This is the probability function for the random variable which is modelled by this binomial distribution. Previously the binomial distributions have had the following probability functions associated with them:

B(5, 0.4): $P(R = x) = {}_5C_x(0.4)^x(0.6)^{5-x}$, and

B(7, 0.4): $P(R = x) = {}_7C_x(0.4)^x(0.6)^{7-x}$

In the general case a random variable which has the binomial distribution B(n, p) has probability function:

$$P(R = x) = {}_nC_x\, p^x(1 - p)^{n-x}$$

EXERCISES

7.1A

1 If 20 per cent of the population is left-handed, find the probability that a group of eight people contains:

a) exactly two left-handers,

b) at least two left-handers,

c) not more than three left-handers.

2 Jane wins, on average, seven games of *Backgammon* in every ten games she plays.

a) Find the probability that she wins exactly 14 games out of 20.

b) Find the probability that she loses exactly two games out of ten.

c) Find the probability that, in a match of six games, she wins more than she loses.

d) What assumptions have been made in the solution of these problems?

3 In a certain school, 60 per cent of the pupils are brought to school by car, 30 per cent travel by public transport and the rest walk or cycle. Find the probability that a Statistics class of ten contains:

a) at least one person who walks or cycles,

b) at least two who travel by public transport,

c) not more than eight who are given a lift.

What assumptions have been made in the consideration of the class?

4 Simon wins, on average, three-quarters of his tennis games. He loses one in ten and draws the rest. If he plays twelve games, find the probability that:

a) he wins at least ten games,

b) he loses not more than one game,

c) he draws exactly two games.

5 Describe the conditions under which a binomial distribution may be used as a model for a random variable. State, with reasons, whether or not a binomial distribution would be an appropriate model for:
   a) the number of days on which rainfall is recorded in Paris during May,
   b) the number of rolls of a cubical die needed to get two successive 6s,
   c) the number of mathematics students in a group of 15 students chosen randomly from a university.

EXERCISES

7.1B

1 In the winter term, 25 per cent of students have a cold at any one time. If nine students are selected at random, find the probability that:
   a) fewer than four have a cold,
   b) no one has a cold.
   Find the most likely number with a cold.

2 State the conditions under which a binomial distribution may be used as a model for a random variable. For each of the following, state if a binomial distribution is an appropriate model. Where you feel it is, state the value of n and p.
   a) In a given constituency, where 30 per cent of the electorate support the Liberals, the number of Liberal supporters in a household with four people on the electoral register.
   b) The number of doubles obtained when a pair of tetrahedral dice are rolled ten times.
   c) The number of 6s obtained when a cubical dice is rolled for five minutes.

3 A cinema manager records the occupancy figures for the last showing of the main film on Saturday evenings over a period of time. From the results he calculates the occupancy rate to be 80 per cent so inferring that the probability that any seat is occupied is 0.80. Use this to calculate the probability that in a row of eight seats:
   a) seven are occupied,
   b) six are occupied,
   c) at least six are occupied,
   d) at most five are occupied.
   Detail the modelling assumptions you have made in calculating these probabilities. Do you consider then to be valid? Explain.

4 State the conditions under which a binomial distribution may be used as a model for a random variable.
   For each of the following situations state whether a binomial distribution is a suitable model. If you think it is appropriate, give values for the parameters, n and p. If you think it is not appropriate, give your reasons.
   a) In a large college where 25 per cent of all students study Statistics, a count is made of the number who study statistics in a seminar group of 30 students.

b) A count is made of the number of green wine gums in a random sample of 40 gums, taken from a batch of 100 000 of which 8000 are green.

c) The output from a machine producing biro refills is known to contain five per cent defective refills, is recorded.

d) From a group of 25 workers of whom ten are in favour of a strike, a random sample of eight is taken and a count made of the number who are in favour of a strike.

5 Chocolate bars disgorged from a vending machine are either plain chocolate or milk chocolate, in the ratio 3 : 2.

a) If three bars are bought, find the probability that at least one is milk chocolate.

b) If four bars are bought, find the probability that at least one is milk chocolate.

c) If six bars are bought, find the probability that at least one is milk chocolate.

d) How many bars need to be bought in order for the chance of obtaining at least one milk chocolate bar to be over 95 per cent?

e) How many bars need to be bought in order for the chance of obtaining at least one milk chocolate bar to be over 99 per cent?

THE RECURRENCE FORMULA FOR THE BINOMIAL DISTRIBUTION

It can be quite tedious producing the probabilities for a binomial distribution using the probability function. For a sequence of probabilities such as:

$$P(R=0) \quad P(R=1) \quad P(R=2) \quad P(R=3) \quad \ldots$$

there is a convenient method which relates the probability of one event to the previous one. Consider B(7, 0.4), where:

$$P(R=0) = {}_7C_0(0.4)^0(0.6)^7 = (0.6)^7$$

and

$$P(R=1) = {}_7C_1(0.4)^1(0.6)^6 = \frac{7!}{1!6!}(0.4)^1(0.6)^6 = 7(0.4)^1(0.6)^6$$

Now, compare P(1) with P(0). What are the differences?

To get P(1) from P(0), we: multiply by 7, multiply by 0.4, divide by 0.6.

$$P(1) = 7 \times \frac{0.4}{0.6} \times P(0)$$

Now, consider P(R = 2):

$$P(R=2) = {}_7C_2(0.4)^2(0.6)^5 = \frac{7!}{2!5!}(0.4)^2(0.6)^5 = \frac{7\times 6}{1\times 2}(0.4)^2(0.6)^5$$

(cancelling 5! in top and bottom, then expanding the 2!).

How can we get P(2) from P(1)? We need to multiply by 6, divide by 2, multiply by 0.4, divide by 0.6. This gives:

$$P(2) = \frac{6}{2}\times\frac{0.4}{0.6}\times P(1)$$

This is the same pattern as for getting P(1) from P(0), since that can be written as:

$$P(1) = \frac{7}{1} \times \frac{0.4}{0.6} \times P(0)$$

The $\frac{0.4}{0.6}$ is common, and the other factor, although it changes, has a common feature. Can you see what this feature is? Try writing P(3) in terms of P(2).

Consider:

$$P(R = 3) = {}_7C_3(0.4)^3(0.6)^4 = \frac{7!}{3!4!}(0.4)^3(0.6)^4 = \frac{7 \times 6 \times 5}{1 \times 2 \times 3}(0.4)^3(0.6)^4$$

$$= \frac{5}{3} \times \frac{0.4}{0.6} \times P(2)$$

Hence, the pattern continues. The factor $\frac{0.4}{0.6}$ is present and the changing factor has the same pattern.

One way of expressing this pattern is to say that the top and bottom add up to eight; the bottom is the value of the new event.

What factors are needed to produce P(4) from P(3), P(5) from P(4), P(6) from P(5), P(7) from P(6)?

In each case, the $\frac{0.4}{0.6}$ will appear, then there will be a factor $\frac{8-x}{x}$, where x is the new value of the random variable. Hence:

$$P(4) = \frac{4}{4} \times \frac{0.4}{0.6} \times P(3)$$

$$P(5) = \frac{3}{5} \times \frac{0.4}{0.6} \times P(4)$$

$$P(6) = \frac{2}{6} \times \frac{0.4}{0.6} \times P(5)$$

$$P(7) = \frac{1}{7} \times \frac{0.4}{0.6} \times P(6)$$

The pattern developed here is known as a recurrence formula for binomial probabilities. In general terms, for the distribution B(7, 0.4):

$$P(x) = \frac{8-x}{x} \times \frac{0.4}{0.6} \times P(x-1)$$

CALCULATOR ACTIVITY

Using the recurrence formula

Try using your calculator to produce the probabilities for B(7, 0.4) using the recurrence formula. Start by producing $(0.6)^7$, then multiply this display by $\frac{0.4}{0.6}$ (you might want to simplify this constant factor) and by $\frac{7}{1}$. Then multiply the new display by $\frac{0.4}{0.6}$ and by $\frac{6}{2}$. Then ...

Continue the process. Check the successive probabilities with those on page 189.

CALCULATOR ACTIVITY

Programming probabilities

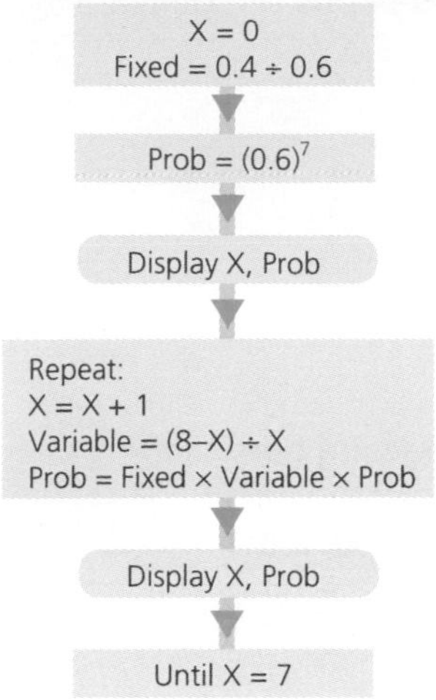
X = 0
Fixed = 0.4 ÷ 0.6
Prob = (0.6)^7
Display X, Prob
Repeat:
X = X + 1
Variable = (8–X) ÷ X
Prob = Fixed × Variable × Prob
Display X, Prob
Until X = 7

If you have a programmable calculator you might like to write a program to display the successive probabilities. This flowchart might be used as a basis for the program.

Is it possible to develop a recurrence formula for any binomial distribution? It ought to be. Consider the distribution B(30, 0.15), where:
$P(0) = (0.85)^{30}$
and
$P(1) = {}_{30}C_1\,(0.15)^1\,(0.85)^{29} = \frac{30!}{1!29!} \times (0.15)^1 \times (0.85)^{29} = \frac{30}{1} \times \frac{(0.15)}{(0.85)} \times P(0)$

This is a promising start. The fixed factor appears to be $\frac{0.15}{0.85}$, the variable factor looks promising with top and bottom adding up to 31 and the bottom being the new value of the random variable. How are these the features seen in the B(7, 0.4) case?

Can you predict the factors required to produce P(2) from P(1), P(3) from P(2), $P(x)$ from $P(x - 1)$?

Using the probability function initially:
$P(2) = {}_{30}C_2\,(0.15)^2 \times (0.85)^{28} = \frac{30!}{2!28!} \times (0.15)^2 \times (0.85)^{28}$
$= \frac{30 \times 29}{1 \times 2} \times (0.15)^2 \times (0.85)^{28}$
hence: $P(2) = \frac{29}{2} \times \frac{0.15}{0.85} \times P(1)$
similarly: $P(3) = \frac{28}{3} \times \frac{0.15}{0.85} \times P(2)$
These follow the pattern seen earlier and $P(x) = \frac{31-x}{x} \times \frac{0.15}{0.85} \times P(x - 1)$

CALCULATOR ACTIVITY

Generating the probability distribution

Generate the probability distribution B(30, 0.15) using the recurrence relation.

You might like to check your probabilities with those given below.

x	$P(R = x)$	x	$P(R = x)$
0	0.0076	9	0.0181
1	0.0404	10	0.0067
2	0.1034	11	0.0022
3	0.1703	12	0.0006
4	0.2028	13	0.0001
5	0.1861	14	0.0000
6	0.1368	15	0.0000
7	0.0828	⋮	
8	0.0420		

***Note*:** All probabilities are given accurate to four decimal places.

Notice that the fixed factor is always related to the constant probability, p. How?

Fixed factor $= \frac{p}{1-p}$

The variable factor is related to the number of possible events, n, and to the new value of the random variable. What is this relationship?

Variable factor $= \frac{n+1-x}{x}$

The recurrence relation for $B(n, p)$ is:

$$P(x) = \frac{n+1-x}{x} \times \frac{p}{1-p} \times P(x-1)$$

Exploration 7.3

Finding recurrence relationships

What are the recurrence relations for the following binomial distributions?

a) $B(5, 0.45)$ **b)** $B(10, \frac{1}{2})$ **c)** $B(15, \frac{1}{4})$ **d)** $B(150, 0.01)$

Using distribution tables

At the end of the book, there are binomial distribution tables. Look at these tables, and identify the section which refers to $B(7, 0.40)$. Locate the table for the specific value of n, then the column corresponding to $p = 0.4$. Here is part of the appropriate extract from the table.

$n = 7$

p	0.4
r	
0	0.0280
1	0.1586
2	0.4199
3	0.7102
⋮	⋮
7	1.0000

Compare the values of the probabilities shown in the table and those given on page 189 for the number of people with blood type A in a group of seven people. Are the probabilities the same?

Perhaps the first thing to note is that the probabilities only coincide when referring to no one in the group of seven having blood of type A. Both values are 0.0280 (to four decimal places). Thereafter the tables are quite distinctly different. Why do you think this is?

Notice that for $r = 1$, the table has the probability 0.1586. However, the probability $P(R = 1)$ is 0.1306. The entry in the table against $r = 1$ represents $P(R = 0) + P(R = 1)$. So the entries in the table represent the sum of the probabilities from $R = 0$ up to $R = r$.

Checking that this is the case for $R = 2$ and $R = 3$ reveals:

$P(R = 0) + P(R = 1) + P(R = 2) = 0.4199$

$P(R = 0) + P(R = 1) + P(R = 2) + P(R = 3) = 0.7102$

which are the entries in the table.

The tables show the values of the **cumulative binomial distribution function**. The words 'cumulative' and 'function' are often dropped when referring to such tables. Using the binomial distribution tables yields the following:

$P(R \le 4) = 0.9037$
$P(R \le 5) = 0.9812$
$P(R \le 6) = 0.9984$
$P(R \le 7) = 1.0000$

when $R \sim B(7, 0.4)$.

The notation: $R \sim B(7, 0.4)$ means that R is a random variable which is binomially distributed with $n = 7$ and $p = 0.4$.

Clearly, using the tables saves time and effort when we are calculating cumulative probabilities. They can also be used to find approximations to individual probabilities, but, both uses are limited in accuracy and, of course, to the (n, p) combinations tabulated.

Example 7.3

a) *Calculate the probabilities of:*

- **i)** *three or fewer people in a group of ten having blood of type O,*
- **ii)** *exactly six people in a group of ten having blood of type O,*
- **iii)** *more than five people in a group of ten having blood of type O,*
- **iv)** *at least seven people in a group of ten having blood of type O.*

b) *It can be assumed that 20 per cent of people in the southern hemisphere have blood of type B. Calculate the probabilities of:*

- **i)** *in a group of 15 such people, five or fewer having blood of type B,*
- **ii)** *in a group of twelve such people, at least four having blood of type B,*
- **iii)** *in a group of eight such people, between three and five (inclusive) having type B blood.*

Solution

a) *The questions in this part refer to the* $B(10, 0.45)$ *distribution (if Exploration 7.2 is correct).*

i) *The event 'three or fewer' is required, hence:*

$P(R \le 3) = 0.2660$

can be read directly from the table for $n = 10$*, the column* $p = 0.45$*, and the row corresponding to 3.*

ii) *From the cumulative probabilities* $P(R \le 6)$ *and* $P(R \le 5)$ *the probability* $P(R = 6)$ *can be found.*

$P(R = 6) = P(R \le 6) - P(R \le 5) = 0.8980 - 0.7384 = 0.1596$

iii) *The event 'more than five' is the complement of the event 'five or fewer'. The tables yield* $P(R \le 5) = 0.7384$.

The probability of the complement is:

$1 - P(R \le 5) = 1 - 0.7384 = 0.2616$

Hence $P(R > 5) = 0.2616$

iv) *The event 'at least seven' corresponds to 'seven, eight, nine or ten'. This event is the complement of 'six or fewer', hence:*

$P(R \ge 7) = 1 - P(R \le 6) = 1 - 0.8980 = 0.1020$

b) ***i)*** *The appropriate probability distribution is* B(15, 0.2)*, and the event is* $R \le 5$*. Hence, from the tables* $P(R \le 5) = 0.9389$.

ii) *This question requires the use of* B(12, 0.2)*. The event 'at least four' corresponds to 'four, five, six, ... or twelve' which is the complement of 'zero, one, two or three'.*

$P(R \ge 4) = 1 - P(R \le 3) = 1 - 0.7946 = 0.2054$

iii) *Here, the binomial distribution required is* B(8, 0.2)*. The probability of the event 'three, four or five' can be found from the probabilities:*

$P(R \le 5)$, $P(R \le 2)$.

$$\begin{aligned} P(R = 3 \text{ or } R = 4 \text{ or } R = 5) &= P(R \le 5) - P(R \le 2) \\ &= 0.9988 - 0.7969 \\ &= 0.2019 \end{aligned}$$

EXERCISES

7.2A

1 A coin is biased and is twice as likely to come down heads as tails. If the coin is tossed eight times, use the recurrence formula to calculate the probability of obtaining fewer than four tails.

2 Use the appropriate recurrence formula to help you draw probability line diagrams for each of the following distributions.
a) B(5, 0.3) **b)** B(8, 0.6) **c)** B(8, 0.4) **d)** B(10, 0.2)
e) B(10, 0.5) **f)** B(12, 0.4)

3 Last year, 55 per cent of all school meals served were meat dishes, 20 per cent were fish and the rest were vegetarian. A group of 18 teachers takes lunch together. Find the probability, using tables, that:
a) at least six vegetarian meals are served,
b) no fish is served,
c) fewer than half the meals are meat.
What assumptions have you made?

4 On average two-fifths of films shown in a cinema are thrillers, and three-quarters of the rest are comedies; the remainder are science fiction. If eight films are selected from the programme at random, find the probability that:
a) at least four are comedies,
b) no more than two are science fiction,
c) the number of thrillers exceeds the number of other films.

5 On average, John correctly solves the newspaper brain teaser seven times out of ten. He attempts each puzzle over a two-week period (not including Sundays). Find the probability that:
a) he solves eight or fewer brain teasers,
b) he solves at least ten brain teasers,
c) he solves exactly seven brain teasers,
d) he solves fewer than six brain teasers.

EXERCISES

7.2B

1 Suppose 30 per cent of school-age children wear spectacles. In a class of 15, what is the most likely number to wear spectacles?

2 Illustrate as probability line graphs the following distributions.
a) B (6, 0.3) **b)** B (7, 0.7) **c)** B (10, 0.45) **d)** B (18, 0.15)
e) B (20, 0.1) **f)** B (8, 0.3)

3 Answer Question 5 of Exercises Set 7.2A for Jane, who is less proficient with brain teasers, and on average, correctly solves only four out of ten.

4 An examination consists of 20 multiple-choice questions each with four alternatives. Only one of the alternatives is correct.
X denotes the number of correct responses obtained by candidates who guess the answers to all 20 questions. Explain why X may be considered to have a binomial distribution. Hence determine:
a) $P(X > 10)$ **b)** $P(X \leq 5)$ **c)** $P(X = 5)$.

5 Each of a group of 20 people flips a coin. What is the probability that:
a) fewer than half of them get a head,
b) at least a quarter of them get a tail,
c) fewer than three-quarters of them get a head?

STATISTICAL TESTS

Suppose that a group of 16 teachers had their blood tested. How many should we expect to have blood of type O? Recall that 45 per cent of the world's population have blood of that type, so it would seem reasonable to expect that about half of the group, say seven or eight, would have type-O blood.

What if none of the group had type-O blood? We might be tempted to conclude that there was a lower likelihood that teachers had blood of that type than in the population at large. But why might we come to this conclusion? It is, of course, quite possible for a group to have no one with that blood type even when the chance for any individual in the group is 0.45. What is the probability? Making the modelling assumptions of the earlier part of the chapter leads to the number of people being a binomial variable, B(16, 0.45).
Hence $P(O) = (0.55)^{16} \approx 0.000\,07$

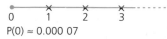

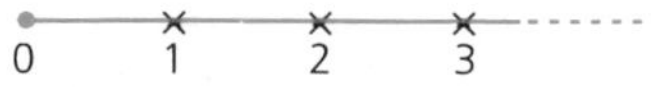

$P(0) \approx 0.000\,07$

An interpretation of this suggests that fewer than one occasion in ten thousand would result in there being no one with blood of type O when there is actually a 45 per cent chance of finding a person with such blood. So, this result is highly unlikely.

It would seem to suggest that type-O blood is less commonly found in teachers than in the general population. Would we draw the same conclusion if there had been one teacher with type-O blood in this group of 16? We have decided already that if none of the 16 had type-O blood, then we would be suggesting that the blood type occurred less frequently. The chance that either one teacher or no teacher is found, in the 16, with type-O blood is:

$$P(0) + P(1) = P(R \leq 1) \approx 0.0010$$

according to the binomial distribution tables for B(16, 0.45).

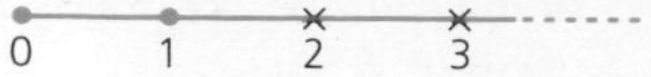

P(≤ 1) ≈ 0.001

This indicates that such an event would occur about once in a thousand times of testing groups of 16 people. Again, it is a very rare occurrence and it would be reasonable to draw the same conclusion.

What conclusion would you draw if you found two type-O teachers in the group? Using, the same model B(16, 0.45), leads to $P(R \leq 2) \approx 0.0066$.

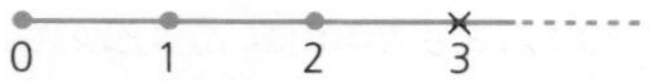

P(≤ 2) ≈ 0.0066

This indicates that on fewer than seven occasions in a thousand would we find two or fewer people with type-O blood. Again, it is something which is rarely likely to occur, and we could reasonably justify concluding that there is a lower incidence of blood type-O found in teachers than in the general population.

Exploration 7.4

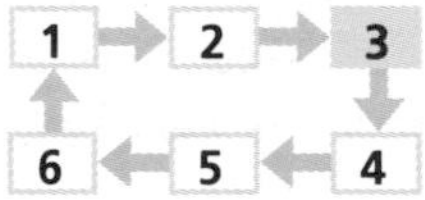

Formulate the mathematical problem

Blood groups among teachers

What conclusions would you reach about the occurrence of type-O blood when:

a) three people are found,
b) four people are found,
c) five people are found,

with type-O blood in a group of 16 teachers?

Interpreting the results

The modelling activity in Exploration 7.4 leads to B(16, 0.45) in each case. The mathematical problem formulated varies.

In part **a)** it was necessary to evaluate the likelihood of observing three or fewer people, since we have already decided that observing zero, one or two people would lead us to draw the same conclusion.

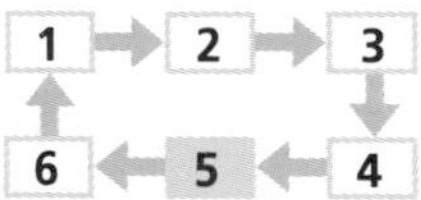

Interpret the solution

$P(R \leq 3) = 0.0281$

This indicates that we observe such a result on fewer than three occasions out of 100. Again, this is an unlikely event, and it seems reasonable to conclude that teachers are less likely to have type-O blood than the general population.

In parts **b)** and **c)**, $P(R \leq 4) \approx 0.0853$;

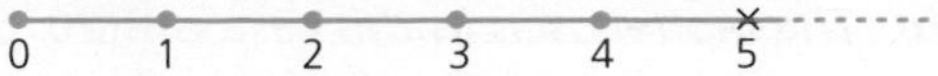

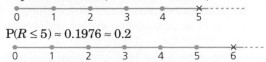

$P(R \leq 5) \approx 0.1976 \approx 0.2$

A reasonable interpretation of the last result is that on about 20 per cent of the occasions we should find five or fewer people with type-O blood. This does not seem to be terribly rare. It might be difficult to justify the conclusion that the teachers are any different from the general population, so which of these results do we take to be significant?

Significance level

The case of four or fewer people is less clear cut and depends on what probability is considered to be unlikely or to be rare. The probability is referred to as the **alpha** or **significance level** of the test. A common significance level is five per cent.

If the significance level is chosen to be five per cent, since $P(R \leq 4) \approx 8\frac{1}{2}\%$ is greater than five per cent, then the conclusion we should reach is that the incidence of four people is not one of the rare events.

Critical region

The group of observations which we consider to be rare events is called the **critical region**. Clearly, the critical region depends on the significance level chosen. For the model being explored here B(16, 0.45) and a significance level of five per cent we have discovered that an observation lying in the critical region $R \leq 3$ leads to the conclusion that the proportion of teachers with type-O blood is lower than in the population at large.

Example 7.4

a) Sixteen teachers are asked about their blood type. What critical region is associated with the conclusion that teachers are less likely to have type-O blood than the population at large when the significance level is:

i) 10% ii) 2% iii) 1%?

b) A researcher wants to address the question, 'Are scientists less likely to have type-A blood than the general population?' The researcher performs blood tests on 20 scientists. What critical region should the researcher use if the test is conducted at:

i) the 10% level, ii) the 2% level?

Solution

a) The model used here is B(16, 0.45).

i) Tables indicate that $P(R \leq 5) \approx 0.2 \approx 20\%$

$P(R \leq 4) \approx 0.085 \approx 8\frac{1}{2}\%$

Hence the critical region at 10% level is $R \leq 4$.

ii) The tables indicate that $P(R \leq 3) \approx 0.028 \approx 3\%$

$P(R \leq 2) \approx 0.007 \approx 1\%$

Hence the 2% critical region is $R \leq 2$.

iii) *The 1% critical region is also* $R \leq 2$.

b) *The model used in this case is* B(20, 0.4).

i) *The distribution tables indicate that* $P(R \leq 5) = 0.1256 \approx 12\frac{1}{2}\%$

$P(R \leq 4) = 0.0510 \approx 5\%$

Hence the 10% critical region is $R \leq 4$.

ii) *The tables indicate that* $P(R \leq 4) = 0.0510 \approx 5\%$

$P(R \leq 3) = 0.0160 \approx 1\frac{1}{2}\%$

Hence the 2% critical region is $R \leq 3$.

The alternative and null hypotheses

What do we mean by the term 'significance level'? It is the maximum probability of getting an observation in the critical region when the model used is appropriate. However, if we get an observation in the critical region, we are going to conclude that the model is not appropriate. So using the significance level inevitably offers the chance of wrongly rejecting the model used.

In statistical tests, the two models involved are the underlying set of assumptions from which the probability distribution is formulated, and the alternative set of assumptions which will be accepted if the test results in an observation in the critical region. These models are called **hypotheses**. The set of assumptions which underpin the model used is called the **null hypothesis**. The alternative set is known as the **alternative hypothesis**.

In common with much of mathematics, a short notation for presenting the concepts of these hypotheses is adopted. The null hypothesis, in the case of the 16 teachers and the type-O blood would be presented as:

$$H_0: p = 0.45$$

This represents 'null hypothesis'.

This represents all the assumptions underlying the binomial model.

The alternative hypothesis would be presented as:

$$H_1: p < 0.45$$

This represents 'alternative hypothesis'.

This indicates that the proportion of teachers with type-O blood is less than in the population at large.

Example 7.5

What are the appropriate hypotheses for Example 7.4b?

Solution

The model used to develop the probability distribution and to establish the critical region is B(20, 0.4). *The set of assumptions which give rise to this model constitute the null hypothesis. The alternative hypothesis concerns itself with scientists being less likely than the general population to have type-A blood. Hence, the two hypotheses can be summarised as:* $\mathrm{H}_0: p = 0.4$; $\mathrm{H}_1: p < 0.4$.

One-tailed tests

What would the alternative hypothesis be if a particular blood type occurred with greater frequency in one group of people than in the general population? Suppose we assume that ten per cent of the general population have type-B blood. What alternative hypothesis could we use if we wished to respond to the question: Is type-B blood found more frequently among Australian mathematicians than in the general population?

The null hypothesis would be: $\mathrm{H}_0: p = 0.10$

and the alternative hypothesis would be: $\mathrm{H}_1: p > 0.10$.

With this alternative hypothesis, we need to consider the probability of getting large numbers of people in a group with type-B blood rather than the small numbers we have considered so far. We need to examine the 'upper tail' of the probability distribution involved. This information can be gained from the distribution tables, but not directly.

Suppose the blood types of a group of 17 Australian mathematicians are recorded. What critical region would be appropriate for a hypothesis test with a five per cent significance level against the alternative hypothesis $\mathrm{H}_1: p > 0.10$?

Examining the distribution function for B(17, 0.1) shows:
$\mathrm{P}(R \le 3) = 0.9174$ and $\mathrm{P}(R \le 4) = 0.9779$.

The first of these implies $\mathrm{P}(R \ge 4) = 1 - 0.9174 = 0.0826 \approx 8\frac{1}{4}\%$ which is more than the significance level of five per cent. Hence, an occurrence of four is not in the critical region.

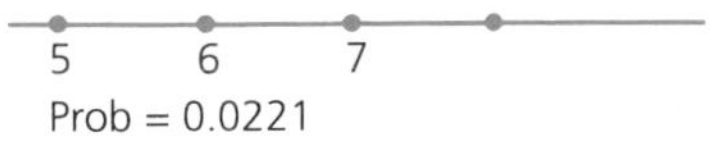

The second of these implies that $\mathrm{P}(R \ge 5) = 1 - 0.9779 = 0.0221 \approx 2\frac{1}{4}\%$ which is less than the significance level. Hence, the critical region is $R \ge 5$.

Example 7.6

Twenty Australian mathematicians had their blood types identified. Eight of them were found to have type-B blood.

a) *Write down appropriate null and alternative hypotheses to respond to the question: Is there sufficient evidence, at the five per cent level of significance, to support the view that Australian mathematicians are more likely to have type-B blood than the general population?*

b) *Calculate the five per cent upper-tail critical region.*
c) *Conduct the test and report your findings.*

Solution

a) *Assuming that in the general population the proportion of people with type B blood is 10% leads to the hypotheses* $H_0: p = 0.1$; $H_1: p > 0.1$.

b) *The appropriate distribution tables are those for the probability model* B*(20, 0.1).*
Examining the table yields:
$P(R \le 3) = 0.8670$
$P(R \le 4) = 0.9568$
$P(R \le 5) = 0.9887$
These cumulative probabilities lead to:
$P(R \ge 4) = 1 - 0.8670 \approx 13.3\%$
$P(R \ge 5) = 1 - 0.9568 \approx 4.3\%$
$P(R \ge 6) = 1 - 0.9887 \approx 1.1\%$
Now 13.3% is larger than the significance level, but 4.3% is smaller than the significance level (and 1.1% is considerably smaller!). Hence, the critical region is $R \ge 5$.

c) *The critical region is* $R \ge 5$*, and in the group there were eight people with type-B blood. This is unlikely to occur under the null hypothesis, hence it is reasonable to conclude that we are more likely to find people with type-B blood among Australian mathematicians than in the general population.*

Two-tailed tests

The two types of alternative hypothesis we have considered so far have been:

- $H_1: p < ...$
- $H_1: p > ...$

Both of these are 'one-tailed', which means the critical region is either:

- a low set of values of the random variable, or
- a high set.

The question that led to the test being done asked:

i) are the occurrences less likely, or
ii) are the occurrences more likely?

A question of the type, 'Is the occurrence different ...?' does not require a specific direction or tendency to be given. This sort of question leads to an alternative hypothesis such as $H_1: p \ne ...$.

The critical region needs to take into account both the upper and lower tails. An hypothesis test with this sort of alternative hypothesis is called a **two-tailed test**. The significance level is generally divided equally between the upper and lower tails.

Suppose that quarter of American scientists have type-O blood. Twenty Mexican scientists are going to have their blood type identified to see if there is any difference between them and the American scientists. A five per cent significance level is chosen for the test. The critical region will consist of a lower tail of up to $2\frac{1}{2}$ per cent and an upper of up to $2\frac{1}{2}$ per cent.

An examination of the tables for the model $B(20, \frac{1}{4})$ yields:
$P(R \leq 1) = 0.0243 \approx 2.4\%$ and $P(R \geq 10) = 1 - 0.9861 = 0.0139 \approx 1.4\%$.

Note: $P(R \leq 2) \approx 9\%$ and $P(R \geq 9) \approx 4\%$ each of which is much larger than the allowed $2\frac{1}{2}$ per cent contribution to the significance level.

Hence, the two parts of the critical region are $R \leq 1$ and $R \geq 10$.

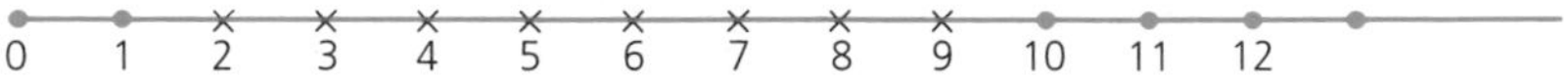

Now suppose that there are eight Mexicans with type-O blood in the group. The result 'eight' does not lie in either part of the critical region, hence the test indicates that there is no difference in the proportion of Mexican and American scientists with type-O blood. If, however, there had been twelve Mexicans with type-O blood, we could justifiably conclude that there was a difference. It is worth noting that we would not be justified in saying that proportionately more Mexicans had that blood type; this is not justified because this was not the question posed and consequently the alternative hypothesis and its critical region did not correspond to such a question.

Example 7.8

Suppose that 30 per cent of Western males have type-A blood. A test is carried out to see if the proportion of Western females with type-A blood is different. A group of 20 females have their blood types identified.

a) *Write down appropriate hypotheses and identify the critical region appropriate to a ten per cent significance level.*

b) *What conclusion would you draw if the test revealed that two females had type-A blood?*

Solution

a) *The hypotheses are* $H_0: p = 0.30$; $H_1: p \neq 0.30$.
The appropriate probability model is B(20, 0.30). *The size of the tails are, at most, 5%. An examination of the distribution table yields:*
$P(R \leq 2) = 0.355 \approx 3.6\%$ *and* $P(R \geq 10) = 1 - 0.9520 \approx 4.8\%$.
($P(R \leq 3) \approx 11\%$ *and* $P(R \geq 9) \approx 11\%$, *both of which are too large.*
Hence, the critical region is $R \leq 2$ *and* $R \geq 10$.

b) *The result 'two' is in the critical region. Hence, it is reasonable to conclude that the proportion of Western females with type-A blood is different from that of Western males.*

EXERCISES

7.3A

1 A coin is tossed three times and shows heads each time.

a) Is there evidence, at the five per cent level, that the coin is biased towards heads?

b) If the coin shows six heads on six tosses, is there evidence, at the five per cent level, that it is biased towards heads?

c) Repeat **b)** at the one per cent level of significance. Comment.

2 An independent weather forecaster predicted that the probability of rain on any day in Wimbledon fortnight (twelve days) was 0.5. In the event it rained on two days out of the twelve.

a) Does this provide evidence, at the five per cent level, that his theory was incorrect?

b) Does this provide evidence, at the one per cent level, that his theory was incorrect?

c) What assumptions were made by the forecaster? Were these realistic assumptions?

3 A driving instructor claims that 80 per cent of his pupils pass the driving test at their first attempt.

a) In a month when ten of his pupils took the test for the first time, only four passed. Does this evidence tend to disprove his claim?

b) Repeat **a)** assuming two passes out of five attempts.

c) Repeat **a)** assuming eight passes out of 20 attempts.

d) Compare **a), b)** and **c)**.

Conduct your tests at the five per cent level.

4 Explain the terms 'critical region', 'significance level' and 'null hypothesis'.

a) A new waterproof material is introduced which is claimed to be better than existing materials. A group of 20 volunteers try out jackets made of the new material. The number, X, of these people who report that the new material is an improvement is recorded. Assume that X is binomially distributed: B (20, p).

A statistical test is proposed in which the hypotheses are:

$$H_0: p = \tfrac{1}{2},\ H_1: p > \tfrac{1}{2},$$

and the critical region is $X \geq 14$. What is the significance level of the test?

b) What is the critical region if the significance level is one per cent?

c) If p is actually 0.7, what is the probability that $X \geq 14$?

d) Why is a one-tailed test preferred to a two-tailed test in this case?

5 A researcher claimed that 30 per cent of chicken sandwiches are contaminated with listeria. A school canteen refuted this claim and sent 20 of its chicken sandwiches to be tested. The canteen manager decided that he would sue the researcher if fewer than three of the 20 sandwiches were contaminated.

a) State suitable hypotheses for the test.

b) What is the critical region?

c) What is the significance level?

d) If in fact 15 per cent of the canteen's sandwiches were contaminated, what is the probability of fewer than three of the 20 being found to have listeria?

EXERCISES

7.3B

1 A national charity organised a raffle and claimed that 20 per cent of the tickets sold would win a small prize. Susan bought twelve tickets but won no prizes.
   a) Does this evidence suggest, at the five per cent significance level, that the number of prizes won was lower than expected?
   b) Repeat **a)** for 30 per cent of tickets sold.
   d) Repeat **b)** for 10 per cent of tickets sold.

2 In previous years, sixth-formers have been equally likely to choose drama or photography as their extra class. This year, 14 out of 20 have chosen drama.
   a) Does this provide evidence of a change in preferences? Use the two per cent significance level.
   b) In previous years, 70 per cent of sixth-formers have chosen drama. This year, only ten out of 20 have done so. Does this provide evidence of a change in preferences? Use the two per cent significance level.

3 A lottery offered a choice of one number from the integers 1 to 25 inclusive. Mary selected the number 7 on 50 consecutive entries but 7 was never the winning number.
   a) Does this evidence suggest, at the five per cent significance level, that the lottery was biased against the number 7?
   b) What if Mary had made 100 consecutive entries?

4 State what you understand by the terms 'critical region' and 'significance level'.
   a) A drug for treating a certain medical condition has side-effects in 40 per cent of patients. A new drug is introduced which it is claimed reduces the risk of side-effects.
   State suitable null and alternative hypotheses to test this claim.
   b) A consultant agrees to use the new drug in treating 20 patients. She decides to accept the claim if fewer than four patients suffer side-effects. State the critical region and find the significance level of the test.
   c) On average the new drug actually causes side-effects in 20 per cent of patients, calculate the probability of fewer than four patients, in a group of 20, suffering from side-effects whilst on the new drug.

5 Under what circumstances would you use a one-tailed test rather than a two-tailed test? An international bank proposes to take over an insurance company. The bank claims that the proposal has the backing of over one half of the company's investors. The directors of the company claim that their investors are evenly split. Independent arbitrators are brought in to test the validity of these claims. They decide to obtain the opinions of 20 investors randomly chosen.
   a) State suitable null and alternative hypotheses.
   b) The number of investors in favour of the proposals is X. The arbitrators decide to accept the bank's opinion of $X \geq 13$. What is the significance level of the test?

c) What is the critical value if the significance level is $2\frac{1}{2}$ per cent?
d) What is the probability of rejecting the company's claim if the actual proportion if its investors in favour of the take-over is 75 per cent?

CONSOLIDATION EXERCISES FOR CHAPTER 7

1 In a large batch of eggs, ten per cent are brown eggs.
a) In a box of six eggs find the probability that at least two are brown.
b) How many eggs must be bought for the probability of getting at least one brown egg to be greater than 90 per cent?

2 Thirty per cent of satsumas have no pips. A bag of satsumas contains six fruits.
a) Find the probability that a bag contains at least three pip-free satsumas.
b) If the first satsuma eaten was pip-free, find the probability that there will be at least two more pip-free satsumas.
c) If the first two satsumas eaten were pip-free, find the probability of at least one more pip-free satsuma.

3 A case of wine contains twelve bottles. A packing carton contains six cases. Seventy per cent of the bottles contain red wine; the rest contain white wine.
a) Find the probability that a case contains at least nine bottles of red wine.
b) Find the probability that all six cases in a carton contain at least nine bottles of red wine.
c) Find the probability that exactly five cases in a carton contain at least nine bottles of red wine.
d) Find the most likely number of cases in a carton which contain at least nine bottles of red wine.

4 A coin is tossed five times and no heads are obtained.
a) Is there evidence, at the one per cent significance level, that the coin is biased against heads?
b) Is there evidence, at the five per cent level, that the coin is biased against heads?
Discuss the usefulness of each of these tests.

5 a) A coin is tossed five times and one head is obtained. Is there evidence at the two per cent level of significance that the coin is biased against heads?
b) Repeat **a)** when the coin is tossed ten times and two heads are obtained.
c) Repeat **a)** when the coin is tossed 15 times and three heads are obtained.
d) Discuss **a), b), c)**.

6 A multiple choice test consists of four sections with three options for each answer. James selects his answers completely at random.

a) Find the probability that he scores no correct answers in section A.
b) Find the probability that he scores at least two correct answers in section A.
c) Find the probability that he scores no correct answers in exactly one section.
d) Find the probability that he scores at least two correct answers in exactly one section.
e) Find the probability that he scores no correct answers in at least two sections.
f) Find the probability that he scores at least two correct answers in at least two sections.

7 A test is applied to a certain type of component produced in a factory. The probability that a randomly chosen component fails is $\frac{1}{5}$. Ten components are tested. Find, correct to three significant figures the probability that:
a) none will fail,
b) exactly two will fail,
c) at most two will fail,
d) at least two will fail.

8 The national pass rate of an examination is 70 per cent. A teacher finds that six out of the twelve children in his class pass. Is there evidence at the five per cent level that this group did significantly worse than average?

9 An electric circuit contains 20 components each of which has a probability of 0.01 of being defective. Calculate the probability that:
a) none of the components is defective,
b) two of the components are defective,
c) at least three of the components are defective.

10 **a)** Give proof that the mean and variance of a random variable, X, which is binomially distributed are np and $np(1-p)$ with the conventional notation.
A disease for which there is yet no known cure has a natural recovery rate of 30 per cent, that is to say that 30 per cent of the patients who suffer from this disease recover. A group of ten patients suffering from the disease are to receive a new drug. The drug will be given more extensive trialling if four or more of this group recover.
b) What is the probability that a drug which in fact has no effect is given further trialling?
c) Find also the probability that the new treatment will be rejected even if it increases the recovery rate to 70 per cent.

11 I suspect that a cubical die is biased against coming up as a 6. I decide to conduct a statistical test.
a) Explain why $\mathrm{H}_1: p < \frac{1}{6}$ is an appropriate alternative hypothesis.
b) I decide to conduct my test at the five per cent level. My experiment consists of rolling the die 16 times and recording the number of 6s which occur. What conclusion should I reach if the number is zero?

c) Suppose, instead, I had chosen to roll the die 26 times. What is the critical region?

12 A trainer claims that one of her squad of gymnasts has improved her performance on the vault . Last season, the gymnast had perfect scores on 70 per cent of her vaults. This season she has had a perfect score on 18 of the 20 vaults she has done. Test the claim of the trainer at the five per cent level.

13 A packet contains 20 seeds, and on average, 75 per cent of this variety of seed germinate. Calculate the probability that 18, 19 or 20 of the seeds will germinate. A new way of packaging seeds becomes available and it is claimed that this results in higher germination rate. What is the critical region if the level of significance is:
a) 10% **b)** 5% **c)** 2%?

14 Metrication has led to changes in packaging. Eggs from suppliers are now sold in boxes of 20. A cynical shop-keeper believes that the eggs in the new boxes are more varied in colour than they used to be. At one time $\frac{2}{3}$ of the eggs, on average, had deep brown shells. His customers now complain about having too many or too few deep brown eggs.
a) State suitable hypotheses to test the belief of the shop-keeper.
b) Calculate the probability of having:
i) fewer than nine, **ii)** more than 17,
deep brown eggs in a box.
c) A box is chosen at random and the number, X, of deep brown eggs noted. Identify the symmetric critical region for the test if it is conducted at:
i) 5% **ii)** 10% **iii)** 1%.
d) He decides to use a non-symmetric critical region:
$X \leq 10$ or $X \geq 19$; what is his significance level?

15 Large batches of biro refills are sampled as part of a quality assurance exercise. Two alternative sampling inspection procedures are proposed.
Proposal 1 Select a random sample of ten refills and reject the batch if more than one defective refill is found.
Proposal 2 Select a random sample of 20 refills and reject the batch if more than three defective refills are found.
For each proposal, calculate the probability that a batch is rejected if:
a) the probability of defective refill is 0.05 (acceptable level),
b) the probability of defective refill is 0.20 (unacceptable level).
Which proposal is better? Why?

16 Charlie has two apparently identical dice. However, one is fair, and the other is biased so that the probability of a 6 is $\frac{1}{3}$.
He decides to try to identify the biased die as follows:

$$H_0: P(\text{score} = 6) = \tfrac{1}{6};\ H_1: P(\text{score} = 6) = \tfrac{1}{3}$$

and reject H_0 in favour of H_1 if, in twelve rolls of a die he observes more than four scores of 6.
He makes a decision by selecting one of the two dice at random and undertaking the above testing procedure once.

a) If he in fact selects the fair die, calculate the significance level of the test.

b) If he selects the biased die, calculate the probability of an incorrect decision.

17 A seed supplier advertises that, on average, 80 per cent of a certain type of seed will germinate. Suppose that 18 of these seeds, chosen at random, are planted.

a) Find the probability that 17 or more seeds will germinate if:

A the supplier's claim is correct,

B the supplier is incorrect and 82 per cent of the seeds, on average, germinate.

Mr Brewer is the advertising manager for the seed supplier. He thinks that the germination rate may be higher than 80 per cent and he decides to carry out a hypothesis test at the ten per cent level of significance. He plants 18 seeds.

b) Write down the null and alternative hypotheses for Mr Brewer's test, explaining why the alternative hypothesis takes the form it does.

c) Find the critical region for Mr Brewer's test. Explain your reasoning.

d) Determine the probability that Mr Brewer will reach the wrong conclusion if:

A the true germination rate is 80%,

B the true germination rate is 82%.

(MEI Question 4, Paper 1, January 1995)

18 Given that X has a binomial distribution in which $n = 15$ and $p = 0.5$, find the probability of each of the following events.

a) $X = 4$

b) $X \leq 4$

c) $X = 4$ or $X = 11$

d) $X \leq 4$ or $X \geq 11$

A large company is considering introducing a new selection procedure for job applicants. The selection procedure is intended to result over a long period in equal numbers of men and women being offered jobs. The new procedure is tried with a random sample of applicants, and 15 of them, eleven women and four men, are offered jobs.

e) Carry out a suitable test at the five per cent level of significance to determine whether it is reasonable to suppose that the selection procedure is performing as intended. You should state the null and alternative hypotheses under test, and you should explain carefully how you arrive at your conclusions.

f) Suppose now that, of the 15 applicants offered jobs, w are women. Find all the values of w for which the selection procedure should be judged acceptable at the five per cent level.

(MEI Question 4, Paper 1, January 1993)

19 Charlotte-Anne claims to have extra-sensory powers. To test the claim she is asked to predict the suit (clubs, spades, hearts or diamonds) of a series of 20 cards chosen at random, with replacement, from a well-shuffled pack.

Assuming that Charlotte-Anne does not, in fact, have extra-sensory powers, but merely guesses at random, find the probability that

a) she gets exactly five of the 20 predictions correct,

b) she gets nine or more predictions correct.

Now suppose that a hypothesis test is to be carried out to examine Charlotte-Anne's claim.

c) Write down suitable null and alternative hypotheses.

d) Determine how many correct predictions out of 20 Charlotte-Anne would need to justify her claim at the one per cent level of significance.

In a hypothesis test on a separate occasion, Charlotte-Anne is given a series of n cards and she correctly predicts the suit of each of them.

e) Find the least value of n which would justify her claim at the five per cent level of significance.

(MEI Question 4, Paper 1, January 1994)

20 State the conditions under which a binomial distribution may be used as a model for a variable.

For each of the variables described below, state whether or not a binomial distribution is a suitable model. If you think a binomial distribution is suitable give appropriate values for the parameters. If you think it is not suitable give one reason.

a) The number of aces in a hand of five cards dealt from a well-shuffled pack of 52 playing cards.

b) The number of wins in five minutes play on a slot machine when 15 per cent of all plays on the machine result in a win.

c) The number of double 6s obtained when a pair of unbiased dice is thrown 20 times.

(NEAB Question 7, Specimen paper 8, 1994)

21 In the Autumn of 1992 the Department of Transport carried out roadside checks on the lighting of randomly-chosen commercial vehicles. The investigation showed that 35 per cent of all commercial vehicles had some fault with their lighting. For a random sample of 15 commercial vehicles, find the probability that:

a) fewer than five vehicles have faulty lights,

b) between three and eight vehicles inclusive have faulty lights,

c) exactly six vehicles have faulty lights.

(NEAB Question 12 (part), Specimen paper 8, 1994)

22 Making use of the appropriate tables, demonstrate clearly that there is greater than a 93 per cent chance that a random observation from the binomial distribution with n equal to 16 and p equal to 0.35 (with the usual notation) lies between 3 and 9 inclusive.

(Oxford Question 1(a), Specimen paper 6, 1996)

23 Over a long period of time it has been found that in Eric's restaurant the ratio of non-vegetarian to vegetarian meals ordered is 3 : 1. During one particular day at Eric's restaurant, a random sample of 20 people contained two who ordered a vegetarian meal.
Carry out a significance test to determine whether or not the proportion of vegetarian meals ordered that day is lower than is usual. State clearly your hypotheses and use a ten per cent significance level.

(ULEAC Question 8 (a), Paper 1, June 1992)

24 The random variable X has the binomial distribution B(10, 0.35). Find $P(X \leq 4)$.

(ULEAC Question 1 (part), Paper 1, June 1993)

25 A manufacturer of balloons produces 40 per cent long ones and 60 per cent round ones, and five per cent of all balloons produced are purple.
Assuming that packets of balloons contain a random selection from the output, calculate the probability that in a packet containing 20 balloons there are:

a) equal numbers of long and round ones,

b) more long ones than round ones.

(ULEAC Question 8 (a) and (b), Paper 1, January 1993)

Summary

In this chapter you have:

- discovered a model for a random variable which represents the number of successful events when a fixed number, n, of events occur, each of which can result in either a success or a failure, the events are independent, and the probability, p, that any event results in a success is constant,
- encountered the $B(n, p)$ notation for the binomial probability distribution,
- discovered the distribution function and tables recording $P(R \leq x)$,
- discovered the processes involved in statistical testing including null and alternative hypotheses, critical regions, significance levels, one- and two-tailed tests.

STATISTICS 8

Sums and differences of distributions

In this chapter we shall:

- *combine independent random variables and discover how the means and variances of the combined variates can be obtained from the individual components,*
- *consider how a binomial variate can be treated as a combination of individual variates,*
- *discover how this combination can be used to predict the mean and variance of a binomial variate.*

UNIFORM AND TRIANGULAR DISTRIBUTIONS

An ordinary cubical die is rolled and the number on the uppermost face is recorded. What probability distribution is appropriate to model the possible outcomes? This is a familiar scenario and a familiar discrete distribution.

x	1	2	3	4	5	6
$P(C = x)$	$\frac{1}{6}$	$\frac{1}{6}$	$\frac{1}{6}$	$\frac{1}{6}$	$\frac{1}{6}$	$\frac{1}{6}$

This is known as a **uniform probability distribution**. It can be shown graphically, as in the diagram.

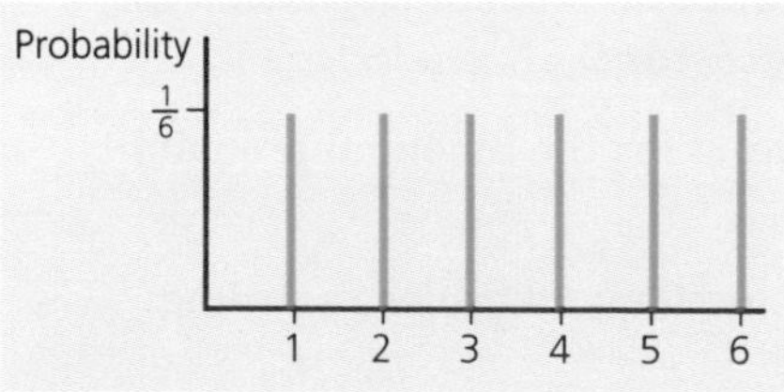

If C is a random variable which is distributed in this way, then:

$$P(C = x) = \tfrac{1}{6} \text{ for } x = 1, 2, \ldots, 6$$

Use your calculator to discover the mean and variance of this distribution of probabilities. You should find:

$$\mu_C = \frac{7}{2} = 3.5 \text{ and } \sigma_C^2 = \frac{35}{12} = 2.917$$

where μ_C and σ_C^2 are the mean and variance of C, the outcome on a cubical die.

Suppose two cubical dice are rolled and the outcomes are added. What are the highest and lowest possible totals? As the lowest outcome on either die is 1, the lowest total is 1 + 1 i.e. 2. Equally, the largest possible total is 6 + 6 i.e. 12.

These results are not startling, rather they are quite comforting. The highest and lowest possible outcomes on a single die are 6 and 1.

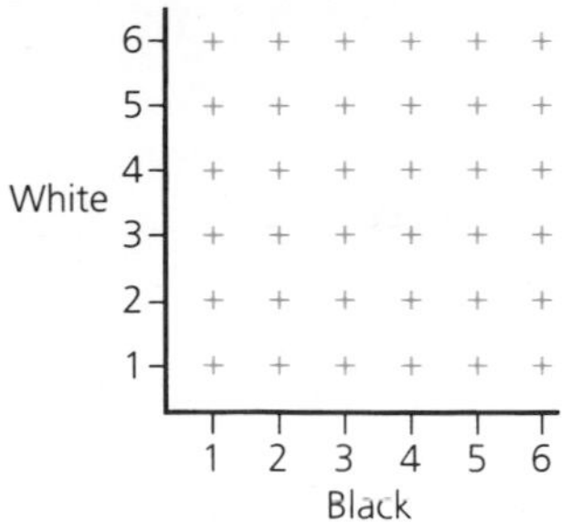

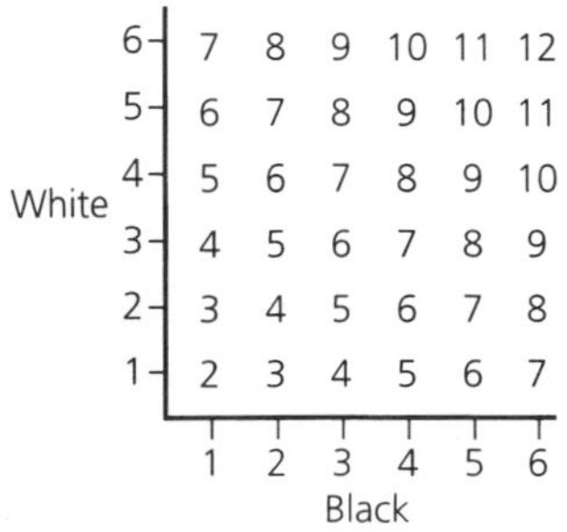

The results for the sum of the outcomes on two dice are simply twice those for a single die. What about the distribution of the probabilities? Is the distribution the same? Is it similar? Is it a uniform distribution?

We have already met a way of identifying all possible outcomes of rolling two dice. We can represent the outcomes as coordinates as in the diagram.

Then at each of the 36 points the total can be found, as shown in this diagram.

From the diagram, the probability distribution can be discovered.

x	2	3	4	5	6	7	8	9	10	11	12
$P(C + C = x)$	$\frac{1}{36}$	$\frac{2}{36}$	$\frac{3}{36}$	$\frac{4}{36}$	$\frac{5}{36}$	$\frac{6}{36}$	$\frac{5}{36}$	$\frac{4}{36}$	$\frac{3}{36}$	$\frac{2}{36}$	$\frac{1}{36}$

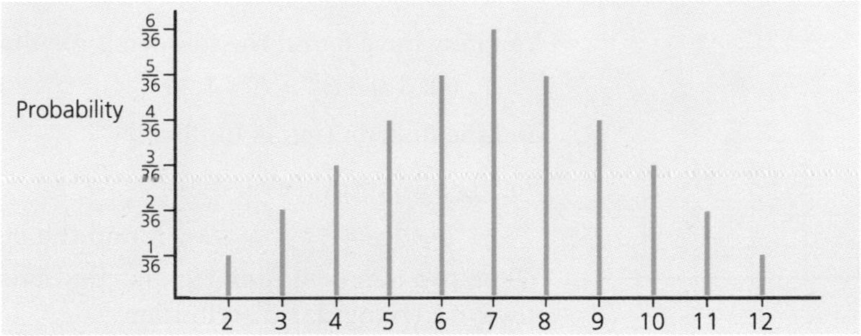

Graphically, this distribution resembles a triangle which is nothing like the shape for a single die. What do you think the mean and the variance of this distribution are? In what way are these measures related to those for a single die? Use your calculator to explore this.

You should find the following values for the mean and variance of the sum of the outcomes on two dice.

$$\mu_{C+C} = 7 \text{ and } \sigma^2_{C+C} = \frac{35}{6}$$

Reflect for a moment on these results: they are, again, comforting. The mean for $C + C = \mu_{C+C} = \mu_C + \mu_C$ is what we might have hoped for. The same is true of the variance.

$$\sigma^2_{C+C} = \sigma^2_C + \sigma^2_C$$

But, what about standard deviation? Recall that the standard deviation is the square root of the variance, and so, for the outcome on a single die:

$$\sigma_C = \sqrt{\frac{35}{12}} \approx 1.7078$$

and for the sum of the outcomes on two dice:

$$\sigma_{C+C} = \sqrt{\frac{35}{6}} \approx 2.4152$$

No simple relationship between the standard deviations is apparent, and in fact there isn't one.

Exploration 8.1 *Outcomes on a tetrahedral die*

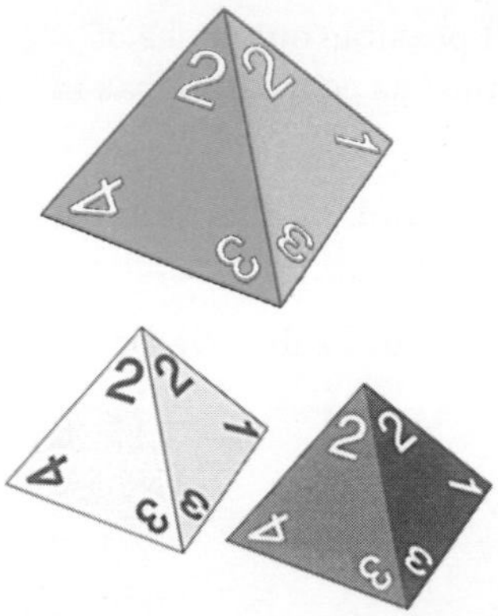

a) Consider the distribution of the possible outcomes on a tetrahedral die with faces numbered 1, 2, 3, 4.

- What model is appropriate?
- What is the mean and the variance of the model distribution?
- What is the shape of the distribution?

b) Consider, now, the distribution of possible outcomes if two tetrahedral dice are rolled.

- What is the variance?
- What is the shape?
- Is there any similarity in the results as seen in the case of cubical dice?

Expected results

You may have found the following results for the tetrahedral die.

$\mu_T = 2.5$ and $\sigma_T^2 = \frac{15}{12} = 1.25$

and the distribution is uniform.

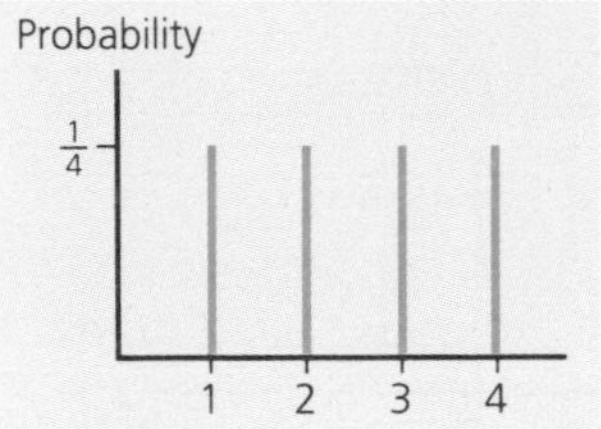

For the two tetrahedral dice

$\mu_{T+T} = 5$ and $\sigma_{T+T}^2 = \frac{15}{6} = 2.5$ and the distribution is triangular.

When two identical uniform distributions are combined, the result is always a triangular distribution.

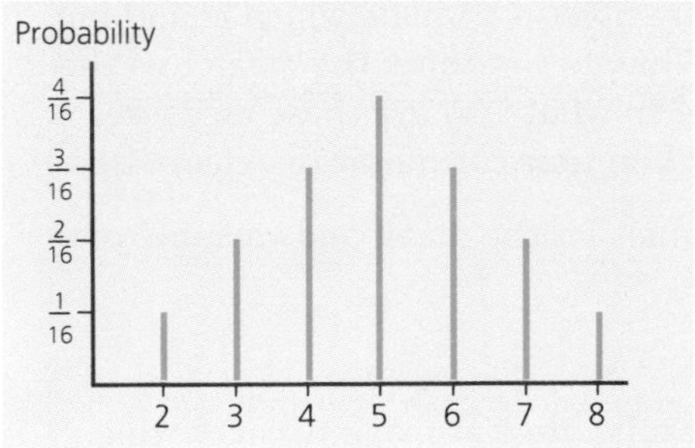

A moment's reflection reveals that these results are in line with those of the cubical dice. The mean for the sum of the outcomes on the two tetrahedral dice is the same as the sum of the means for the individual dice.

$$\mu_{T+T} = \mu_T + \mu_T$$

The same relationship holds for the variance: the variance of the sum of the outcomes on two dice is the same as the sum of the variance of the outcomes on each dice.

$$\sigma_{T+T}^2 = \sigma_T^2 + \sigma_T^2$$

Combining different uniform distributions

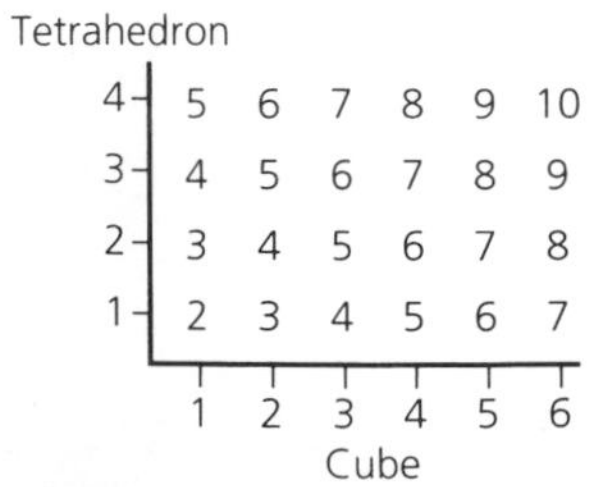

Imagine the set of possible outcomes when a tetrahedral die and a cubical die are rolled and the total is recorded. The diagram shows the outcomes.

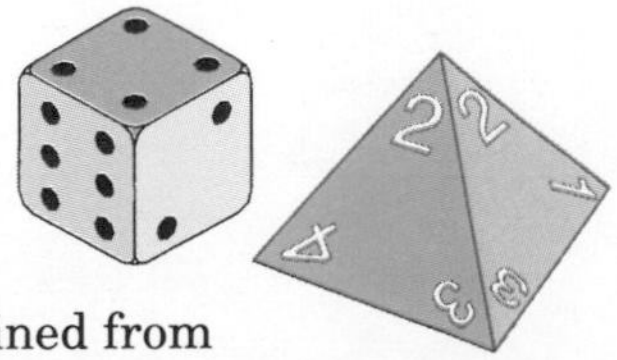

The distribution table can be readily obtained from this diagram. There are 24 distinct combinations which are equally likely, hence the probabilities are:

x	2	3	4	5	6	7	8	9	10
$P(C+T=x)$	$\frac{1}{24}$	$\frac{2}{24}$	$\frac{3}{24}$	$\frac{4}{24}$	$\frac{4}{24}$	$\frac{4}{24}$	$\frac{3}{24}$	$\frac{2}{24}$	$\frac{1}{24}$

Then what is the mean outcome? What about the variance? Use your calculator to find out.

You should find:

$$\mu_{C+T} = 6$$

$$\sigma^2_{C+T} = \frac{50}{12} = 4.167$$

How are these are related to the means and variances of the outcomes for the individual dice? The relationship is all that might be hoped for.

$$\mu_{C+T} = \mu_C + \mu_T = 3.5 + 2.5$$

and:

$$\sigma^2_{C+T} = \sigma^2_C + \sigma^2_T = \tfrac{35}{12} + \tfrac{15}{12}$$

These may be summarised as:

'the mean of the sum is the sum of the means'

and:

'the variance of the sum is the sum of the variances'.

A graphical display of the distribution of probabilities is shown in this diagram.

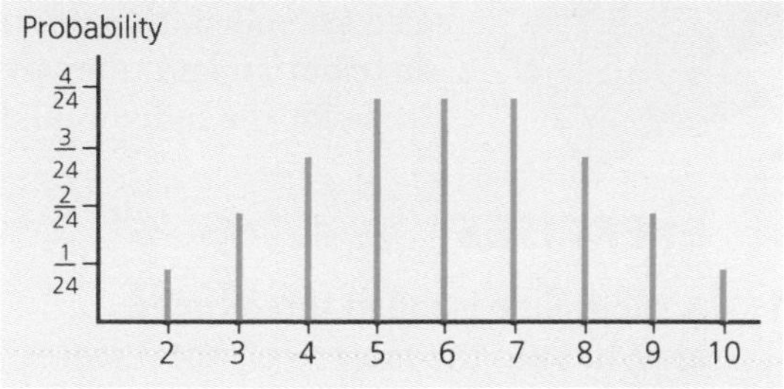

It is clear from this display that the sum of outcomes on two different dice does not necessarily result in a triangular shape to the distribution.

Example 8.1

A disc is marked with a 1 on one side and a 2 on the other. Let X be the random variable which models the number on the visible face of the disc when it is flipped and lands. Calculate the mean and variance of X.

The disc and a tetrahedral die are thrown together. The sum of the outcomes is recorded. Let Z be the random variable which models this sum. Write down the probability distribution for Z and calculate its mean and variance. How are these related to those for X and T, where T is the random variable which models the number on the face of the tetrahedral die?

Solution

An appropriate model for X is:

x	1	2
$P(X = x)$	$\frac{1}{2}$	$\frac{1}{2}$

Hence:

$\mu_X = 1\frac{1}{2}$ *and* $\sigma_X^2 = \frac{1}{4}$

The probability distribution for Z is:

Z	2	3	4	5	6
P(Z = z)	$\frac{1}{8}$	$\frac{2}{8}$	$\frac{2}{8}$	$\frac{2}{8}$	$\frac{1}{8}$

Hence:

$\mu_Z = 4$ *and* $\sigma_Z^2 = 1$

Since:

$\mu_T = 2.5$ *and* $\sigma_T^2 = 1.25$

it can be seen that:

$\mu_Z = \mu_X + \mu_T$ *and* $\sigma_Z^2 = \sigma_X^2 + \sigma_T^2$

Exploration 8.2 Combining three distributions

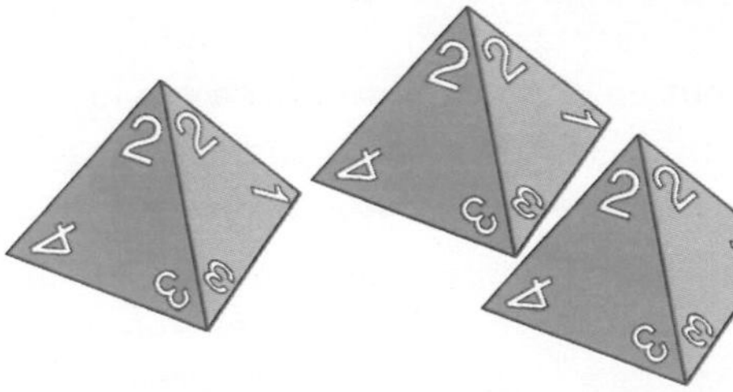

Imagine you now have three tetrahedral dice each with vertices numbered 1, 2, 3 and 4.

The total of the three outcomes is recorded when the dice are rolled. Explore the resulting probability distribution and compare its mean and variance with those for the individual die.

EXERCISES

8.1A

1 A cubical die is rolled three times.
 a) What are the most likely total scores?
 What are the least likely total scores?
 b) State the mean total score, and its variance.

2 A combination lock has three digits, □ □ □ which can each take values of 0, 1, 2, ..., 9
 a) How many different ways are there of setting the combination?
 b) If the three digits are added together, find the mean and variance of their sum.
 c) Find the mean and variance of the sum of the settings of a similar lock with four digits.

3 A computer program generates a random digit R from 0 to 4 inclusive followed by an (independent) random digit S from 5 to 9 inclusive. It then shows the total of the two random digits, $T = R + S$.
 a) Calculate the mean and variance of R.
 b) Calculate the mean and variance of S.
 c) State the mean and variance of T.

By listing the possible values of T, and from first principles, verify your answer to **c)**.

It is quite possible to test those conjectures using the results we discovered in Chapter 7, *The binomial probability model*, about the binomial distribution. Remember that if $X \sim B(5, 0.3)$ then the probability function is:

$$P(X = x) = {}_5C_x\ 0.3^x\ 0.7^{5-x}$$

This produces the following tabulation.

x	0	1	2	3	4	5
$P(X = x)$	0.168 07	0.360 15	0.308 70	0.132 30	0.028 35	0.002 43

Use expectation to calculate the mean and variance.

Exploration 8.6

Mean and variance of Bernoulli variates

- Use an argument based on linear combination of Bernoulli variates to predict the mean and variance of the binomial variate $X \sim B(5, 0.4)$.
- Check your predictions using the appropriate probability function and expectation.

Mean and variance of the binomial distribution

In the previous section, we saw how a binomial variate could be considered to be the sum of n independent Bernoulli variates with the same probability of 'success'. Symbolically this can be written as:

$$X \sim B(n, p)$$

and this is equivalent to:

$$X \sim \text{Ber}(p) + \text{Ber}(p) + \ldots + \text{Ber}(p)$$

This equivalence leads readily to the mean and variance of a general binomial variate.

Mean $\quad \mu_{B(n,p)} = \mu_{\text{Ber}} + \mu_{\text{Ber}} + \ldots + \mu_{\text{Ber}}$ (n means)

$$= p + p + \ldots + p = np$$

Variance $\quad \sigma^2_{B(n,p)} = \sigma^2_{\text{Ber}} + \sigma^2_{\text{Ber}} + \ldots + \sigma^2_{\text{Ber}}$

$$= p(1-p) + p(1-p) + \ldots + p(1-p) = np(1-p)$$

Example 8.3

A survey indicates that 35 per cent of university students smoke cigarettes at least once a day. Describe how the number of smokers in a group of 18 students can be modelled using a sum of Bernoulli variates. What is the mean and variance of the number of smokers in groups of 18 university students?

Solution

A student either does or does not smoke at least one cigarette per day. This is a Bernoulli trial and the survey suggests:

$$P(\text{smoke}) = 0.35$$

The number of smokers in a group of 18 students can be modelled as the sum of 18 Bernoulli variates each having probability equal to 0.35. If X is the number of smokers, then:

$X \sim B(18, 0.35)$

Hence, the mean of X is:

$18 \times 0.35 = 6.3$

and, the variance of X is:

$18 \times 0.35 \times 0.65 = 4.095$

EXERCISES

8.3A

1 Thirty per cent of the members of a certain profession have an A-Level in Statistics. A random sample of ten members is obtained and the number, X, with an A-Level in Statistics is noted. Use a binomial distribution to model the occurrence of X. Estimate the probability that:
a) $X < 2$ **b)** $X =$ the expected value, μ **c)** $X > \mu + \sigma$.

2 A fair coin is tossed eight times.
a) Find the mean and variance of the number of heads obtained.
b) Repeat part **a)** for a biased coin for which the probability of getting a head is four times the probability of getting a tail.

3 Forty per cent of sixth-formers in a certain county study A-Level Mathematics. In a sixth-form group of 20, find the probability that:
a) exactly the expected number study A-Level Mathematics,
b) more than the expected number study A-Level Mathematics.
c) Calculate the variance of the number who do not study A-Level Mathematics.

4 In a sixth-form college, 45 per cent are in the Science department, 35 per cent in the Humanities department and the rest in the Languages department.
If a group of ten students is chosen at random,
a) calculate the mean and variance of the number in the Science department,
b) calculate the mean and variance of the number in the Languages department,
c) find the probability that more than half the group are in the Humanities department.

5 In a binomial distribution, the mean = 10.2 and the variance = 1.53.
a) Calculate n, p and q.
b) If p represents the probability of passing an exam, calculate the probability that more than the expected number pass the exam.

EXERCISES

8.3B

1 Sixty per cent of the members of a profession have an A-Level in Mathematics. A random sample of ten members is obtained. The number, X, with A-Level Mathematics is recorded. Use a binomial distribution to model the occurrence of X. Estimate the probability that:
a) $X > 2$ **b)** $X = \mu$, the expected value of X **c)** $X < \mu + \sigma$.

2 In a particular area it has been established that 75 per cent of cars are fitted with some form of alarm system. A random sample of twelve cars from the area is selected. Deduce the distribution of the number of cars in the sample fitted with some form of alarm. Hence calculate the probability that the 'expected' number of cars are fitted with some form of alarm.

3 A test consists solely of questions requiring a 'Yes' or 'No' answer. State a distribution for the answer to a single question.

The test contains 18 questions.

a) For a candidate who simply guesses the answer to each question, state the distribution which describes his number of correct answers. Hence calculate the probability that he answers less than half the questions correctly.

b) Repeat **a)** for a candidate who has a 65 per cent chance of answering each question correctly.

4 On any given night the probability that each single room in a hotel is occupied is 0.80, and the probability that each double room is occupied is 0.74.

a) The hotel has five single rooms. If X denotes the number of single rooms occupied, deduce its probability model. Hence determine the mean and variance of X. Calculate the probability that on a given night, the mean number of single rooms are occupied.

b) The hotel has 15 double rooms. Explain how the number of double rooms occupied, Y, may be constructed as a sum of Bernoulli variables.
State the mean, μ and variance, σ^2 of Y. Hence calculate the probability that Y is greater than $(\mu + \sigma)$.

5 **a)** The random variables $X_i \sim \text{Ber}(0.5)$. State the distribution of:

i) $Y = X_1 + X_2 + X_3$

ii) $Z = X_4 + X_5 + X_6 + X_7$

iii) $T = \sum_{i=1}^{7} X_i$

Hence show that $\mu_T = \mu_Y + \mu_Z$ and $\sigma_T^2 = \sigma_Y^2 + \sigma_Z^2$.

b) If $X_1, X_2, X_3 \sim \text{Ber}(0.3)$ and $X_4, X_5, X_6, X_7 \sim \text{Ber}(0.8)$ show that, although no distribution can be deduced for T (as defined in **iii)** above), the results for μ_T and σ_T^2 still hold.

CONSOLIDATION EXERCISES FOR CHAPTER 8

1 Three fair dice are thrown and 20 points are scored if all three dice show the same number; one point is scored if two of the dice show the same number.

a) Calculate the mean and variance of the number of points scored.

b) Calculate the mean and variance of the total number of points scored if the dice are thrown together four times.

2 Consider a binomial distribution.

a) Calculate the variance, when $n = 10$, for values of p from:

$p = 0.9$
$p = 0.8$
$p = 0.7$ etc. down to
$p = 0.1$.

Discuss your results.

b) Repeat for $n = 15$. Discuss.

3 X is a discrete random variable. If $4X$ has a mean of 12 and a variance of 3.2, calculate:

a) the mean and variance of X,

b) the mean and variance of $5X$.

4 X and Y are independent random variables.

The variance of $(X + Y) = 7$.
The variance of $(3X - 2Y) = 33$.

Calculate the variance of X and the variance of Y.

5 A random variable, X, has the following distribution.

r	-1	0	1	2
$P(X = r)$	0.2	0.2	0.4	0.2

Find:

a) $E(X)$ **b)** $\text{Var}(X)$.

Hence find:

c) $\text{Var}(2X)$ **d)** $\text{Var}(-6X)$ **e)** $\text{Var}(3X + 7)$ **f)** $\text{Var}\left(\frac{X}{2} - 4\right)$

6 The random variable X has a mean μ and a variance σ^2. Find in terms of μ and σ^2 the mean and variance of:

a) $5X$ **b)** $X - \mu$ **c)** $\dfrac{X}{\sigma}$ **d)** $10 - 3X$.

7 A six-sided die is biased so that the probability of scoring n (for $n = 1$ to 6) is proportional to n. Determine the probabilities of getting each of the values 1 to 6 inclusive and find the mean and variance of this probability distribution.

What would be the mean and variance if the score showing were doubled?

8 A player throws a die with faces numbered 1 to 6 inclusive. If the player gets a 6 he throws the die a second time and in this case his score is the sum of 6 and the second number: otherwise his score is the number obtained. The player has no more than two throws.

Let X be the random variable denoting the player's score. Write down the probability distribution of X, and determine $E(X)$.

9 If $X \sim B(n, p)$, write down expressions for μ_X and σ_X^2 in terms of n and p. You are told that $\mu_X = 1$, find the values of n and p if:

a) $\sigma^2 = 0.8$ **b)** $\sigma^2 = 0.9$ **c)** $\sigma^2 = 0.95$ **d)** $\sigma^2 = 0.995$.

10 The number of 'inners' scored by 150 archers at an archery contest is given in the table (each archer had five arrows to shoot at the target).

Number of inners	0	1	2	3	4	5
Number of archers	19	44	42	19	15	11

a) Calculate the mean number of hits per archer.
b) Assume that the number of hits may be modelled by a binomial distribution. What are the values of n and p?
c) Calculate the frequencies of each number of hits given by the binomial distribution.
d) Do you think the binomial is a good model?

11 Five coins are flipped and the number of heads recorded. This experiment is conducted 200 times, the results are as shown in the table.

Number of heads	0	1	2	3	4	5
Frequency	47	81	52	16	3	1

a) Calculate the mean number of heads.
b) Compare this with the binomial model for a 'fair' coin.

12 A survey conducted by a local education authority requires schools to complete a questionnaire about a sample of their pupils. The sample is defined as all pupils born on the 5th of any month, so that the probability of a randomly chosen pupil being in the sample is $\frac{12}{365}$, or about 0.0329.
The A-level year in a secondary school consists of 100 pupils.
a) Find the probability that none of these pupils appears in the sample.
The school has 1500 pupils in total.
b) Write down the expectation and the standard deviation of the total number of pupils sampled in a school of this size.

(MEI Question 4 (part), Paper S2, June 1993)

13 An envelope contains ten stamps, seven being 20p stamps and the other three being 25p stamps. Four stamps are to be chosen at random from the envelope.
a) Let T denote the total value of the four stamps when they are chosen without replacement. Find the sampling distribution of T and deduce its mean and variance.
b) Let S denote the total value of the four stamps when they are chosen with replacement. By considering the distribution of the number of 20p stamps that are chosen, or otherwise, find the mean and the variance of S.

(WJEC Question 11, Specimen, Paper A3, 1994)

14 The discrete random variable X has the probability distribution shown in the following table.

r	-1	0	1
$P(X = r)$	$\frac{1}{4}$	$\frac{1}{2}$	$\frac{1}{4}$

a) Write down the expectation of X, and find the variance of X.

A second random variable Y had the same distribution as X, and the two random variables are independent.

b) List all possible values of $X + Y$. Show that $P(X + Y = -2) = \frac{1}{16}$, and find $P(X + Y = 0)$.

c) Show in the complete probability distribution for $X + Y$.

d) Verify that, for these random variables,

$$\text{Var}(X + Y) = \text{Var}(X) + \text{Var}(Y).$$

(MEI Question 2, Paper S2, 1993)

15 **a)** A regular tetrahedron is a solid with four faces, all identical. What is the probability, if it is tossed in the air, that it will land on a given face?

A die is made by numbering the four faces 1, 2, 3, 4. The random variable X represents the number on the face on which the tetrahedron lands. The mean of X is 2.5. Calculate the variance of X.

b) A circular disc is marked with a 1 on one side and a 2 on the other. Y is the random variable representing the number showing on the visible face of the disc when it is tossed and lands. Demonstrate that the mean and variance of Y are $1\frac{1}{2}$ and $\frac{1}{4}$ respectively.

c) The disc and the dice are tossed together. The sum of the outcomes is recorded. Z is the random variable representing independent sums of X and Y. Write down the probability distribution for Z, and use this to calculate its mean and variance. In what way are these parameters related to those for X and Y?

(Oxford Question 5, Paper 3, 1991)

Summary

- If $X \sim B(n, p)$ then

  $\mu_X = np$

  $\sigma_X^2 = np(1-p)$

- The mean of the sum is the sum of the means.

  $\mu_{C+T} = \mu_C + \mu_T$

- The variance of the sum is the sum of the variances.

  $\sigma_{C+T}^2 = \sigma_C^2 + \sigma_T^2$

- The mean of the difference is the difference of the means.

  $\mu_{C-T} = \mu_C - \mu_T$

- X is the sum of n independent Bernoulli variates each with probability p.

- The mean and variance of simple sums of random variables bear a simple relation to the individual variates. In particular:

  $\mu_{X+Y+Z} = \mu_X + \mu_Y + \mu_Z$

  $\sigma_{X+Y+Z}^2 = \sigma_X^2 + \sigma_Y^2 + \sigma_Z^2$

- The mean and variance of a linear combination of variates are

  $\mu_{aX+bY} = a\mu_X + b\mu_Y$

  $\sigma_{aX+bY}^2 = a^2\sigma_X^2 + b^2\sigma_Y^2$

STATISTICS

9 The Poisson distribution

In this chapter we shall:

- *discover a probability distribution which models the occurrence of random events,*
- *validate the model in its use to approximate the occurrence of rare events.*

MODELLING RANDOM EVENTS

Gardeners are proud of the state of their gardens, and especially of the state of their grass – their lawn. They invest much time and effort into the appearance of their lawn. One of the reasons why so much effort is required is that weeds keep appearing. They crop up anywhere in the lawn. There doesn't seem to be any way of predicting where the next weed will appear. There's no particular part of the lawn favoured by the weeds nor is any part avoided. And, during the growing season, they seem to appear at about the same rate everywhere. Even in some of the most carefully attended lawns, the weeds, or in particular, daisies, appear, at a rate of three daisies per square metre of lawn each week.

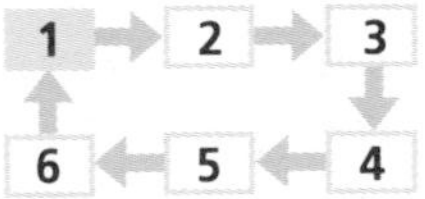

Specify the real problem

The gardener of a well-tended lawn decides to call in an expert to offer advice about the number of daisies she might find in any part of the lawn, to see if it is possible to reduce the amount of weedkiller that she is using.

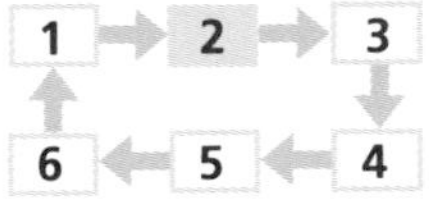

Set up a model

Imagine that you are the lawn expert. How might you set about modelling the appearance of the daisies?

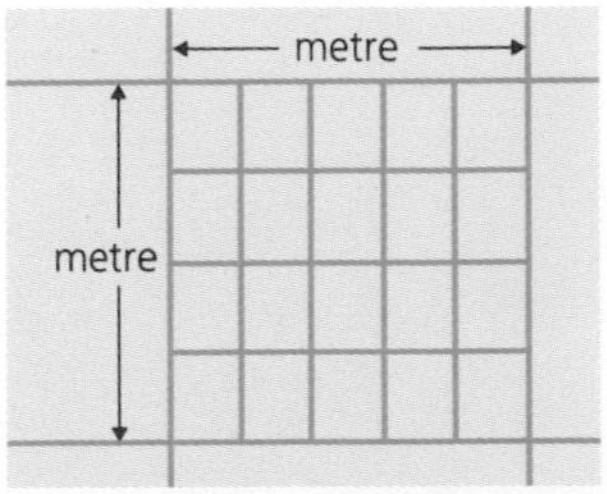

One possible approach is to imagine that a typical square metre of lawn is divided into a number of equal subdivisions. Let's start with 20 subdivisions.

What is the probability that none of the 20 subdivisions contains a daisy?

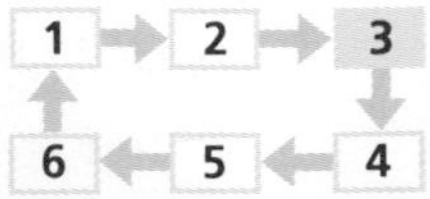

Formulate the mathematical problem

Suppose that a subdivision is big enough to contain one daisy but no more than one. We could then model the appearance of a daisy as a Bernoulli trial (see Chapter 8, *Sums and differences of distributions*). The 'success' corresponds to a daisy appearing in the subdivision. The average number of daisies in a square metre (i.e. in 20 Bernoulli trials) is three. Hence, the probability of a 'success' is $\frac{3}{20}$, and the probability of a 'failure' is $\left(1-\frac{3}{20}\right)$.

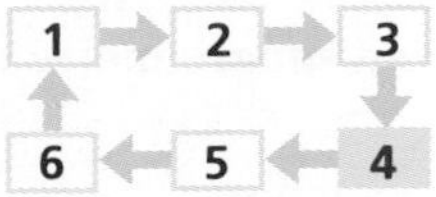

Solve the mathematical problem

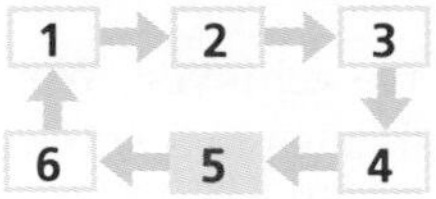

Interpret the solution

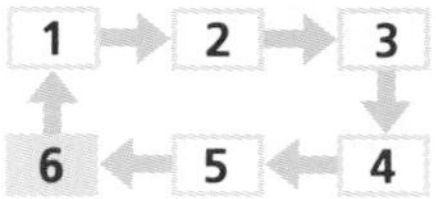

Compare with reality

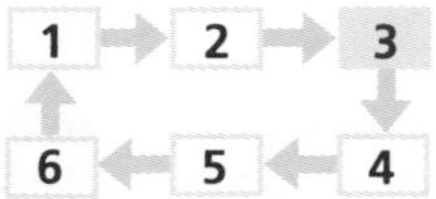

Formulate the mathematical problem

This leads to the conclusion:

$$\text{P(no daisies in 20 subdivisions)} = \left(1-\tfrac{3}{20}\right)^{3} \approx 0.03876$$

assuming that daisies occur independently. This assumption of independence enables us to use the multiplication law of independent events (as we saw in Chapter 4, *Probability*, page 112).

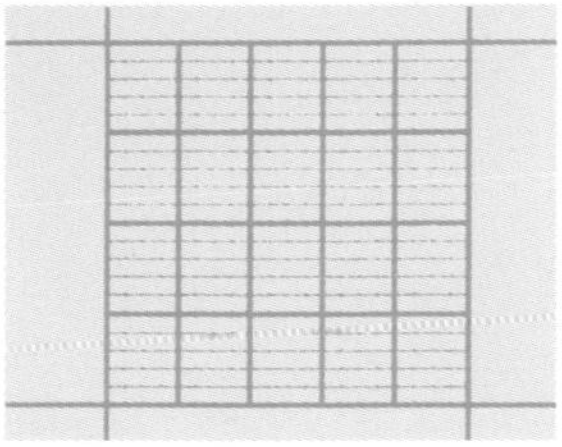

Is it realistic to suppose that a patch of ground, $\frac{1}{20}\,\text{m}^2$ in area, is so small that it could not contain more than one daisy? Perhaps we need to subdivide the square metre of lawn further? We might divide each subdivision further into 20 equal parts, giving 400 new subdivisions in each square metre.

What is the probability that a square metre of lawn does not contain a daisy?

By developing an argument similar to the one above, we have 400 Bernoulli trials with a probability of 'success' equal to $\frac{3}{400}$. Hence:

$$\text{P(failure)} = 1-\tfrac{3}{400}$$

This leads to:

$$\text{P(no daisies in a square metre)} = \left(1-\tfrac{3}{400}\right)^{400} \approx 0.049\,23$$

Clearly, this is a different result from that achieved by considering just 20 Bernoulli trials.

CALCULATOR ACTIVITY

Evaluating the model

Explore the probability of getting no daisies in a square metre of lawn by using the model set up above, but with more subdivisions. You might try dividing each of the new subdivisions into 20 equal parts and repeating this process.

Recurrence formula

The sort of results you might expect are given in the table.

No. subdivisions	P(success)	P(failure)	P(no daisies in a square metre)
20	$\frac{3}{20}$	$1-\frac{3}{20}$	0.038 760
400	$\frac{3}{400}$	$1-\frac{3}{400}$	0.049 227
8000	$\frac{3}{8000}$	$1-\frac{3}{8000}$	0.049 759
160 000	$\frac{3}{160\,000}$	$1-\frac{3}{160\,000}$	0.049 786
3 200 000	$\frac{3}{3\,200\,000}$	$1-\frac{3}{3\,200\,000}$	0.049 787
64 000 000	$\frac{3}{64\,000\,000}$	$1-\frac{3}{64\,000\,000}$	0.049 787
1 000 000 000	$\frac{3}{1\,000\,000\,000}$	$1-\frac{3}{1\,000\,000\,000}$	0.049 787

So the likelihood of there being no daisies in a given square metre of lawn settles down to approximately 0.049 787.

Note: Some calculators give false results for this probability when there are more than 1000 million subdivisions.

It is worth reflecting, at this point, on what we have achieved. We have a model for the likelihood of there being no daisies in a square metre *when* there are, on average, three daisies. It will be useful to give this probability a label:

$$P_{Av=3}(0)$$

We have arrived at a value for $P_{Av=3}(0)$ using the limiting process:

$$P_{Av=3}(0) = \text{the limit of } \left\{\left(1-\tfrac{3}{n}\right)^n\right\} \text{ as } n \text{ increases}$$
$$\approx 0.049\,787 \text{ (to six decimal places)}$$

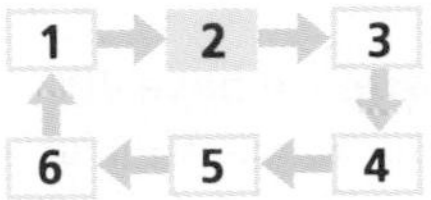

Set up a model

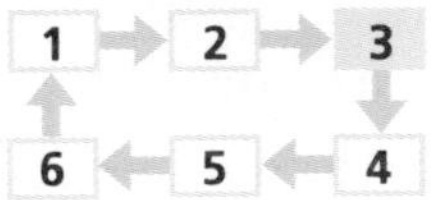

Formulate the mathematical problem

But we set out to model the *appearance* of daisies, so we ought to be trying to model the probability of finding one daisy, two daisies, three daisies, etc.

Let's return to the point where we set up 20 equal subdivisions of a square metre of lawn. We modelled the situation as 20 Bernoulli trials where the probability of success was $\frac{3}{20}$. We now want to know the probability of there being exactly one daisy, exactly two daisies, exactly three daisies, etc. in this square metre.

In Chapter 7, *The binomial probability model*, we met the binomial distribution and the associated recurrence formula:

$$P(x) = \frac{p}{1-p} \times \frac{n+1-x}{x} \times P(x-1)$$

The situation that we are now considering could be modelled as $B(20, \frac{3}{20})$. Hence we could use the recurrence formula to work out the probability of there being exactly one daisy from our model for the probability of finding no daisies at all.

$$P(1) = \frac{\frac{3}{20}}{1-\frac{3}{20}} \times \frac{20+1-1}{1} \times P(0)$$

where $P(0)$ is $P_{Av=3}(0)$. As the model is refined, try considering, say, 400 subdivisions and using the recurrence formula for the binomial model $B(20, \frac{3}{400})$ which produces:

$$P(1) = \frac{\frac{3}{400}}{1-\frac{3}{400}} \times \frac{400+1-1}{1} \times P(0)$$

Further refinement will produce:

$$P(1) = \frac{\frac{3}{8000}}{1-\frac{3}{8000}} \times \frac{8000+1-1}{1} \times P(0)$$

and so on. The limit of this refinement process will lead to the probability we are seeking:

$$P_{Av=3}(1) = \text{limit}\left\{\frac{\frac{3}{n}}{1-\frac{3}{n}} \times \frac{n+1-1}{1}\right\} P_{Av=3}(0)$$

CALCULATOR ACTIVITY

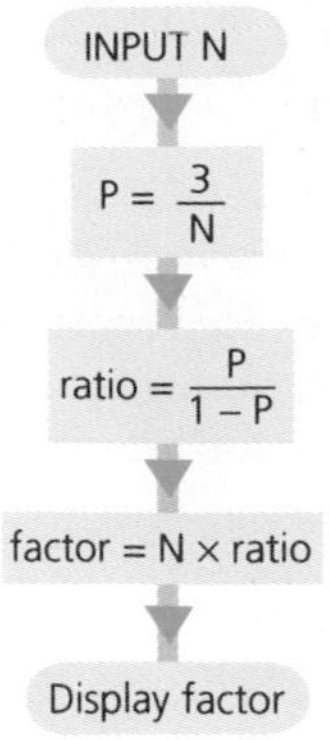

The value of the factor

Explore the value of the factor, in brackets, in the recurrence formula:

$$\left\{\frac{\frac{3}{n}}{1-\frac{3}{n}} \times n\right\}$$

as n increases.

If you have a programmable calculator, you could use a program based on this flowchart.

The sort of results you might achieve are shown in the table below.

Values of n	Value of factor $\left\{\frac{\frac{3}{n}}{1-\frac{3}{n}} \times n\right\}$
20	3.53
400	3.023
8000	3.0011
160 000	3.000 06
3 200 000	3.000 003
64 000 000	3.000 000 1
1 000 000 000	3.000 000 009

Note: This program does not seem to give false results even for values as large as a ten thousand million.

- What do you think the limiting value of the factor is?
- In what way is this limiting value related to the average number of daisies per square metre?

Interpreting the results

It seems that the limit value is the same as the mean, i.e. 3. This suggests that the probability of exactly one daisy is:

$$P_{Av=3}(1) = 3P_{Av=3}(0) \approx 0.149\,361 \text{ (to six decimal places.)}$$

The recurrence formula can be used to find the probability of finding exactly two daisies:

$$P_{Av=3}(2) = \text{limit}\left\{\frac{\frac{3}{n}}{1-\frac{3}{n}} \times \frac{n+1-2}{2}\right\}P_{Av=3}(1)$$

and the probability of exactly three daisies:

$$P_{Av=3}(3) = \text{limit}\left\{\frac{\frac{3}{n}}{1-\frac{3}{n}} \times \frac{n+1-3}{3}\right\}P_{Av=3}(2)$$

and the probability of exactly four daisies:

$$P_{Av=3}(4) = \text{limit}\left\{\frac{\frac{3}{n}}{1-\frac{3}{n}} \times \frac{n+1-4}{4}\right\}P_{Av=3}(3)$$

etc.

Before setting out to discover these probabilities, it is worth considering the similarities and differences between the successive factors. The part which changes in the limiting process is the part which contains n, i.e. the $\frac{\frac{3}{n}}{1-\frac{3}{n}} \times (n+1-x)$ part for $x = 2, 3, 4, \ldots$

CALCULATOR ACTIVITY

The limiting value

Explore the limiting value of these factors for $x = 2, x = 3, x = 4, \ldots$. You could try adapting your previous program to allow for an input of x as well as n.

- What do you think the limit of this factor is?
- Does it depend on the value of x?

Expected results

The results you might have achieved are summarised in the table below where the value of this factor is given for $x = 2, 3, 4$ and 5 and for a range of values of n up to one thousand million.

Value of n	New value of random variable			
	$x = 2$	$x = 3$	$x = 4$	$x = 5$
20	3.35	3.18	3	2.82
400	3.015	3.008	3	2.992
8000	3.000 8	3.000 4	3	2.999 6
160 000	3.000 04	3.000 02	3	2.999 98
3 200 000	3.000 002	3.000 000 9	3	2.999 999 1
64 000 000	3.000 000 9	3.000 000 05	3	2.999 999 95
1 000 000 000	3.000 000 006	3.000 000 003	3	2.999 999 997

This suggests that in every case the limit of the factor:

$$\frac{\frac{3}{n}}{1-\frac{3}{n}} \times (n+1-x)$$

is 3 i.e. this factor gets closer to the average number of daisies per square metre the more extensive the subdivision of a square.

So, if you now interpret this result, the model suggests the following relationship between successive probabilities.

$$\mathrm{P}_{\mathrm{Av}=3}(1) = 3 \times \mathrm{P}_{\mathrm{Av}=3}(0)$$

$$\mathrm{P}_{\mathrm{Av}=3}(2) = \tfrac{3}{2} \times \mathrm{P}_{\mathrm{Av}=3}(1)$$

$$\mathrm{P}_{\mathrm{Av}=3}(3) = \tfrac{3}{3} \times \mathrm{P}_{\mathrm{Av}=3}(2)$$

$$\mathrm{P}_{\mathrm{Av}=3}(4) = \tfrac{3}{4} \times \mathrm{P}_{\mathrm{Av}=3}(3)$$

$$\mathrm{P}_{\mathrm{Av}=3}(5) = \tfrac{3}{5} \times \mathrm{P}_{\mathrm{Av}=3}(4)$$

etc. This suggests that the recurrence formula for probabilities of successive numbers of events for this model is:

$$\mathrm{P}_{\mathrm{Av}=3}(x) = \tfrac{3}{x} \times \mathrm{P}_{\mathrm{Av}=3}(x-1)$$

Probability distribution

You, the lawn expert, are now in a position to advise the gardener about the likelihood of her seeing various numbers of daisies in square metres of her lawn. We saw at the beginning of this chapter that $P_3(0) = 0.049\,787$ (this notation is less cumbersome than $P_{Av=3}(0)$).

Then we saw that:

$$P_3(1) = 3 \times P_3(0) \approx 0.149\,361$$
$$P_3(2) = \tfrac{3}{2} \times P_3(1) \approx 0.224\,042$$
$$P_3(3) = \tfrac{3}{3} \times P_3(2) \approx 0.224\,042$$
$$P_3(4) = \tfrac{3}{4} \times P_3(3) \approx 0.168\,031$$
$$P_3(5) = \tfrac{3}{5} \times P_3(4) \approx 0.100\,819$$

Extending the process:

$$P_3(6) = \tfrac{3}{6} \times P_3(5) \approx 0.050\,409$$
$$P_3(7) = \tfrac{3}{7} \times P_3(6) \approx 0.021\,604$$
$$P_3(8) = \tfrac{3}{8} \times P_3(7) \approx 0.008\,102$$
$$P_3(9) = \tfrac{3}{9} \times P_3(8) \approx 0.002\,701$$
$$P_3(10) = \tfrac{3}{10} \times P_3(9) \approx 0.000\,810$$

etc.

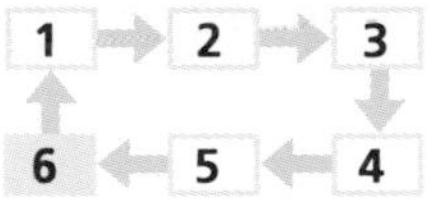

Compare with reality

How would you report your discoveries to the gardener? Suppose the garden is marked out in 100 metre squares. You might summarise your findings, briefly, by saying that she could expect to find five of the metre squares with no daisies: 'about 92 of them with between one and six daisies, and about three of the metre squares with more than six daisies.'

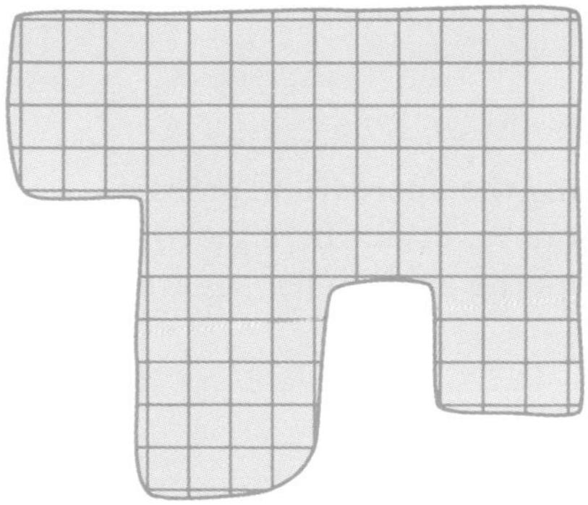

If you now reflect on the recurrence formula, it is quite possible to express the probability of the appearance of any given number of daisies in terms of the probability of none: $P_3(0)$. Try to express $P_3(4)$ in terms of $P_3(0)$. Consider the following argument.

$$P_3(1) = 3 \times P_3(0)$$

$$P_3(2) = \tfrac{3}{2} \times P_3(1) = \tfrac{3}{2} \times \{3 \times P_3(0)\} \quad \text{or} \quad \frac{3^2}{2} P_3(0)$$

Then:

$$P_3(3) = \tfrac{3}{3} \times P_3(2) = \tfrac{3}{3} \times \left\{\frac{3^2}{2} \times P_3(0)\right\} = \frac{3^3}{3 \times 2} \times P_3(0)$$

and so:

$$P_3(4) = \tfrac{3}{4} \times P_3(3) = \tfrac{3}{4} \times \left\{\frac{3^3}{312} \times P_3(0)\right\} = \frac{3^4}{4 \times 3 \times 2} \times P_3(0)$$

There seems to be an interesting pattern emerging in both the top and the bottom of the factor. The top appears to be:

3^x i.e. (average number)^x

But how can we rewrite the bottom of the factor? $4 \times 3 \times 2$ is the same as 4!, so it appears that:

$$P_3(4) = \frac{3^4}{4!} \times P_3(3)$$

which suggests that the probability function is:

$$P_3(x) = \frac{3^x}{x!} \times P_3(0)$$

Exploration 9.1

Check the validity

Check the validity of this probability function by evaluating $P_3(x)$ for $x = 1, 2, 3, \ldots, 10$ using the function and comparing with the values found using the recurrence formula.

Poisson probability distribution

The model developed in this chapter has centred on the distribution of the number of daisies which might be found in metre squares of a lawn where there were, on average, three daisies per square metre. How would the probability function differ had there been an average of two daisies, or one daisy, or just $\frac{1}{2}$ a daisy per square metre?

This table summarises the probability function that would result from a similar modelling process.

Average	Probability function
3	$P_3(x) = \frac{3^x}{x!} \times P_3(0)$
2	$P_2(x) = \frac{2^x}{x!} \times P_2(0)$
1	$P_1(x) = \frac{1^x}{x!} \times P_1(0)$
$\frac{1}{2}$	$P_{\frac{1}{2}}(x) = \frac{\frac{1}{2}^x}{x!} \times P_{\frac{1}{2}}(0)$

Each of the functions rests on finding the appropriate probability of there being no daisies in a metre square, $P_\lambda(0)$ where the symbol, λ, (the Greek letter lambda) represents the average per square metre.

CALCULATOR ACTIVITY

Relating probabilities

a) Use your calculator to evaluate $P_1(0)$ where:
$P_1(0) = \text{limit of } \left(1 - \frac{1}{n}\right)^n$

b) Use your calculator to discover the relationship between:
$P_2(0)$ and $P_1(0)$
$P_3(0)$ and $P_1(0)$
$P_{\frac{1}{2}}(0)$ and $P_1(0)$.

c) Can you predict the relationship between $P_\lambda(0)$ and $P_1(0)$?

Expected results

You may have found that:

$P_1(0) \approx 0.367\,879$ (to six d.p.)

and:

$P_2(0) \approx 0.135\,335$

which is the square of $P_1(0)$ i.e.

$P_2(0) = \{P_1(0)\}^2$

and you know already that:

$P_3(0) \approx 0.049\,787$

so your calculator discoveries may have led you to:

$P_3(0) = \{P_1(0)\}^3$

further:

$P_{\frac{1}{2}}(0) = 0.606\,531 = \{P_1(0)\}^{\frac{1}{2}}$

It would seem reasonable to suggest that there is a power relationship between $P_\lambda(0)$ and $P_1(0)$.

$P_\lambda(0) = \{P_1(0)\}^\lambda$

As a result of this, all probabilities of events for which the occurrence may be modelled in this way can be expressed in terms of $P_1(0)$. The value of $P_1(0)$ is:

$P_1(0) = 0.367\,879\,441\,2$ (to ten d.p.).

It is closely related to the natural number $e = 2.718\,281\,828\,5$ (to ten d.p.).

$P_1(0) = \frac{1}{e}$ or $= e^{-1}$

Hence, since $P_\lambda(0) = \{P_1(0)\}^\lambda$ then:

$P_\lambda(0) = \{e^{-1}\}^\lambda = e^{-\lambda}$

The probability function for this model is:

$$P_\lambda(x) = \frac{\lambda^x e^{-\lambda}}{x!}$$

This probability model was proposed by the 19th-century French mathematician, Simeon Poisson. It bears his name and is the probability model is known as the **Poisson distribution**. He proposed that this modelled the number of random events which occur in a fixed interval of a continuous medium.

In the example about the lawn, the random event is the appearance of a daisy and the continuous medium is the lawn, with the fixed interval being a square metre per week.

Poisson also required the events to occur singly (i.e. no more than one daisy can appear at the same place in the garden at one time) and the occurrence to be at a uniform rate (e.g. in the garden we assumed that it was the growing season and the daisies grew at a weekly rate of three per square metre, on average). This assumption implies that there is no tendency for daisies to appear more densely in one part of the garden than another. It also assumes that there is no change of season. Clearly, in a different season, daisies are not going to grow at the same rate – they might not grow at all. Poisson also required the events to be independent (i.e. the presence of a daisy in one place is not going to attract others appearing, nor prevent others appearing nearby).

Using the Poisson distribution

Example 9.1

A Geiger counter is used in recording radioactive events. Each radioactive event arriving at the counter shows as a number. The number of radioactive events recorded in a room is, on average, two events every five seconds. This is called the ***background count****.*

a) *Which probability distribution is appropriate to model the number of radioactive events, which are recorded every five seconds, in the room?*

b) *Use the appropriate probability function to find:*

i) the probability of exactly three events in five seconds,

ii) the probability of more than three events in five seconds.

Solution

a) *Radioactive events can be regarded as 'random events' and here, the Geiger counter records the number of events occurring in five-second intervals. These are 'fixed intervals' (5 s), in a 'continuous medium' (time). These events are recorded one at a time, i.e. singly, and may be assumed to occur independently and uniformly (two per five second interval, on average).*

This leads to the Poisson distribution with an average of 2.

b) *The probability model is:*

$$P_2(x) = \frac{2^x e^{-2}}{x!}$$

i) This requires you to find $P_2(3)$ *i.e.*

$$P_2(3) = \frac{2^x e^{-2}}{3!} \approx 0.1804$$

ii) Here, you need to find:

$P_2(4) + P_2(5) + P_2(6) + \ldots$ *etc.*

This can most readily be found by evaluating the probability of the 'complementary event' i.e. to find $P_2(0) + P_2(1) + P_2(2) + P_2(3)$.

$$P_2(0) + P_2(1) + P_2(2) + P_2(3) = e^{-2} + 2e^{-2} + \frac{2^2}{2!}e^{-2} + \frac{2^3}{3!}e^{-2} = 0.8571$$

Hence, the required probability is:

$1 - 0.8571 = 0.1429$

EXERCISES

9.1 A

1 The number of vehicles passing a traffic check-point every ten seconds is, on average, 5.3. Use the Poisson distribution to find the probability that in a ten-second iterval **a)** 4 vehicles pass, **b)** no vehicles pass, **c)** at least one vehicle passes, **d)** at most two vehicles pass the check point.

2 The random variable, X, has a Poisson distribution with mean 1.8. Evaluate:
a) $P(X = 1)$ **b)** $P(X \leq 2)$ **c)** $P(X \geq 1)$.

3 The number of broken eggs in a crate follows a Poisson distribution. If the probability of no broken eggs in a crate is $\approx$ 0.049 787, estmate the probability of:
a) one broken egg in a crate, **b)** two broken eggs in a crate.

4 A car-hire firm owns two vehicles capable of taking a wheelchair. On average there is one request each day for the daily hire of one of these vehicles. On a particular day, find the probability that:
a) no such vehicle is demanded,
b) both vehicles are in use.

5 Two car-hire firms, A and B, each own two vehicles capable of taking a wheelchair. On average each firm receives one request each day for the daily hire of one of these vehicles. On a particular day, find the probability that:
a) no such vehicle is demanded,
b) firm A hires out both of its vehicles but firm B hires out neither of its vehicles,
c) exactly one of the four vehicles is demanded,
d) exactly two of the vehicles are demanded,
e) firm A cannot satisfy demand for such vehicles,
f) neither firm can satisfy demand for such vehicles.

6 The carpet in a hotel's corridors is 2 m wide. During production, small flaws occur at a mean rate of 2.3 per 20 m^2 area. Determine the probability that in 20 m^2 of the carpet there are:
a) more that two flaws,
b) exactly three flaws,
c) at most four flaws,
d) between three and five flaws, inclusive.

7 The number of cars travelling north under a motorway bridge during the hours of darkness may be modelled as a Poisson variate having a mean rate of 3.26 every ten seconds. Calculate the probability that in a ten-second interval during the hours of darkness, the number of cars passing under the bridge travelling north is:
a) exactly three, **b)** at least two, **c)** more than four.

EXERCISES

9.1 B

1 The random variable, Y, may be modelled as a Poisson variate with parameter $\lambda = 2.1$. Calculate:
a) $P(Y = 2)$ **b)** $P(Y \geq 2)$ **c)** $P(Y < 4)$ **d)** $P(2 < Y \leq 4)$.

2 Investigations have revealed that spelling mistakes occur in a local newspaper at an average rate of 1.37 per 100 words of text and that their occurrence may be modelled by a Poisson distribution. Calculate the probability that in 100 words of text, the number of spelling

mistakes is:

a) exactly two, **b)** more than one,
c) at most two, **d)** between one and four, exclusive.

3 The number of faulty paving bricks in large batches has been shown to follow a Poisson distribution with mean 4.3. Evaluate the probability that in one of these batches the number of faulty bricks is:

a) exactly four, **b)** at most three,
c) at least two, **d)** between three and six, inclusive.

4 Explain why a Poisson distribution may be appropriate as a model for the *number* of faults per 20 m length of carpet.
Assume X may be modelled by a Poisson distribution with mean 0.98. Find the probability that in a randomly-selected 20 m length of this carpet:

a) there are no small faults, **b)** there are at most two small faults.

5 On average, there are two rotten apples in a box. Find the probability of finding:

a) exactly four rotten apples, **b)** at least two rotten apples,
c) not more than four rotten apples in a box.

6 A large computing facility is being designed to handle enquiries from all over the country. If enquiries are expected to arrive randomly at a mean rate of five per minute, determine the probability that:

a) no enquiry arrives in a minute,
b) at least two enquiries will arrive in a minute,
c) there will be two or three enquiries in a minute.

7 A field is sown with barley seed after the previous year's crop of wheat has been ploughed in. Unfortunately, the ploughing does not totally eliminate the wheat. Some wheat seeds remain and germinate. Assume that the average number of wheat seeds which subsequently germinate is 2.3 per square metre. Calculate the probability that a randomly-chosen square metre contains:

a) exactly one wheat seed,
b) two or three wheat seeds,
c) at least four wheat seeds which germinate.

VALIDATING THE MODEL

On page 238 we found the recurrence formula: $P_3(x) = \frac{3}{x} P_3(x-1)$

for the Poisson probability distribution where the average is 3. A convenient notation for a random variable, X, with occurrence that may be modelled in this way, is:

$$X \sim \text{Poi}(3)$$

In general, if a random variable, X, may be modelled with a Poisson distribution with average λ, then this is written:

$$X \sim \text{Poi}(\lambda)$$

The recurrence formula for this probability distribution is:

$$P_\lambda(x) = \frac{\lambda}{x} P_\lambda(x-1)$$

Just as in the case of a random variable which is binomially distributed, the recurrence formula is a very convenient way of calculating successive probabilities. We saw this on page 239, when we were calculating the probability of 0, 1, 2, 3, ..., 10 daisies in a metre square.

Example 9.2

Buttercups occur at an average rate of 1.8 per square metre per week. A gardener picks the buttercups that he finds in his garden once a week. What is the probability that he will collect:

a) *up to five buttercups,*

b) *more than eight buttercups,*

from an area of one square metre in a week?

Solution

If X is the number of buttercups found in a square metre in a week then it is reasonable to model X as a Poisson variate:

$X \sim \text{Poi}(1.8)$

The recurrence formula for X is:

$$P_{1.8}(x) = \frac{1.8}{x} P_{1.8}(x-1)$$

and:

$P_{1.8}(0) = e^{-1.8} = 0.165\,299$

a) *Hence, the required probability is:*

$P_{1.8}(X \le 5) = P_{1.8}(0) + P_{1.8}(1) + P_{1.8}(2) + \ldots + P_{1.8}(5)$

and:

$$P_{1.8}(1) = 1.8 \times P_{1.8}(0) \approx 0.297\,538$$

$$P_{1.8}(2) = \frac{1.8}{2} \times P_{1.8}(1) \approx 0.267\,784$$

$$P_{1.8}(3) = \frac{1.8}{3} \times P_{1.8}(2) \approx 0.160\,671$$

$$P_{1.8}(4) = \frac{1.8}{4} \times P_{1.8}(3) \approx 0.072\,302$$

$$P_{1.8}(5) = \frac{1.8}{5} \times P_{1.8}(4) \approx 0.026\,029$$

Adding these probabilities produces: $P_{1.8}(X \le 5) = 0.989\,622$

b) *To find $P_{1.8}(X > 8)$, the property of complementary events can be used to good effect. Thus:*

$P_{1.8}(X > 8) = 1 - P_{1.8}(X \le 8)$

We have already calculated the probabilities for 0, 1, 2, 3, 4 and 5, so all that needs to be done is to calculate:

$$P_{1.8}(6) = \frac{1.8}{6} \times P_{1.8}(5) \approx 0.007\,809$$

$$P_{1.8}(7) = \frac{1.8}{7} \times P_{1.8}(6) \approx 0.002\,008$$

$$P_{1.8}(8) = \frac{1.8}{8} \times P_{1.8}(7) \approx 0.000\,452$$

Adding the total of these to $P_{1.8}(X \le 5)$ gives:

$P_{1.8}(X \le 8) = 0.999\,890$

Hence: $P_{1.8}(X > 8) = 1 - 0.999\,890 = 0.000\,11$

The distribution function

In Chapter 7, *The binomial probability model*, we encountered the cumulative probability distribution function and the associated tabular values. The same sort of tables exist for the Poisson probability model. Entries in these tables provide us with probabilities such as:

$$P_{1.8}(X \leq 5),\ P_{1.8}(X \leq 8),\ P_2(X \leq 3),\ P_3(X \leq 10)$$

etc. When there is no requirement for large numbers of decimal places, the tables provide a quick and easy way of obtaining probabilities for Poisson events.

Exploration 9.2

Using the Poisson distribution

a) Use the Poisson distribution tables to write down, to four places of decimals, the probabilities:

- $P_{1.8}(X \leq 5)$
- $P_{1.8}(X \leq 8)$
- $P_2(X < 3)$
- $P_3(X \leq 10)$.

b) Check your answers to part **a)**, using the appropriate recurrence formula.

Example 9.3

A national newspaper has, on average, 3.6 misprints per page. Assume that the number of misprints on a page can be modelled as Poi(3.6). *Obtain, accurate to three places of decimals, the probabilities of the following events.*

a) *A page has fewer than six misprints.*

b) *A page is free from misprints.*

c) *A page has more than three misprints.*

d) *A page has exactly five misprints.*

e) *A page has between three and eight (inclusive) misprints.*

Solution

The degree of accuracy required allows the tables to be used.

a) $P_{3.6}(X < 6) = P_{3.6}(X \leq 5) = 0.8441 = 0.844$ *(three d.p.)*

b) $P(\text{page is free from misprints}) = P_{3.6}(X = 0) = 0.0273$
$= 0.027$ *(three d.p.)*

c) $P(\text{more than } 3) = 1 - P_{3.6}(X \leq 3) = 1 - 0.5152 = 0.485$ *(three d.p.)*

d) $P(\text{exactly } 5) = P_{3.6}(X \leq 5) - P_{3.6}(X \leq 4) = 0.8441 - 0.7064$
$= 0.1377 = 0.138$ *(three d.p.)*

e) $P(\text{between 3 and 8 inclusive}) = P_{3.6}(X \leq 8) - P_{3.6}(X \leq 2)$
$= 0.9883 - 0.3027$
$= 0.6856 = 0.686$ *(three d.p.)*

Mean and variance

When we developed our model for the daisies, we started by assuming that there were, on average, three daisies per square metre. Does the model actually have a mean value equal to three? In other words, is the expected value equal to three?

Remember from Chapter 5, *Expectation*, that the expected value of a random variable, X, was seen to be the sum:

$$\mu = \sum_{\text{all } x} x\text{P}(X = x)$$

In the case of this Poisson model, the possible values for the random variable are $x = 0, 1, 2, 3, 4, \ldots$ etc. This implies that the mean can be found from summing all products from $x = 0$.

$$\mu = \sum_{x=0} x\text{P}(X = x)$$

CALCULATOR ACTIVITY

Find the mean number of daisies

Try one of the following suggestions, using your calculator to discover the mean number of daisies according to the probability model you developed.

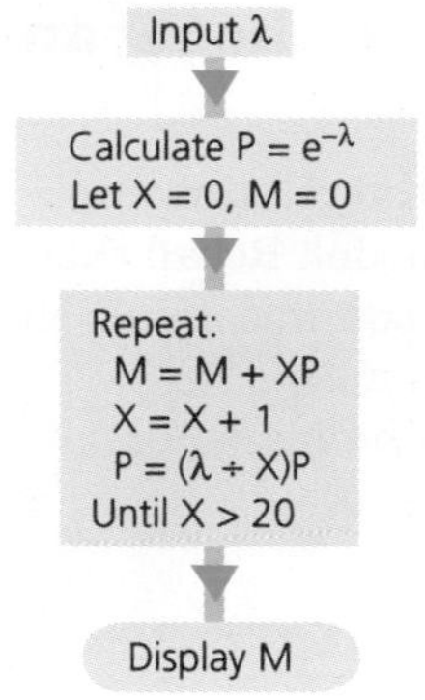

a) Use a program based on this flowchart which makes use of the recurrence formula for Poisson probabilities. This program displays the expected value to an accuracy of about eight places of decimals for Poi(λ) when λ is less than four.
To improve the accuracy, or to explore the expected value for larger values of λ, you need to change the condition of the 'Repeat ... Until' loop. If you change the stopping condition to 'Until $X > 40$' the program will display the expected value with a greater accuracy for λ up to the value 12.

b) You could use your calculator in 'Statistics' mode if it accepts probability distributions. See the advice offered in part **a)** on the limit needed on the value of X. If you are content with fewer than eight decimal places of accuracy, you could reduce the limit on X from 20 to 15 or fewer.

c) If you do not have an appropriate 'Statistics' mode calculator, you may be able to use its memory to evaluate the sum:

$$0\times e^{-3}+1\times 3e^{-3}+2\times\frac{3^2}{2!}e^{-3}+3\times\frac{3^3}{3!}e^{-3}+4\times\frac{3^4}{4!}e^{-3}\times\ldots+15\times\frac{3^{15}}{15!}e^{-3}+\ldots$$

A bit of factorisation might reduce the effort required on keying in this sum:

$$3c^{-3}\left\{1+3+\frac{3^2}{2!}+\frac{3^3}{3!}+\frac{3^4}{4!}+\ldots+\frac{3^{14}}{14!}\right\}$$

Expected results

You should find that the mean of the Poisson model for the daisies is the same as the quoted average of three. This is a somewhat reassuring result since it is what you would want of the model! It is generally true that the mean of the Poisson model, Poi(λ), is the same as its parameter, λ.

Mean: $\mu = \lambda$

Example 9.4

A random variable, X, is believed to follow a Poisson distribution, $\text{Poi}(\lambda)$. *It is known that the probability,* $P(X = 0)$, *is 0.1. What is the mean of the distribution?*

Solution

We know that the probability, $P(X = 0)$, *is* $e^{-\lambda}$. *But the information says that this is 0.1. So we need to find* λ *when* $e^{-\lambda} = 0.1$.

Using natural logs (since this is the inverse function):

$\log(e^{-\lambda}) = \log(0.1)$

Therefore $-\lambda \approx -2.303$ *or,* $\lambda \approx 2.30$

Hence the mean is about 2.30.

This can be checked by evaluating $e^{-2.30}$.

CALCULATOR ACTIVITY

Relationship between the variance and the mean of the Poisson model

Develop the previous calculator activity to discover the relationship between the variance and the mean of the Poisson model. Recall that the variance, σ^2, can be found from the 'mean of the squares *minus* the square of the mean':

$$\sigma^2 = \left\{\sum_{\text{all } x} x^2 P(X = x)\right\} - \mu^2$$

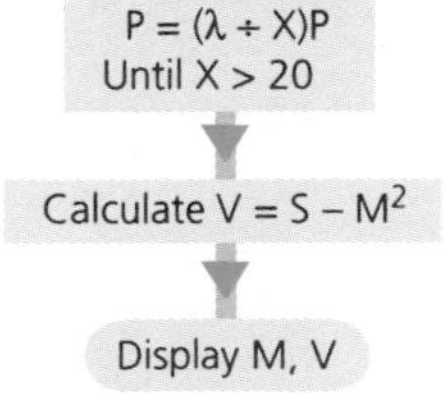
Input λ
Calculate P = e^{-λ}
Let X = 0, M = 0, S = 0
Repeat:
M = M + XP
S = S + X²P
P = (λ ÷ X)P
Until X > 20
Calculate V = S – M²
Display M, V

a) The previous flowchart can be adapted as shown.
This program produces both the mean and variance.

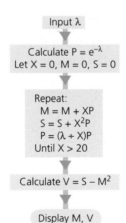

b) If you have an appropriate statistics calculator, you should be able to put in the same probability distribution as in the previous Calculator activity and extract the standard deviation. If you square the standard deviation you should get the variance.

c) Using the memory of a calculator you need to evaluate:

$$0^2 \times e^{-3} + 1^2 \times 3e^{-3} + 2^2 \times \frac{3^2}{2!}e^{-3} + 3^2 \times \frac{3^3}{3!}e^{-3} + \ldots + 15^2 \times \frac{3^{15}}{15!}e^{-3}$$

Factorising gives:

$$3e^{-3}\left\{1 + 2 \times 3 + 3 \times \frac{3^2}{2!} + 4 \times \frac{3^3}{3!} + 5 \times \frac{3^4}{4!} + \ldots + 15 \times \frac{3^{14}}{14!}\right\}$$

Once you evaluate this, you need to subtract the square of the mean.

The sort of results which emerge from a programmable calculator are summarised in the table that follows, for various values of the parameter λ and various limits for X.

λ	Limit on X	Mean	Variance
3	20	3	2.999 999 999
1.2	20	1.2	1.2
2	20	2	2
3.5	20	3.499 999 999	3.499 999 999
4	20	3.999 999 992	3.999 999 992
4	30	4	4
8	40	8	8
10	40	10	10
12	40	12	11.99 999 98

These results suggest that the variance of the Poisson model is the same as its mean.

Variance: $\sigma^2 = \lambda$

EXERCISES

9.2 A

1 If $X \sim \text{Poi}(4)$, use the Poisson tables to find the following.
a) $P(X < 4)$ **b)** $P(X > 3)$ **c)** $P(X = 2)$ **d)** $P(3 < X < 7)$ **e)** $P(2 < X \leq 8)$

2 A Mathematics student makes an average of three arithmetic errors on each homework assignment. Find the probability that, on a particular assignment, he made:
a) more than three errors, **b)** fewer than five errors,
c) a number of errors between two and six inclusive.

3 A Poisson variable has a mean of 5. Find the probability that:
a) $X < 5$ **b)** $X > 5$ **c)** $X \geq 5$

4 A Poisson variable is such that $P(X = 5) = 0.8\ P(X = 3)$. Find its mean and standard deviation.

5 A Poisson variable is such that $P(X = 3) = P(X = 4)$. Find $P(X = 5)$.

6 Sales of a particular pre-packed sandwich from a small shop follow a Poisson distribution with a mean of 9.4 per day. Calculate that probability that a particular day, the number of pre-packed sandwiches sold:
a) is fewer than eight, **b)** exceeds ten,
c) is between five and 15 exclusive.

7 The daily number, X, of items of crockery, broken in a hotel's kitchen may be modelled by a Poisson random variable with average 1.5.

x	$P(X = x)$	$f = 10\,000 \times P(X = x)$
0	0.2231	2231
≥ 8	0.0002	1
		1000

Complete the above table using $\lambda = 1.5$. Hence calculate the mean and variance of X. Approximate ≥ 8 by exactly 8. Comment on your two calculated values.

EXERCISES

9.2 B

1 A Poisson variable has a mean of 3.5. Find the probability that:
a) $X \geq 3$ **b)** $X < 3$ **c)** $X \leq 6$

2 The random variable, Y, has a Poisson distribution with parameter $\lambda = 4.35$. Use the recurrence formula to evaluate $P(Y = 0), \ldots, P(Y = 6)$. Hence state:
a) $P(Y \leq 4)$ **b)** $P(Y \geq 6)$ **c)** $P(3 \leq Y \leq 6)$ **d)** $P(1 < Y < 4)$.

3 A Poisson variable is such that $P(X = 3) = 0.6P(X = 2)$. Find its mean and variance.

4 If $X \sim \text{Poi}(\lambda)$ and $P(X = 4) = 2\,P(X = 3)$, calculate the value of λ and hence $P(X = 5)$.

5 Show that if $P(Y = 5) = 1.8P(Y = 3)$ then $\lambda = 6$. Hence calculate $P(Y = 6)$.

6 The discrete random variable X is known to follow a Poisson distribution with mean 5.15. Use the recurrence formula to calculate $P(X = 0), \ldots, P(X = 8)$. Hence determine:
a) $P(2 \leq X \leq 7)$ **b)** $P(X \leq 6)$ **c)** $P(X > 5)$ **d)** $P(3 < X \leq 8)$.

7 A store's weekly number of sales of a particular make of hairdrier may be modelled by a Poisson random variable with mean 6.4. Use Poisson distribution tables to calculate the probability that sales in a particular week:
a) exceed eight,
b) are fewer than four,
c) are between five and ten inclusive,
d) are less than eight but more than two.

If, at the beginning of the week, the store's stock of the hairdrier is ten, what is the probability that demand will exceed supply? Determine the minimum number of hairdriers that should be in stock at the start of the week if the probability of failing to meet demand is to be:
e) at most ten per cent
f) at most five per cent
g) at most one per cent.

ADDITIVE PROPERTY OF POISSON VARIATES

The lawn, which was the subject of expert investigation in the beginning of this chapter, sometimes has dandelions randomly appearing. They occur at a rate of about 0.6 per square metre per week. The growing seasons for daisies and dandelions overlap. If the only weeds which appear in the garden are daisies and dandelions, what is the weekly average number of weeds per square metre during the overlap period?

If we assume that the daisies and dandelions appear independently, it would seem reasonable to say that the weekly average number of weeds is (3 + 0.6) i.e. 3.6. This result would be in line with the results established in Chapter 8, *Sums and differences of distributions*. Let X be the random variable 'the number of daisies in a square in one week' and let Y be the random variable 'the number of dandelions in a square metre in the same week'. Then, Z, which is the total number of weeds in a square metre in the week in question, can be expressed as the sum of X and Y.

$$Z = X + Y$$

hence:

$$\mu_Z = \mu_X + \mu_Y$$

What would the variance of Z be?

Again, Chapter 8, *Sums and differences of distributions* suggests that:

$$\sigma_Z^2 = \sigma_X^2 + \sigma_Y^2$$

and the results from earlier in this chapter indicate that:

$$\sigma_X^2 = \mu_X = 3 \quad \text{and} \quad \sigma_Y^2 = \mu_Y = 0.6$$

This indicates that the variance of Z is the same as the mean of Z. This is one feature of the Poisson model we have developed in this chapter. Why might a Poisson model be appropriate for the number of weeds in a metre square?

It seems reasonable to suggest that the conditions which are met by the daisies and the dandelions separately: random, independent, uniform rate, single occurrence; are also met by them jointly as 'weeds'.

The argument presented here indicates that if X and Y are independent Poisson variables, then this sum, $X + Y$, is also a Poisson variate.

Example 9.5

In a physics laboratory, Geiger counter records indicate that the average count is one radioactive event every second. In a biology laboratory, the average count is four every five seconds. Use an appropriate probability model to calculate:

a) *the average count per five seconds in the physics laboratory,*
b) *the probability of a count of at least four radioactive events, in a five-second interval:*
  i) *in the physics laboratory,*
  ii) *in the biology laboratory,*
  iii) *in total in the two laboratories.*

Give your answers to three places of decimals.

Solution

a) *The Poisson model is appropriate. This model requires that the random events occur at a constant rate. This implies that if the rate of occurrence is one per second, then in five seconds the average will be five.*
b) *This question requires you to calculate:*
$P(X \geq 4)$
when X is a Poisson variate.

i) *In this part* $X \sim \text{Poi}(5)$. *The distribution tables indicate:*
$P(X < 4) = P(X \leq 3) = 0.2650$
Hence:
$P(X \geq 4) = 1 - P(X < 4) = 0.735$

ii) *Here,* $X \sim \text{Poi}(3)$. *From the tables:*
$P(X \geq 4) = 1 - P(X \leq 3) = 1 - 0.6472 \approx 0.353$

iii) *The random variable is the sum of the counts in each of the laboratories. The average count is therefore going to be (5 + 3) i.e. 8. Hence, for this part of the question* $X \sim \text{Poi}(8)$. *The tables indicate:*
$P(X \geq 4) = 1 - P(X \leq 3) = 1 - 0.0424 \approx 0.959$

Example 9.6

A skilled word-processor operator produces documents with an average of 0.3 errors per page. What is the probability that a document consisting of 15 pages has:

a) *no error,*
b) *at least one error,*
c) *fewer than five errors,*
d) *at least one error on each of five pages and none on the others?*

Solution

It is reasonable to assume that the errors are random events. These occur in a page i.e. in a fixed interval of a continuous medium. If it is also possible to assume that errors occur independently, singly and at a constant rate, a Poisson distribution is appropriate to model the number of errors which occur on a page. $\text{Poi}(0.3)$ *is the particular model for the errors on the page.*

a) *The model for the number of errors in a document of 15 pages is* $\text{Poi}(4.5)$. *Then:*
$P(0) = e^{-4.5} \approx 0.0111$

b) $P(\text{at least one error}) = 1 - P(\text{no error}) \approx 0.9889$.

c) $P(\text{fewer than five errors}) = P(\leq 4) \approx 0.5321$.

d) *This question is quite demanding. Consider the situation being modelled. Suppose the first five pages have at least one error on each and there are no errors on any of the remaining ten pages.*

≥ 1	≥ 1	≥ 1	≥ 1	≥ 1	0	0	...	0

The probability that this occurs is:
$\{P(\geq 1)\}^5 \{P(0)\}^{10}$
where $P(\geq 1)$ *refers to the chance that a single page has at least one error. Hence:*
$P(\geq 1) = 1 - e^{-0.3} \approx 0.259\,18$
and:
$P(0) = e^{-0.3} \approx 0.740\,82$
Hence, the probability that the first five pages each have at least one error and the rest have none is:
$(0.259\,18)^5(0.740\,82)^{10}$
However, this is one way where five pages each have at least one

error and none of the rest have any errors. There are ${}_{15}C_5$ *possible combination of pages which produce such a document. Hence, the probability required is:*
${}_{15}C_5(0.25918)^5\,(0.74082)^{10} \approx 0.1749$

An alternative perspective on this part of the question identifies the conditions which give rise to a binomial probability model, in particular, the B(15, 0.259 18) and the requirement is to find P(5).

Hypothesis tests

Our gardener from the start of this chapter notes one week that a particular metre square in the south-east corner of the garden contained seven daisies. She asks the expert if this means that the daisy population is more dense in this part of the garden. How would you set about answering the query and what advice would you offer the gardener?

Statistically speaking, the gardener is asking her lawn expert to conduct an **hypothesis test**. From Chapter 7, *The binomial probability model*, we know that we need to set up an appropriate model in order to determine probabilities. The modelling assumptions are those that lead to the Poisson model Poi(3), which assumes that daisies occur at the same rate throughout the lawn including the south-east corner. This model is summarised as the **null hypothesis**, H_0: $\mu = 3$.

The question asks whether daisies occur at a greater rate in the south east corner, this is the alternative hypothesis, H_1: $\mu > 3$.

Since there is a distinct direction in H_1, the test to be conducted is one-tailed. The direction indicates that 'large' observations would be considered unusual under the null hypothesis.

Suppose we conduct the test at the five per cent level of significance, so that if the observed number of daisies lies in an upper tail of five per cent then it would be considered as 'significant' under the null hypothesis. The distribution tables for Poi(3) indicate that:

$$P(X \geq 7) = 1 - P(X \leq 6) = 1 - 0.9665 = 0.0335$$

Hence:

$$P(X \geq 7) = 3.35\% < 5\%$$

which suggests that an observation of seven or more will occur in less than five per cent of the metres squares on average.

It is appropriate to report to the gardener that this is an unusually high number of daisies. It is not impossible when the average is actually three, but it is reasonable to conclude that the mean occurrence of daisies in the south-east corner of the garden is higher than three.

Briefly, the expert might say that the daisies are more dense there than elsewhere.

EXERCISES

9.3 A

1 Faults occur in the manufacture of yarn at random and at a mean rate of 0.6 per 1000 m. Determine the probability that in a 1000 m roll there are:
a) no faults,
b) more than two faults.
Find also the probability that in a roll of 10 000 m the number of faults is:
c) exactly six,
d) fewer than ten,
e) between two and eight, inclusive.

2 The independent random variables, X, Y, Z, have Poisson distributions with means 2.6, 3.4 and 1.5 respectively. Evaluate:
a) $P(X + Y = 6)$
b) $P(X + Y \geq 8)$
c) $P(X + Z < 4)$
d) $P(X + Y + Z > 10)$
e) $P(Y + Z \leq 5)$
f) $P(3 \leq X + Z \leq 7)$.

3 Accidents occur on a process plant at an average of 2.1 per week. Determine the probability of:
a) more than three accidents in a week,
b) exactly two accidents in each of two successive weeks,
c) exactly four accidents in a fortnight,
d) more than eight accidents in a four week period.
Explain why your answers to **b)** and **c)** differ.

4 A newspaper of 24 pages contains on average 36 misprints.
a) On the leader page find the probability of:
  i) at least three misprints,
  ii) not more than four misprints.
b On the two sports pages together, find the probability of:
  i) no misprints at all,
  ii) only one misprint,
  iii) two misprints in all.

5 A PC computer network which operates five days per week is subject to random crashes at an average rate of 0.9 per day. Use the Poisson distribution to estimate the probability of:
a) exactly five crashes in a week,
b) fewer than six crashes in a fortnight,
c) between three and eight crashes inclusive in a fortnight,
d) more than five crashes in a fortnight.

6 A newspaper of 24 pages contains, on average, 36 misprints. Find the probability of:
a) no misprints on the front page,
b) no misprints on any of the three Arts pages,
c) that there is exactly one misprint in the Arts section,

d) that there are exactly two misprint in the Arts section,
e) that no page in the newspaper is free of misprints.

7 The number of emergency admissions each day to a hospital is found to have a Poisson distribution with mean 2.
a) Evaluate the probability that on a particular day there are no emergency admissions.
b) At the beginning of one day the hospital has five beds available for emergencies. Calculate the probability that this will be insufficient to meet the demand for that particular day.
c) Calculate the probability that there will be a total of three admissions during a period of two consecutive days.
d) How many beds should b available at the beginning of a day for the hospital to be at least 95 per cent sure of coping with the demand for beds?

8 Explain briefly, referring to your practical work if possible, the role of the null hypothesis and of the alternative hypothesis in a test of significance.
Over a long period, John has found that the bus taking him to school arrives late, on average, nine times per month. In the month following the start of new summer schedules, John finds that his bus arrives late 13 times. Assuming that the number of times that the bus is late has a Poisson distribution, test, at the five per cent level of significance, whether the new schedules have, in fact, increased the number of times on which the bus is late. State clearly your null and alternative hypotheses.

9 If, in question 3, there were five accidents in a particular week, would this suggest that the average weekly rate had increased? (Use a five per cent level of significance.)

10 Small bubbles in a particular type of glass occur at random. When the production process is operating satisfactorily, the mean number is 0.2 per square metre. A random sample of 20 pieces of glass, each 1 m^2 in area, reveals a total in seven bubbles. Investigate the claim that the process is not operating satisfactorily as it is producing too many small bubbles.

EXERCISES

9.3 B

1 The demand for a certain spare part occurs randomly at an average rate of four per week. Find the probability that the demand is for:
a) at least five in a week,
b) exactly four in a week,
c) at most six in a fortnight,
d) exactly four in each of two successive weeks.
What is the minimum number of this spare part which must be in stock at the beginning of each week so that the probability of failing to meet demand during the week is less than one per cent?

2 The independent Poisson random variables, R, S and T have parameters, 0.8, 2.6 and 5.6 respectively. Calculate the probability that:

a) $R + S > 3$ **b)** $R + T \leq 6$ **c)** $2 < S + T < 10$
d) $R + S + T \geq 5$ **e)** $R + S + 7 = 10$.

3 For the period 6.30 a.m. to 8.30 a.m. the demand for a particular daily paper at a small newsagents may be modelled by a Poisson variate with mean 12.8. Calculate the probability that the demand for the paper:

a) exceeds five during the period 6.30 a.m. to 7.00 a.m.
b) is fewer that ten during the period 7.00 a.m. to 8.00 a.m.
c) is between five and 15, inclusive, during the period 6.30 a.m. to 8.00 a.m.
d) exceeds five during the period 7.15 a.m. to 7.45 a.m. and between 8.00 a.m. and 8.30 a.m.

4 On average a switchboard receives six calls in an hour.

a) Find the probability there are no calls in a five-minute period.
b) Find the probability there are no calls in a ten-minute period.
c) Find the probability of there being no calls between 4.05 p.m. and 4.10 p.m. given that there were no calls between 4.00 p.m. and 4.05 p.m.
d) Find the probability of there being no calls between 4.10 p.m. and 4.15 p.m. given there were no calls between 4.00 p.m. and 4.10 p.m.,
e) What do you deduce from your answers to **c)** and **d)**?

5 A garage showroom has a demand for cars which is known to have a Poisson distribution with an average of 1.2 per day. Use the tables to answer the following questions.

a) What is the probability that four cars are sold in one day?
b) What is the probability that more than five cars are sold in a three-day period?
c) How many cars should the showroom have at the beginning of a five-day working week to be at least 95 per cent sure of meeting that week's demand?

6 X is a Poisson variable with mean 1.3, and Y is a Poisson variable with mean 2.7. Find the mean and standard deviation of:

a) $X + Y$ **b)** $Y - X$ **c)** $2X$ **d)** $3Y - X$.

7 The number of students dropping out of a university course each year follows a Poisson distribution with a mean of 6.4. Last year, only two students dropped out. Does this provide evidence, at the five per cent significance level, of a change in behaviour?

8 The number of serious accidents on a certain stretch of road averages three each year. Last year there were seven accidents on this stretch. Does this provide evidence, at the five per cent significance level, that the situation has deteriorated?

9 A traffic survey has established that the number of vehicles passing a school's entrance has a Poisson distribution with mean 9.6 per minute. Following the operating of a new alternative route, five vehicles pass the school's entrance in a one-minute interval. Test the hypothesis that the opening of the alternative route has made no difference to the mean number of vehicles passing the school's entrance.
Suggest how the test could be improved.

CONSOLIDATION EXERCISES FOR CHAPTER 9

1 The number of wrong notes hit by a pianist in a certain concerto is a Poisson variable with a mean of 1.6. If he plays the concerto twice, find the probability that:
a) he plays the concerto perfectly both times,
b) he plays the concerto perfectly once only,
c) he hits two wrong notes altogether,
d) he hits three wrong notes altogether,
e) he hits no more than two wrong notes in either concerto.

2 The number of double-yolked eggs in a crate is a Poisson variable with mean 1.5.
a) In one crate of eggs, find the probability of finding:
  i) more than three such eggs,
  ii) no more than two such eggs.
b) In two crates of eggs, find the probability of finding:
  i) exactly two such eggs altogether,
  ii) no more than two such eggs in either crate.

3 The number of faulty batteries in a box is a Poisson variable with a mean of 2.8.
a) Find the probability that a box is free of faulty batteries.
b) If there are three boxes, find the probability that exactly two of them are faultfree.
c) If there are five boxes, find the probability that not more than two are faultfree.

4 On average there are three orange Smarties in each packet. Find the probability of finding.
a) fewer than three orange Smarties in a packet,
b) more than three orange Smarties in a packet,
c) fewer than three orange Smarties in each of two packets,
d) exactly three orange Smarties in total, if two packets are bought.

5 A Poisson variable is such that its mean is equal to its standard deviation. Find the possible value of P(3).

6 On average an office receives three faxes an hour. Find the probability that they receive:
a) exactly two faxes in an hour,
b) exactly one fax in half an hour,
c) at least one fax in 15 minutes,
d) not more than two faces in a two-hour period.

7 The number of defective escalators each day in an underground system is a Poisson variable with a mean of 3 and the number of defective lifts in the system is a Poisson variable with a mean of 2. Find the probability that, on a given day:
a) there are one defective escalator and one defective lift,
b) there are at least two defective escalators and at least one defective lift,
c) there are not more than three defective machines altogether,
d) the system is operating perfectly.

8 A footballer has just been transferred from a Division 3 club to one in the Premier division. He has a scoring record of 0.8 goals per match.
a) Assuming that he maintains this record in the Premier division what is the chance that he will score two goals in a given game? State your assumptions.
b) What is the probability that he will have scored more than eight goals after twelve games?
c) Another striker with a scoring rate of 0.7 goals per game joins the Premier division club. The two strikers play together in a match. What is the probability they score four goals between them?

9 Every five minutes, two taxis, on average, pass a particular spot. Buses come past, on average, one every ten minutes. No other vehicles pass this spot. Find the probability that:
a) no buses or taxis pass in a ten-minute period,
b) exactly three vehicles pass in a five-minute period,
c) more taxis than buses pass in ten minutes.

10 Two independent observations, x_1 and x_2, are taken from a Poisson distribution with mean μ in order to test the hypothesis H_0: $\mu = 5$ against H_1: $\mu < 5$. The critical region selected is $x_1 + x_2 \le 4$.
a) State the distribution of the random variable $Y = X_1 + X_1$.
b) Find the size of the critical region.

11 The independent random variables X and Y have Poisson distributions with means λ and μ respectively. The random variable $Z = X + Y$. Show that:
a) $P(Z = 0) = e^{-(\lambda + \mu)}$
b) $P(Z = 1) = (\lambda + \mu)\,e^{-(\lambda + \mu)}$.
Determine $P(Z = 2)$ and hence deduce the distribution of Z.
Verify your deduction by evaluating $P(Z = 3)$ directly, and by using appropriate combinations of probabilities for X and Y.

12 Explain why, if m is the modal value of a Poisson distribution with mean λ, then $P(X = m) \ge P(X = m - 1)$ and $P(X = m) \ge P(X = m + 1)$. By making use of the recurrence formula, deduce that $\lambda - 1 \le m \le \lambda$. State the mode when $\lambda = 5.2$. Use the Poisson tables to confirm your answer.

13 The independent Poisson random variables X and Y have means of 2.5 and 4.5 respectively. Obtain the mean and variance of the random variables:

a) $X - Y$ **b)** $2X + 5$.

For each of these random variables give a reason why the distribution is **not** Poisson.

14 The annual number of accidents at a certain road junction may be modelled by a Poisson distribution with mean 8.5. Improvements intended to reduce the accident rate are made at the junction. To test their effectiveness it is planned to monitor the number of accidents at the junction in the following year.

a) State suitable null and alternative hypotheses for the test,

b) If the improvements are to be considered effective when fewer than five accidents occur in the year, determine the significance level of the test.

c) If the improvements reduce the mean annual number of accidents at the junction to three, what is the probability of having five or more accidents?

15 A car-hire firm finds that the daily demand for its cars follows a Poisson distribution, with mean 3.6.

a) What is the probability that on a particular day the demand will be:

i) two or fewer,

ii) from three to seven (inclusive),

iii) zero?

b) What is the probability that ten consecutive days will include two or more on which the demand is zero?

c) Suggest reasons why daily demand for car hire may not follow a Poisson distribution.

(AEB Question 6 (part), Paper 9, Winter 1994)

16 In an area of heathland, squares of equal area were chosen at random. The number of plants of a particular species in each square follows a Poisson distribution with mean 3.8.

What is the probability that the number of plants of this different species in a particular square is:

a) five or fewer,

b) exactly four,

c) eight or more?

(AEB Question 5 (part), Paper 9, Summer 1994)

17 Data files on computers have sizes measured in megabytes. When files are sent from one computer to another down a communications link, the number of errors has a Poisson distribution. On average, there is one error for every ten megabytes of data.

a) Find the probability that a three-megabyte file is transmitted:

i) without error, **ii)** with two or more errors.

b) Show that a file which has a 95 per cent chance of being transmitted without error is a little over half a megabyte in size.

A commercial organisation transmits 1000 megabytes of data per day.

c) State how many errors per day they will incur on average.

(MEI Question 3 (part), Paper S2, January 1993)

18 A single observation is to be taken from a Poisson distribution with mean μ and used to test H_0: $\mu = 6$ against H_1: $\mu < 6$. The critical region is chosen to be $x \leq 2$. Find the size of the critical region.

(ULEAC Question 2 (part), Specimen paper T3, 1994)

19 A shop sells a particular make of radio at a rate of four per week on average. The number sold in a week is thought to have a Poisson distribution.

a) Using a Poisson distribution, find the probability that the shop sells at least two in a week.

b) Find the smallest number that can be in stock at the beginning of a week in order to have at least a 99 per cent chance of being able to meet all demands during that week.

c) Comment on the applicability of a Poisson distribution.

(ULEAC Question 2, Specimen paper T1, 1994)

20 The number of items of type A and type B sold in a week by a shop have independent Poisson distributions with means 2.4 and 3.6 respectively.

a) Calculate the probability that the shop will sell three items of type A and five items of type B in a week.

b) Find the probability that the shop will sell a total of fewer than eight of these items in the week.

(NEAB Question 8 (part), Specimen paper 4, 1996)

21 Measurements of the time intervals between successive arrivals of telephone calls at an office exchange were taken. The first 100 time intervals were recorded and the following grouped frequency distribution was obtained.

Time interval (x minutes)	Frequency
$0 < x \leq 0.5$	39
$0.5 < x \leq 1.0$	23
$1.0 < x \leq 2.0$	23
$2.0 < x \leq 3.0$	9
$3.0 < x \leq 6.0$	6

a) Draw a histogram to illustrate this distribution.

b) Calculate, showing your working, estimates of the mean and the standard deviation of the distribution.

c) State, with a reason, whether or not a Poisson distribution may be used as a model for the distribution.

(JMB Question 5, Paper II, June 1992)

22 Inspection of 100 metres of curtain material revealed a total of four flaws. Using a suitable Poisson distribution as a model, test at the two per cent significance level the null hypothesis that the mean number of flaws per 10 metres of the material is equal to 1, against the alternative that it is less than 1.

(WJEC Question 3, Specimen paper A3, 1994)

23 The number of accidents which occur on a particular section of road in a one-week period is modelled by a Poisson distribution with mean 1.2. Find the probability that there is more than one accident on this section of the road in a particular week.

(UCLES Question 7, Specimen (Linear) paper 4, 1994)

24 Analysis of the scores in a football match in a local league suggests that the total number of goals scored in a randomly-chosen match may be modelled by the Poisson distribution with parameter 2.7. The numbers of goals scored in different matches are independent of one another.

a) Find the probability that a match will end with no goals having been scored.

b) Find the probability that four or more goals will be scored in a match.

One Saturday afternoon, eleven matches are played in the league.

c) State the expected number of matches in which no goals are scored.

d) Find the probability that there are goals scored in all eleven matches.

(MEI Question 2, Specimen paper S2, 1994)

Summary

- The Poisson distribution models the number of random events which occur in a fixed interval of a continuous medium.
- $P_{\lambda}(x) = \dfrac{\lambda^x e^{-\lambda}}{x!}$

  mean of Poisson distribution $\mu = \lambda$

  variance of Poisson distribution $\sigma^2 = \lambda$

STATISTICS

10 Samples and populations

In this chapter we shall investigate:

- *random sampling,*
- *systematic, stratified, quota and cluster sampling,*
- *sampling from distributions.*

We shall now find out how to explore characteristics of data where we are unable to gain access to all the data. The exploration process can be refined to improve its accuracy, to improve its efficiency, to make it easier to use; however there are drawbacks. These refinements are introduced in this chapter. Finally, we shall consider how models for data may be simulated using related processes.

SUBJECTIVE AND RANDOM SAMPLING

Have a look at the diagram and, in no more time than it takes you to finish this sentence, select one value from the display which you feel is representative of all 200 numbers.

	A	B	C	D	E	F	G	H	I	J	K	L	M	N	O	P	Q	R	S	T
1	8.8	6.6	9.2	6.0	11.1	15.3	7.0	6.0	6.0	6.8	7.9	12.2	6.9	8.1	9.8	11.1	6.2	9.0	7.1	7.7
2	6.4	8.2	6.6	7.0	9.8	6.1	12.2	6.8	6.6	6.9	9.5	8.6	8.0	6.0	7.8	6.6	11.7	7.1	11.7	12.2
3	17.4	8.0	12.8	13.8	9.5	7.4	7.7	16.6	5.8	6.5	7.0	19.3	9.9	7.9	11.4	10.8	8.3	6.4	5.7	16.2
4	8.4	7.9	7.5	7.2	11.0	19.5	6.2	9.3	13.1	15.5	7.1	9.4	10.8	10.2	10.7	13.4	7.2	12.0	9.4	9.4
5	13.6	6.8	8.2	12.6	13.1	14.2	8.4	8.4	5.9	9.0	7.8	10.1	11.8	7.2	7.2	6.7	17.2	9.7	8.9	10.4
6	18.9	11.5	6.7	6.3	7.8	5.5	6.2	13.3	14.7	15.2	13.3	7.2	6.4	8.9	9.6	10.4	6.5	13.0	10.7	7.3
7	12.1	7.7	10.7	16.0	8.5	10.8	10.3	11.4	9.6	11.9	10.7	7.7	15.7	7.3	11.1	8.5	8.1	8.4	8.1	8.6
8	8.7	8.3	6.3	6.7	9.6	8.1	6.1	11.0	10.9	6.0	15.7	8.9	6.0	8.3	15.5	8.7	8.7	17.8	11.6	5.8
9	6.5	18.5	6.5	8.6	6.7	6.8	7.9	6.1	12.2	11.2	6.8	15.8	9.6	8.6	10.5	9.2	8.7	13.2	6.2	7.8
10	9.1	11.0	8.1	8.6	7.6	9.0	7.7	8.5	11.6	8.3	10.3	8.6	7.4	7.9	6.2	8.6	9.3	7.7	12.1	15.0

You might think that was a pretty tall order i.e. you might think you were being asked to carry out a virtually impossible task. In a way, you would be right. However, think back to Chapter 2, *Summary statistics 1*, which was about summary values, about averages. The difference between what you were asked to do above, and what you did in Chapter 2, was that in Chapter 2 you had more time, and you could involve every item of the data in your summary value.

There are many occasions where we need to know something about the characteristics of data but our access is restricted in some way.

Consider a company that manufactures fireworks. Fireworks need a fuse that must burn for a sufficiently long time to allow the organiser to be out of the way when the firework goes off, or explodes, or ... .

A responsible manufacturer would want to use only those fuses with a sufficiently long burn time. So, how do they check the burn time of all fuses? The only sure way is to set every fuse alight and time it – but then they wouldn't have any fuses left to put on the fireworks!

They need a reliable method of **sampling** from the entire stock of fuses available. This entire stock is known as the **population** of fuses. The firework manufacturer needs to obtain a **representative sample** of fuses i.e. a small collection of fuses, having the characteristics of the whole stock. Why would taking a sample of just one fuse and testing this not be of much benefit? Clearly, this would give no indication of the variation there might be in the times. A larger sample is needed.

Return to the data in the diagram above and select a **sample of size 6** i.e. a sample of six items of data, that you think is **representative** of the population of 200. (Carry out your selection without doing any calculations.)

There are many thousands of possible samples, including:

5.5 6.5 7.5 8.5 18.5 19.5

In what way do you think this sample is representative of the population? If you scan the population you will see that it only contains numbers in this range, so it is representative in one sense.

In what ways may the sample not be representative of the population? The mean of the sample may not be the same as the population mean, the variances may differ, the sample does not have a unique mode, the distributions of frequencies are not the same, ...

Reflect on your own sample. In what ways is it representative, and in what ways may it not be representative of the population?

It is rather unrealistic to expect a sample of size 6 to be wholly representative of a population, even a population of such limited size. It is more reasonable to require that the sampling method is not subjective and introduces no bias in the selection. One method of achieving this is called **random sampling**. This process gives every member of the population the same chance of being in the sample and does not exclude any sections of the population at any stage of the process.

One approach at random sampling is through **random numbers** such as those found on some calculators, on computers or in tables. This involves setting up a one-to-one correspondence between the random numbers and the members of the population i.e. there is a unique pairing between each member of the population and a unique collection of random numbers. A fairly obvious approach, in the case of the population of 200 opposite, is to use three-digit random numbers and the one-to-one correspondence, as shown here.

Random number	Member of population
001	1st
002	2nd
003	3rd
…	…
…	…
200	200th

The correspondence is unique and each member has an equal chance of being selected. Selecting a sample in this way requires a list of the population. Such a list is known as a **sampling frame**. The list must be complete, up to date, and contain only the population.

CALCULATOR ACTIVITY

The random number facility

Use the random number facility on your calculator to select a sample of size 6 from the population listed above. It might be convenient to number the members of the population so that column A contains the 1st to 10th members; column B contains the 11th to 20th members, etc.

Evaluating the results

What difficulties did you encounter in the calculator activity? Suppose your calculator produced the following random numbers:

0.588 0.269 0.117 0.002 0.824 0.715 0.513

If the method suggested is followed, 0.588 and 0.269 would correspond to the 588th and 269th members of the population. But these do not exist, so these random numbers would be ignored. Many other random numbers would also be ignored, so clearly this is an inefficient way of using random numbers.

Efficiency

There are many ways of improving the efficiency and trying to use all of the random numbers.

Casting out

In this method, the random numbers giving rise to 201, 202, 203, etc. have the 200s removed. This would produce the following results.

Random number	Remove	Position in sampling frame	Member
269	200	69th	7.9
588	400	188th	11.6
824	800	24th	7.5
117	0	117th	7.7

Grouping

An equivalent procedure which works very well if a computer or calculator is available, but is rather awkward to use manually, is to group the random numbers into, in this case, fives. This means we allow 0.001, 0.002, 0.003, 0.004, 0.005 to correspond to the first entry in the sampling frame, and 0.006, 0.007, ..., 0.010 to correspond to the second entry in the sampling frame, etc. We need to consider which random numbers will correspond to the 200th entry, since it seems as if only 0.996, 0.997, 0.998, 0.999 are available once the others have been assigned. So the 200th member of the population is associated with only four random numbers, unlike the five for every other. (A way of overcoming this is to include 0.000.)

Try to identify which members of the population would be selected using the 'grouping' approach.

Pure random sampling

You may have found that grouping was not very easy. One way of simplifying it is to divide the three-digit random number by the group size, and use the next integer (unless the result is already a whole number). This is illustrated in this table.

Random number	Divide by group size	Position in sampling frame	Member
588	588 ÷ 5 = 117.6	118th	8.9
269	269 ÷ 5 = 53.8	54th	19.5
117	117 ÷ 5 = 23.6	24th	7.5
002	2 ÷ 5 = 0.4	1st	6.6
824	824 ÷ 5 = 164.8	165th	17.2
715	715 ÷ 5 = 143	143rd	11.4
513	513 ÷ 5 = 102.6	103rd	7.0

One very real reservation about this and the previous procedure is that we have to treat the random number 000 (i.e. 0.000) in a special way and convert the random decimals from our calculator into integers. This can be overcome by using a function know as **the integer part of** ... or INT(x).

To get 588 from the random decimal 0.588 we have effectively multiplied it by 1000; the process involved here requires us to divide the result by 5. A short-cut to this is to do the multiplication (× 1000)) and division (÷ 5) in one move, i.e. to multiply the calculator random decimal by 200. For example:

$$\text{INT}(200 \times 0.588) = \text{INT}(117.6) = 117$$

and $\text{INT}(200 \times 0.715) = \text{INT}(143) = 143$

and $\text{INT}(200 \times 0.002) = \text{INT}(0.4) = 0$

but this does not give quite the same results. The major concern is that 0.002 corresponds to a position 'zero' in the sampling frame. This is a little difficult to work with, so a convenient alteration to the procedure is to *add one* to the result of the multiplication. This leads to the results in the following table.

Random decimal	Conversion	Position in sampling frame
0.588	INT(200 × 0.588 + 1)	118th
0.269	INT(200 × 0.269 + 1)	54th
0.117	INT(200 × 0.117 + 1)	24th
0.002	INT(200 × 0.002 + 1)	1st
0.824	INT(200 × 0.824 + 1)	165th
0.715	INT(200 × 0.715 + 1)	144th
0.513	INT(200 × 0.513 + 1)	103rd

There is only one difference between the positions here and those found previously. This is illustrated below where 0.000 is not treated any differently.

Random decimal	Correspondence in sampling frame
0.000, 0.001. .... 0.004	1
0.005, 0.006, ..., 0.009	2
0.100, ..., 0.104	3
...	...
...	...
0.995, 0.996, 0.997, 0.998, 0.999	200

Now each member of the population has an equal chance of being in the random sample.

Sampling populations of any size

The method just described is readily adaptable to populations of other sizes. For instance, if the population has 82 members then:

INT(82 × random decimal + 1)

will generate appropriate positions in the sampling frame.

The number of places in the random decimal affects the randomness of this process. The greater the number of digits in the decimal the better. If you are dependent on a calculator which produces random three-digit decimals, it is possible to extend these to six places, nine places, ... For example, the two three-digit decimals 0.588, 0.269 can be computed as:

$0.588 + (0.269 \div 1000)$

to give the six-digit decimal 0.588 269.

Random decimals with six places should be sufficient to sample from sampling frames with tens of thousands of elements.

The above discussion indicates that an efficient way of using calculator random decimals to sample from populations of size N is to use:

$\text{INT}(N \times (r + r \div 1000) + 1)$

where r is a three-digit random decimal.

In most practical situations, sample sizes are quite small, so duplication is not allowed in a sample. This means that if the same position in the sampling frame is selected more than once, then it is ignored on the second and subsequent occasions. In short, no repeats are allowed in small samples. For example, when using the 'casting out' manual technique of selecting a sample from the data on page 262, the first few random decimals which appear may be:

0.517 0.317 0.991.

These correspond to:

Random number	Remove	Position in frame	Member
517	400	117	7.7
317	200	117	7.7
991	800	191	7.7

The random numbers 517, 317 correspond to the same position in the sampling frame, hence 317 is ignored and only the member of the population selected by 517 is allowed. The random number 991 corresponds to a different position and so the corresponding member of the population is accepted in the sample, even though it is the same in value.

Exploration 10.1

A sample of size 10

The following is a collection of calculator random decimals.

0.431	0.214	0.143	0.596	0.475
0.423	0.270	0.606	0.313	0.309
0.928	0.378	0.851	0.238	0.734
0.200	0.190	0.525	0.933	0.278
0.064	0.821	0.209	0.888	0.808

Obtain a random sample of burn times of size 10 using:

a) casting out,
b) grouping,
c) Pure random sampling using:
i) three decimal places, **ii)** six decimal places.

Calculate the mean burn time for each sample.

Example 10.1

A shopkeeper, each year, sells 200 packets of sparklers. Each packet contains five sparklers.

Last year she had many complaints that the sparklers were difficult to light. She receives her supply of 200 packets to sell this year and decides to use a sample of five per cent of the packets to establish the proportion of sparklers which are difficult to light.

a) *How would you advise the shopkeeper to undertake this task?*
b) *The shopkeeper obtains a sample of ten packets and finds the number of sparklers in each packet that are difficult to light.*

The results are:
2, 1, 5, 2, 2, 2, 2, 3, 1, 0.
What proportion of the sparklers do you estimate to be difficult to light?

Solution

a) *The shopkeeper needs to establish a sampling frame. She could do this by numbering the packets: 1, 2, 3, ... , 200. Having done this she needs to obtain a random sample. She wants a sample of five per cent of the packets, so she needs ten packets in her sample. A random sample may be obtained using a random number generator e.g. programming a calculator to identify which packets to include in the sample. The calculator could be programmed to produce the following:*
INT(200 × (random decimal) + 1)
until ten distinct numbers are obtained. Then the ten packets of sparklers can be selected. Having selected the packets, she needs to open each of them and test the contents, to see how many are difficult to light. The problem, of course is that once alight they are impossible to stop, hence the sparklers will be destroyed.

b) *The sample indicates that 2 + 1 + 5 + ... + 1 + 0 i.e. 20 sparklers were difficult to light. The ten packets each had five sparklers. So, 20 sparklers out of 50 were difficult to light. This suggests that an estimate of the proportion is:*

$\frac{20}{50}$ *or* $\frac{2}{5}$ *or* 0.4

EXERCISES

10.1A

1 A population has N members. A calculator produces the following random decimals.

0.687, 0.802, 0.070, 0.802, 0.212, 0.974. 0.562, 0.297, 0.179, 0.395

A sample of size five is selected from a sampling frame listing the population. The position in the frame of each member of the sample is found using:

$$\text{INT}(N(r_1 + r_2 \div 1000) + 1)$$

where r_1, r_2 are consecutive three-digit random numbers. Identify which members of the frame would be selected if:
a) $N = 400$ **b)** $N = 5000$ **c)** $N = 147$ **d)** $N = 60000$ **e)** $N = 200000$.

2 Describe how you would use random numbers to select a random sample of 15 members from a population of 438 members.

3 State which of the following are subjective samples and which may be random samples.
a) a sample of patients chosen as participants in a drug test
b) the winners in a lottery
c) the top Year 10 mathematics set
d) the children whose surnames begin with B in a primary school

4 A partnership of GPs would like to obtain a sample of patients on their register to investigate their level of satisfaction with the management of the practice. Discuss the problems they would face in simply assigning random numbers to each name on their register.

5 Indicate an appropriate sampling frame for each of the following.
a) a sample to survey local attitudes to the building of a new local supermarket on green belt land
b) children's attitudes in a secondary school to a shortened lunch break
c) change of a region's telephone dialling code
d) proposed merger of two pharmaceutical companies

EXERCISES

10.1B

1 A sample of 20 is to be taken from a school with 470 students. Describe how you would use random numbers to select the sample. Explain how you would ensure efficiency in your selection process.

2 It is required to sample the population of a certain town. Discuss the suitability of the following sampling frames.
a) names in the telephone directory
b) names on the electoral roll
c) membership of the local civic society

3 Identify, where possible, an existing sampling frame for each of the following populations.
a) households with telephones in the Birmingham area
b) employees at a factory
c) lions in Africa
d) guests in a large hotel
e) visitors to a free art exhibition

4 For each of the following investigations, detail the population from which the sample should be taken.
a) proportion of people displaying an allergy to a new treatment for hay fever
b) proportion of winners on the football pools
c) number of households with outstanding gas accounts
d) number of cars without a valid MOT certificate

5 A research study into prostate problems in men involved sending a questionnaire to 552 men aged over 50 who were registered with a general practice.
a) Suggest one advantage and one disadvantage of this sampling method.
b) Of the men contacted, 358 returned the questionnaire. Of these, 273 reported that they had undergone treatment for prostate problems. Obtain an estimate of the proportion of men aged over 50 who experience prostate difficulties.
c) Criticise the method of obtaining the estimate in **b)** and suggest an alternative method.

SYSTEMATIC, STRATIFIED, QUOTA AND CLUSTER SAMPLING

So far we have considered reasons why it is sometimes necessary to use a sampling procedure to obtain information about the population from which the sample is taken. Sometimes, data are obtained from an entire population, even as large as the population of people in a country. This is a **census.** Although it has the distinct advantage of presenting a complete picture of the population, a census is time-consuming and expensive to carry out. It is out of date before completion. The accuracy of the data decreases with time that elapses after the data are collected. The amount of data collected is usually so extensive that it is virtually impossible to analyse every bit of it. Often, census information is actually sampled after it has been collected in order to identify characteristics effectively.

Stratified sampling

Simple random sampling does not guarantee a **representative** sample. In random sampling any collection is possible. An approach which tries to enhance this aspect without reintroducing subjectivity is **stratified sampling**. Sub-samples are taken randomly from distinct sections or **strata** of data. The size of the sub-samples are in proportion to the size of the strata in the population.

Suppose we return to the firework scenario. The shopkeeper wants to improve her sampling procedure, so she decides to use a stratified random sampling method. Her supplier tells her that he classifies the fuses as short, medium and long. These have burn times, t seconds, such that $t < 7$, $7 \le t < 10$, $t \ge 10$, respectively. The fuse manufacturer says that the fuses are produced in the ratio 1 : 2 : 2.

A stratified sample would contain burn times in these ratios. Why is it not possible to have a stratified sample of size six? Clearly, the sample size should be a multiple of five.

The shopkeeper decides to take a stratified sample of ten fuses to estimate the mean burn time. So, she needs *two* short, *four* medium and *four* long burn time fuses in her sample. The sampling is to be done randomly. One approach is to generate random numbers until she has selected the two short-burn fuses, then to generate random numbers until she has selected the four medium burn times etc. This is rather inefficient. How might she improve efficiency? She could generate the random numbers, select the fuse and not reject any unless the stratum is complete.

Suppose the random numbers generated are those in the table. The fuse times selected and strata to which they belong would also be recorded in the table.

Random decimal	Position	Time	Action	Stratum
0.641	129	9.6	Keep	M
0.894	179	13.2	Keep	L
0.509	102	9.5	Keep	M
0.956	192	12.2	Keep	L
0.288	58	8.1	Keep	M
0.066	14	7.9	Keep	M
0.839	168	8.7	Reject	
0.957	192		Reject	
0.202	41	11.1	Keep	L
0.292	59	6.8	Keep	S
0.434	87	10.9	Keep	L
0.849	170	8.6	Reject	
0.935	188	6.2	Keep	S

The stratified random sample contains:

Time	Stratum
6.8, 6.2	short
9.6, 9.5, 8.1, 7.9	medium
13.2, 12.2, 11.1, 10.9	long

The mean burn time is 9.55 seconds.

Exploration 10.2

Mean burn times

Obtain a stratified random sample of fuse burn times of size 10 using your calculator to generate random decimals. Calculate the mean burn time for your sample.

Quota sampling, cluster sampling

Randomising the sampling process is highly desirable. However when the population under investigation is large or spread widely, e.g. in investigating consumer spending habits or seeking opinion from owners of cafés, it has cost and time implications. Subjective alternatives are sometimes used for pragmatic reasons. **Quota sampling** is a subjective alternative to stratified sampling.

Sub-samples must be in the appropriate proportion but the selection of the member for the sub-sample is left up to the sampler. Often the method used is on a 'first come, first in' basis.

Cluster sampling is another approach to gauging characteristics of a large, widely-spread population. A small area, often geographical, where typical members of the population are to be found is identified. Then, every member of the population is recorded in this cluster. In the case of the firework display, you might choose a firework party nearby and record the fuse burn time of all the fireworks let off at the party. Why might this sample not be representative of the population?

Exploration 10.3

Quota sampling burn times

Conduct a five per cent quota sampling survey of the burn times, starting by going down column A. Calculate the mean time of your sample. In what way do you anticipate that your sample differs from the sample obtained by anyone else conducting this survey in this way?

Systematic sampling

A very convenient form of sampling which gives every member of a population the same chance of being included *before* the process begins, but one which excludes large sections of the population the moment the process begins, is **systematic sampling**. If the sample is to contain five per cent of a population, then systematic sampling would involve selecting every 20th member of the population. If a ten per cent sample is sought then every tenth member of the population is selected. Clearly, this process effectively excludes 19 out of every 20, or nine out of every ten in the case of the ten per cent sample. However, if the first item to be included in the sample is chosen randomly then every member of the population has an equal chance of being included.

Exploration 10.4

Systematic sampling

Obtain a five per cent systematic sample of burn times using a randomly-generated start point. (You need not restrict your start point to one of the first 20 times since you can always go back to the beginning.)

Consider the situation where 21 students carry out Exploration 10.4. How likely is it that there will be at least two identical samples? Can you think of any other reasons why systematic sampling is undesirable?

EXERCISES

10.2A

1 The principal of a college asks members of a Statistics class to suggest an appropriate sampling method for seeking student opinion on the refurbishing of the main hall.

Proposal 1: Select every 20th student from the college's alphabetical listing of all students.
Proposal 2: Select students at random from each years listing and in proportion to the number on the list.
Proposal 3: Simply select students as they enter the dining hall.
Proposal 4: Select students as they arrive at college, but ensure representative proportions with respect to year and to gender.

In each case identify the sampling method and list its advantages and disadvantages. Which proposal would you recommend to the principal?

2 A school has 210 sixth-formers, 30 doing arts foundation courses, 80 doing science A-Levels and the rest doing humanities A-Levels.

Describe how you would obtain a ten per cent stratified sample. Explain how this differs from a simple random sample.

3 **a)** Explain briefly what is meant by a random sample.
A college has 2580 students. A survey into student attitudes to work, using a questionnaire, is being carried out. The questionnaire is to be given to a sample of 250 students. The following three methods of selecting the sample are proposed:
A: Ask for volunteers.
B: Ask every fifth student entering the college on a certain morning.
C: Use the alphabetical list of students in the college, taking every tenth student on the list.
b) Discuss briefly the advantages and drawbacks of each of the proposed methods of selection.
c) Explain in detail a method which could be used to obtain a random sample of 250 students from the population of the college.

4 Explain why it is often necessary to sample, and state the properties that are considered essential for a statistically acceptable sample.

A district councillor decides to investigate the attitudes of residents on a large housing estate towards the proposed building of a new light industrial estate. If a sample of residents is to be selected, state, with reasons, the method of sampling you would recommend the councillor to use.

5 A town has a population of 60 000. The water company is interested in how many households own dishwashers.
a) What are the advantages of a sampling method of enquiry over a census?
b) Describe briefly how you would select a suitable sample in this situation. Your answer should include details of the sampling frame, the type of sample to be used and how you would calculate your estimate based on your sample information.

6 Return to the data on page 262 and present them in a two-part stem-and-leaf display. Estimate the mean burn time. Compare with the means obtained in the various sampling processes discussed in this chapter. How accurate do you feel the fuse manufacturer is in his ratio 1 : 2 : 2 for short : medium : long burn times?

EXERCISES

10.2B

1 It is required to assess opinions about the management of a town centre shopping mall. Describe which method of sampling is used in each of the following proposals, and discuss the advantages and disadvantages of each.
a) Give a questionnaire to the driver of every tenth car in the nearby shoppers' car park.
b) Give a questionnaire to people emerging from the supermarket check-out, taking account of their sex and age.

c) Give a questionnaire to anyone walking about the shopping mall who will accept one.
d) Send a questionnaire to local residents selected from the electoral roll using random numbers.

2 Distinguish between:
a) random and subjective sampling,
b) random and representative sampling,
c) random and systematic sampling.

3 A school has tickets for 48 of its students to go to an international art exhibition. There are 1200 students at the college. There are the following numbers in each year group.

Year	7	8	9	10	11	12/13
Number	200	225	224	226	150	175

If the 48 are chosen from all 1200 students, describe how you would select a representative group using:
a) pure random sampling, **b)** stratified sampling,
c) quota sampling.
State which of these methods is the most appropriate.

4 Describe, briefly, how you would obtain a representative sample of the opinion of the 5374 students at a tertiary college on having a totally non-smoking campus.

5 The total numbers of goals scored in 100 hockey matches are recorded in the table.

3	4	2	5	3	4	6	4	5	4	5	6	5	4	5	3	4	5	2	5
6	1	6	4	5	4	4	4	6	4	4	4	5	3	5	5	5	6	5	3
4	3	5	2	4	3	4	1	5	6	4	6	5	5	4	5	5	6	5	3
5	2	6	3	5	6	2	4	5	5	4	6	5	4	4	5	4	6	3	4
5	5	4	6	2	5	4	3	5	2	3	6	6	3	5	3	4	6	4	6

a) Represent the data in a stem-and-leaf diagram.
b) Calculate the mean number of goals per match.
c) **i)** Use a simple random sampling method to select a sample of size 5. Calculate the mean of the sample.
ii) Repeat this process until you have ten sample means. Calculate the mean of the sample means.
d) **i)** Use a systematic sampling method to select a sample of size five. Calculate the mean of the sample.
ii) Repeat this process until you have ten sample means. Calculate the mean of the sample means.
e) **i)** Use your stem-and-leaf diagram to identify three strata in the data. Use a stratified sampling procedure to select a sample of size 5. Calculate the mean of the sample.
ii) Repeat until you have ten sample means. Calculate the mean of the sample means.
f) Comment on the results in **b)**, **c)**, **d)** and **e)**.

6 The diagram represents the number of sparklers which prove to be difficult to light in packets of five, sold by a corner shop.

	A	B	C	D	E	F	G	H	I	J	K	L	M	N	O	P	Q	R	S	T
1	2	3	1	2	3	1	2	4	1	2	0	2	3	1	2	3	1	2	4	1
2	2	0	2	3	1	2	3	1	2	4	1	2	0	2	3	1	2	3	1	2
3	4	1	3	0	2	3	1	2	3	1	2	4	2	3	0	2	3	1	2	3
4	1	2	5	2	3	1	2	3	1	2	3	1	2	0	2	3	1	2	3	1
5	2	4	1	2	0	2	3	1	2	3	1	2	4	1	2	0	2	3	1	2
6	3	1	2	4	1	3	0	2	3	1	2	3	1	2	4	1	3	0	2	3
7	1	2	3	1	2	4	2	3	1	2	3	1	2	3	1	2	0	2	3	1
8	2	3	1	2	4	1	2	0	2	3	1	2	3	1	2	4	1	2	0	2
9	3	1	2	3	1	2	4	1	2	0	2	3	1	2	3	1	2	4	1	3
10	0	2	3	1	2	3	1	2	4	2	3	0	2	3	1	2	3	1	2	5

The shopkeeper decided to use a systematic sampling procedure by opening every 20th packet and setting the contents aside for testing. He decided to use random numbers to determine which of the 20 packets he would start with. How might he do this?

Carry out the following process which represents a simulation of the experience of the shopkeeper.

a) Obtain a five per cent systematic sample by randomly choosing a starting point from the row labelled 1. Record the mean of the sample.

b) Repeat the process in **a**).

c) Examine the columns of data and offer a criticism of this process of sampling.

SAMPLING FROM PROBABILITY DISTRIBUTIONS

Reflect on the 'sparklers' data in question 6 of Exercises Set 10.2B. If these came from the same source as the sparklers as the shopkeeper's in Example 10.1, then 40 per cent of the sparklers are difficult to light. What probability distribution is appropriate to model the number of the sparklers which are difficult to light in packets of five?

Provided the assumptions which led to the binomial model are valid then B(5, 0.40) would be appropriate. This suggests that the data in question 6 are from this distribution. You may have noticed elements of a pattern in the systematic sample you obtained in tackling question 6 of Exercise 10.2B. Is this pattern inevitable or is it merely a feature of the sparkler production? One way of tackling questions like this is to simulate observations of a random variable i.e. to obtain random samples from probability distributions.

A common way of doing this makes use of the cumulative distribution function appropriate to the random variable. Then it uses random decimals to represent a cumulative probability. The observation selected is the one which corresponds to this cumulative probability. This process is analogous, in the sparkler example, to having the entire production, say 10 000 packets of sparklers, in a long line.

The packets are arranged so that all those with none that are hard to light are at the front of the line, followed by those with one, then those with two etc. Then random numbers between 1 and 10 000 are generated, to select packets.

Sampling from binomial and Poisson distributions

The cumulative distribution for the random variable X where $X \sim B(5, 0.4)$, is shown in the table below.

x	0	1	2	3	4	5
$P(X \le x)$	0.077 76	0.336 96	0.682 56	0.912 96	0.989 76	1.000 00

Since there are five places of decimals in this display of the distribution function, it would be appropriate to use random five-place decimals to obtain observations of X. Suppose the following random decimals are generated.

0.306 90, 0.592 52, 0.041 96, 0.958 58

The observations corresponding to these random decimals can be found by locating the decimals on a diagram like this, which is equivalent to the cumulative distribution.

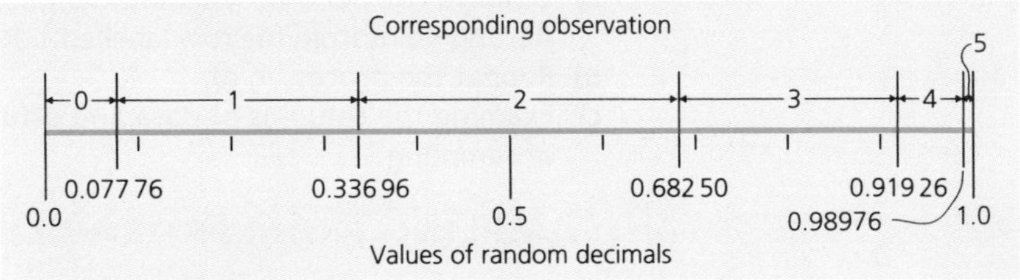

Hence:

0.306 90 corresponds to 1
0.592 52 corresponds to 2
0.041 96 corresponds to 0
0.958 58 corresponds to 4.

Exploration 10.5

Observations

Use you calculator or the random number tables in this book to obtain five observations of the random variable X where:

a) $X \sim B(5, 0.4)$,
b) $X \sim B(10, 0.2)$, (if you use the distributions tables then four-place decimals)
c) $X \sim Poi(3.2)$, (again four-place decimals are appropriate in using the tables)
d) $X \sim Poi(7.9)$.

Sampling from other distributions

Where distributions tables are not available, we can use our knowledge of the probability function and hence the cumulative probability function to obtain random samples.

Imagine you are working in a firework factory and your job is to ensure that the 'volcanoes' are standing on their base rather than lying on their side as they travel along the production line. On average, 20 per cent of the volcanoes lie on their side. Whenever you see a volcano on its side, you stand it upright and then count the number of volcanoes up to the next one lying on its side, and so on. Why might a geometric distribution be an appropriate model for this count?

From the information available, Geo(0.2) would seem to be appropriate. The distribution function can be tabulated.

x	1	2	3	4	5	
$P(X \le x)$	0.2	$1 - 0.8^2$ $= 0.36$	$1 - 0.8^3$ $= 0.488$	$1 - 0.8^4$ $= 0.5904$	$1 - 0.8^5$ $= 0.672\,32$	etc. etc.

So, a random decimal less than 0.2 corresponds to the value 1,
a random decimal between 0.2 and 0.36 corresponds to the value 2,
a random decimal between 0.36 and 0.488 corresponds to the value 3, etc.

Exploration 10.6

Finding observations

Find the observations on X, where $X \sim \text{Geo}(0.2)$, which correspond to the following decimals.

0.126 57, 0.339 93, 0.988 83, 0.515 64, 0.782 98

Example 10.2

Simulate the outcome of rolling a tetrahedral die three times using these random numbers.
482 51 735 46 821 43

Solution
The distribution function can be given in linear form as shown.

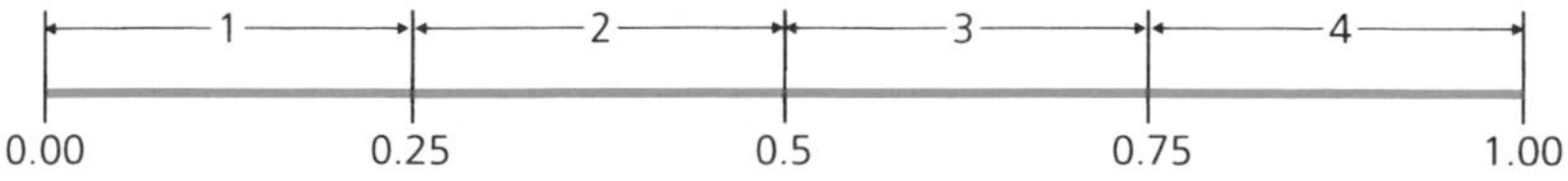

Then considering the random numbers as five-digit random decimals:
0.482 51 corresponds to the outcome 2,
0.735 46 corresponds to the outcome 3,
0.821 43 corresponds to the outcome 4.

EXERCISES

10.3A

1 A trial consists of simultaneously rolling twelve fair cubical dice and counting the number of sixes obtained.

Use random numbers and the table of the appropriate distribution function to simulate an experiment which consists of 20 such independent trials.

2 The weekly number of accidents on a particular stretch of road is known to be a Poisson variate with a mean rate of 1.6. Use random numbers and distribution function tables to simulate 26 weeks' accident figures.

Calculate the mean and variance of your simulated values. Compare with the values you would expect.

3 The number of Smarties in a mini pack takes one of the values 9, 10 and 11, all of which are equally likely.

a) Andrea always eats exactly one box of Smarties a day. Simulate the daily number of Smarties eaten by Andrea over a 28-day period.

b) Brian always eats exactly two boxes of Smarties a day. Simulate the daily number of Smarties eaten by Brian over a 28-day period.

In each case compare the simulated totals eaten over the period with that expected.

4 As part of a factory's quality control procedures, random samples of ten items are selected from large batches and a count made of the number which are defective.

a) When the process is operating adequately two per cent of items are defective. Simulate the counts recorded from sampling 20 batches. Use your results to then estimate the proportion defective.

b) Repeat **a)**, when the process is producing 15 per cent defective items due to an unidentified malfunction.

EXERCISES

10.3B

1 Use random numbers to simulate the scores obtained when a fair six-sided die is rolled 36 times.

Repeat the simulation when the die is biased such that:
$P(1) = P(2) = P(3) = P(4) = P(5) = \frac{1}{9}$ and $P(6) = \frac{4}{9}$

In each case compare your simulated results with those you would expect.

2 A trial consists of tossing two fair coins simultaneously and counting the number of heads, H, observed.

Specify completely the distribution of H.

Using random number tables, simulate 50 observations of H. Estimate the probability of a head in a single toss of one of the coins. Comment.

3 The weekly number of word-processors sold by Shop A may be modelled by a Poisson distribution with mean 5. Independently, the

weekly number of word-processors sold by Shop B may also be modelled by a Poisson distribution, but with mean 4.

a) Simulate 20 weeks' sales for shop A.
b) Simulate 20 weeks' sales for Shop B
c) Simulate 20 weeks' sales for Shops A and B combined

Compare the combined results for **a)** and **b)** with those obtained in **c)**.

4 A fair tetrahedral die is repeatedly rolled until a score of 4 is observed. If R denotes the number of rolls required, specify completely the distribution of R. Copy and complete the following probability table for R.

r	1	2	3	4	5	6	7	≥ 8
$P(R = r)$								

Hence, using random numbers, simulate 48 observations of R. Using your results construct an empirical probability table for R and compare it with the theoretical probability table.

CONSOLIDATION EXERCISES FOR CHAPTER 10

1 Explain what is meant by:

a) a stratified sample,
b) a cluster sample,
c) a quota sample.

For each case outline a situation where it may be considered to be the most appropriate sampling method.

2 **a)** Describe how you would use the random number 48 722 obtained from a table of random digits arranged in groups of five digits to obtain a single sample from a population of 3027 members.

b) Use random numbers from tables to select a five per cent stratified sample from a population of size 360 divided into three strata of sizes 40, 120 and 200 respectively.

3 The daily demand for a particular spare part from a garage's stores is known to be a random event with an average of 2.5.

a) Use random numbers and appropriate distribution tables to simulate 50 days' demands.
b) If there are five of these spare parts in stock at the beginning of each day, use your results to estimate the probability of being out of stock.
c) Compare this estimate with that given by the appropriate probability model.

4 As part of a traffic investigation, a count, X, is made of the number of cars passing a checkpoint between each successive alternative type of motor vehicle.

a) Given that 80 per cent of vehicles passing the check point are cars, identify a possible probability model for X, stating any further assumptions that are necessary.
b) Construct a probability table for X and hence use random numbers to simulate 30 observed values of X.

5 Describe how you could obtain representative samples of the following.
a) the opinion of students, in a school, on government education policy
b) the distribution of daisies in a hockey pitch
c) the diameter of trees in a 20 acre woodland
d) the make of car driven by teachers in primary schools in Cornwall
e) the ownership of telephone answering machines in private households in Yorkshire
f) the size of carrots sold in supermarkets
g) the size of farms in Wales

6 Describe two methods by which you might be able to determine the pattern of membership of professional associations by teachers. Carry out one of these and report your findings. In what way might your findings differ from those you may have obtained using your other method?

7 Describe a method which you could use to determine ownership of computers by households in a town of about 25 000 people. Explain why you chose your approach rather than other methods.

8 Use the random decimals 0.4751, 0.8230, 0.1308 to obtain three observations for each of the following random variables.
a) $X, X \sim \mathrm{B}(12, 0.4)$
b) $Y, Y \sim \mathrm{Poi}(4.8)$
c) $Z, Z \sim \mathrm{Ber}(0.6)$

9 Use the random decimals:

0.5862, 0.1037, 0.0049, 0.8540, 0.9500

to obtain five observations from a random variable which is distributed $\mathrm{Geo}(0.1)$.

10 Describe how you could use computer-generated random decimals to simulate outcomes of rolling:
a) a tetrahedral die,
b) a cubical die,
c) an icosahedral die.

11 Describe how you might conduct a survey of the opinions of concerned people about the wearing of uniforms in school.

12 Describe how you could conduct a survey into the readability of daily newspapers.

13 Give a brief explanation of and an example of the use of:
a) a census,
b) a sample survey.

(ULEAC Question 3, Statistics paper 2, June 1994)

14 Write brief notes on:
a) simple random sampling,
b) quota sampling.
Your notes should include a description of each method, and an advantage and a disadvantage associated with it.

(ULEAC Question 5, Statistics paper 2, January 1993)

15 a) Explain briefly:
i) why it is often desirable to take samples,
ii) what you understand by a sampling frame.
b) State two circumstances when you would consider using:
i) clustering, ii) stratification,
when sampling from a population.
c) Give two advantages and two disadvantages associated with quota sampling.

(ULEAC Question 10, Statistics paper 2, June 1992)

16 a) Explain briefly:
i) why it is often desirable to take samples,
ii) what you understand by a sampling frame.
b) State a situation in which you would consider using:
i) a systematic sample,
ii) a stratified example,
when sampling from a population. Give a specific example in each case.
c) Give one advantage and one disadvantage associated with stratified sampling.

(ULEAC Question 5, Specimen paper T1, 1994)

17 a) Give one advantage and one disadvantage of using:
i) a census, ii) a survey.
b) It is decided to take a sample of 100 from a population consisting of 5000 elements. Explain how you would obtain a simple random sample without replacement from this population.
c) An electrical company repairs very large numbers of television sets and wishes to estimate the mean time taken to repair a particular fault.
It is known from previous research that the standard deviation of the time taken to repair this particular fault is 2.5 minutes.
The manager wishes to ensure that the probability that the estimate differs from the true mean by less than 30 seconds is 0.95.
Find how large a sample is required.

(ULEAC Question 10, Paper S2, June 1993)

18 Explain briefly the difference between a sample survey and a census.
A GCSE student, embarking upon a statistics project, wishes to investigate the proportion of left-handed students at his school.
Describe how the student could obtain a random sample of 70 pupils from his school of 380 girls and 320 boys using:
a) simple random sampling,
b) stratified random sampling.

State one possible advantage of stratified random sampling over simple random sampling.

(NEAB Question 10 (part), AS Statistics paper, June 1994)

19 Members of a public library may borrow up to four books at any one time. The number of books borrowed by a member on each visit to the library is a random variable X, with the following probability distribution.

X	0	1	2	3	4
probability	0.24	0.12	0.20	0.28	0.16

a) Find the mean and the standard deviation of X.
This distribution is unknown to the librarian. In order to estimate the mean number of books borrowed by members she decides to record the number of books borrowed by a sample of 40 members using the library. She chooses the first member of the sample by selecting a random integer, r, between 1 and 5 inclusive. She then includes in her sample the rth member to leave the library one morning and every fifth member to leave after that until her sample of 40 is complete. Thus if $r = 3$ she chooses the third, eighth, 13th, ... 198th members leaving the library as her sample.

b) **i)** Is each of the first 200 people leaving the library equally likely to be included in the sample?
ii) Does the sample constitute a random sample of the first 200 people leaving the library? Give a reason.
iii) Comment on whether the sample will provide a useful estimate of the mean number of books borrowed by members.

A list of names of 8950 members of the library is available.

c) Describe how random sampling numbers could be used to select a random sample (without replacement) of 40 of these names.

(AEB Question 5 (part), Paper 9, Winter 1994)

20 Following a spell of particularly bad weather, an insurance company received 42 claims for storm damage on the same day. Sufficient staff were available to investigate only six of these claims. The others would be paid in full without investigation. The claims were numbered 00 to 41 and the following suggestions were made as to the method used to select the six. In each case six different claims are required, so any repeats would be ignored.

Method 1	Choose the six largest claims.
Method 2	Select two-digit random numbers, ignoring any greater than 41. When six have been obtained choose the corresponding claims.
Method 3	Select two-digit random numbers. Divide each one by 42, take the remainder and choose the corresponding claims (e.g. if 44 is selected claim number 02 would be chosen).
Method 4	As 3, but when selecting the random numbers ignore 84 and over.
Method 5	Select a single digit at random, ignoring 7 and over. Choose this and every seventh claim thereafter (e.g. if 3 is selected, choose claims numbered 03, 10, 17, 24, 31 and 38).

Comment on each of the methods, including an explanation of whether it would yield a random sample or not.

(AEB Question 4, Specimen paper 2, 1994)

21 A village has 450 households. The landlord of the local pub wants to obtain the opinion of the villagers about the new licensing laws which allow greatly extended opening hours. He decides to conduct a ten per cent survey of the households. Describe briefly but clearly how he might select the households to survey. Illustrate your description by using the following random digits to select the three households in his sample.

083 95 947 60 261 32

(Oxford Question 6 (part), Specimen Paper S6, 1994)

22 The random number facility on a calculator produced the number 0.657. Use this to obtain a sample from a Poisson distribution with mean 5, carefully explaining the steps you take. You are advised to make full use of the recommended tables.

(Oxford Question 11 (part), Paper 3, 1991)

Summary

In this chapter we have discussed:

- the process of simple random sampling
- the use of the integer part function INT(x) in converting random decimals into selections
- the importance of sampling frame
- how sampling can be made representative
- how to simulate random variables.

STATISTICS

11 Continuous random variables

In this chapter we shall investigate:

- *how to model the occurrence of continuous random variables,*
- *how the summation process for discrete variables is paralleled by integration for continuous variables,*
- *how a review of the Poisson distribution leads to a new model of random events.*

MODELLING CONTINUOUS RANDOM VARIABLES

Exploration 11.1

Probability density function

A large retail DIY store sells planks of wood which it saws to the length the customer requires. The manager of the timber department is not prepared to sell any length less than 1.2 m. As a result she often has off-cuts of timber left over, which are put in a waste container. One of the customers of the store is the coordinator for Technology of a consortium of primary schools. He thinks there would be a use for these off-cuts but would like to know what lengths to expect in the container.

Imagine you are approached to offer the coordinator advice. You decide to obtain a random sample of off-cuts.

Length, x m	$0 \le x < 0.2$	$0.2 \le x < 0.4$	$0.4 \le x < 0.5$
Number off-cuts, f	15	14	5

Length, x m	$0.5 \le x < 0.6$	$0.6 \le x < 1.0$	$1.0 \le x < 1.2$
Number off-cuts, f	8	21	12

The lengths of your sample of 75 off-cuts are recorded in a table, and displayed in this histogram.

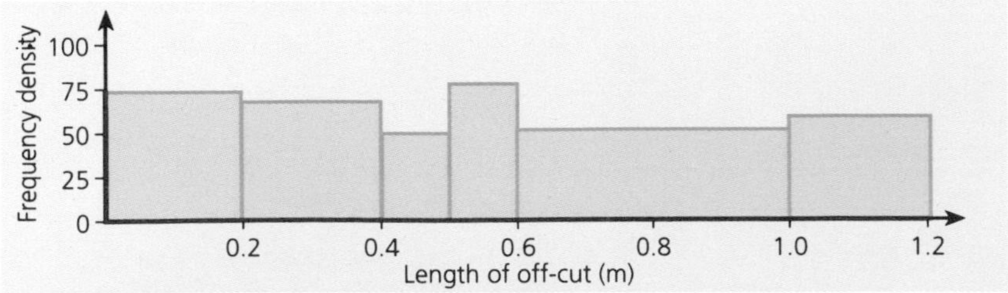

The coordinator asks you if this is typical of the lengths he might find and if you could tell him what proportion of the lengths are less than $\frac{3}{4}$ m. What do you think your reply is?

Planning an approach

You might talk to the coordinator about the nature of inherent random variability in data. You might also want a more appropriate presentation of the data, to respond to the question about proportion. The histogram concerns itself with **frequency** whereas **relative frequency** is more directly appropriate to **proportion**. Try presenting the data in a relative frequency histogram.

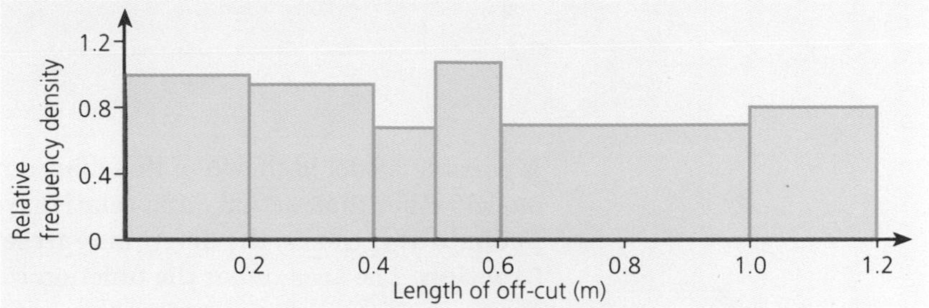

The relative frequency histogram is shown in this diagram. The Technology coordinator comments that the diagram looks no different from the first frequency histogram and asks how you can use it to answer his question on proportion. What is the main difference between the two diagrams?

You might point out that frequencies can be found from the first histogram, whereas the relative frequency histogram allows relative frequencies or proportions to be found from the areas. Suppose you invite the coordinator to estimate the proportion of off-cuts with lengths less than $\frac{3}{4}$ m. What do you think the estimate should be?

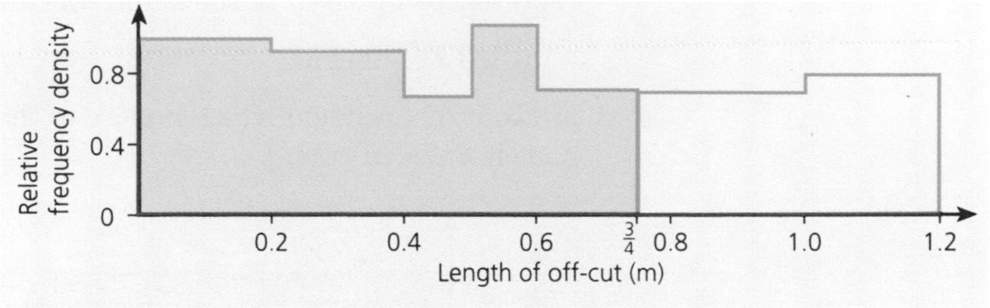

The area shaded in this diagram is an estimate of the proportion. This area consists of:

$$0.2 \times 1.0 + 0.2 \times 0.933 + 0.1 \times 0.667 + 0.1 \times 1.067 + 0.15 \times 0.7 = 0.665$$

So, a reasonable estimate of the proportion is 66.5 per cent or about $\frac{2}{3}$.

Reflecting on your discussion with the coordinator, you might mention that the relative frequency histogram you have used to estimate proportions is based on one random sample. You could suggest that you would prefer to obtain many more random samples so that you would be sure of the shape of the resulting relative frequency histogram.

If the sample of 75 lengths is representative of the off-cuts, then the relative frequency histogram is likely to be a smoother version of the top graph on page 285. What model do you suggest might fit the **smoothed relative frequency density?**

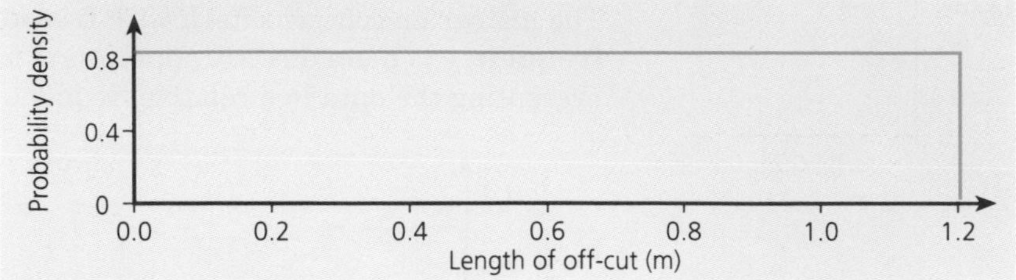

A possible model is shown in this diagram. Note that, since this is a model rather than actual data, relative frequency has been replaced by **probability**. Hence, the diagram represents a **probability density function**. The area under the function represents probability in the same way that area represents relative frequency in a histogram.

Since all the off-cuts of timber have lengths between 0.0 and 1.2 m, what is the area under this probability density function? The area represents the probability of all possibilities, hence its value should be equal to 1. The region is rectangular and its area can be found from 'length × breadth'. Hence, the density function, $f(x)$, is:

$$f(x) = \frac{1}{1.2} \approx 0.833 \quad 0 \le x \le 1.2$$

Note: The word 'probability' is not always used in this context.

What does this model suggest is the proportion of off-cuts with length less than $\frac{3}{4}$ m? The proportion is modelled by the probability represented by the area shaded in this diagram. This is:

$$0.833 \times \frac{3}{4} \approx 0.625$$

or about $62\frac{1}{2}$ per cent which is close to the value obtained from the sample of 75 off-cuts.

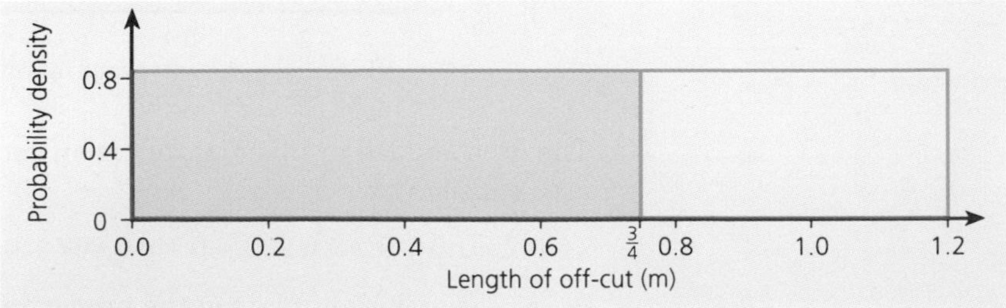

Exploration 11.2

Using the density function

Use the density function of the diagram showing the smoothed relative frequency density, above, to estimate the following proportions.

a) $0.3 \le x \le 0.6$ **b)** $x \ge 0.9$ **c)** $x < 0.8$

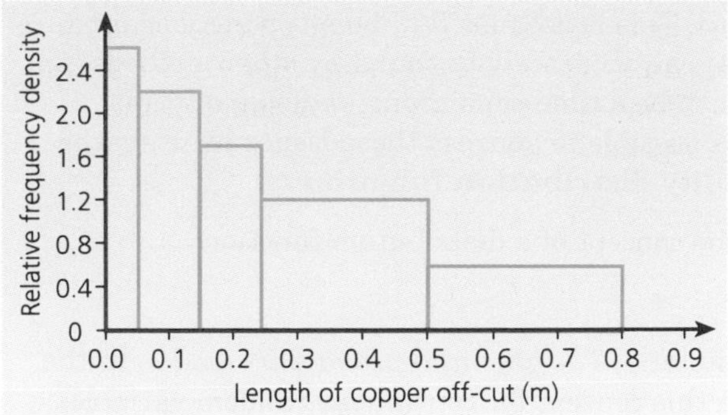

Distribution function

The plumbing department of the store sells copper pipe cut to any required length. There is no 'minimum length' policy. However, at the end of every day, the employee responsible for cutting copper pipe puts lengths he feels are unsaleable into a waste container. An examination of the container revealed a relative frequency histogram of the lengths as shown here.

Exploration 11.3

Using the relative frequency histogram

Use the relative frequency histogram above to calculate the following proportions of pipe lengths, x m.

a) $0 \le x \le 0.05$ **b)** $0.25 \le x \le 0.5$ **c)** $0.05 \le x \le 0.25$

A model for the probability density function of the length of copper pipe off-cuts is given in this diagram. The model proposed is a straight line joining the points with coordinates (0, 2.5) and (0.8, 0). Calculate the area between the density function and the horizontal axis. This area is equal to 1. Why?

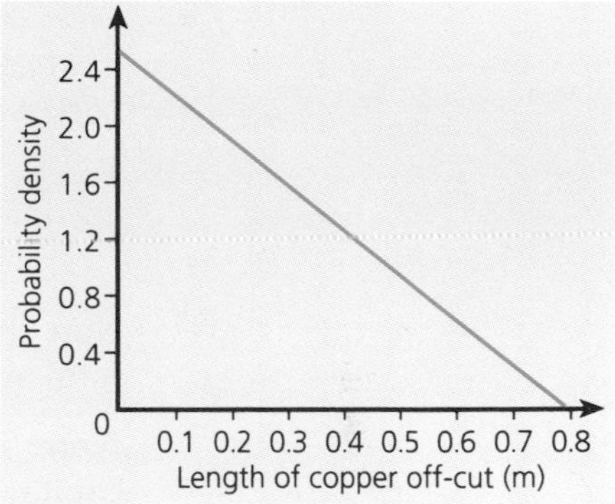

Interpreting the results

Since the density function is a straight line, it is relatively simple to find its equation. Using $y = mx + c$, the density function, $f(x)$, is:

$$f(x) = -\tfrac{2.5}{0.8}x + 2.5$$

and simplifying this gives:

$$f(x) = 2.5 - 3.125x \quad 0 \le x \le 0.8$$

The area between the density function and the horizontal axis can be found by integrating $f(x)$ between $x = 0$ and $x = 0.8$.

$$\begin{aligned}\text{Area} &= \int_0^{0.8}(2.5 - 3.125x)\,dx \\ &= \left[2.5x - 3.125 \times \tfrac{1}{2}x^2\right]_0^{0.8} \\ &= (2.0 - 1.0) - 0 = 1.0\end{aligned}$$

Integration is used to find the area under a density function. Integration is used to find areas and hence to find probabilities with many density functions. This is because few density functions produce shapes with areas that can accurately be found by other methods. Evaluating integrals can be a time-consuming, tedious, difficult process. However, it is possible to address these issues by using the **cumulative probability distribution function**.

You may recall that the concept of a distribution function:

$$F(x) = P(X \le x)$$

was introduced in Chapter 7, *The binomial probability model*. In the same way, we can use this concept for continuous random variables. There is a difference between the approach adopted for a discrete random variable, such as one which is binomially distributed, and the variables under discussion now. That difference is in the way that the cumulative probabilities are evaluated. Rather than adding up probabilities of consecutive events, we use integration. For example, the probability that the length of an off-cut of copper pipe is less than x m is the area under the density function between 0 and x m. This area can be found, using integration, as:

$$P(X \le x) = \int_0^x (2.5 - 3.125x)\,dx$$

$$= \left[2.5x - 3.125 \times \tfrac{1}{2}x^2\right]_0^x$$

$$= 2.5x - \tfrac{3.125}{2}x^2$$

Note that this is the same integral as was needed to show that the total area under the curve was equal to 1.

Note: The standard notation for the distribution function is to use a capital letter, F(x).

Now that the form of the distribution function for the length of copper off-cuts has been found, there should be no need to do any more integration to evaluate probabilities. Some quite high level mathematics has been used to develop F(x) and it is reasonable to question the validity of the proposed model. In Exploration 11.3, you calculated proportions of off-cuts in particular ranges of lengths. What are the estimates of those proportions suggested by the model?

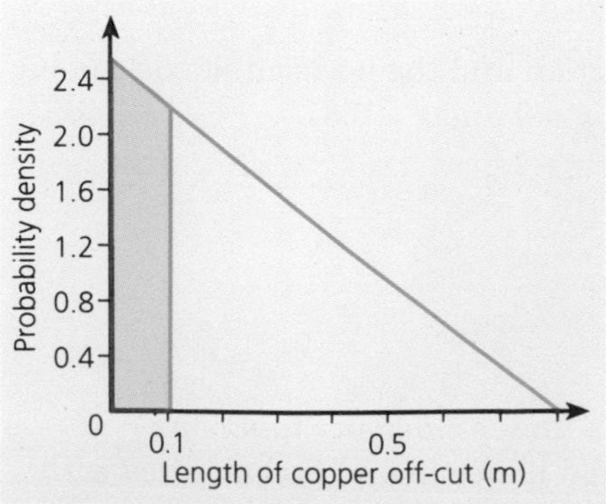

a) The first proportion you calculated can be estimated by:

$$P(0 \le x \le 0.05)$$

which is available directly from the distribution function as:

$$F(0.05) = 2.5x - \tfrac{3.125}{2}x^2 \quad \text{for } x = 0.05$$

$$= 2.5 \times 0.05 - \tfrac{3.125}{2} \times 0.05^2$$

$$\approx 0.121$$

which compares with the proportion, obtained from the relative frequency histogram, $2.6 \times 0.05 \approx 0.13$.

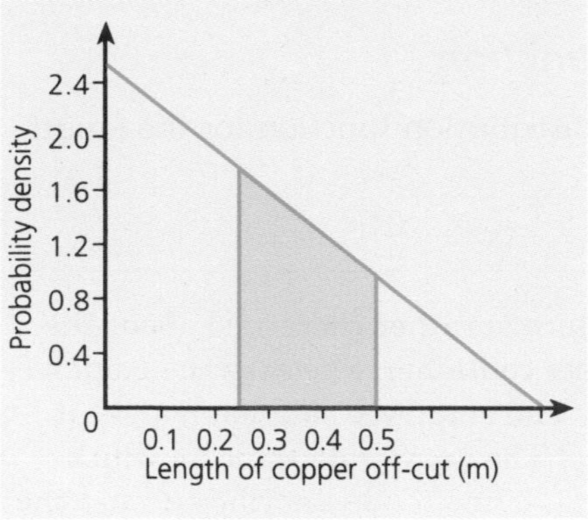

b) The second proportion:

$$P(0.25 \le x \le 0.5)$$

can also be found using the distribution function as a difference between two cumulative proportions.

$$\begin{aligned} P(0.25 \le x \le 0.5) &= P(x \le 0.5) - P(x \le 0.25) \\ &= F(0.5) - F(0.25) \\ &\approx 0.8594 - 0.5273 \\ &= 0.332 \end{aligned}$$

This compares with, $1.2 \times (0.5 - 0.25) = 0.3$, from the histogram.

c) The third proportion can be found in a similar way:

$$\begin{aligned} P(0.05 \le x \le 0.25) &= F(0.25) - F(0.05) \\ &= 0.5273 - 0.1211 \\ &\approx 0.406 \end{aligned}$$

The relative frequency was $2.2 \times (0.15 - 0.05) + 1.7 \times (0.25 - 0.15) = 0.39$.

The table below summarises the observed relative frequencies and the probabilities expected by the proposed model.

Length, x m	Relative frequency	Probability
$0.0 \le x \le 0.05$	0.13	0.12
$0.25 \le x \le 0.5$	0.3	0.33
$0.05 \le x \le 0.25$	0.39	0.41

There appears to be reasonable agreement between the model and the observed data. What do you think the probability of the length of an off-cut of pipe being less than –0.5 m? Clearly, it is impossible for the length of the pipe to be negative. This has implications for the value of the distribution function for negative values of x. The distribution function, under these circumstances, must be zero, i.e:

$F(x) = 0$ when $x < 0$

According to the density function model proposed, what is the probability of getting an off-cut of length less than 1.5 m? Since no off-cuts are longer than 0.8 m, this probability must be 1, i.e:

$F(x) = 1$ when $x > 0.8$

These considerations lead to the **full definition** of the distribution function for the length of copper pipes.

$$F(x) = \begin{cases} 0 & \text{when } x < 0 \\ 2.5x - \frac{3.125}{2}x^2 & \text{when } 0 \le x \le 0.8 \\ 1 & \text{when } x > 0.8 \end{cases}$$

Exploration 11.4

Defining the distribution function

Try giving a full definition of the distribution function for the length of off-cuts of wood.

Median and mode

The general manager of the DIY store is rather concerned about the quantity of pipe going into the waste container. She asks the employee what the average length of pipe is. The employee said he thought it was a bit less than $\frac{1}{4}$ m. Is he right? The answer depends on which average he was giving.

The **mode** is the length which occurs most often. The relative frequency histogram suggests that lengths less than 0.05 m occur relatively more often than other lengths, since this is where the relative frequency density is greatest. The proposed model, in the form of the probability density function, has a similar feature. In the model the density function takes its largest value at $x = 0$. This suggests that zero is the mode. It does not appear that this is the average being quoted by the employee.

The **median** is the middle length i.e. the length of pipe such that half the pipes are longer and half are shorter. This suggest that:

$$\mathrm{P}(X \leq \text{median}) = \tfrac{1}{2}$$

or, in terms of the distribution function:

$$\mathrm{F}(\text{median}) = \tfrac{1}{2}$$

This leads to the quadratic equation:

$$2.5 \times \text{median} - \tfrac{3.125}{2} \times (\text{median})^2 = \tfrac{1}{2}$$

The equation can be rearranged into:

$$1.5625 \times (\text{median})^2 - 2.5 \times \text{median} + 0.5 = 0$$

What is the solution of the equation? Most quadratic equations have two solutions and this is no exception. Its solutions are:

$$\text{median} \approx 1.366 \text{ or } 0.234$$

but only one of these values is possible. Since no pipe length is more than 0.8 m, the 1.366 m length is not possible. Hence:

$$\text{median} \approx 0.234 \text{ m}$$

This suggests that the employee *could* have been referring to the median. A glance at the table on page 289 suggests that this is not an unreasonable value for the median. The table indicates that the relative frequency for lengths between 0 m and 0.25 m is (0.13 + 0.39) and is just over $\frac{1}{2}$.

Example 11.1

A refined model is proposed for the distribution of lengths of the copper pipe off-cuts. The density function is:

$$f(x) = \frac{a}{(x+1)^2} - 1.5625 \quad \text{when } 0 \le x \le 0.8$$

a) *What is the value of a?*

b) *Obtain the full definition for the distribution function, $F(x)$.*

c) *Use your distribution function to estimate the following probabilities.*

i) $P(0 \le x \le 0.05)$

ii) $P(0.25 \le x \le 0.5)$

iii) $P(0.05 \le x \le 0.25)$

d) *i)* *Write down the mode.*

ii) *Show that the median is one of the roots of the equation:*

$1.5625x^2 - 3x + 0.5 = 0$

and hence identify the median of the model.

Solution

a) *If $f(x)$ is a density function for this continuous random variable, then the integral of $f(x)$ over the range $x = 0$ to $x = 0.8$ must equal 1. Hence:*

$$\int_0^{0.8} \left(\frac{a}{(x+1)^2} - 1.5625 \right) dx = 1$$

Integrating produces:

$$\left[-\frac{a}{(x+1)} - 1.5625x \right]_0^{0.8} = 1$$

Evaluating, by putting in the limits:

$$\left(-\frac{a}{1.8} - 1.5625 \times 0.08 \right) - \left(-\frac{a}{1} \right) = 1$$

$$\Rightarrow -\frac{a}{1.8} - 1.25 + a = 1$$

Rearranging produces:

$a = 5.0625$

b) *There are three parts to the full definition of $F(x)$.*

For $x < 0$: $F(x) = 0$

For $x > 0.8$: $F(x) = 1$

For x between 0 and 0.8:

$$F(x) = \int_0^x \left(\frac{5.0625}{(x+1)^2} - 1.5625 \right) dx$$

$$= \left[-\frac{5.0625}{(x+1)} - 1.5625x \right]_0^x$$

$$= \left(-\frac{5.0625}{(x+1)} - 1.5625x\right) - \left(-\frac{5.0625}{1}\right)$$

$$= 5.0625 - 1.5625x - \frac{5.0625}{(x+1)}$$

c) i) $P(0 \le x \le 0.05) = F(0.05) \approx 0.163$

ii) $P(0.25 \le x \le 0.5) = F(0.5) - F(0.25)$

$\approx 0.9063 - 0.6219$

≈ 0.284

iii) $P(0.05 \le x \le 0.25) = F(0.25) - F(0.05)$

$\approx 0.6219 - 0.1629$

≈ 0.459

d) i) *A sketch of the density function may help to identify the mode. A graphics calculator might help to produce a sketch like this.*

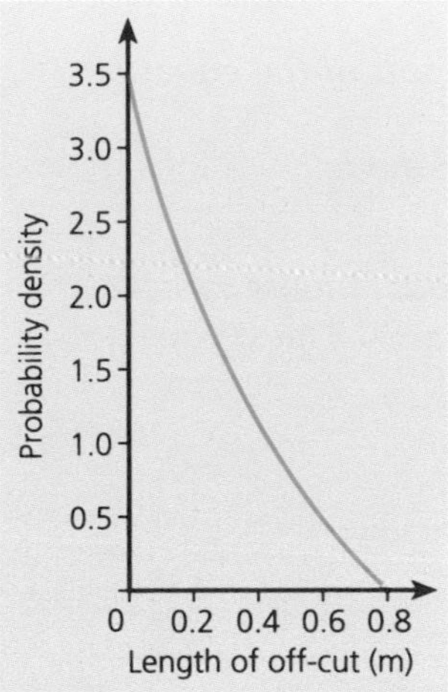

This suggests that the mode is 0.0 metres.

ii) *The median is the value of* x *which satisfies:*

$F(x) = \frac{1}{2}$

Hence:

$$5.0625 - 1.5625x - \frac{5.0625}{1+x} = 0.5$$

Subtracting 0.5 gives:

$$4.5625 - 1.5625x - \frac{5.0625}{1+x} = 0$$

Multiplying by $(1 + x)$ *gives:*

$4.5625(1 + x) - 1.5625x(1 + x) - 5.0626 = 0$

Expanding and collecting like terms:

$3x - 1.5625x^2 - 0.5 = 0$

Rearranging gives the quadratic equation:

$1.5625x^2 - 3x + 0.5 = 0$

as required.

The solutions to these equations are $x = 1.736$ *or* 0.1844.

Hence the median is 0.184 *metres.*

EXERCISES

11.1A

1 The table shows the take-home pay for a group of 1000 people.

Pay : £ x	Frequency
$0 \le x < 25$	41
$25 \le x < 50$	43
$50 \le x < 75$	51
$75 \le x < 125$	142
$125 \le x < 200$	231
$200 \le x < 275$	212
$275 \le x < 350$	175
$350 \le x < 425$	105

a) Illustrate these data in a relative frequency histogram. Use your histogram to estimate the proportion of people whose pay lies between £150 and £250.
b) Devise a model for your histogram in the form of a probability density function.
c) Use your model to estimate the probability of someone having take-home pay between £150 and £250.

2 The relative frequency of occurrence of a random variable, X, is modelled by the probability density function:

$$f(x) = \begin{cases} a(1-x) & \text{when } 0 \le x \le 1 \\ 0 & \text{for all other values of } x \end{cases}$$

where a is constant.
a) Calculate the value of a.
b) Derive the cumulative distribution function of X.
c) Determine the median of X.
d) Calculate the probability that X lies between 0.3 and 0.7.

3 The money, £x, set aside each week by a small company for future investment is modelled using the density function, $f(x)$, where:

$$f(x) = \begin{cases} k(1-0.001x) & \text{when } 0 \le x \le 1000 \\ 0 & \text{for all other values of } x \end{cases}$$

a) Calculate the value of the constant k.
b) Sketch the density function and state the modal amount of money set aside each week.
c) Obtain the distribution function, $F(x)$, and use it to estimate:
i) $P(0 \le x \le 200)$ **ii)** $P(x > 400)$ **iii)** the median.

4 A random variable, X, has probability density function:

$$f(x) = \begin{cases} kx^2(1-x) & \text{when } 0 \le x \le 1 \\ 0 & \text{for all other values of } x \end{cases}$$

a) Determine the value of the constant k.
b) Sketch $f(x)$ and state the mode of X.
c) Calculate the distribution function $F(x)$.
d) Using your calculator, determine the median.

5 A continuous random variable, T, has probability density function:

$$f(t) = \begin{cases} at & \text{when } 0 \le t \le 8 \\ 8a & \text{when } 8 < t \le 9 \\ 0 & \text{elsewhere} \end{cases}$$

a) Sketch $f(t)$.
b) Determine the value of the constant a.
c) Obtain the distribution function $F(t)$.
d) Determine the median value of T.
e) Calculate the probability $P(t > 7)$.

6 A model for the relative frequency of a continuous random variable, X, is proposed as:

$$f(x) = \begin{cases} ax & \text{when } 0 < x \le 2 \\ a(4-x) & \text{when } 2 < x \le 4 \\ 0 & \text{elsewhere} \end{cases}$$

a) Sketch the graph of $f(x)$ and state the mode.
b) Determine the value of the constant a.
c) Obtain the distribution function $F(x)$.
d) Estimate the relative frequency for the event: $1 < X < 3$.

7 Observations are made on a random variable, X. The table records 170 observations.

Observation	Frequency
$0 \le x < 2$	4
$2 \le x < 4$	20
$4 \le x < 6$	26
$6 \le x < 8$	48
$8 \le x < 10$	72

a) Illustrate these data in a relative frequency histogram.
b) Use your histogram to estimate the proportion of observations lying between 3 and 7.
c) Devise a model for the histogram in the form of a density function.
d) Use your model to estimate the probability that an observation lies between 3 and 7.

EXERCISES

11.1B

1 The table shows the age, in years, of a group of 4000 sports cars owned by members of an international enthusiasts' club.

Age, x years	Number of cars
$16 \le x < 18$	450
$18 \le x < 21$	499
$21 \le x < 24$	507
$24 \le x < 27$	588
$27 \le x < 30$	601
$30 \le x < 32$	463
$32 \le x < 34$	389
$34 \le x < 36$	503

a) Display the information in a relative frequency histogram.
b) Use your histogram to estimate the proportion of cars which are between 25 and 35 years old.
c) Determine a probability density function to model your histogram.
d) Use your model to estimate the probability of a randomly-selected car being between 35 and 25 years old.

2 A group of 165 people were asked to record the time, t hours, since they last ate a meal. The results are shown in the table below.

Time, t hours	Number of people
$0 \le t < 2$	41
$2 \le t < 4$	35
$4 \le t < 8$	52
$8 \le t < 12$	29
$12 \le t < 16$	8

a) Display the information in a relative frequency histogram.
b) Use your histogram to estimate the proportion of people who had last eaten between five and ten hours ago.
c) Devise a density function to model the relative frequency.
d) Use your model to estimate the probability of a randomly-selected person having last eaten between five and ten hours ago.

3 A random variable, T, may be modelled using the density function:

$$f(t) = \begin{cases} b(1+t) & \text{when } 1 \le t \le 3 \\ 0 & \text{for all other values of } t \end{cases}$$

a) Sketch the graph of $f(t)$.
b) Determine the value of the constant, b.
c) Obtain the distribution function, $F(t)$.
d) Calculate the probability $P(t > 2)$.

4 A random variable, X, is modelled using the density function:

$$f(x) = \begin{cases} b & \text{when } 0 \le x \le 20 \\ \dfrac{b(40-x)}{20} & \text{when } 20 < x \le 40 \\ 0 & \text{elsewhere} \end{cases}$$

a) Sketch the graph of $f(x)$ and determine the value of b.
b) Calculate the median of x.

5 A survey is conducted of makers of 'home-produced' jam. The quantity, X, of jam left over after filling all the 500 g jars available is recorded. The following model for the density function is proposed.

$$f(x) = \begin{cases} k(1-0.002x) & \text{when } 0 \le x \le 500 \\ 0 & \text{elsewhere} \end{cases}$$

a) Determine the value of the constant, k.
b) Sketch the density function and state the modal value.
c) Obtain the distribution function, $F(x)$.
d) Use $F(x)$ to determine:
i) $P(0 \le x \le 100)$ **ii)** $P(100 \le x \le 200)$
e) Write down the median and confirm your answer using the distribution function.

6 I use the following density function to model my journey time, t hours, to work.

$$\mathrm{f}(t)=\begin{cases} kt^2 & \text{when } 0.1 \le t \le 0.25 \\ 0.25k(0.5-t) & \text{when } 0.25 < t \le 0.5 \\ 0 & \text{elsewhere} \end{cases}$$

a) Sketch $\mathrm{f}(t)$ and determine the value of the constant, k.
b) Write down the mode.
c) Estimate the probability that my journey takes:
i) more than 15 minutes, **ii)** between 12 and 24 minutes.

7 The monthly fuel consumption, X, in thousands of litres, of a road haulage company may be modelled by the density function:

$$\mathrm{f}(x)=\begin{cases} \dfrac{k}{x^2} & \text{when } 2 \le x \le 4 \\ 0 & \text{elsewhere} \end{cases}$$

a) Sketch $\mathrm{f}(x)$ and determine the value of the constant, k.
b) Determine the distribution function $\mathrm{F}(x)$.
c) Use $\mathrm{F}(x)$ to estimate the probabilities:
i) $\mathrm{P}(x > 3.5)$ **ii)** $\mathrm{P}(x < 2.5)$.
d) Determine the median and quartiles of the fuel consumption; hence write down the quartile spread.
e) Find the level of consumption exceeded, on average, in only one month in ten.

EXPECTED VALUE

We met the concept of the expected value of a random variable in Chapter 5, *Expectation*. The focus there was on discrete variables and it was shown that the expected value was the same as the mean:

$$\text{mean} = \mathrm{E}(X) = \sum_{\text{all } x} x\,\mathrm{P}(x)$$

for discrete variables.

In the case of continuous variables, the mean is similarly defined as the expected value. The difference is that summation is replaced by integration, and probability is replaced by probability density:

$$\text{mean} = \mathrm{E}(X) = \int_{\text{all } x} x\,\mathrm{f}(x)\,\mathrm{d}x$$

for continuous variables.

In the case of the copper pipe off-cuts, the employee may have given his general manager his estimate of the mean. Recall that the density function for the length of the off-cuts shown on page 287 is:

$$\mathrm{f}(x)=\begin{cases} 2.5-3.125x & 0 \le x \le 0.8 \\ 0 & \text{elsewhere} \end{cases}$$

Hence, the mean, μ, for this model is found by integration.

$$\text{mean length of pipe, } \mu = \int_0^{0.8} x(2.5 - 3.125x)\,dx$$

$$= \int_0^{0.8} \left(2.5x - 3.125x^2\right)dx$$

$$= \left[2.5 \times \frac{x^2}{2} - 3.125 \times \frac{x^3}{3}\right]_0^{0.8}$$

$$= 0.267$$

Thus, this average, the mean, is a bit more than $\frac{1}{4}$ m.

Exploration 11.5

Using integration to find the mean length

Try using integration to find the mean length of timber off-cut for the model shown in the diagram on page 286.

Variance

You may recall that one of the measures, variance, associated with spread can be defined in expectation terms:

$$\text{Var}(X) = \text{E}(X^2) - \mu^2$$

For a continuous random variable calculation of the expected value of the square, $\text{E}(X^2)$, involves integration:

$$\text{E}(X^2) = \int_{\text{all } x} x^2 \text{f}(x)\,dx$$

Hence, the variance of the lengths of pipe off-cuts according to the model is:

$$\text{variance} = \int_0^{0.8} x^2(2.5 - 3.125x)\,dx - 0.267^2$$

$$= \left[2.5 \times \frac{x^3}{3} - 3.125 \times \frac{x^4}{4}\right]_0^{0.8} - 0.267^2$$

$$= 0.1067 - 0.0711 = 0.0356$$

Exploration 11.6

Calculating the variance

Try calculating the variance of the lengths of timber off-cuts according to the model of the histogram on page 286.

Example 11.2

The DIY store employs a customer information assistant on Saturdays. The assistant offers customers advice about the location of goods in the store. The general manager is interested in the time the assistant spends answering customers' enquiries. A model for the time is proposed. Overleaf is a diagrammatic representation of the model.

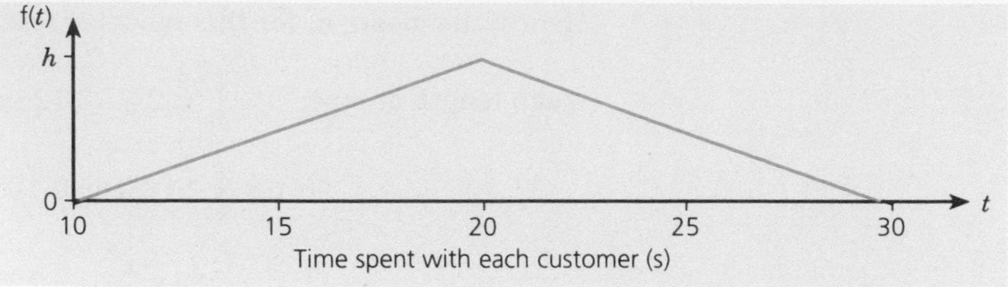

a) *Study the diagram and, without any calculation, estimate each of the following averages.*
i) modal time *ii) median time* *iii) mean time*
b) *Calculate the value of h.*
c) *i) Show that for $10 \le t \le 20$, the probability density function, $f(t)$, has the equation: $f(t) = 0.01t - 0.1$.*
ii) Find the form of $f(t)$ for $20 \le t \le 30$.
d) *Use your expressions for $f(t)$ to determine:*
i) the mean, *ii) the standard deviation,*
of the constant times as indicated by the model.

Solution

a) *i) The modal time is where the density function takes its greatest value. This is for 20 seconds.*
ii) The density function is symmetric about $t = 20$. This indicates that the area under the function is divided into two equal parts at $t = 20$. Hence, the median is 20 seconds.
iii) The symmetry of the density function suggests that the mean may also be 20 seconds.
b) *The area under the function must be equal to 1. The shape is a triangle and its area can be found from 'half base × height'. Hence, the area is:*
$\frac{1}{2} \times 20 \times h = 1$
$\Rightarrow h = 0.1$
c) *i) Between $t = 10$ and $t = 20$, the density function is linear passing through points with coordinates (10, 0) and (20, 0.1). The gradient of the line representing $f(t)$ is:*

$$\frac{0.1}{20 - 10} = 0.01$$

Hence, $f(t)$ is of the form:
$f(t) = 0.01t$ *+ constant*
The constant can be found from either of the points through which the line passes, e.g. finding it for (10, 0) suggests:
$0 = 0.01 \times 10$ *+ constant*
which implies that the constant is –0.1. Check by finding the constant for the line through (20, 0.1). Hence:
$f(t) = 0.01t - 0.1$ *when* $10 \le t \le 20$.
ii) Between $t = 20$ and $t = 30$, $f(t)$ is linear and passes through points (20, 0.1) and (30, 0). The gradient is:

$$\frac{-0.1}{30 - 20} = -0.01$$

Hence, $f(t)$ *is of the form:*

$f(t) = -0.01t +$ *constant*

Ensuring that the line passes through (20, 0.1), say, implies that:

$0.1 = -0.01 \times 20 +$ *constant*

This suggests that the constant is 0.3. Hence:

$f(t) = -0.01t + 0.3$ *when* $20 \le t \le 30$.

d) i) *The mean can be calculated from:*

$$\int_{\text{all } t} t\,f(t)\,dt$$

The range of values of t is from $t = 10$ *to* $t = 30$. *But, the form of t changes at* $t = 20$ *and this must be taken into account in the integral. So:*

$$\begin{aligned}\text{mean} &= \int_{10}^{20} t(0.01t - 0.1)\,dt + \int_{20}^{30} t(0.3 - 0.01t)\,dt \\ &= \left[0.01 \times \frac{t^3}{3} - 0.1 \times \frac{t^2}{2}\right]_{10}^{20} + \left[0.3 \times \frac{t^2}{2} - 0.01 \times \frac{t^3}{3}\right]_{20}^{30} \\ &= 8.333 + 11.667 \\ &= 20\end{aligned}$$

as anticipated in ***a)****.*

ii) *The standard deviation may be found from the variance.*

$$\begin{aligned}\text{variance} &= \int_{10}^{20} t^2(0.01t - 0.1)\,dt + \int_{20}^{30} t^2(0.3 - 0.01t)\,dt - 20^2 \\ &= \left[0.01 \times \frac{t^4}{4} - 0.1 \times \frac{t^3}{3}\right]_{10}^{20} + \left[0.3 \times \frac{t^2}{2} - 0.01 \times \frac{t^4}{4}\right]_{20}^{30} - 400 \\ &= 141.667 + 275 - 400 \\ &= 16.667\end{aligned}$$

Hence, the standard deviation is $\sqrt{16.667} \approx 4.08$.

Rectangular distribution

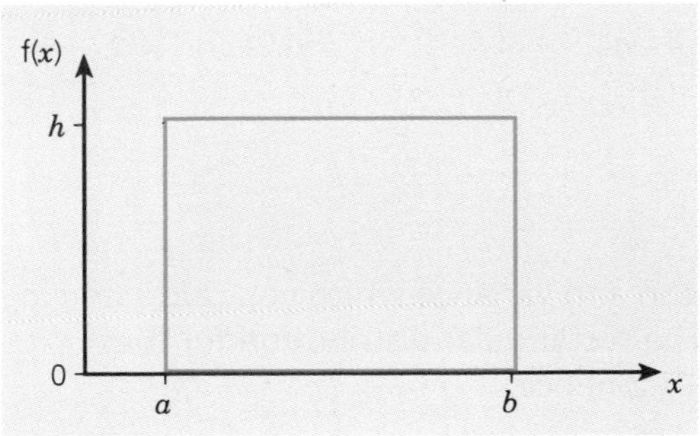

The model proposed for the length of timber off-cuts is an example of a **rectangular probability distribution**. It is characterised by a density function which resembles a rectangle.

Consider a random variable which is rectangularly distributed on $a \le x \le b$. The density function is a constant, h, for this set of values of x. The area of the rectangle must be 1, hence:

$$h \times (b - a) = 1$$

which implies that the **density function** is:

$$f(x) = h = \frac{1}{b-a} \text{ when } a \le x \le b.$$

The mean value of X is:

$$\begin{aligned}
\mathrm{E}(x) &= \int_a^b x\mathrm{f}(x)\mathrm{d}x \\
&= \int_a^b x \times \frac{1}{b-a}\mathrm{d}x \\
&= \left[\frac{1}{b-a}\times\frac{x^2}{2}\right]_a^b \\
&= \frac{1}{b-a}\left(\frac{b^2}{2}-\frac{a^2}{2}\right) \\
&= \frac{1}{2(b-a)}\left(b^2-a^2\right) \\
&= \frac{1}{2(b-a)}(b-a)(b+a) \\
&= \tfrac{1}{2}(b+a)
\end{aligned}$$

The variance is:

$$\begin{aligned}
\int_a^b x^2 \times \frac{1}{b-a}\mathrm{d}x - \left(\frac{b+a}{2}\right)^2 &= \left[\frac{1}{b-a}\times\frac{x^3}{3}\right]_a^b - \frac{(b+a)^2}{4} \\
&= \frac{1}{3(b-a)}\left(b^3-a^3\right)-\frac{(b+a)^2}{4} \\
&= \frac{1}{3(b-a)}(b-a)\left(b^2+ab+a^2\right)-\frac{b^2+2ab+a^2}{4} \\
&= \frac{b^2+ab+a^2}{3}-\frac{b^2+2ab+a^2}{4} \\
&= \frac{4\left(b^2+ab+a^2\right)-3\left(b^2+2ab+a^2\right)}{12} \\
&= \frac{b^2-2ab+a^2}{12} \\
&= \frac{(b-a)^2}{12}
\end{aligned}$$

Exploration 11.7 *Check your results*

Check your values for the mean and variance which you calculated in Explorations 11.4 and 11.5. The rectangular distribution for the timber off-cuts has $a = 0$, $b = 1.2$ hence:

mean $= \dfrac{1.2+0}{2} = 0.6$ metres

and variance $= \dfrac{(1.2+0)^2}{12} = 0.12$

Note: $X \sim \mathrm{Rect}(a, b)$ indicates that the random variable, X, is rectangularly distributed on $a \le x \le b$.

Triangular distribution

Recall the discrete uniform distribution and the triangular distribution from Chapter 8, *Sums and differences of distributions*. Adding together two independent observations of a uniformly-distributed variate results in a triangularly-distributed variate. This is also true for continuous variates. The relationship between the means of the distributions and the relationship between the variances of the distributions also hold.

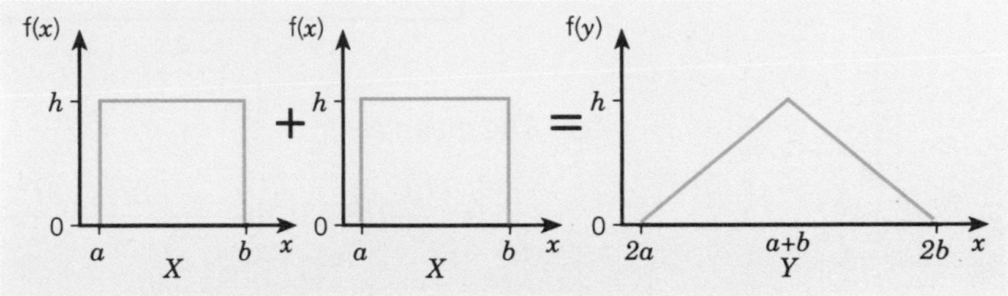

If $X \sim \text{Rect}(a, b)$ and X_1, X_2 are independent observations on X, then the distribution of $Y = X_1 + X_2$ is a triangular distribution symmetric about $y = (a + b)$ as shown in the diagram.

The mean of Y is μ_Y where:

$$\mu_Y = \mu_X + \mu_X = \tfrac{1}{2}(a+b) + \tfrac{1}{2}(a+b) = (a+b)$$

and the variance of Y is $\sigma_Y{}^2$ where:

$$\sigma_Y{}^2 = \sigma_X{}^2 + \sigma_X{}^2 = \frac{(b-a)^2}{12} + \frac{(b-a)^2}{12} = \frac{(b-a)^2}{6}$$

Example 11.3

A model for the distribution of length of time, in minutes, it takes a customer of the DIY store to pay for purchases has density function:

$f(t) = At(8-3t)$ *when* $0 \le t \le 2\tfrac{2}{3}$

a) *Calculate the value of A.*

b) *Sketch a graph of the density function.*

c) *Calculate the mean time to pay for purchases, as indicated by the model.*

d) *What is the standard deviation of the time?*

e) *Estimate the probability that it takes a customer longer than two minutes to pay.*

Solution

a) *The area under the density function must be 1. Hence:*

$$\int_0^{2\frac{2}{3}} At(8-3t)\,dt = 1$$

Expanding and integrating gives:

$$A\left[4t^2 - t^3\right]_0^{2\frac{2}{3}} = 1$$

Evaluating:

$\frac{256}{27} A = 1 \Rightarrow A = \frac{27}{256}$

b)

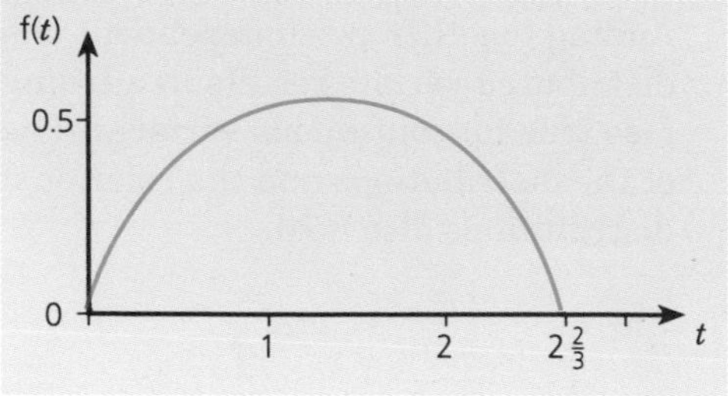

c) *The mean is:*

$$\int_0^{2\frac{2}{3}} t \times \frac{27}{256} t(8-3t)\mathrm{d}t = \frac{27}{256}\left[\frac{8t^3}{3} - \frac{3t^4}{4}\right]_0^{2\frac{2}{3}}$$
$$= 1.333$$

***Note:** We might have found this value without integration, basing our argument on symmetry.*

d) *To calculate the standard deviation, we need to find the variance.*

$$\int_0^{2\frac{2}{3}} t^2 \times \frac{27}{256} t(8-3t)\mathrm{d}t - (1.333)^2 = \frac{27}{256}\left[4t^4 - \frac{3t^5}{5}\right]_0^{2\frac{2}{3}} - (1.333)^2$$
$$= 2.1333 - 1.7778$$
$$\approx 0.3556$$

Hence the standard deviation of times is $\sqrt{0.3556} \approx 0.596$ *minutes.*

e) $$\mathrm{P}(T > 2) = \int_2^{2\frac{2}{3}} \frac{27}{256} t(8-3t)\mathrm{d}t$$
$$= \frac{27}{256}\left[4t^4 - t^3\right]_2^{2\frac{2}{3}}$$
$$\approx 0.156$$

EXERCISES

11.2A

1 The density function for a random variable, X, is given by:

$$\mathrm{f}(x) = \begin{cases} \frac{k}{x^3} & \text{when } 3 \le x \le 6 \\ 0 & \text{elsewhere} \end{cases}$$

a) Calculate the value of the constant, k.
b) Calculate the mean and the variance of X.
c) Calculate the median of X.
d) What is the probability that each of two observations of X are greater than 5?
e) Three observations on X are made. Calculate the probability that exactly one observation is less than 4 and the other two are greater than 4.

2 The distribution of lengths, in cm, of off-cuts of wood is modelled as a rectangular distribution, Rect(5, 10).

a) Write down the mean length and determine the variance of the length.
b) Three randomly chosen off-cuts are measured. Calculate the probability that at least two of them are more than 8 cm.

3 The random variable, X, is distributed rectangularly:

$$X \sim \text{Rect}\left(-\frac{1}{2}\pi, \frac{1}{2}\pi\right)$$

a) Calculate the mean of X.
b) Calculate the variance of X.
c) Determine the distribution function, $F(x)$.

4 A random variable, T, is modelled using the density function:

$$f(t) = \begin{cases} k(7-2t) & \text{when } 0 \le t \le 3 \\ 0 & \text{elsewhere} \end{cases}$$

a) Calculate the value of k.
b) Determine the median of T.
c) Calculate the mean and variance of T.
d) Three independent observations are made on T. Estimate the probability that at least one of these is less than 2.

5 A random variable, X, is distributed rectangularly:

$X \sim$ Rect(–1, 1)

a) Write down the density function, $f(x)$.
b) Calculate the mean and variance of X.
c) Describe the distribution of the sum, Y, of two independent observations on X.
d) Determine the density function, $f(Y)$.
e) Estimate the probability, $P(y > 0.5)$.

6 A random variable, X, has density function:

$$f(x) = \begin{cases} kx(4-x) & \text{when } 0 \le x \le 2 \\ 0 & \text{elsewhere} \end{cases}$$

a) Determine the value of the constant, k.
b) Show that the mean is 1.25.
c) Calculate the variance of X.
d) Estimate the proportion of observations on X which will be within one standard deviation of the mean.

EXERCISES

11.2B

1 A random variable, T, has density function:

$$f(t) = \begin{cases} (2k)^{-1} & \text{when } 0 \le t \le 2k \\ 0 & \text{elsewhere} \end{cases}$$

a) Determine the value of the mean in terms of k.
b) Calculate the variance as a function of k.

2 A random variable, X, is rectangularly distributed:

$X \sim \text{Rect}(a, b)$

a) Write down the density function, $f(x)$.
b) Determine the distribution function, $F(x)$.

3 The density function for a random variable, X, is given by:

$$f(x) = \begin{cases} kx(4-x) & \text{when } 0 \le x < 4 \\ 0 & \text{elsewhere} \end{cases}$$

a) Use calculus to obtain the value of the constant, k.
b) Determine the mean of X.
c) Calculate the standard deviation of X.
d) Estimate the probability, $P(x < 1)$.

4 A random variable, X, is modelled as being rectangularly distributed:

$X \sim \text{Rect}(a^2, 4a^2)$

a) Write down the density function for X.
b) X represents the area of a square which is randomly generated by a computer. What is the mean area of the random square?
c) Calculate the probability that the length of the side of a square is less than $1.5a$.

5 A model for the distribution of mass, W, in grams, of grade II strawberries has density function:

$$f(w) = \begin{cases} k(30w - w^2 - 200) & \text{when } 10 \le w < 20 \\ 0 & \text{elsewhere} \end{cases}$$

where k is constant.

a) Calculate the value of k and sketch the graph of $f(w)$.
b) Calculate the mean mass of grade II strawberries.
c) Calculate the standard deviation of the mass.
d) Four strawberries are chosen at random. Calculate the probability that each weighs over 13 g.

6 A university lecturer estimates that most of her students will spend between two and five hours on a particular piece of coursework. Very few spend less than two hours and none spends in excess of five hours. She models the time, T, in hours spent using the density function:

$$f(t) = \begin{cases} kt & \text{when } 0 \le t < 2 \\ \frac{2}{3}k(5-x) & \text{when } 2 \le t \le 5 \\ 0 & \text{elsewhere} \end{cases}$$

a) Find the value of k.
b) Sketch the graph of the density function.
c) Calculate the mean and variance of T.
d) Determine the probability that a student spends more than four hours on the coursework.
e) Estimate the proportion of students who spend less than an hour on the work.

MATHEMATICAL MODELLING ACTIVITY THE EXPONENTIAL DISTRIBUTION

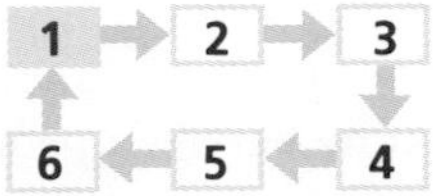

Specify the real problem

Specify the real problem

The DIY store's general manager is concerned about the queues that build up at the checkouts. She decides that she needs to collect information about the pattern of arrivals of customers. An A-level student who works at the store on Saturdays, said that the Poisson distribution was a useful model for the number of people arriving at the store in a fixed period of time. Under what circumstances is this an appropriate model?

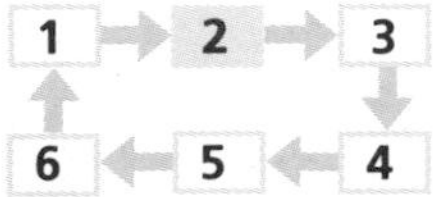

Set up a model

Set up a model

Records of the number of people entering the store on Saturdays indicate that there are, on average, 1500 during the ten hours that the store is open for trading. The manager offers this information to the student. He says that this is equivalent to an average of 150 people per hour or 2.5 people per minute.

The manager says she needs to know something about the time gap between arrivals.

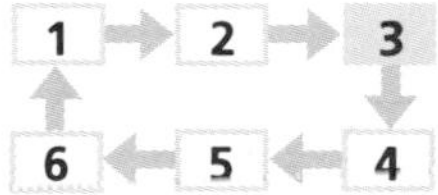

Formulate the mathematical problem

Formulate the mathematical problem

The student says about 92 per cent of the times between arrivals will be one minute or less. The manager asked him how he reached that conclusion.

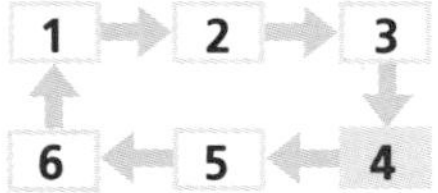

Solve the mathematical problem

Solve the mathematical problem

The student told her that the Poisson model indicates the probability of no person arriving in a given one minute period is $e^{-2.5}$ which is about eight per cent.

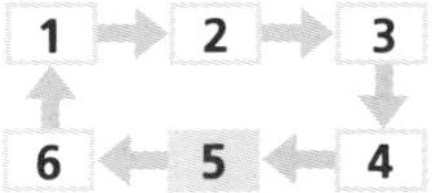

Interpret the solution

Interpret the solution

He says this is therefore the chance of waiting for more than a minute for the next person to arrive in the store. Hence, 92 per cent of times between arrivals will be less than one minute. The manager says she thinks she understands, but she really wants to know more about the distribution of the possible time gap rather than just the likelihood of the time being less than one minute.

The student says that the Poisson model can still help. If there are 2.5 people, on average every minute, then in a two-minute period there would be an average of five. This suggests that in a period of t minutes, there would be an average of $2.5t$ people. Then, if the manager follows the same argument as above:

P(time gap between arrivals $\leq t$ minutes) = 1 – P(no arrival in t minutes)
$= 1 - e^{-2.5t}$

He says that this can be shortened to:

$$P(T \leq t) = 1 - e^{-2.5t}$$

where T is the random variable which models the time between people arriving at the store.

This random variable, T, is a continuous variate, and the probability, $P(T \leq t)$, is a cumulative probability which implies that the distribution function for T is:

$$F(t) = 1 - e^{-2.5t} \quad \text{when } t \geq 0$$

It is quite possible to use the distribution function to determine probabilities. For example, the manager asks what the chance is that there will be less than 20 seconds between arrivals. The student says that this is equivalent to asking for the probability:

$$P(T \leq \tfrac{1}{3}) = 1 - e^{-\frac{2.5}{3}} \approx 0.565$$

And the chance that the time gap is somewhere between one and two minutes? The student says this is:

$$\begin{aligned} P(1 \leq T \leq 2) &= F(2) - F(1) \\ &= \left(1 - e^{-5}\right) - \left(1 - e^{-2.5}\right) \\ &= (1 - 0.0067) - (1 - 0.0831) \\ &\approx 0.0075 \quad \text{i.e. } 7.5\% \end{aligned}$$

The manager asks if it is possible to use this distribution function to find the mean time gap indicated by the model. The student tells her that the density function is needed for that. Normally a distribution function, $F(t)$, is found by integrating a density function, $f(t)$. However, in this case the distribution function is available. So to get the density function required for the mean it is necessary to differentiate $F(t)$.

$$\begin{aligned} f(t) &= \frac{d}{dt}F(t) \\ &= \frac{d}{dt}\left(1 - e^{-2.5t}\right) \\ &= 2.5e^{-2.5t} \end{aligned}$$

Hence, the density function for the time gap is:

$$f(t) = 2.5e^{-2.5t} \quad \text{when } t \geq 0$$

thus the mean can be found:

$$\text{mean} = \int_0^\infty t \times 2.5e^{-2.5t}\,dt$$

using integration by parts:

$$\int_0^\infty t \times 2.5e^{-2.5t}\,dt = \left[-te^{-2.5t}\right]_0^\infty - \left[\frac{1}{2.5}e^{-2.5t}\right]_0^\infty$$

Clearly, it is not really possible to replace t by 'infinity', however, putting a large value in place of infinity gives:

$$\left[-te^{-2.5t}\right]_0^{\infty}-\left[\frac{1}{2.5}e^{-2.5t}\right]_0^{\infty}=[0-0]-\frac{1}{2.5}[0-1]=\frac{1}{2.5}$$

Hence, the mean time is 0.4 minute or 24 seconds.

The time gap considered here is an example of a continuous random variable which has an **exponential distribution**. The modelling assumptions:

- random events
- occurring singly and independently
- in a continuous medium
- at a uniform rate

are necessary as they are required for the discrete Poisson distribution. The random variable of the exponential distribution is the interval of continuous medium between each random event.

Where the random events occur at the uniform rate, λ per unit continuous medium, then if X is the variate measuring the gap between events:

distribution function: $F(x)=1-e^{-\lambda x} \quad x \geq 0$

density function: $f(x)=\lambda e^{-\lambda x} \quad x \geq 0$

mean of X: $\mu=\dfrac{1}{\lambda}$

Example 11.4

The DIY store sells tape measures supplied by one manufacturer. The tapes are supplied in various lengths: 1 m, 2 m, 5 m, 10 m and 50 m. The manufacturer claims that the number of flaws in any 250 m of the material used in making the tape measure is less than 1.

a) *Estimate the probability:*
   i) *of a 10 m tape being free from flaws,*
   ii) *of having at least one flaw in a 50 m tape,*
   iii) *of having no tape measure with a flaw in a box of 100 2 m tapes.*

b) *A DIY enthusiast, who bought a 5 m tape complains to the manager that it was flawed. The manager asks to see it, but the customer has disposed of it. The manager is somewhat suspicious and seeks statistical advice from the student. What advice would you offer?*

Solution

a) *In order to estimate probabilities it is necessary to set up a model. It seems reasonable to consider 'flaws in tape material' as random events in a continuous medium. Provided the assumptions of singularity, independence and uniform rate of occurrence are appropriate, then an exponential model is relevant. The claim of* *the manufacturer might be interpreted in a 'worst case scenario' as indicating that the average number of flaws is one per 250 m. In other words, the rate of occurrence is* $\frac{1}{250}$ *flaw per m, i.e. 0.004 flaw per metre. Thus, the distribution function is:*

$$F(x) = 1 - e^{-0.004x},\ x \geq 0$$

i) P(*10 m tape being free from flaws*)
$= P(X > 10) = e^{-0.004 \times 10} \approx 0.961$

ii) P(*at least one flaw in a 50 m tape*)
$= P(x \leq 50) = 1 - e^{-0.004 \times 50} \approx 0.181$

iii) *It will help to estimate the probability of a 2 m tape being free from flaws.*
This is $e^{-0.0004 \times 2} \approx 0.992\ 03$.
Hence, P(*all 100 tapes free from flaws*) $\approx (0.992\ 03)^{100} \approx 0.449$.

b) *The probability that a 5 m tape is unflawed is 0.9802. Hence, the chance that a tape is flawed is about two per cent. So the student may well report that, under the assumption that flaws occur at an average rate of one per 250 metres, it is unlikely that a 5 m tape will contain a flaw. The manager's suspicions may be appropriate, depending on how many 5 m tapes have been sold.*

CONSOLIDATION EXERCISES FOR CHAPTER 11

1 A random variable is modelled using an exponential distribution with mean equal to 1.

a) Obtain the cumulative distribution function $F(x)$.
b) By appropriate use of $F(x)$, determine:
  i) the median,
  ii) the probability that a random observation is between the median and the mean.

2 The life, X hours, of a high-intensity projector bulb may be modelled by an exponential distribution where the distribution is:

$$f(x) = \begin{cases} 1 - e^{-\frac{x}{10}} & \text{when } x \geq 0 \\ 0 & \text{otherwise} \end{cases}$$

a) Calculate the median life.
b) Calculate the probability that a bulb lasts between five and ten hours.
c) Derive the density function.
d) Calculate the mean and variance of the life of a bulb.

3 The time, T days, between successive notifications of a rare disease may be modelled by an exponential distribution. The mean time is observed to be 80 days. Calculate the probability that the time between two successive notifications is:

a) less than 30 days, **b)** between 50 and 100 days.

4 The number of breakdowns of a computer in a month may be modelled as a Poisson variate with mean equal to 2. Assume that a month has 30 days.

a) Calculate the probability that there is no further breakdown until the 16th day, given that there was a breakdown on the first.
b) Determine the probability that the time between breakdowns is:
i) less than ten days, **ii)** between ten and 20 days,
iii) between 20 and 30 days.
c) Given that there is no breakdown during the first ten days of one month, calculate the probability that there will be a breakdown during the next ten days.

5 The time, T, between the arrival of successive vehicles on a country road may be modelled using an exponential variate with $\lambda = 0.01$ per second.

a) Calculate the mean and variance of T.
b) A heavily-laden pedestrian tries to cross the road. She sets off just as one vehicles passes. Assume that, unhindered, she would take 50 seconds to cross the road. Calculate the probability that she succeeds before another vehicle appears.

6 The probability that at least t metres of curtain material is free from flaws is $e^{\frac{-t}{a}}$.

a) Find the density function for the flaw-free length of material.
b) Prove that the mean length is a metres.
c) If the mean length is 15 m, what is the probability that there is a 30 m length free of flaws?

7 The density function for a random variable, X, is:

$$f(x) = \begin{cases} \frac{1}{2} - \frac{1}{8}x & \text{when } 0 \le x \le b \\ 0 & \text{otherwise} \end{cases}$$

a) Calculate the value of b.
b) Find the mean and variance of X.
c) Sketch the density function and state the modal value of X.
d) Determine the median and the quartiles of X.

8 A random variable, T, may be modelled using the density function:

$$f(t) = \begin{cases} k(1 - t^2) & \text{when } -1 \le t < 1 \\ 0 & \text{otherwise} \end{cases}$$

a) Calculate k and sketch the graph of $f(t)$.
b) Determine the distribution function for T.
c) Calculate the probabilities:
i) $P(X > \frac{1}{2})$ **ii)** $P(-\frac{1}{2} < X < \frac{1}{2})$
d) Determine the mean and variance of X.

9 A random variable, X, may be modelled using the rectangular distribution:

$X \sim$ Rect(5,10)

a) Write down the mean and variance of X.
b) Determine the mean and variance of the sum of two independent observations on X.
c) Calculate the probability that the sum of two independent observations on X is more than 12.

10 Assume that $T \sim$ Rect(0, 1) and that t_1, t_2 are independent observations on T.

a) Calculate the probabilities:
i) $P(t_1 + t_2 < 0.5)$ **ii)** $P(t_1 + t_2 > 1.8)$
b) Determine the mean, variance and density function for $t_1 - t_2$.

11 Lengths of off-cuts of copper piping in a plumber's yard are rectangularly distributed between 5 cm and 30 cm. State the mean and variance of the lengths of off-cuts.

Calculate the probability that, out of four randomly chosen off-cuts, exactly two are more than 20 cm long.

(AEB Question 5(b), Applied Statistics 2, Summer 1994)

12 The number of telephone calls per minute received by the switchboard is known to have a Poisson distribution with mean 15.

Calculate the probability that the time between two successive calls is:
a) at most 6 seconds, **b)** between 5 and 10 seconds.

(NEAB Question 4, Specimen paper 9, 1995)

13 Workers on a large industrial estate have journey times to work, in hours, which are modelled by the random variable X with probability density function:

$f(x) = 20x^3(1 - x), \quad 0 \le x \le 1$

a) Sketch the graph of the probability density function.
b) Hence state, with an explanation, the likely location of the industrial estate in relation to local housing.
c) Find the expectation and the standard deviation of the workers' journey times *in minutes*.
d) Obtain the cumulative distribution function, and use it to verify that the median journey time is a little over 41 minutes.

(MEI Question 4, Paper S3, January 1994)

14 The lifetime, X hours, of a certain type of cathode ray tube in a given environment has an exponential distribution with mean lifetime 100 hours.

a) Write down the probability density function of X and obtain the cumulative distribution function X.
b) Find the probability that a tube chosen at random lasts less than 50 hours.

(ULEAC Question 4, Specimen paper T1, 1995)

15 The continuous random variable X has the probability density function given by:

$$f(x) = \begin{cases} \dfrac{1}{x} & 1 \le x \le a \\ 0 & \text{otherwise} \end{cases}$$

a) Show that $a = \mathrm{e}$.
b) Find, in terms of e, the mean and variance of X, simplifying your answers.
c) Show that the probability that the value of X lies within one standard deviation of the mean is approximately 0.59.

(NEAB Question 13, Specimen paper 2, 1995)

16 The amount, X kg, of impurities per 10 kg batch of mineral ore is a continuous random variable with probability density function f given by:

$$f(x) = \begin{cases} \frac{1}{8}(x-5) & 5 \le x \le 9 \\ 0 & \text{otherwise} \end{cases}$$

a) Find the mean of X.
b) Show that, in the interval $5 \le x \le 9$, the distribution function F of X is given by $F(x) = \frac{1}{16}(x-5)^2$.
State the values of $F(x)$ for $X < 5$ and for $X > 9$.
c) Find the median value of X.

The value, £Y, of a batch of ore after processing is given by:
$Y = 5 - \frac{1}{2}X$
Find, in either order:
d) the distribution function, G of Y,
e) the probability density function, g of Y.

(NEAB Question 10, Paper 2, June 1993)

17 The continuous random variable, X, has a rectangular (or uniform) distribution on the interval $0 \le x \le 4$.

a) Write down the probability density function for X, and sketch its graph.
b) Find the cumulative distribution function $F(x) = P(X \le x)$.

A second random variable, Y, is related to X by the equation $Y = \sqrt{X}$.
c) Show that $P(Y \le y) = P(X \le y^2) = \frac{1}{4}y^2$. Using this, the cumulative distribution function for Y, find the probability density function for Y. State the range of values which Y takes.
d) Show that the median value of Y is the square root of the median value of X.

(MEI Question 2, Paper S3, June 1993)

18 In an attempt to economise on her telephone bill, Debbie times her calls and ensures that they never last longer than four minutes. The length of the calls, T minutes, may be regarded as a random variable with probability density function:

$$f(t) = \begin{cases} kt & 0 < t \le 4 \\ 0 & \text{otherwise} \end{cases}$$

where k is a constant.

a) **i)** Show that $k = 0.125$.

ii) Find the mean and standard deviation of T.

iii) Find the probability that a call lasts between three and four minutes.

iv) What is the probability that, of five independent calls, exactly three last between three and four minutes?

Calls are charged at a rate of 6p per call plus 4p for each complete minute that the call lasts.

b) **i)** Copy and complete the following table.

Length of call, in minutes	Probability	Cost, in pence
0–1		
1–2		
2–3		
3–4		18

ii) Find the mean and standard deviation of the cost, in pence, of a call.

(AEB Question 8, Paper 9, Summer 1994)

19 The length, in metres, of off-cuts of wood found in a timber yard can be modelled by a continuous uniform distribution with density function, $f(x)$, defined as:

$$f(x) = \begin{cases} \dfrac{1}{k} & 0.2 < x < 0.8 \\ 0 & \text{otherwise} \end{cases}$$

Write down the value of k and the mean length. Calculate the variance of the length.

The triangular distribution may be used to model the sum of the lengths of two randomly chosen off-cuts. Sketch the density function. Calculate the probability of the sum of two lengths being more than 1.2 m.

(Oxford Question 3, Specimen 63, 1995)

20 Lengths of off-cuts of copper piping, collected after the installation of central heating in houses, are rectangularly distributed between 10 cm and 50 cm.

Calculate the probability that the length of a randomly selected off-cut:

a) is precisely 25 cm

b) is more than 25 cm.

Calculate the probability that, out of five randomly selected off-cuts, exactly two are more than 25 cm long.

(NEAB Question 2, Specimen paper 9, 1995)

21 A continuous random variable, X, has probability density function f given by:

$$f(x) = \begin{cases} \frac{1}{\pi}(1+\cos x) & 0 \le x \le \pi \\ 0 & \text{otherwise} \end{cases}$$

a) Use the method of integration by parts to show that:

$$E(X) = \frac{\pi}{2} - \frac{2}{\pi}$$

b) Find the distribution function of X for all values of x.

c) Show that the 60th percentile, t, of the distribution of X satisfies the equation:

$$t = 0.6\pi - \sin t$$

d) Show that $1 < t < 1.1$ and use the iterative formula:

$$t_{n+1} = 0.6\pi - \sin t_n$$

to find the value of t correct to two decimal places.

(NEAB Question 13, Specimen paper 2, 1995)

Summary

- The probability density function is used as a model for relative frequency.
- If f(x) is a probability density function then:

$$\int_{\text{all } x} f(x)\,dx = 1$$

$$\text{mean} = E(X) = \int_{\text{all } x} x\,f(x)\,dx$$

$$\text{variance} = \text{Var}(X) = \int_{\text{all } x} x^2\,f(x)\,dx - \mu^2$$

- The rectangular probability distribution is:

$$f(x) = \frac{1}{b-a} \quad \text{for } a \le x \le b$$

It has mean $E(X) = \frac{1}{2}(a+b)$ and variance $\frac{1}{12}(b-a)^2$

- The exponential probability distribution is:

$$f(x) = \lambda e^{-\lambda x} \quad x \ge 0$$

It has a mean $E(X) = \frac{1}{\lambda}$

STATISTICS

12 Validating models – the chi-squared distribution

In this chapter we shall:

- *reflect on the models discussed so far and find out how well they fit observed data,*
- *investigate an intuitive way of measuring the quality of the fit by considering the difference between the frequencies observed and those expected by the model.*

Exploration 12.1

Observed frequencies

You need two cubical dice, each numbered 1, 2, 3, 4, 5, 6. Roll the dice and record the number of **even** numbers (this will be 2, 1 or 0) in the outcome. Now repeat this process until you have recorded the outcomes for 20 rolls of the two dice. Record your results in a table like this.

No. of evens	Observed frequency
0	
1	
2	

Display your results in a frequency block graph.

EXPECTED FREQUENCIES

Let X represent the random variable 'the number of even numbers displayed when two dice are rolled'. What probability distribution is appropriate to model the relative frequency of the outcomes? We can assume that the dice are symmetrical and each face is equally likely to appear uppermost. Hence the probability that an even number results is $\frac{1}{2}$ for each die and for each roll. We can also assume that the outcome on one die is independent of the outcome on the other. The number of evens seen on each roll of the pair of dice may thus be modelled by a binomial distribution:

$$X \sim \mathrm{B}(2, \tfrac{1}{2})$$

This model suggest the following probabilities:

$$P(X=0) = {}_2C_0\left(\tfrac{1}{2}\right)^0\left(\tfrac{1}{2}\right)^2 = \tfrac{1}{4}$$

$$P(X=1) = {}_2C_1\left(\tfrac{1}{2}\right)^1\left(\tfrac{1}{2}\right)^1 = \tfrac{1}{2}$$

$$P(X=2) = {}_2C_2\left(\tfrac{1}{2}\right)^0\left(\tfrac{1}{2}\right)^2 = \tfrac{1}{4}$$

How well does this model fit your observed data?

Compare with reality

A first attempt at answering this might be to compare these probabilities with the frequencies you observed. But, if you do, you are not comparing like with like. The quantities you are comparing should be like quantities, either both frequencies or both probabilities. In this chapter we concentrate on comparing frequencies. We need to convert the probabilities suggested by the model into **expected frequencies**.

The model suggests that if a quarter of outcomes result in no evens, a quarter of the 20 rolls of the dice are expected to result in no evens. The model suggests the following results.

X	Expected frequency in 20 rolls
0	5
1	10
2	5

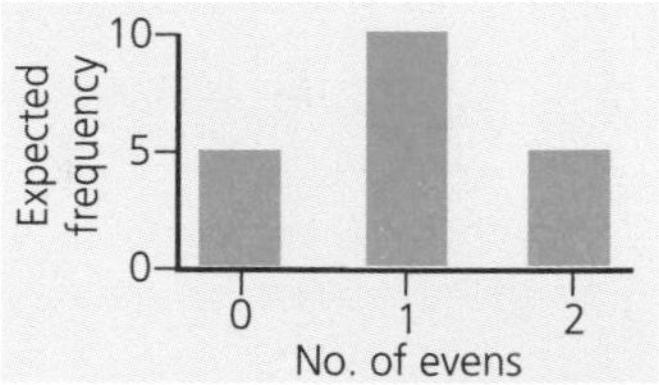

A graphical display of these expected frequencies is shown here. Compare these expected frequencies with your observed frequencies.

A measure of the quality of fit

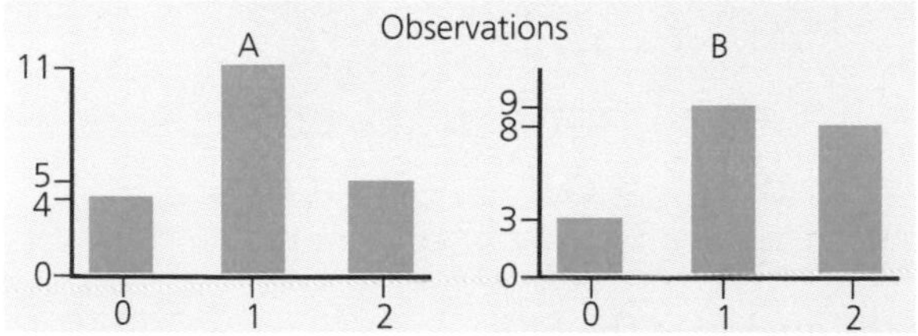

We are now in a better position to address the issue of how well a model fits observed results. Consider the two collections of observations displayed in this diagram. Do you think that:

- either
- neither
- both

are reasonably modelled by the display above?

Intuitively, it seems reasonable to say that observations A are very well modelled, but that the fit of the model is less good for observations B.

Descriptions of the fit such as 'good', 'poor' or 'perfect' are **qualitative**. In this chapter we are interested in **quantifying** the fit of providing a *measure of the quality of the fit* provided by a model. A simplistic approach to this is to find the total differences between the observed frequency and the frequency expected by the model. For observations A, this is shown here.

Outcome	$X = 0$	$X = 1$	$X = 2$
Observations A: O	4	11	5
Expected frequencies: E	5	10	5
Difference: $O - E$	−1	1	0

The total of the differences is zero. Try calculating the total of the differences between observations B and the expected frequencies. You should find that the total of the differences is also zero despite the fit appearing to be poorer. Unfortunately, the result will always be zero, regardless of the quality of the fit. The total of the positive differences will always be balanced by the total of the negative differences.

To overcome this we can square the differences. However, there is a further consideration. In observations A, the model expected five occurrences of $X = 0$ whereas four were observed, and ten occurrences of $X = 1$ when eleven were observed. In each case, the squared difference:

$$(O - E)^2 = 1$$

is the same. But this squared differences is a much larger proportion of the expected frequency in the case $X = 0$ than in the case $X = 1$. It is $\frac{1}{5}$ i.e. 20 per cent, compared with $\frac{1}{10}$ i.e. ten per cent. This **relative difference** should be taken into consideration.

A measure of the quality of fit can be found by the following process.

1 Find the difference: $(O - E)$

2 Square the differences: $(O - E)^2$

3 Express these as a proportion of the expected frequency: $\frac{(O-E)^2}{E}$

4 Find the total of these relative differences: $\sum \frac{(O-E)^2}{E}$

Applying this process to observations A produces this table.

Outcome	O	E	$(O - E)^2$	$\frac{(O-E)^2}{E}$
$X = 0$	4	5	$(-1)^2$	$\frac{1}{5} = 0.2$
$X = 1$	11	10	$(1)^2$	$\frac{1}{10} = 0.1$
$X = 2$	5	5	$(0)^2$	$\frac{0}{5} = 0.0$

Hence, the measure of the quality of the fit is:

$$\sum \frac{(O-E)^2}{E} = 0.3$$

Exploration 12.2 *Measure of fit for observations B*

Find the measure of fit of the model to the observation B.

The results you may have found are shown below.

Outcome	O	E	$(O - E)^2$	$\frac{(O-E)^2}{E}$
$X = 0$	3	5	$(-2)^2$	0.8
$X = 1$	9	10	$(-1)^2$	0.1
$X = 2$	8	5	$(3)^2$	1.8

This produces:

$$\sum \frac{(O-E)^2}{E} = 2.7$$

Interpreting the measure of fit

It is worth reflecting on these measures. The binomial model seemed to fit observations A better than it fitted observations B. The measures of fit are 0.3 and 2.7 respectively. These measures are associated with the difference between the observations and the model. When the difference is large then the measure is large. The reverse is also true – when the difference is small the measure is small. A simple interpretation of the measure of fit is possible.

Value of $\sum \frac{(O-E)^2}{E}$	$\Rightarrow$	Quality of fit
small	$\Rightarrow$ small difference $\Rightarrow$	good
large	$\Rightarrow$ large difference $\Rightarrow$	poor

Exploration 12.3

An experiment

You need two cubical dice again. Roll the two dice and record the number of square numbers (i.e. 1 or 4) in the outcome.

- Repeat the process until you have recorded the results of 50 rolls of the two dice.
- Record your results in a frequency table.
- Try to explain why the binomial model B(2, $\frac{1}{3}$) is appropriate to model the relative frequencies.
- Use the probabilities of the model to find the expected frequencies of observing 0, 1, 2 squares.
- Obtain the measure of the quality of the fit.
- Interpret the measure.

Note: The expected frequencies are 22.222…, 22.222…, 5.555… in this case. It is reasonable to use these to three decimal places of accuracy. Although they are taken to represent frequencies, we do not use integers because these are the average frequencies expected by the model. Sticking to integers introduces errors which can cause confusion in the interpretation of the measure of fit.

Empirical distribution of the measure of fit

The table above provides a very simple way of interpreting the measure of fit for a particular model. A large value indicates a poor fit. The larger the value, the poorer the fit – but how large is large? We need to discover the distribution of possible values of this measure of fit. From this we ought to be able to determine which values might be considered to be large. One way of doing this is to generate 20 observations of a random variable, X, where:

$$X \sim B(2, \tfrac{1}{2})$$

as in Exploration 12.1. Having obtained the observed frequencies, we need to calculate the corresponding measure of fit:

$$\sum \frac{(O-E)^2}{E}$$

This process needs to be repeated, say 50 to 100 times. Clearly, using dice could be time consuming. You could use a spreadsheet, a computer program or a programmable calculator. The following activity provides a way of automating the process.

CALCULATOR ACTIVITY

Simulating the distribution of $\sum \frac{(O-E)^2}{E}$ for the B(2, 0.5) model

Observations of X can be simulated using 0s and 1s. The 1 corresponds to an even outcome on the die, and the 0 corresponds to an odd outcome. The 0s and 1s are generated in pairs so that on average they occur in equal proportions. The observation of the random variable, X, is the sum of the pair of digits.

Pair of digits	Sum	Value of X
(0, 0)	0	0
(0, 1) or (1, 0)	1	1
(1, 1)	2	2

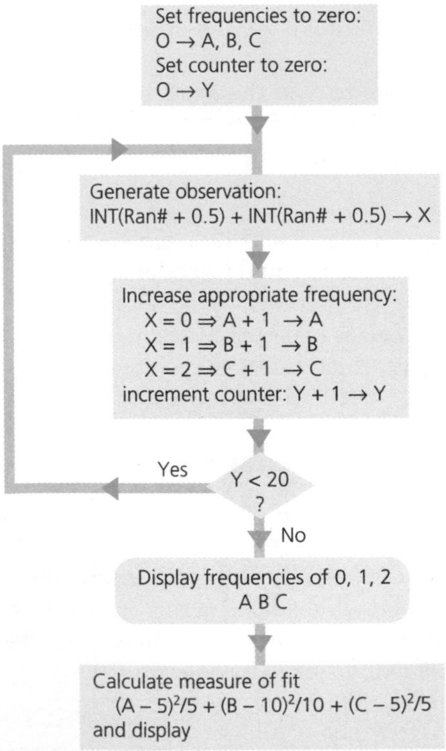

Random decimals occur uniformly in the range 0.000 to 0.999, and if 0.5 is added to them, numbers in the range 0.500 to 1.499 are produced. Taking the integer part of this produces the numbers 0 and 1 in equal proportions. So:

$$\text{INT}(r + 0.5)$$

where r is random decimal, produce 0, 1 with equal likelihood. An observation of X can be produced by adding two of these independently produced numbers:

$$X = \text{INT}\,(r_1 + 0.5) + \text{INT}(r_2 + 0.5)$$

The flowchart shows a program to produce 20 observations of X. The observed frequencies of $X = 0$, $X = 1$, $X = 2$ are stored in memories A, B, C. The value of the measure of fit is produced in the final section.

Run this program, recording the outcome each time, between 50 and 100 times. Now display the distribution of values of the measure of fit you have obtained. A stem-and-leaf display is very effective.

The sort of results you might obtain are shown below. This display represents the values of the measure of fit for 84 sets of 20 observations of X.

Unit = 0.1

0	4 3 4 3 3 3 8 0 3 3 3 3 3 8 3 3 8 3 0 4 3 8
1	2 2 1 5 1 1 9 5 0 9 2 9 2 9 2 9 9 2 2 2 9 6 1 1 9 1 6
2	7 7 4 7 7 4 7 4 7 7 7 4
3	6 6 6 2 6 6 6 6 6 2
4	8 8 4 8 8 4
5	1 1 9
6	8 8
7	2
8	9

The 84 values span 0.0 to 8.9. Their distribution is highly skewed and positively skewed. Each value has arisen as the measure of fit of the binomial model B(2, 0.5) to collections of observations generated in the same way. The aim here is to determine what is meant by a 'large' value of the measure of fit. We might think that a value greater than 8.9 (since this is the largest obtained in this experiment) would be large but we should be wrong. However, it is worth reflecting on the concept of hypothesis testing and significance.

Now we shall test the null hypothesis: 'A model provides a good fit for observed frequencies' against the alternative hypothesis: 'It is not a good fit'. In brief:

H_0: model fits; H_1: model doesn't fit.

Using the measure we have discovered in this chapter, we can state the parallel hypothesis:

H_0: $\sum \frac{(O-E)^2}{E}$ is small; H_1: $\sum \frac{(O-E)^2}{E}$ is large.

If you use a five per cent significance level, you are prepared to take a five per cent chance of wrongly rejecting the null hypothesis.

Five per cent of the $\sum \frac{(O-E)^2}{E}$ values in the stem-and-leaf diagram above are larger than 5.9, and the next largest value is 6.8. It is reasonable, on this evidence, to conclude that the five per cent critical value lies somewhere between 5.9 and 6.8.

Exploration 12.4

Identifying critical values

Try identifying the critical value according to the distribution you obtained in the Calculator activity.

Discovering the critical value

Before we can obtain a definitive five per cent critical value for this measure, we need to explore various observations of random variables, expected frequencies and associated measures of fit.

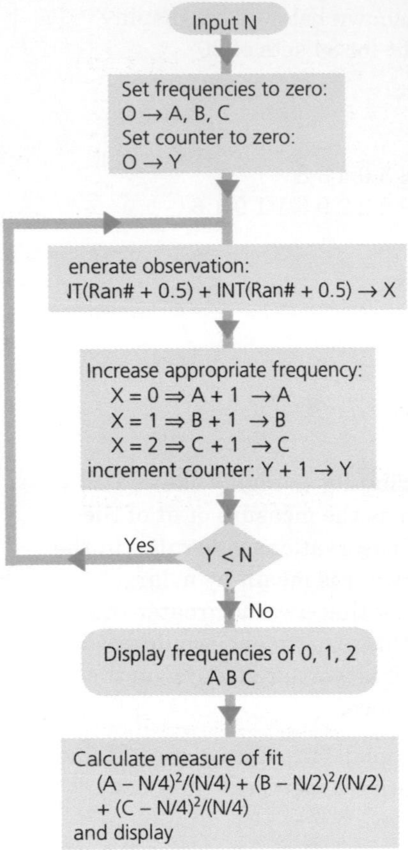

The stem-and-leaf diagram above was based on sets of 20 observations of a random variable, X, modelled by B(2, 0.5). We might ask if sets of 30, 40, 50, 100 observations of X would have provided a substantially different distribution of the measure of fit. How might you modify the flowchart in the Calculator activity on page 318 to accomplish this? You need to input the number of observations required, then to replace the 20 by this input.

You might like to adjust your program to take this change into account.

The sort of results you might get by generating sets of 50 observations of X are shown in the stem-and-leaf display here. The values of 120 measures of fit are presented. Compare this distribution with the one above. The range is similar and the distribution is also highly positively skewed.

Unit = 0.1

0	883376470003736633672877387006663272338 72
1	6300903935639653633033363 3
2	096410292691612221211602 66
3	5252585255282
4	5887
5	344
6	44401
7	
8	5
9	9

What value of the measure of fit does this distribution suggest should be treated as 'large'? Since there are 120 values of the measure of fit, the five per cent critical region would contain the six largest values: 9.9, 8.5, 6.4, 6.4, 6.4, 6.1. Hence, the five per cent critical value lies between 6.0 and 6.1 which is similar to the conclusion reached when using the binomial distribution to model different circumstances.

Critical values for other models

Perhaps a different distribution and a distinctly different critical region will be indicated when using different binomial models? For instance, the model proposed in Exploration 12.3 was B(2, $\frac{1}{3}$) and this seems an appropriate model when recording the number of 'squares' obtained on rolling two cubical dice. It is useful to explore the distribution of measures of fit obtained under these circumstances.

Using dice to generate the observed frequencies is time consuming. The following Calculator activity develops the method already introduced for B(2, 0.5) and can be adapted for a programmable calculator, computer program or spreadsheet.

CALCULATOR ACTIVITY

Simulating the distribution of $\sum \frac{(O-E)^2}{E}$ for B(2, p)

In the previous Calculator activity, you used the integer function, random decimals and 0.5 to simulate observations of X ~ B(2, 0.5). A simple modification of this allows observation of Y:

$Y \sim B(2, p)$

to be simulated.

$Y = \text{INT}(r_1 + p) + \text{INT}(r_2 + p)$

■ Adapt your program by replacing

INT (Ran# + 0.5) + INT (Ran# + 0.5) → X

by

INT (Ran# + $\frac{1}{3}$) + INT (Ran# + $\frac{1}{3}$) → X

What else needs to be changed in the program? The expected values are not $\frac{1}{4}$N, $\frac{1}{2}$N, $\frac{1}{4}$N, so these need to be calculated. What is the probability that $X = 0$?

$$P(X = 0) = {}_2C_0 \left(\frac{1}{3}\right)^2 \left(\frac{2}{3}\right)^2 = \frac{4}{9}$$

So, the expected frequency for $X = 0$ is
$\frac{4}{9} \times N$

Similarly, the expected frequency for:
$X = 1$ is $\frac{4}{9} \times N$ $X = 2$ is $\frac{1}{9} \times N$

This means that the final box in the flowchart is changed to:

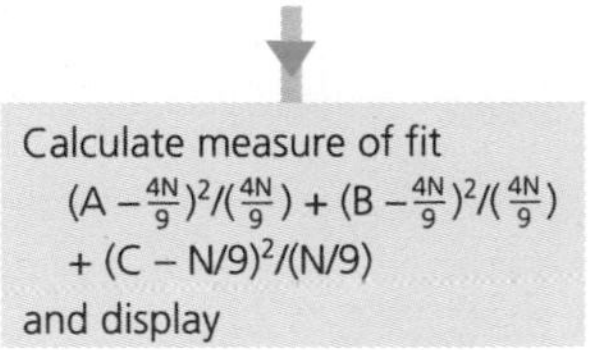

Run your modified program to obtain values of the measure of fit and identify your five per cent critical value.

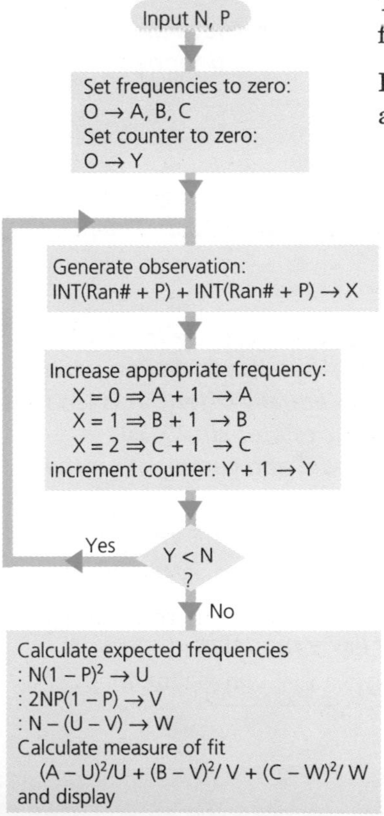

■ Consider how you might adapt your program to consider the distribution of measure of fit associated with the model:
$X \sim B(2, p)$

You will need to input p and to modify the variable X and finally to modify the expected frequencies used in calculating the measure of fit.

Run your program to obtain measures of fit associated with a variety of B(2, p) models.

On page 322 there are stem-and-leaf displays of measures of fit obtained when using the B (2, $\frac{1}{3}$) and B(2, 0.675) models. There are 120 values of the measure obtained in each case.

Calculate measure of fit
$(A - \frac{4N}{9})^2/(\frac{4N}{9}) + (B - \frac{4N}{9})^2/(\frac{4N}{9})$
$+ (C - N/9)^2/(N/9)$
and display

Input N, P

Set frequencies to zero:
$0 \rightarrow A, B, C$
Set counter to zero:
$0 \rightarrow Y$

Generate observation:
INT(Ran# + P) + INT(Ran# + P) → X

Increase appropriate frequency:
$X = 0 \Rightarrow A + 1 \rightarrow A$
$X = 1 \Rightarrow B + 1 \rightarrow B$
$X = 2 \Rightarrow C + 1 \rightarrow C$
increment counter: $Y + 1 \rightarrow Y$

Yes $Y < N$
?

No

Calculate expected frequencies
: $N(1 - P)^2 \rightarrow U$
: $2NP(1 - P) \rightarrow V$
: $N - (U - V) \rightarrow W$
Calculate measure of fit
$(A - U)^2/U + (B - V)^2/V + (C - W)^2/W$
and display

0	0460214248280410049818408644650458405510488
1	833332515815354415545913325153334
2	7294902060946672
3	174388452357
4	8300249
5	316
6	0225
7	8
8	5

0	318975691864778992733391190673535634913442737877637170164
1	544484029125154344477457442
2	52595989241
3	11499321823
4	0217
5	53
6	204
7	61
8	
9	6

These distributions of the measure of fit are similar to those you have discovered already. In each case, *the five per cent critical value is about 6.0.*

Example 12.1

Each of two boys had a coin and was flipping it. They recorded the number of heads showing each time after each had flipped his coin. The results are shown in the table.

Number of heads	Number of times
0	11
1	32
2	17

a) *One boy proposed the following model: 'There are three possibilities i.e. no head, one head, or two heads, so you would expect to see each of these outcomes equally often'. This led him to suggest that the expected frequencies are: 20, 20, 20. Calculate a measure of fit for this model and comment on the result.*

b) *Propose your own model and test how well it fits the observed frequencies.*

Solution

a) *The measure of fit for the equally likely model is:*

$$\sum \frac{(O-E)^2}{E} = \frac{(11-20)^2}{20} + \frac{(32-20)^2}{20} + \frac{(17-20)^2}{20}$$
$$= 11.7$$

Appropriate hypotheses are:
H_0: equal likelihood is an appropriate model;
H_1: the model is not appropriate
The critical value for the measure of fit under these circumstances is 6.0, but the calculated value, 11.7, is much larger. Hence, the model is a poor fit. It is reasonable to conclude that the proposed model is inappropriate.

b) *A binomial model B(2, 0.5) is a reasonable proposal. This would have expected frequencies 15, 30, 15. Hence, the measure of fit would be:*

$$\frac{(11-15)^2}{15}+\frac{(32-30)^2}{30}+\frac{(17-15)^2}{15}=1.47$$

This measure is smaller than the critical value of 6.0. Hence, it is reasonable to conclude that the binomial distribution is a reasonable fit for these frequencies.

The chi-squared statistic and degrees of freedom

Suppose a group of 20 infant school children is engaged in a mathematical counting activity. In groups of four, the children are flipping coins. Each child in a group has a coin and flips it. The group then counts the number of heads showing and records it. The group repeats the process until they have recorded a total of 16 numbers. This means that the class as a whole has recorded 80 numbers. These numbers are always 0, 1, 2, 3 or 4.

What probability distribution is appropriate to use as a basis to model the frequency of occurrence of these numbers? The recorded numbers can be taken as observations which may be modelled by a binomial distribution:

$$X \sim \mathrm{B}(4, 0.5)$$

What are the expected frequencies for $X = 0, 1, 2, 3, 4$? They are shown in the following table.

Number of heads, X	0	1	2	3	4
Expected frequency	5	20	30	20	5

The class teacher, who studied mathematics at college, collected the class results and calculated the measure of fit of the binomial model. The classes frequencies were as shown below.

Number of heads, X	0	1	2	3	4
Observed frequency	3	27	25	23	2

What value of the measure of fit should the class teacher obtain for these frequencies? The measure of fit is:

$$\sum\frac{(O-E)^2}{E} = \frac{(3-5)^2}{5}+\frac{(27-20)^2}{20}+\frac{(25-30)^2}{30}+\frac{(23-20)^2}{20}+\frac{(2-5)^2}{5}$$
$$= 6.333$$

Having consulted tables of critical values, she concluded that the model was appropriate. This seems to contradict what we have discovered in this chapter. However, there is a fundamental difference between this model and all the other models explored so far in the chapter. The number of possible values of the random variable is greater. Hence, the number of **frequency cells** is larger.

In each of the B(2, p) cases we have explored the number of frequency cells has been 3. In this B(4, $\frac{1}{2}$) case, the number of cells is 5. Each difference between an observed frequency and an expected frequency makes a positive contribution to the measure of fit. Hence, where there are more frequency cells, the measure of fit is likely to be larger. So the infant school teacher may be right. The five per cent critical value for five cells may well be larger than the 6.0 value which we have discovered in this chapter.

How can we investigate the distribution of $\sum\frac{(O-E)^2}{E}$ appropriate to the B(4, $\frac{1}{2}$) model? A suitable method could be a simulation using a programmable calculator and a computer spreadsheet.

CALCULATOR ACTIVITY

Simulating the distribution of $\sum\frac{(O-E)^2}{E}$ for B(4, $\frac{1}{2}$)

Each time the program outlined below is run, 80 observations of a random variable, X, are generated. The value of $\sum\frac{(O-E)^2}{E}$ is calculated assuming that B(4, $\frac{1}{2}$) is an appropriate model for X.

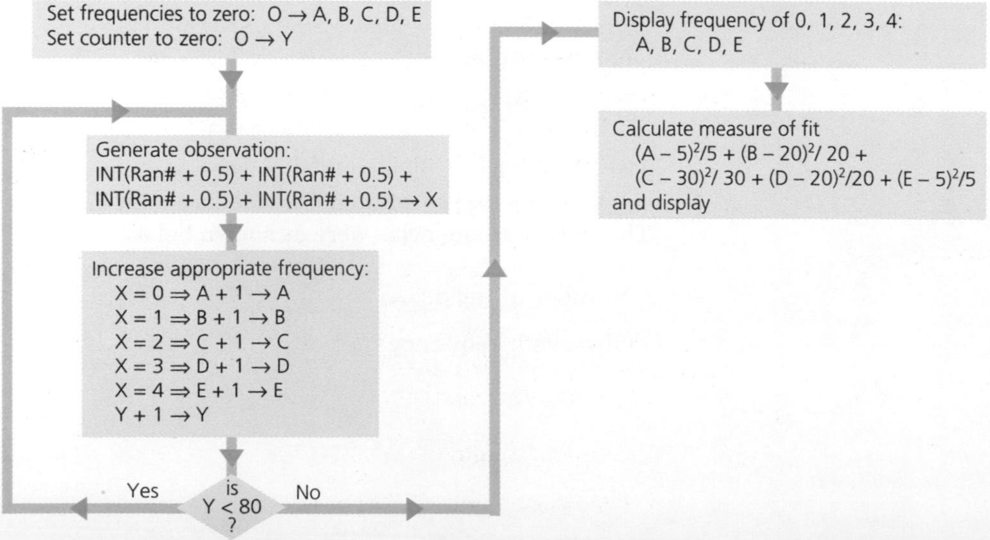

Run the program about 100 times and record the value of the measure of fit each time. Show the resulting 100 values using a stem-and-leaf display. Determine the five per cent critical value from your empirical distribution.

0	3 4 4 5 6 6 6 7
1	0 1 1 1 3 3 4 4 4 5 5 5 7 9 9 9
2	0 0 1 1 2 3 3 3 5 5 5 6 6 6 6 7 8 8 9 9
3	0 0 1 1 2 2 2 3 3 4 5 6 6 8 9 9 9
4	1 3 4 4 5 5 6 6 7 7 7 9
5	0 0 2 2 2 3 4 5 5 7
6	6 7 7 7 7 7
7	2 4 4
8	2 4 7
9	8
10	4
11	
12	
13	
14	2
15	
16	
17	1
18	
19	3

The data above were obtained using the program in the last calculator activity. The distribution is different from the previous ones, in that it is less skewed and the values obtained are generally greater. The five per cent critical value lies between 8.7 and 9.8. This is substantially different from the 6.0 obtained in the case of three cells. The teacher seems to be justified.

Chi-squared

Critical values of the measure of fit are found using a statistic called χ^2 (chi-squared), which has **degrees of freedom** associated with it. The degrees of freedom are closely related to the number of frequency cells. One way of identifying the degrees of freedom can be illustrated by returning to the boys in Example 12.1.

The model they proposed turned out to be inappropriate but that does not really matter. To what extent were they free to propose a model for the observed frequencies? Their experiment produced 60 observations of a random variable which could take the values 0, 1, 2, which led to three frequency cells. The content of these cells is the topic of the model. One boy proposes a model in which there are A occurrences of $X = 0$ and B occurrences of $X = 1$ are expected.

Value of random variable, X	0	1	2
Frequency expected by model	A	B	?

How many occurrences of $X = 2$ are expected? The answer must be $60 - (A + B)$. The experimental outcome imposes this on the model regardless of what the boy might want since the *total of the frequencies* for both model and observed data must be the same. So the boy is free to fill only two of the frequency cells; the content of the third is constrained by the data. Hence, the number of degrees of freedom is $3 - 1 = 2$ for this experiment.

Definition

The number of degrees of freedom of the χ^2 statistic is the number of frequency cells *minus* the number of constraints imposed by the data on the model.

Notation

The Greek letter for n i.e. ν (nu) is used to represent the number of degrees of freedom of the χ^2 statistic. It is often written as a subscript, hence χ^2_4 represents a chi-squared statistic with $\nu = 4$ (i.e. four degrees of freedom).

Exploration 12.5

Exploring χ^2

Explore the χ^2_ν critical value tables at the end of the book. Identify the upper-tail five per cent critical value for:

a) χ^2_2 **b)** χ^2_4.

(You should find these are 5.991 and 9.488 respectively.)

Example 12.2

A medical researcher investigated the blood type found in 1500 people. He chose 300 families each with five people and recorded the number of people with type-O blood in each family. His results are shown in the table.

No. of people with type O	0	1	2	3	4	5
No. of families	18	81	85	70	38	8

a) *Assume that a binomial distribution is appropriate to model the occurrence of people with type-O blood. If 45 per cent of the world's population have blood of this type, use the binomial model to determine the expected frequencies of occurrence of 0, 1, 2, 3, 4 people with type-O blood in families of five people.*

b) *Explain why it is not necessary to use the model to determine the expected frequency of five people.*

c) *Obtain the value of the measure of fit:*

$\sum \frac{(O-E)^2}{E}$ *for the data.*

d) *Conduct an appropriate hypothesis test at the five per cent level of significance and make your conclusions clear.*

Solution

a) *The binomial model is B(5, 0.45) and so the expected frequencies are:*

$X = 0 \quad 300 \times (0.55)^5 = 15.099$

$X = 1 \quad 300 \times 5(0.45)(0.55)^4 = 61.767$

$X = 2 \quad 300 \times 10(0.45)^2(0.55)^3 = 101.073$

$X = 3 \quad 300 \times 15(0.45)^3(0.55)^2 = 82.695$

$X = 4 \quad 300 \times 5(0.45)^4(0.55) = 33.831$

b) *We know that there were a total of 300 families investigated, so the expected frequencies must come to 300. This means that the final frequency cell is constrained to contain:*
300 – (15.099 + 61.767 + ... + 33.831) = 5.535

c)

Observed	18	81	85	70	38	8
Expected	15.099	61.767	101.073	82.695	33.831	5.535
$\frac{(O-E)^2}{E}$	0.5574	5.9888	2.5560	1.9489	0.5137	1.0978

Hence,

$$\sum \frac{(O-E)^2}{E} = 12.6626$$

d) *Appropriate hypothesis are:*

H_0: binomial model is a good fit for these data;
H_1: the model does not fit.

Approximate five per cent critical value for the $\sum \frac{(O-E)^2}{E}$ statistic can be found by consulting the chi-squared tables. The number of degrees of freedom in this case is:

	number of cells	*minus*	*constraints*	
i.e.	*6*	*–*	*1*	*= 5*

and the five per cent χ^2_5 (upper-tail) is 11.070. Since the test statistic, 12.6626, is larger then the critical value, there is a large difference between observed and expected frequencies. We can conclude that the binomial model is not appropriate for these data.

EXERCISES

12.1A

1 Observations made on a discrete random variable, X, are shown in the table.

x	0	1	2	3	4
Observed frequency	35	76	62	21	6

a) Assume that the binomial distribution, B(4, 0.3), is an appropriate model for these data. Determine the frequencies expected by this model.

b) Explain why it is not necessary to use the model to determine the expected frequency of observation of the value $x = 4$.

c) Obtain the value of the measure of fit, $\sum\frac{(O-E)^2}{E}$, and conduct a χ^2 test at the five per cent level of significance.
d) What are your conclusions?

2 Two dice are rolled 216 times. The number of sixes appearing after each roll is recorded. The results are shown in the table.

Number of sixes	0	1	2
Observed frequency	128	78	10

a) Obtain the frequencies expected by the binomial model B(2, $\frac{1}{6}$).
b) Carry out a chi-squared test at the five per cent level of significance making your hypotheses and conclusion clear.

3 A survey was conducted with 500 economists. Each was asked five questions about economic indicators. The numbers of positive responses are recorded in the table.

No. of positive responses	0	1	2	3	4	5
No. of economists	7	22	128	167	133	44

a) Explain what assumptions would be needed in order to model the number of positive responses using a binomial distribution.
b) Determine the frequencies expected in using the binomial distribution, B(5, 0.6), to model the positive responses.
c) Conduct a chi-squared test at the ten per cent level of significance to determine if this binomial model fits the observed data. Make your conclusions clear.
d) What conclusion would you have drawn if the test had been conducted at the five per cent level?

4 A charity fund-raiser makes four telephone calls every hour he works. He recorded the number of calls which produced a donation each hour over a period of two months. These are shown in the table.

No. of successful calls	0	1	2	3	4
No. of hours	25	85	97	58	15

a) What modelling assumptions are needed to use a binomial distribution in this situation?
b) Assume that the distribution, B(4, 0.4) is appropriate and obtain expected frequencies to fit the observed frequencies.
c) Conduct a test at the five per cent level to judge how good a fit this model is. Report your findings.
d) Repeat the procedure using the model B(4, 0.5).

5 Some students are conducting an experiment where they drop 20 coins on a large sheet of graph paper. They record the number of coins which land clearly in a grid square of the paper. They repeat this process 50 times.

No. coins landing clearly	≤ 5	6	7	8	≥ 9
Frequency	9	7	17	8	9

a) Explain why a binomial distribution may be an appropriate model for the number of coins landing clearly each time.
b) Assume that the distribution, B(20, 0.33), is appropriate and obtain the frequencies expected by this model for the classes indicated in the table.
c) Conduct a chi-squared test at the two per cent level to determine if this model fits. Report your findings.

EXERCISES

12.1 B

1 A student throws four coins and records the number of coins landing heads. She repeats this until she has a total of 160 recordings. The results are shown in the table.

Number of heads	0	1	2	3	4
Frequency	5	32	68	42	13

a) Explain why the binomial model, B(4, 0.5), may be appropriate to provide a fit for these data.
b) Conduct an appropriate test at the five per cent level of significance and report your findings.

2 On each of 200 days, a potter throws four plates and records the following data.

No. of exhibition plates produced	0	1	2	3	4
No. of days	9	35	69	71	16

a) Assume that the quality of the potter's work is such that 60 per cent of his plates are of exhibition standard. Use an appropriate model to determine expected frequencies.
b) Conduct a chi-squared test at the one per cent level of significance to determine if your model provides a reasonable fit for the data. Report your findings.

3 A calculator is programmed to produce random digits, from the range 0 to 9, in pairs. The number of even digits in each pair is noted until 50 pairs have been produced. The results are shown in the table.

No. of even digits	0	1	2
No. of pairs	13	17	20

a) Determine an appropriate model to fit these data.
b) Obtain the frequencies expected by your model and conduct an appropriate hypothesis test at the two per cent level of significance. Report your findings.
c) What conclusion would you have reached had you conducted your test at the:
i) ten per cent level **ii)** 0.5 per cent level?
d) Is there evidence in these data to suggest that the calculator is biased in its production of random digits?

4 The binomial model, B(8, 0.2), is proposed to fit the 200 observations in the table.

x	0	1	2	3	≥4
Frequency	31	65	60	31	13

a) Determine the frequencies expected by the model.
b) Conduct a chi-squared test at the ten per cent level of significance and report your findings.
c) Which one of the following descriptions would you use to describe the quality of the fit?
very poor, poor, reasonable, good, very good

5 A wholesale fruit merchant claims that at least 70 per cent of all the satsumas she sells are free from pips. A consignment of 3000 bags each containing five satsumas produces the following data.

Number free of pips	0	1	2	3	4	5
Number of bags	3	64	371	963	1091	508

a) Use an appropriate model to determine the expected frequencies.
b) Test the fit of the model at five per cent level and report your findings.
A consignment of 6000 bags produced the data below.

Number free of pips	0	1	2	3	4	5
Number of bags	6	128	742	1926	2182	1016

c) Are the two observed distributions similar in appearance?
d) Test the fit of the binomial distribution, B(5, 0.7), as a model for these data.
e) Comment on the results of **b)** and **d)**.

GOODNESS OF FIT

The χ^2 statistic may often be referred to as the **chi-squared goodness of fit**. We shall now develop its use in judging the fit of other models, where the expected frequencies give rise to concerns and where additional constraints are imposed.

Expected frequency convention

First, we consider the situation where each of a group of twelve children rolls a ten-faced die. The faces are numbered 0, 1, 2, ... , 9 and the number of zeros which result is counted. The experiment is repeated until the children have recorded a total of 50 counts. We need an appropriate model which could be used to determine the expected frequencies of occurrence of the possible counts of zeros.

The random variable, X, is the number of zeros recorded each time. We could use a binomial model:

$$X \sim B(12, 0.1)$$

The frequencies expected by this model are:

Outcome	$50 \times P(X)$		E
$X = 0$	$50 \times (0.9)^{12}$	=	14.122
$X = 1$	$50 \times 12(0.1)(0.9)^{11}$	=	18.829
$X = 2$	$50 \times 66(0.1)^2(0.9)^{10}$	=	11.506
$X = 3$	$50 \times 220(0.1)^3(0.9)^9$	=	4.262
$X = 4$	$50 \times 495(0.1)^4(0.9)^8$	=	1.065
$X = 5$	$50 \times 792(0.1)^5(0.9)^7$	=	0.189
$X = 6$	$50 \times 924(0.1)^6(0.9)^6$	=	0.025

etc.

The expected frequencies have been given to three decimal places. Clearly it is possible to continue calculating expected frequencies for the remaining possible outcomes. The frequencies are getting smaller each time. Suppose that the results obtained by the children are as given in this table.

Outcome, X	0	1	2	3	4	5	≥ 6
Observed, O	14	19	12	4	0	0	1
Expected, E	14.122	18.829	11.506	4.262	1.065	0.189	0.027

Are the expected frequencies close to the observed frequencies?

It seems that there is very little difference between any corresponding pair of O and E. Is this view born out by the measure of fit?

$$\sum \frac{(O-E)^2}{E} \approx 36.36$$

This seems to indicate an enormous difference between the observed and expected frequencies. What gives rise to this large value? The differences in the first four cells:

$$\sum_{X=0}^{X=3} \frac{(O-E)^2}{E} \approx 0.04$$

contribute about 0.1 per cent of the total measure of fit. This seems to agree with the closeness between the observed and expected frequencies. The problem lies with the remaining cells.

In these cells, the expected frequencies are comparatively small, at 1.065, 0.189, 0.025, … and the measure of fit, $\sum \frac{(O-E)^2}{E}$, involves dividing these small quantities. The contributions from these cells are 1.065, 0.189, 35.064. To avoid wrongly rejecting the null hypothesis more often than anticipated by the significance level, it is necessary to *amalgamate frequency cells* so that the expected frequency is no longer small. It is generally agreed that *expected frequencies should not be less than 5*.

In this case, the sum of the expected frequencies for the first three outcomes, $X = 0, X = 1, X = 2$ comes to:

$$\sum_{X=0}^{X=2} E = 44.456$$

This means that the total for all the remaining outcomes:

$$\sum_{X=3}^{X=12} E = 50 - 44.456 = 5.544$$

is just over 5. So, to conduct a chi-squared goodness-of-fit test for these data based on fitting the model, B(12,0.1), it is necessary to reduce the number of frequency cells to just four.

Outcome, X	0	1	2	3 or more
Observed, O	14	19	12	5
Expected, E	14.122	18.829	11.506	5.543

Then the measure of fit:

$$\sum \frac{(O-E)^2}{E} \approx 0.077$$

is much more in line with the very reasonable view that a binomial model fits the observed frequencies. To confirm this, we consult the chi-squared statistic for the five per cent upper tail critical value for three degrees of freedom, i.e. 7.815.

There are three degrees of freedom because the number of frequency cells has been reduced to four. The total frequency must be 50 and this imposes the additional constraint on the model.

Generally, frequency cells should be amalgamated unless all expected frequencies are at least 5. This is not an unbreakable rule. Somewhat greater flexibility in testing models can be obtained by adopting the following convention. Amalgamate frequency cells until at least 80 per cent of the expected frequencies are at least 5 and no expected frequency is less than 2.

Fitting other models

The process of amalgamating cells opens the way to testing the fit of models such as the **Poisson distribution** or the **geometric**

distribution. These distributions allow for exceedingly large values of the random variable which may never be observed in practical situations.

Suppose the headteacher of the infant school gave all 55 children in Year 2 a piece of written work. The numbers of errors made by the children are summarised below.

No. of errors, X	0	1	2	3	4	5	6	7	8
No. of children, O	4	12	15	13	6	4	0	1	0

The headteacher's experience suggests that children make 2.5 errors, on average, doing this piece of work and that the errors occur randomly. Does a Poisson model provide a reasonable fit for these frequencies?

The Poisson model Poi(2.5) indicates that the expected frequencies are:

X	0	1	2	3	4	5	6	7	8...
E	4.515	11.287	14.108	11.757	7.348	3.674	1.531	0.547	0.171...

The cells which might be amalgamated are those where the expected frequencies are small i.e. the cells corresponding to $X = 0, X = 5, X = 6$, etc. There is no need to amalgamate the cell corresponding to $X = 0$, particularly if the cells corresponding to $X = 5, X = 6$, etc are amalgamated. This gives the following table:

X	0	1	2	3	4	≥ 5
O	4	12	15	13	6	5
E	4.515	11.287	14.108	11.757	7.348	5.985

and the value of the measure of fit:

$$\sum \frac{(O-E)^2}{E} \approx 0.745$$

This compares favourably with the five per cent critical value of the statistic with five degrees of freedom, 11.070. It is reasonable to conclude that the Poisson model provides a good fit for the observed frequencies.

Additional constraints

Suppose that, in the infant school just considered, the headteacher was new. This might be the first time that the piece of writing had been set. The average would not be known, so the only way that a reasonable average could be obtained is by using the average from the data. This imposes a further constraint on the model and reduces further the degrees of freedom.

The mean, $\overline{x}$, number of errors per child for the data given initially is $\frac{132}{55} \approx 2.4$. Using the model Poi(2.4) produces the following expected frequencies.

X	0	1	2	3	4	≥ 5
O	4	12	15	13	6	5
E	4.989	11.975	14.370	11.496	6.897	5.273

The measure of fit works out to be 0.551. This time the number of degrees of freedom is:

6	−	(1	+	1)	=	4
frequency cells	less	(total to be 55	+	mean obtained from data)		

Hence, the appropriate critical value of consider is the χ^2_4 statistic i.e. 9.488. We conclude that there is no reason to reject the Poi(2.4) model.

Exploration 12.6

Mind the gap

You need eight 2p coins, a large sheet of paper ruled with lines 4 cm apart, and some means of randomly projecting the coins onto the paper. A coin is described as a clear if, having landed on the paper, no part of it touches or covers any of the ruled lines. Your experiment consists of randomly projecting the eight coins onto the paper and counting the number of clears. Then repeat this process until you have a total of 40 counts. Now tackle the following model fitting processes.

- Decide what probability distribution you feel is appropriate to model the occurrence of the number of clears when eight coins are randomly dropped on the paper. Determine the appropriate parameters from your experimental data and conduct a chi-squared goodness-of-fit test. Report on your findings.
- Test the appropriateness of the model B(8, $\frac{1}{3}$) as providing a basis to model your observed frequencies. Report your findings.

Example 12.3

The fabric department of a large department store sells material from which curtains are made. The material is supplied to the store in rolls containing 50 m. When the departmental manager notices that there is no more than 3 m left on a roll, she checks the length. If she finds there is less than 1 m remaining, the material is discarded. If there is between 1 m and 3 m the material is sold at a reduced price as a 'remnant'. A model of the length of material offered for sale as remnant material is proposed. The density function, f(x), of the model is:

$$f(x)=\begin{cases} x-1 & 1\leq x\leq 2 \\ 3-x & 2\leq x\leq 3 \\ 0 & \text{elsewhere} \end{cases}$$

a) *If 40 remnants are measured, determine the expected number which will be found in the following length classes.*
$1.0 \le x \le 1.25$ $1.25 \le x \le 1.5$
$1.5 \le x \le 1.75$ $1.75 \le x \le 2.0$
$2.0 \le x \le 3.0$

b) *An examination of 40 remnants reveals the following data.*

Class	$1.0 \le x \le 1.25$	$1.25 \le x \le 1.5$	$1.5 \le x \le 1.75$	$1.75 \le x \le 2.0$	$2.0 \le x \le 3.0$
Observed	3	4	6	6	21

Carry out a goodness-of-fit test at the five per cent level to determine if there is sufficient evidence to support the proposed model.

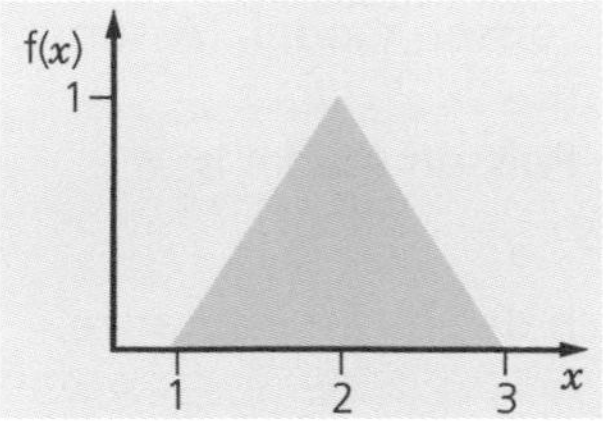

Solution

a) A sketch of the density is shown. It is apparent that the graph of the function is symmetric about $x = 2$. The probability, determined by the model, that a remnant will lie in a given range can be found from consideration of symmetry or by integration e.g.

$$P(1.0 \le x \le 1.25) = \int_{1.0}^{1.25} f(x)\,dx$$

$$= \left[\tfrac{1}{2}x^2 - x\right]_{1.0}^{1.25} = 0.031\,25$$

which is $\frac{1}{32}$. Multiplying this by 40 produces the expected frequency of 1.25.

Other expected frequencies can be found as:

Class	$1.0 \le x \le 1.25$	$1.25 \le x \le 1.5$	$1.5 \le x \le 1.75$	$1.75 \le x \le 2.0$	$2.0 \le x \le 3.0$
Expected	1.25	3.75	6.25	8.75	20

b) The first two cells should be amalgamated since the expected frequencies are low. This results in the following table.

Class	$1.0 \le x \le 1.5$	$1.5 \le x \le 1.75$	$1.75 \le x \le 2.0$	$2.0 \le x \le 3.0$
E	5	6.25	8.75	20
O	7	6	6	21

The value of the measure of fit:

$$\sum \frac{(O-E)^2}{E} = 1.724$$

There are four frequency cells and one constraint (the total frequency must be 40), hence, the number of degrees of freedom for the chi-squared statistic is 3. The five per cent upper-tail critical value is 7.815. Clearly, the test result is much less than this, which suggests that the model is reasonable.

EXERCISES

12.2 A

1 A Geiger counter is a device for recording the number of radioactive particles emitted by an appropriate source. The numbers of emissions recorded by a Geiger counter in each of 100 one-second intervals are shown in the table.

Number recorded	0	1	2	3	4	≥ 5
Frequency	20	19	29	13	13	6

a) Explain briefly why a Poisson distribution may be appropriate to model the number of particles recorded in one second.
b) Determine the mean number of emissions in one second. Use this to derive the frequencies expected by the Poisson model. (Assume, for the mean, that 5.5 is representative of the '≥ 5' class.)
c) Test at the five per cent level whether the Poisson model is appropriate. Report your findings carefully.

2 The number of fraudulent claims received each day by an insurance company is shown in the table.

Number of frauds	0	1	2	3	≥ 4
Frequency	29	38	21	10	2

a) Assume that the mean number of fraudulent claims that the company has usually received is 1.1 per day. Determine an appropriate probability model for these data.
b) Carefully conduct a chi-squared goodness-of-fit test at the two per cent level.
c) Interpret your findings.

3 The librarian of a college library wanted to know if the number of books borrowed each day (of a six-day week) was uniformly distributed. She collected the following records.

Day	Mon	Tue	Wed	Thu	Fri	Sat
No. borrowed	308	292	243	229	226	202

a) Determine the daily number of books expected by a uniform model.
b) Conduct a goodness-of-fit test at the two per cent level.
c) Interpret your findings.

4 A model is proposed for a continuous random variable. The proposed density function is given below.

$$f(x) = \begin{cases} \frac{(6-x)}{18} & \text{when } 0 \leq x < 6 \\ 0 & \text{otherwise} \end{cases}$$

Test if this model is an appropriate fit for the 100 items of data in the table. Conduct your test at the five per cent level.

Class	$0 \le x < 1$	$1 \le x < 2$	$2 \le x < 4$	$4 \le x < 6$
Frequency	36	29	22	13

5 A manuscript in an unknown hand is discovered. It is suggested that it is the work of a well-known 19th-century novelist. It is known that 23 per cent of the novelist's sentences contain fewer than 20 words and 13 per cent contain at least 40 words. The unidentified manuscript contains 171 sentences under 20 words, 110 sentences with at least 40 words and there are 519 remaining sentences.

a) Write down suitable null and alternative hypotheses for a goodness-of-fit test.
b) Calculate the expected number of sentences, in the three classes identified, under the null hypothesis.
c) Carry out the test at ten per cent and interpret your outcome.

6 When recording for a radio play, the probability that an actor will get his lines exactly right on a given 'take' is 0.7. The following data were obtained in the studio during the recording of *King Lear*.

No. of takes needed for a speech	1	2	3	4	≥ 5
Frequency	149	48	29	10	4

a) Set up an appropriate mathematical model for the number of takes required for a speech.
b) Use your model to determine the frequencies expected.
c) Carry out a goodness-of-fit test at the one per cent level. Interpret the outcome carefully.
d) Determine the probability from the data and repeat the test.

EXERCISES

12.2 B

1 In 400 ten-minute intervals, a taxi firm received the following numbers of calls.

No. of calls	0	1	2	3	4	5	6	7	≥ 8
No. of intervals	3	11	56	74	70	73	48	40	25

a) Carry out a goodness-of-fit test at the five per cent significance level to determine if the Poisson distribution with mean 4.5 is a suitable model for the data.
b) Repeat the test for Poi(4).
c) Describe, but do not carry out, how you would modify your procedure if you used the mean obtained from the given data.

2 The numbers of letters of complaint received by a major radio station each day over a period of 200 days are shown in the table.

No. of complaints	0	1	2	3	4	5	≥ 6
No. of days	34	78	60	20	5	3	0

Use a chi-squared test, with a five per cent level of significance, to determine if a Poisson distribution is an appropriate model for these data.

3 A successful local rugby team has five entrances to its grounds. The numbers of spectators using each entrance one Saturday are recorded in the table.

Entrance	N	S	E	W	M
No. of spectators	620	685	644	639	767

Use an appropriate test to determine if there is sufficient evidence to support the view that the same number of people, on average, use each entrance.

4 There are four common blood types: O, A, B and AB. The proportions of people in one north European country in each of these blood-type groups are assumed to be 48 : 40 : 8 : 4. The records of 200 pre-school children reveal the data shown in the table.

Blood type	O	A	B	AB
No. of children	113	62	13	12

Do these data support the hypothesis that the distribution of blood type among pre-school children is different from the assumed distribution?

5 A model is proposed for the weight distribution of trout in a fish farm. The density function of the model is:

$$f(x) = \begin{cases} \frac{1}{36} w(6-w) & \text{when } 0 \leq w \leq 6 \\ 0 & \text{elsewhere} \end{cases}$$

A sample of 80 fish is obtained and their weights are recorded. The table contains a summary of the data.

Class of weight	$0 \leq x < 1$	$1 \leq x < 2$	$2 \leq x < 4$	$4 \leq x < 6$
No. of fish	6	14	39	21

a) Use the model to determine the expected number of fish in each weight class.

b) Conduct a goodness-of-fit test at the ten per cent level to determine if the model is appropriate for these data. Make your conclusions clear.

6 A student flips a coin and records the number of flips it takes to get a head. He repeats this until he has 100 recordings. The results are shown in the table.

No. of flips to get a head	1	2	3	4	≥ 5
Frequency	48	32	16	4	0

a) Determine an appropriate model to obtain expected frequencies.
b) Conduct a chi-squared test to judge the appropriateness of your model. Report your findings at:
i) ten per cent **ii)** five per cent **iii)** two per cent levels.

CONTINGENCY TABLES AND INDEPENDENCE

The concept of **independence** is important in developing probability models. Recall that if two events, A, B are independent then the probability, P(A and B), that they both occur is the product, $P(A) \times P(B)$, of their individual probabilities.

Consider a survey carried out with 100 primary school children. There are 60 boys and 40 girls in the survey. The children are asked to state their first choice among various makes of car 'British' or 'European' and 'Other country' of origin.

The assumption that gender and preference are independent would indicate for instance that the proportion of boys among those who state a preference for 'British' cars is the same as the proportion $\frac{60}{100}$ of boys taking part in the survey. Equally, $\frac{60}{100}$ of those stating 'European' as their preference would be boys.

Suppose that 50 of the children state a preference for European cars. How many of these 50 would you expect to be boys? Assuming independence leads to $\frac{60}{100} \times 50$ i.e. 30 boys preferring European cars. This leaves 20 (50 – 30) girls expected to have chosen European.

If 30 children state a preference for British cars, how many of these do you expect to be boys, and how many does this suggest will be girls? The independence model suggest that the expected number of boys is $\frac{60}{100} \times 30 = 18$ and the expected number of girls must be $30 - 18 = 12$.

These expected values can conveniently be summarised in a table known as a **contingency table**.

		Preference			
		European	**British**	**Other**	**Subtotal**
Gender	**Boy**	$\frac{60}{100} \times 50 = 30$	$\frac{60}{100} \times 30 = 18$		60
	Girl	$50 - 30 = 20$	$30 - 18 = 12$		40
	Subtotal	50	30	20	100

How many boys and how many girls do you expect to choose cars of 'other' country of origin? Since 48 boys are already accounted for out of 60, there must be twelve expected and this indicates that eight girls are expected to prefer 'other' manufacturers.

A summary of the numbers of children expected to be found in each possible cell is shown below.

Expected frequencies			
	European	**British**	**Other**
Boy	30	18	12
Girl	20	12	8

The preferences actually stated by the children are shown in the 'observed frequencies' table below.

Observed frequencies			
	European	**British**	**Other**
Boy	25	20	15
Girl	25	10	5

We can use the measure of fit developed so far to judge whether there is evidence to support the independence model assumed to derive the expected frequencies above.

$$\sum_{all\ cells} \frac{(O-E)^2}{E} = \frac{(25-30)^2}{30} + \frac{(25-20)^2}{20} + \frac{(20-18)^2}{18} + \frac{(10-12)^2}{12} + \frac{(15-12)^2}{12} + \frac{(5-8)^2}{8}$$
$$\approx 4.514$$

Again, we can approximate the critical values of this statistic using the chi-squared distribution. However, first we need to find the number of degrees of freedom of the model. Refer back to the table of expected frequencies. Only 30 and 18 are determined by the assumed model, the remaining cell frequencies are constrained by column sub-totals, row sub-totals and the total frequency.

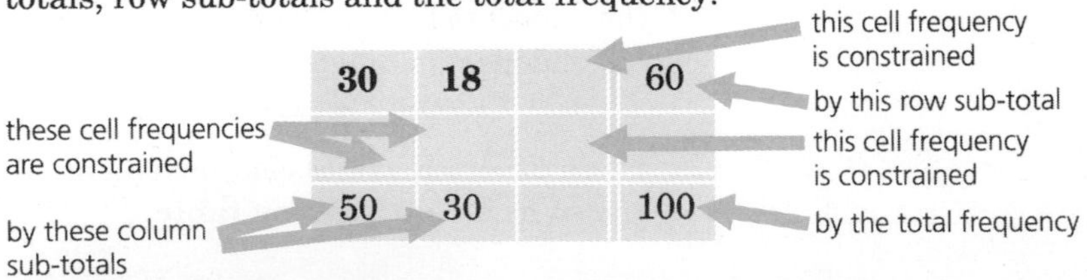

Hence, the number of degrees of freedom is:

3 × 2	−	(2	+	1	+	1)	=	2
cells in the table		column		row		total		

The five per cent critical value for a χ^2_2 statistic is 5.991. The test statistic for these data is 4.514, which is smaller than the critical value indicating a reasonable fit. Hence, the independence of gender and preference of car manufacturer is supported by these data.

Contingency tables and association

The model we used to determine the expected frequencies in the 'gender and car manufacturer' study was independence. When that model is used and the **measure of fit** turns out to be **large**, indicating that the model does not fit, the appropriate conclusion is that the row factor and the column factor are *not independent*. When the factors are not independent, they are said to be **associated**. Hence, the process explored previously can be used to determine whether two factors are associated.

Consider a survey conducted among all the overseas students at a large English university. The survey included a question in which the students were asked which of three car styles they preferred. They were asked to choose between hatchbacks, estates and sports versions. The students were identified by their year of study: first year, second year, third year or postgraduate.

There were 343 students altogether. The largest group was made up of the 108 first-year students. There were 87 second years, 88 in the third year and 60 postgraduates. Use the assumption that choice of style and year of study are independent, to determine the proportion of students in the first year who choose hatchbacks.

The assumption of independence implies that the proportion of first years who choose hatchbacks is the same as in the survey i.e. $\frac{108}{343}$. Similarly, of those choosing estates the proportion of first years is expected to be $\frac{108}{343}$. The proportion of second-year students among those choosing hatchbacks is expected to be $\frac{87}{343}$ and so on.

The actual numbers of students who choose each car style are given in the table.

	Hatchback	Estates	Sports
No. of students	185	102	56

We can use the information we have to explain how these six expected frequencies have been obtained.

		Expected Car Choice			
		Hatchback	**Estate**	**Sports**	**Sub-total**
Year of Study	**1st**	58.251	32.117		108
	2nd	46.924	25.872		87
	3rd	47.464	26.169		88
	Postgraduate				60
	Subtotal	185	102	56	343

The 58.251 arises as the appropriate proportion, $\frac{108}{343}$, of 185, similarly $\frac{108}{343} \times 102 = 32.117$, and $\frac{87}{343} \times 185 = 46.924$ etc.

How many postgraduate students are expected to have chosen hatchbacks? The answer is the number that gives the sub-total 185 i.e.

$$185 - (58.251 + 46.924 + 47.464) = 32.361$$

Similarly, the number of postgraduates choosing estates is expected to be 17.942. The number of each year choosing sports cars can be found by making the row sub-totals correct. The result of this process is shown in the table of expected frequencies.

		Expected frequencies		
		Hatchback	Estate	Sports
Year of Study	1st	58.251	32.117	17.632
	2nd	46.924	25.872	14.204
	3rd	47.464	26.169	14.367
	Postgraduate	32.361	17.842	9.797

The actual results of the survey are shown below in the table of observed frequencies.

		Observed frequencies		
		Hatchback	Estate	Sports
Year of Study	1st	71	26	11
	2nd	52	23	12
	3rd	38	32	18
	Postgraduate	24	21	15

Is there evidence of association between year of study and choice of car style? The table of expected frequencies has been derived from the assumption that the factors are independent, hence the appropriate hypotheses are:

H_0: year of study and choice of style are independent;
H_1: year of study and choice of style are associated.

The measure of fit, $\sum \frac{(O-E)^2}{E}$, for these data based on the independence hypothesis is:

$$\frac{(71-58.251)^2}{58.251} + \frac{(26-32.117)^2}{32.117} + \ldots + \frac{(15-9.797)^2}{9.797} \approx 17.247$$

This appears to be a large value, however, to decide this you need to sort out the degrees of freedom for this contingency table. What are the degrees of freedom? Reflect on the table of expected car choice, once the model determined the content of the six cells shown, the data imposed constraints on all the other cell frequencies. Hence, there are six degrees of freedom.

The five per cent upper-tail critical value of the χ_6^2 statistic is 12.592. The calculated measure of fit, 17.247, is larger than this, hence there is evidence to reject the null hypothesis.

In other words, these data indicate that there is association between year of study and choice of car style.

Nature of association

We can use a simple process to identify the nature of the association where there is evidence of this existence. We start by identifying the sign of the association – this is the same as the sign (positive or negative) of the differences $(O - E)$ in each frequency cell.

Sign of $(O - E)$			
	Hatchback	Estate	Sports
1st	+	–	–
2nd	+	–	–
3rd	–	+	+
Postgraduate	–	+	+

The next stage is to identify where the greatest relative difference between the observed and expected frequencies occurs. We do this by expressing the measure of fit $\frac{(O-E)^2}{E}$ for each cell as a proportion or percentage of the overall measure of fit, $\sum \frac{(O-E)^2}{E}$. Combining these percentage differences with the signs leads to this table.

	Hatchback	Estate	Sports
1st	+16.2%	–6.8%	–14.5%
2nd	+3.2%	–1.8%	–2.0%
3rd	–10.9%	+7.5%	+5.3%
Postgraduate	–12.5%	+3.2%	+16.0%

The greatest differences are in first-years choosing hatchbacks and postgraduates choosing sports cars. These show a high positive association which can be interpreted as indicating that first-year students are attracted by hatchback cars and postgraduates prefer sports cars.

Large negative associations indicate that third-year students and postgraduates shun hatchbacks and that first-year students don't like sports cars.

Yates' correction

Critical values of the distribution of the measure of fit, $\sum \frac{(O-E)^2}{E}$, are approximated by the critical values of the chi-squared statistic with the appropriate number of degrees of freedom. When there is only *one degree of freedom* this approximation is considered to be poor. Under these circumstances it is possible to improve the approximation by using a correction to the measure known as **Yates' correction**.

This correction involves replacing $(O - E)^2$ by:

$$\{|O - E| - \tfrac{1}{2}\}^2 \quad \text{or} \quad \{\text{ABS}(O - E) - \tfrac{1}{2}\}^2$$

The measure $\sum \frac{\left[ABS(O-E)-\frac{1}{2}\right]^2}{E}$ is approximately a χ_1^2 statistic.

Example 12.4

One hundred primary school children were asked whether they had arrived at school by motorised transport or non-motorised. The results are given in the table. Is there evidence, at the ten per cent level of significance, that there is an association between mode of transport and gender?

	Motorised	Non-motorised
Boy	44	19
Girl	24	13

Solution

The row and column totals for this contingency table are as follows.

	Motorised	Non-motorised	
Boy			63
Girl			37
	68	32	100

The null hypothesis for this test is that the factors are independent. This means that the number of boys expected to have used motorised transport is $\frac{63}{100} \times 68 = 42.84$. *Once this cell frequency is entered, all others are constrained by the data. Hence, the number of degrees of freedom for this 2 × 2 contingency table is 1. This means that Yates' correction should be used.*

The expected frequencies are listed here.

	Motorised	Non-motorised
Boy	42.84	20.16
Girl	25.16	11.84

Hence, the measure of fit is:

$$\frac{(|44 - 42.84| - \frac{1}{2})^2}{42.84} + \frac{(|19 - 20.16| - \frac{1}{2})^2}{20.16} +$$

$$\frac{(|24 - 25.16| - \frac{1}{2})^2}{25.16} + \frac{(|13 - 11.84| - \frac{1}{2})^2}{11.84} \approx 0.086$$

The ten per cent upper-tail critical value is 2.706, the measure for this data is considerably less than this critical value. There is no reason to reject the independence model. Hence, there is insufficient evidence to support the view that mode of transport and gender are associated.

EXERCISES

12.3 A

1 An audit of a school library revealed the following data concerning the popularity of 'novels' and 'non-fiction' books. The table shows that there were 659 novels which were never borrowed, 101 non-fiction books which were borrowed very often, and so on.

		Popularity			
		Never	Rarely	Often	Very often
Classification	Novel	659	816	343	222
	Non-fiction	628	427	214	101

a) How many books are accounted for in the table?
b) What is the ratio of novel to non-fiction, and the ratio of the four popularity categories?
c) Assume that the book classification and the popularity category are independent.
How many novels are expected to be in the 'Never' borrowed category? How many novels does the independence model expect to be in the 'Rarely' and in the 'Often' categories?
d) Find the number of degrees of freedom of a chi-squared test.
e) Carry out the test and report your findings fully.

2 A publisher of an evening and a morning daily newspaper uses different production teams for the two editions. The publisher wants to establish if the number of reported misprints is independent of the edition. The following data are collected over a ten-week period.

		Misprints				
		0	1	2	3	≥ 4
Edition	Morning	13	17	15	7	8
	Evening	8	11	24	8	9

Use an appropriate chi-squared test to find whether there is evidence of independence or association between the edition and the number of misprints.

3 A survey of attitudes to smoking was conducted at a college. Students, academic staff and support staff were interviewed. Their opinions on the introduction of a designated smoking area are recorded in the table.

		Opinion		
		In favour	Undecided	Against
Interviewee	Academic staff	28	42	10
	Support staff	9	11	40
	Students	26	24	10

a) Is there evidence of association between interviewee and opinion?
b) Report your findings fully and identify the nature of association, if any.

4 A washing machine manufacturer uses a four-stage classification of quality of the output. There are three types of machine manufactured which are distinguished by their maximum spin speed. The following data represent one day's production.

		Classification			
		Perfect	Near perfect	Repairable	Reject
Machine type	600 spin	18	7	4	6
	800 spin	30	10	12	8
	1200 spin	37	13	4	1

a) Assume that output quality and machine type are independent and use an appropriate model to determine the number of machines in each cell expected by the independence model.
b) Find the proportion of cells where the expected frequency is less than 5. Do any of the cells have an expected frequency less than 2?
c) Conduct an appropriate test at ten per cent to judge whether there is evidence of independence between quality and machine type.
d) Report your findings fully.

5 A survey conducted of 1250 passengers on long-distance railway journeys revealed the following data.

		Class of ticket held	
		First	Standard
Gender of passenger	Female	95	405
	Male	165	585

Use an appropriate chi-squared test to find out if there is evidence of association between gender of passenger and class of ticket held:
a) with Yates' correction, **b)** without Yates' correction.

6 The data below represent a random sample of hospital patient records.

		Length of hospitalisation	
		Short stay	Long stay
Costs met by:	Private patient	38	27
	National Health	52	133

Use a test based on the chi-squared distribution to find out whether there is evidence of an association between the length of stay in hospital and who met the cost. Report your findings fully.

EXERCISES

12.3 B

1 A group of 450 randomly-selected married couples were questioned about their daily alcohol consumption. The data are recorded below.

		Gender of partner	
		Female	Male
Alcohol consumption relative to maximum recommended	None	46	34
	Lower	264	151
	Equal to	83	202
	Higher	57	63

a) Assume that alcohol consumption and gender are independent. How many female partners does this assumption expect to consume:
i) no alcohol, **ii)** some but lower,
iii) an amount equal to the maximum recommended?

b) In using a chi-squared test, how many degrees of freedom does the statistic associated with this 4×2 table have?

c) Find the value of the measure of fit of an 'independence' model for the data and carry out the appropriate test at 0.5 per cent.

d) Report your findings fully, mention the two most deviant cells as far as the model proposed is concerned.

2 A group of 100 A-level students consisted of 55 boys and 45 girls. During one week, when each recorded the time spent watching television, 23 boys and 15 girls watched for more than ten hours. Is there sufficient evidence in these data to support the view that there is no association between gender and time spent watching television? (Make it clear whether you use Yates' correction.)

3 An analysis of the passenger list of the tragic voyage of the *Titanic* revealed the data shown.

		Ticket class		
		First	Second	Third
Outcome of tragedy	Survived	203	118	178
	Perished	122	167	528

Does the data support the view that the class of ticket held by a passenger had a significant effect on the outcome of the tragedy? Identify clearly any association you find.

4 A survey of 192 international companies revealed the following data about their size and profits.

Profitability	No.of employees				
	Under 500	500 to 1000	1000 to 5000	5000 to 10000	$\geq$ 10 000
Increased	2	6	39	17	24
Decreased	4	9	58	14	19

a) Assume that the size of the company and its profitability are independent. Determine how many companies this model expects in each of the ten possible cells.

b) What proportion of cells have an expected frequency of less than 5?

c) Use an appropriate chi-squared test to test the hypothesis that profitability is affected by the size of the company.

d) Report your findings fully.

5 A survey was conducted among 300 randomly-selected people in a town to gauge opinion about diverting traffic from the town. Three options are offered: a tunnel taking traffic under the town, a new surface dual carriageway, to do nothing. The interviewees were also asked where they lived: in the town, on the outskirts, elsewhere. The results are shown in the table.

		Option		
		Tunnel	Surface	No change
Living	In town	31	20	49
	Outskirts	34	34	32
	Elsewhere	72	25	3

Investigate these data to determine if the preferred option is associated with home location. Report your findings fully.

6 Students studying part-time for a degree were questioned about whether they were self-funded or received support from their employer. The results are shown below.

		Gender	
		Male	Female
Funding	Self	34	67
	Employer	91	14

Conduct an appropriate chi-squared test to find out if gender of student and funding are independent. Make it clear if you use Yates' correction and report your findings fully.

CONSOLIDATION EXERCISES FOR CHAPTER 12

1 A local council has records of the number of children and the number of households in its area. It is therefore known that the average number of children per household is 1.40. It is suggested that the number of children per household can be modelled by the Poisson distribution with parameter 1.40. In order to test this, a random sample of 1000 households is taken, giving the following data.

Number of children	0	1	2	3	4	≥ 5
Number of households	273	361	263	78	21	4

a) Find the corresponding expected frequencies obtained from the Poisson distribution with parameter 1.40.

b) Carry out a χ^2 test, at the five per cent level of significance, to determine whether or not the proposed model should be accepted. State clearly the null and alternative hypothesis being tested and the conclusion which is reached.

(MEI Question 4, Specimen S3, 1994)

2 A fruit farmer's apples are graded on a scale from A to D before sale. Lengthy past experience shows that the percentage of apples in the four grades are as follows.

Grade	A	B	C	D
Percentage	29	38	27	6

The farmer introduces a new treatment and applies it to a small number of trees to see if it affects the distribution of grades. The apples produced by these trees are graded as follows.

Grade	A	B	C	D
No. of apples	79	94	58	19

a) Write down suitable null and alternative hypothesis for a chi-squared test.

b) Calculate the expected number of apples in each grade under the null hypothesis.

c) Carry out the chi-squared test at the five per cent level of significance. State the conclusions of the test clearly.

(MEI Question 3, Paper S3, June 1994)

3 Two schools enter their pupils for a particular public examination and the results obtained are shown below.

	Credit	**Pass**	**Fail**
School A	51	10	19
School B	39	10	21

By using an approximate χ^2 statistic, assess at the five per cent level whether or not there is a significant difference between the two schools with respect to the proportions of pupils in the three grades. State your null and alternative hypothesis.

(ULEAC Question 5, Specimen paper T2, 1994)

4 A golfer is practising on a putting green. She deduces that the number of strokes required at each hole can be modelled by the discrete random variable X defined as follows.

$$P(X = r) = k\left(\tfrac{2}{3}\right)^r \qquad r = 1, 2, 3, \ldots$$

a) Show that $k = \frac{1}{2}$. In order to test her model, the golfer records the number of strokes required at each of 54 holes. These figures are summarised in the following frequency table.

No. of strokes	1	2	3	4	5	6+
No. of holes	20	16	14	3	1	0

b) Calculate the expected frequencies according to the golfer's proposed model. Carry out a suitable test at the five per cent level to determine whether the model is a good one or not. State your hypothesis and your conclusions carefully.

(MEI Question 1, Paper S3, June 1993)

5 The following table is the result of analysing a random sample of the invoices submitted by branches of a large chain of book shops.

	Novel	Textbook	General interest
Hardback	24	10	22
Paperback	66	10	18

Using an approximate χ^2 statistic assess, at the five per cent level of significance, find whether or not there is any association between the type of book sold and its cover. State clearly your null and alternative hypothesis.

(ULEAC Question 5, Paper S2, June 1994)

6 All entrants to a particular science course at a university are required to study French or Russian. The numbers of students of each sex choosing each language are shown in the following table.

	French	Russian
Male	39	16
Female	21	14

Use a χ^2 test (including Yates' correction) at the five per cent significance level to test whether choice of language is independent of sex.

(AEB Question 3, Specimen paper 2, 1994)

7 A statistics conference, lasting four days, was held at a university. Lunch was provided and on each day a choice of a vegetarian or a meat dish was offered for the main course. Of those taking lunch, the uptake was as follows.

	Tuesday	Wednesday	Thursday	Friday
Vegetarian	17	24	21	16
Meat	62	42	38	22

a) Use the χ^2 distribution at the five per cent significance level to test whether the choice of dish for the main course was independent of the day of the week.

On each day a choice of fruit, ice-cream or apple pie was offered for dessert. A contingency table was formed showing the number making each choice on each day and $\sum \frac{(O-E)^2}{E}$ was calculated to be 3.7. (No classes were grouped together.)

b) Test, at the five per cent significance level, whether the choice of dessert was independent of the day of the week.

c) There was a total of 80 participants at the conference entitled to lunch each day. Test, at the five per cent significance level, whether the number of participants taking lunch was independent of the day of the week.

d) Describe briefly any variations in the choice and attendance at lunch over the four days.

(AEB Question 2, Paper 9, Winter 1994)

8 During hockey practice, each member of a squad of 60 players attempted to hit a ball between two posts. Each player had eight attempts and the numbers of successes were as follows.

3 4 8 1 0 3 3 4 4 2 6 7 3 3 2 5 5 5 8 1 3 5 6 1 3 4 4 4 1 0
5 3 6 0 6 7 4 3 5 7 0 1 2 6 1 8 0 0 3 0 4 4 1 3 5 0 8 1 8 8

a) Form the data into a ungrouped frequency distribution.

b) Use the χ^2 distribution at the five per cent significance level to test whether the binomial distribution is an adequate model for the data.

c) State, giving a reason, whether the data support the view that the probability of success is the same for each player.

(AEB Question 1, Paper 9, Summer 1994)

9 The numbers of incidences of different categories of crime over a certain period are recorded in the table below.

Category of crime	A	B	C	D	E	F	G	H
No. of incidences	23	19	14	20	21	28	26	37

a) Write down the mean number of incidences per category, and illustrate the data in an appropriate graphical form.

b) One theory concerning these categories of crime is that their incidences are equal. Write down hypotheses which might be used in statistically testing the data against this theory.

c) Conduct a chi-squared goodness of fit test to determine a response to the theory and make your conclusions clear at the five per cent level.

(Oxford Question 7, Specimen paper S1)

10 **a)** The data in the following table are the result of counting radioactive events in five-second intervals.

No. of events	0	1	2	≥ 3
No. of observations	5	14	13	8

Show that the number of events in a five-second interval is 1.7 (taking the group with frequency 8 to have a mean of 3.5).

b) Write down the probability of 0, 1, 2, ≥ 3 events for Poisson distribution with mean 1.7. Hence obtain to one decimal place the expected frequencies.

c) Use the chi-squared goodness of fit test to assess whether it is reasonable to claim that the data come from a Poisson distribution. Make your method clear and conduct your test at the ten per cent level.

(Oxford Question 4, Specimen paper S2, 1994)

11 The table shows the numbers of men and women recruited into the various UK services in 1987/88.

	Men	Women
Navy	4601	580
Army	19865	1146
Air Force	5728	885

Use a χ^2 test to see whether there is evidence of any association between a person's gender and the service into which they are recruited. State your null hypothesis clearly, and test at the one per cent level.

(Nuffield Question 2, Specimen, 9870/43, 1994)

12 **a)** When two varieties of tomato plant, tall cut-leaf and dwarf potato-leaf, are crossed there are four possible phenotypes: tall cut-leaf (A), tall potato-leaf (B), dwarf cut-leaf (C), dwarf potato-leaf (D). Mendel's laws of inheritance state that these four phenotypes should appear in the ratios 9 : 3 : 3 : 1, respectively. The following table shows the results of a particular investigation into phenotype numbers from crossings of the two varieties of tomato plant.

Phenotype	A	B	C	D
Frequency	931	288	293	104

Test the claim that these results are consistent with Mendel's laws.

b) On an open road where traffic is flowing freely, the time interval, in seconds, between successive red cars is recorded precisely, with the results as tabulated below.

Time interval (s)	0–	15–	30–	45–	60–	90– 120
Frequency	70	54	32	20	17	7

Show that an estimate of the mean time between successive red cars is 30 seconds. Hence test the hypothesis that the time interval between successive red cars on this road follows an exponential distribution.

(NEAB Question 2, Paper 2, June 1994)

13 The results of a student's investigation were as follows.

	Left-handed	**Right-handed**
Boys	5	27
Girls	13	25

Use a χ^2 test (including Yates' correction), at the five per cent significance level, to test the hypothesis that there is no association between the sex of a student and the incidence of left-handedness.

(NEAB Question 10 (part), AS, June 1994)

Summary

- The chi-squared statistic can be used as a test for goodness of fit.

  $$\chi^2_\upsilon = \sum \frac{(O-E)^2}{E}$$

  where υ is the number of degrees of freedom.

  $\nu = n - \text{con}$

  n = number of frequency cells

  con = number of constraints on the model

- In a contingency table, the expected entry in cell (i, j) is:

  $$\frac{(\text{sum of row } i) \times (\text{sum in column } j)}{\text{total number}}$$

  For a contingency table with r rows and c columns the number of degrees of freedom is $(r-1)(c-1)$.

- When there is only one degree of freedom the χ^2_1 statistic is:

  $$\sum \frac{\left[\text{ABS}(O-E) - \frac{1}{2}\right]^2}{E}$$

This is called **Yates' correction**.

STATISTICS

13 Bivariate data

In this chapter we investigate:

- *how to present linked pairs of data and how to explore the nature of the link between the two variables,*
- *one way of modelling a linear relationship,*
- *how cyclical variations in data may be discovered.*

SCATTERGRAPHS AND COVARIANCE

All the random variables we have studied so far have been single variables or **univariates**. We shall now study **two-variable** or **bivariate data**. An example of bivariate data is given here.

Time (minutes)		
Conditions:	Dry	Wet
Runner	X	Y
A	35.5	37.4
B	40.4	40.5
C	38.8	39.6
D	39.9	39.8
E	38.0	37.1
F	37.2	36.1
G	34.7	36.7
H	37.9	36.8

The data are the times taken by eight young runners in two 10 km races. The races were run on the all-weather track at their school under different weather conditions. What are the mean times in the dry and in the wet for these runners?

$\bar{x} = 37.8$ and $\bar{y} = 38.0$

This seems to indicate that the runners take about $\frac{2}{10}$ of a minute more in the wet than in the dry. Is there any other comment you could make about the times?

It is clearly possible to identify the way the times are spread out. The standard deviations of the times are:

$s_x \approx 1.853$ and $s_y \approx 1.580$

This suggests that there is more of a spread of times in the dry.

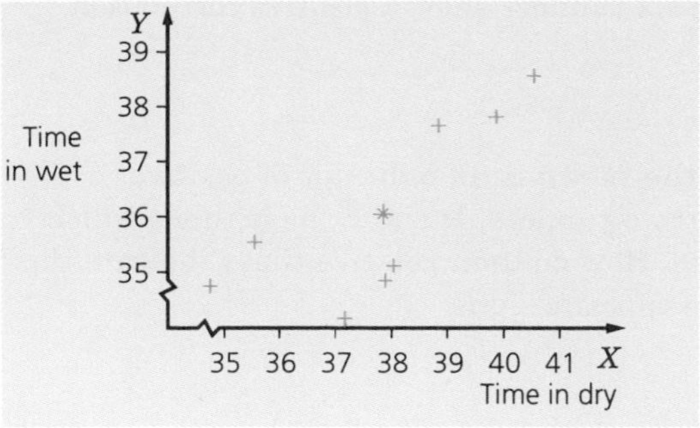

In this elementary analysis we have treated the data as two collections of single variables. However, there are two times given for each runner. This allows us to make a valid pairing of the times and so treat the data as bivariate. Then we can display the data graphically in a scattergraph or scatter diagram.

What does this display suggest about the time a runner takes in the dry and the time taken in the wet?

The display suggests a link or **positive correlation** between the times. This means that longer dry times are associated with longer wet times and shorter dry times are associated with shorter wet times.

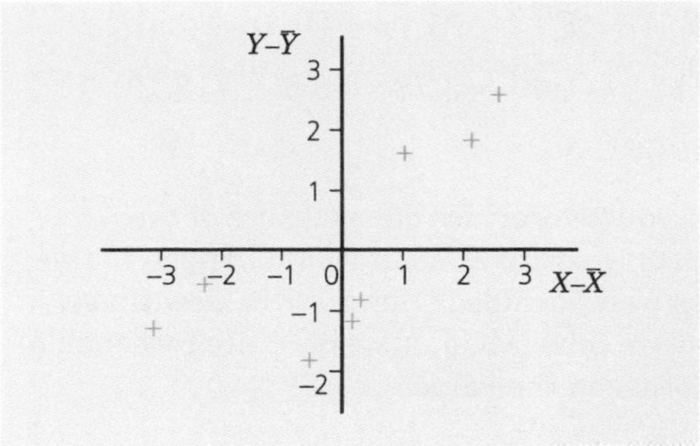

The point (37.8, 38.0), marked differently on the scattergraph is the **mean point**, i.e. the point with the two means as its coordinates. Notice that virtually all the points representing the actual data are either *above and to the right* or *below and to the left* of the mean point. This is a feature of bivariate data which are positively correlated. This graph shows the data related to the respective means. It was produced from the data in the table below in which the respective means were subtracted from the individual times.

Times relative to means		
Runner	$X - \bar{X}$	$Y - \bar{Y}$
A	–3.1	–1.3
B	–2.3	–0.6
C	–0.6	–1.9
D	+0.1	–1.2
E	+0.2	–0.9
F	+1.0	+1.6
G	+2.1	+1.8
H	+2.6	+2.5

The runners at the points which are above and to the right of the mean are F, G and H. What do the relative times of these three runners have in common? The pair of relative times for each of F, G and H are positive. Now consider the runners whose points are below and to the left of the mean point. What do their relative times have in common? These are A, B and C and in each case the pairs of relative times are negative.

The times for each of these six runners show a positive correlation. The 'same sign' nature,

(+, +) or (− , −)

of their times **relative to the mean** is an indicator of positive correlation. The remaining two runners, D and E, have times which are not positively correlated. How do their relative times indicate this? We can see that these have opposite signs:

(+, −)

If the product of a runner's relative times is formed then that product has the same sign as the nature of the correlation shown in the runner's times. The products of the relative times for each runner is shown below.

Runner	A	B	C	D	E	F	G	H	Total
Product $(X-\bar{X})\times(Y-\bar{Y})$	+4.03	+1.38	+1.14	−0.12	−0.18	+1.6	+3.78	+6.5	+18.13

The total, 18.13, of these products provides one measure of the correlation. This total, +18.13 is positive. What would happen to the total if more runners' times were included? Suppose the new times, generally, also showed positive correlation, it seems quite reasonable that there would be an increase in the value:

$$\sum_{\text{all runners}}\left\{\left(x-\bar{x}\right)\left(y-\bar{y}\right)\right\}$$

It would be sensible to take into account the number of runners by finding the average amount of correlation. This leads to:

$$\frac{\Sigma\left\{\left(x-\bar{x}\right)\left(y-\bar{y}\right)\right\}}{n}$$

where n is the number of runners, or in general, the number of data pairs. This measure is known as the **covariance** of X and Y.

There are two common notations in use:

$$\text{cov}(x, y) = s_{xy}$$

$$= \frac{1}{n}\sum_{\text{all }(x,y)}\left\{\left(x-\bar{x}\right)\left(y-\bar{y}\right)\right\}$$

So the covariance of the times for the two 10 km races is:

$$s_{xy} = \frac{18.13}{8}$$

$$\approx 2.266$$

Exploration 13.1 *Covariance*

a) i) Calculate the mean times, $\bar{u}$, $\bar{v}$ for the following collection of data.

U (s)	2082	2120	2232	2274	2280	2328	2394	2424
V (s)	2202	2244	2166	2208	2226	2376	2388	2430

ii) Calculate the times relative to their respective means.
iii) Demonstrate that the covariance of U and V is 8158.5 and comment on its value.
iv) Plot a scattergraph of these data and compare with the previous scattergraph.

b) The same eight runners, A to H take part in a 10 km race on grass. The weather conditions were dry. These are their results.

Runner	A	B	C	D	E	F	G	H
Time, z, (minutes)	37.5	39.1	38.5	38.8	38.0	37.6	36.7	37.8

i) Calculate the mean, $\bar{z}$ of these race times.
ii) Calculate, the covariance of the all-weather 'dry' time and the grass track 'dry' time. Comment on the value.
iii) Draw a scattergraph showing the dry weather times for each of the eight runners on the two types of surface. Comment on the link between the times.

Product moment correlation coefficient

Let us look back to the data in Exploration 13.1 part **a)**. This is the same as the original data for the runners' times, the only changes that have been made are in the units used i.e. seconds as opposed to minutes, and in the order of the runners: G, A, F, H, E, C, D, B. This means that these data should show exactly the same correlation, but the two covariances are different.

$$\text{cov}(x, y) \approx 2.266 \ldots$$
$$\text{cov}(u, v) \approx 8158.5$$

This is decidedly unsatisfactory. The measure of correlation should be independent of the units of measurement used but changing units from minutes to seconds seems to change the spread of the data, the standard deviations found in above were:

s_x 1.853 $\quad$ s_y 1.580

and

s_u 111.2 $\quad$ s_v 94.77

Try dividing the covariance by each of the relevant standard deviations.

You should find:

$$\frac{s_{xy}}{(s_x s_y)} = \frac{2.266}{(1.853 \times 1.580)} \approx 0.774$$

$$\frac{s_{uv}}{(s_u s_v)} = \frac{8158.5}{(111.2 \times 94.77)} \approx 0.774$$

so these measures agree with each other. Clearly, this is a far more satisfactory measure of correlation. It is known as **Pearson's product moment correlation coefficient** (PMCC). It is usually represented by r.

Notation: $r = \dfrac{s_{xy}}{(s_x s_y)}$

Exploration 13.2

Calculating r

Calculate the product moment correlation coefficient for the dry weather 10 km race times, i.e. calculate:

$$r = \frac{s_{xz}}{(s_x s_z)}$$

Comment on the result.

(You should find $r \approx 0.967$ which suggests a more positive correlation between the 'dry' times than between the times for the all-weather track.)

An alternative approach

The way in which the covariance is calculated:

$$\tfrac{1}{n}\Sigma\{(x - \bar{x})(y - \bar{y})\}$$

is fraught with potential for 'rounding' errors and subsequent errors in the value of the product moment correlation coefficient. You may recall the way in which variance was introduced in Chapter 3, *Summary statistics 2*.

$$s_x^{\,2} = \tfrac{1}{n}\Sigma(x - \bar{x})^2$$

which was shown to be exactly the same as:

$$s_x^{\,2} = \tfrac{1}{n}\Sigma x^2 - \bar{x}^2$$

i.e. the mean of the squares minus the square of the mean. The covariance can be calculated in an exactly similar way:

$$s_{xy} = \tfrac{1}{n}\Sigma xy - \bar{x}\,\bar{y}$$

i.e. **the mean of the products minus the product of the means.**

The advantage of this is that the original data are used rather than the relative data and that the means we need only enter the calculation at the end.

In the case of the all weather track data:

$$\begin{aligned}\Sigma xy &= 35.5 \times 37.4 + 40.4 \times 40.5 + \ldots + 37.9 \times 36.8 \\ &= 11\,509.33\end{aligned}$$

Hence the covariance is:

$$s_{xy} = \frac{11509.33}{8} - 37.8 \times 38.0$$
$$\approx 2.266$$

Exploration 13.3

Mean of products minus product of means

a) Calculate the covariance, s_{xz}, using this alternative method.

b) **i)** Calculate the covariance, s_{yz}.

ii) Show that the correlation coefficient between the all-weather 'wet' times and the 'dry' grass times is 0.867.

CALCULATOR ACTIVITY

Bivariate data

Will your calculator accept bivariate data? Will it enable you to evaluate:

a) covariance, **b)** the product moment correlation coefficient, r? Find out.

Interpretation of r

Pearson's product moment correlation coefficient, r is a sound measure of correlation. It will always lie between –1 and +1. The closer it is to these values, the greater the degree of correlation, or more strictly, **linear correlation** in the data. If r is close to zero, it is unlikely that there is any linear correlation in the data. The sign of the coefficient is an indication of the positive or negative nature of the correlation.

Example 13.1

The eight runners who have featured in this chapter also recorded times for a 10 km race carried out on grass in wet weather. Their times are shown below.

Runner	A	B	C	D	E	F	G	H
Wet grass time, W (minutes)	38.5	40.6	40.3	40.4	39.9	42.3	39.6	43.2
Wet all weather time, Y (minutes)	37.4	40.5	39.6	39.8	37.1	36.1	36.7	36.8

(The wet all-weather track times are also reproduced for convenience.)

a) *Plot a scattergraph for these data, comment on the correlation exhibited in them.*

b) *Calculate the product moment correlation coefficient for these data and interpret its value.*

Solution

a)

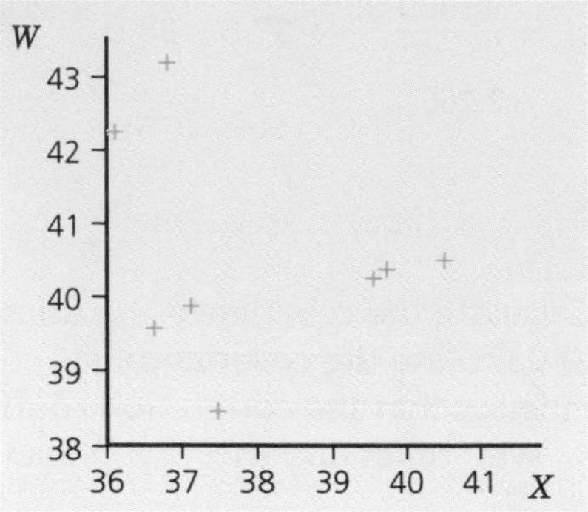

The runners who are fast in the wet on the all-weather track seem to be slow in the wet on grass. But the slower all-weather track runners do not seem to be noticeably fast or slow on grass in similar conditions.

b) $\Sigma yw = 12\,338.4$, *hence the covariance is:*

$s_{yw} = -0.5$

and the correlation coefficient is

$$r \approx \frac{-0.5}{(1.5796 \times 1.4)} \approx -0.266$$

The sign indicates a negative correlation in the data. However, the value, 0.226, is not close to 1, so it is unlikely that there is any linear correlation.

EXERCISES

13.1 A

1 Explain what Pearson's product moment correlation coefficient measures. Why is this a better measure of correlation than covariance?

x, km	69	65	62	61	59	70	71	75
y, $	0.52	0.25	0.55	0.20	0.62	0.47	0.25	0.37

a) Calculate the covariance of the data in the table. What inference are you able to draw from the value you have calculated?

b) Calculate the product moment correlation coefficient for these data and comment on its value.

c) The variables u, v are defined as:
$u = x - 65$, $v = 100(y - 0.40)$.
Determine the value of $\text{cov}(u, v)$ and compare this with the covariance you found in **a)**.

d) What is the product moment correlation coefficient of the transformed data? Comment.

e) Plot a scattergraph of the data and comment on the relationship between unit cost, $\$y$, and length, x km.

2 **a)** Plot a scattergraph of the data in the table below.

Date: x	10	11	12	13	14	15	16	17	18	19
Milk yield: y	1183	1159	1288	1254	1269	1274	1316	1318	1399	1322

b) Do the data exhibit correlation?
c) Calculate the mean:
i) date, **ii)** milk yield, **iii)** product.
d) Use your calculation to estimate the value of Pearson's product moment correlation coefficient for these data. Comment.

3 The data in question **2** were recorded by a farm manager in February. The milk yield was the total volume of milk from her herd of Jersey cows produced on that date. The data for 20–28 February are shown below.

Date: x	20	21	22	23	24	25	26	27	28
Milk yield: y	1227	1258	1245	1278	1193	1156	1200	1201	1151

a) What is the nature, if any, of the correlation exhibited in the data?
b) Justify your comment in **a)** by calculating the product moment correlation coefficient for these data.
c) Plot the milk yield for the 19 days; comment on the correlation.
d) Calculate the product moment correlation coefficient for the 19 data pairs. Comment.

4 The data in the table represent the total live births recorded for a period of twelve months.

Month: x	1	2	3	4	5	6	7	8	9	10	11	12
Live births: y (1000s)	63.5	73.0	67.4	69.7	65.9	67.2	62.9	63.6	63.1	59.2	59.9	61.7

a) Plot a scattergraph for these data. Identify the data pair which may be termed an outlier. Comment on the correlation of the data.
b) Calculate the product moment correlation coefficient for these data:
i) excluding the outlier, **ii)** including the outlier.
c) Comment on the results of your calculations.

5 The total hours of sunshine in an Eastern city are recorded over a period of 20 days in the table.

Day	1	2	3	4	5	6	7	8	9	10	11	12	13	14	15	16	17	18	19	20
Sunshine	9.3	8.9	8.6	10.3	8.6	8.1	7.6	8.0	7.1	7.9	7.6	7.5	7.3	7.7	7.3	7.4	6.8	6.9	4.9	6.1

a) Which two items of data may be regarded as outliers?
b) Calculate the covariance for the hours of sunshine:
i) including the outliers, **ii)** excluding the outliers.
c) Comment on the correlation shown in these data.

6 The data represent the recorded crime rate (per thousand households) over a period of a year. The crime rate is calculated on the last day of each month. Use a code to represent the month and calculate the product moment correlation coefficient for the data. What interpretation do you put on its value?

Month	Jan	Feb	Mar	Apr	May	June	July	Aug	Sep	Oct	Nov	Dec
Crime rate	45.5	43.5	43.9	47.0	54.7	64.3	71.9	76.0	77.0	76.9	76.4	75.4

7 Explain why each of the following statements may be incorrect.

a) When the product moment correlation coefficient is close to zero the two variables are not associated.

b) If the product moment correlation coefficient is 0.975 then an increase in one variable causes an increase in the other.

c) Flowers in a garden are watered and the correlation coefficient between the petal size and the amount of water given is 0.073, so it follows that there is no relation between petal size and quantity of water given to the flower.

8 The proportion of people registered as unemployed each quarter over a period of five years is given in the table.

Year	1				2				3			
Quarter	1	2	3	4	1	2	3	4	1	2	3	4
% unemployed	10.5	10.1	9.5	8.9	8.4	7.9	7.5	7.0	6.5	6.2	5.9	6.1
Year	4				5							
Quarter	1	2	3	4	1	2	3	4				
% unemployed	6.8	7.6	8.4	8.9	9.2	9.4	9.6	10.2				

Calculate the product moment correlation coefficient and interpret its value. Draw a scattergraph of the data and reflect on your interpretation.

EXERCISES

13.1 B

1 The yield per hectare of a crop is supposed to depend on the amount of rainfall in the month preceding its harvest. Nine areas are monitored and the values of yield (y) and the corresponding rainfall (x) are recorded. The results are given in the table below.

x	14.7	10.3	18.6	13.4	14.7	13.6	16.7	11.7	12.1
y	8.3	10.2	15.3	6.4	11.8	12.0	13.4	11.9	9.9

a) Plot a scatter diagram of the data and comment on whether the diagram supports the theory that there is a linear association between y and x.

b) Calculate the covariance of x and y.

c) You are given that the standard deviations of x and y are $s_x = 2.4160$ and $s_y = 2.5165$. Use these together with part **b)** to calculate the product moment correlation coefficient.

d) Does your answer to **c)** support your answer to **a)**? Give reasons.

2 When students start secondary school they are given an IQ test to determine their level of ability. The table below gives the scores for the test together with the end of year test score in mathematics.

IQ test (x)	335	201	430	281	385	231	300	130	190	320
Maths test (y)	58	52	76	64	74	52	65	43	47	68

a) Draw a scattergraph of these data and estimate the value of r, the product moment correlation coefficient.

b) Calculate the covariance of x and y.

c) Calculate the value of the product moment correlation coefficient.

d) State whether there seems to be a linear relationship between the two test scores.

3 Seven students are surveyed to find the number of points they got on average at GCSE and at A-level. The results are given below.

Average points at GCSE (x)	6.1	5.8	6.3	5.9	5.8	5.9	6.2
Average points at A Level (y)	7.4	6.7	7.9	4.3	6.6	6.8	7.3

a) Draw a scatter diagram of these data.

b) Calculate s_{xy}, the covariance.

c) Find r, the product moment correlation coefficient and state whether you think there is a positive correlation between GCSE scores and A-level scores.

4

Student	A	B	C	D	E	F	G	H	I	J
French mark (x)	56	50	72	67	31	50	65	40	80	61
German mark (y)	60	50	67	75	64	56	73	48	76	62

a) Plot a scattergraph for the above data and identify the outlier. Suggest reasons to account for it.

b) Calculate the product moment correlation coefficient for these data:
i) including the outlier, **ii)** excluding the outlier.
Comment.

5 State in each case whether you would expect a positive correlation, a negative correlation or no correlation.

a) total weight of passengers on a bus and its fuel consumption in miles per gallon

b) number of hours piano practice and number of wrong notes

c) shoe size and mean number of hours sleep

d) ice-cream sales and sales of sunglasses

6 A student calculated the product moment correlation coefficient between height above sea level and mean July temperature for a number of towns.

Heights were given in feet and temperatures in degrees Fahrenheit. She wanted to find the product moment correlation coefficient between heights in metres and temperatures in degrees Celsius. If there are 3.28 feet per metre, and if temperature °C = $\frac{5}{9}$(°F – 32), how should she correct her calculations?

7 Use diagrams to help to provide an explanation of the terms:

a) perfect correlation, **b)** zero correlation.

EMPIRICAL CRITICAL VALUES FOR PEARSON'S PRODUCT MOMENT CORRELATION COEFFICIENT

When is a value of the product moment correlation coefficient close to unity and when is it not? The answer to this question is very important. We know that if a value of r is close to unity, then there is evidence in the data of a linear correlation between the variates. But at this point, it is not possible to be clear about what we mean by 'close to unity'.

One approach to this is to set up bivariate populations which do not show linear correlation and then obtain random samples for which the value of r is calculated. If this process is repeated a large number of times it ought to be possible to obtain an empirical distribution for r.

A simple way of generating a bivariate distribution is to use pairs of random decimals as (x, y) observations. We could generate eight pairs and then calculate the corresponding value of the product moment correlation coefficient. This gives one value of r. Try it yourself. Repeat this process many (say 100, or more) times and display the resulting distribution of the sample product moment correlation coefficient for eight pairs of data.

CALCULATOR ACTIVITY

Bivariate data – Sample size 8

a) Try programming your calculator to run a program based on this flowchart. It is intended to produce a value of the product moment correlation coefficient for a sample of size 8 from a bivariate population in which there is no correlation.

Record the output each time the program is run. Use a stem-and-leaf display with a unit equal to 0.01.

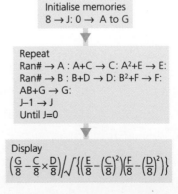

Initialise memories
8 → J: 0 → A to G

↓

Repeat
Ran# → A : A+C → C: A²+E → E:
Ran# → B : B+D → D: B²+F → F:
AB+G → G:
J–1 → J
Until J=0

↓

Display
$\left(\frac{G}{8}-\frac{C}{8}\times\frac{D}{8}\right)\Big/\sqrt{\left\{\left(\frac{E}{8}-\left(\frac{C}{8}\right)^2\right)\left(\frac{F}{8}-\left(\frac{D}{8}\right)^2\right)\right\}}$

b) Once you have a 100 or more values of r recorded, you will have a reasonable impression of the distribution of r for eight pairs of data. Describe the distribution and try to identify ten per cent two-tailed (or five per cent one-tailed) critical values.

This is the output obtained when the program of the calculator activity above was run 131 times.

Unit = 0.01

–9	
–8	0
–7	8 4 8
–6	2 5 2 3 2
–5	6 7 3 2 6
–4	2 2 8 4 2 7
–3	9 1 1 9 2 4 3
–2	9 0 9 5 5 5 4 8 5
–1	9 6 8 4 7 8 6 9 1 4 7
–0	2 8 0 4 7 5 2 0 0 0 2 0 1
0	9 7 2 3 0 2 1 5 9 0 1 0 7 0
1	1 7 4 0 7 4 5 6 7 2 1 7
2	9 1 9 1 2 2 7 8 6 4 2
3	0 2 8 9 3 8 2 4 3 9
4	5 0 0 3 3 9 7 8
5	7 7 3 9 3 4 8
6	8 5 1 0 2 3
7	6 4
8	3
9	

To use this empirical distribution to determine critical values for r is perhaps optimistic. However, the following three examples show how we might use the display to obtain critical values.

a) H_0: product moment correlation coefficient = 0;
H_1: product moment correlation coefficient > 0

The null hypothesis reflects the way the distribution was obtained i.e. no linear correlation. The alternative hypothesis reflects an interest in positive linear correlation. H_1, therefore, indicates that you should be considering the upper tail. The five per cent upper-tail critical value has a depth of $0.05 \times (131 + 1) = 6.6$ which suggests that it is between 0.63 and 0.62.

b) H_0: product moment correlation coefficient = 0;
H_1: product moment correlation coefficient < 0

The interest in this as in the lower tail and the five per cent critical value is between –0.63 and –0.62. This agrees closely with the upper-tail value.

c) H_0: product moment correlation coefficient = 0;
H_1: product moment correlation coefficient $\neq$ 0

The alternative hypothesis addresses the issue of whether the data show linear correlation without any concern over positive or negative. Examining the tails of the distribution suggests that the five per cent critical values are between 0.74 and 0.68, and between –0.78 and –0.74. Think back to the 10 km race times in the light of these critical values. For instance, is there sufficient evidence in the data to support the view that race times on all-weather tracks in wet and dry conditions are positively correlated? The hypotheses relevant to test this are:

H_0: product moment correlation coefficient = 0;
H_1: product moment correlation coefficient > 0

and the sample of eight pairs of race times had a correlation coefficient of about 0.774. This value is closer to +1 than the critical value, 0.62 to 0.63, found in part **a**). It is reasonable to conclude at the five per cent level, that the null hypothesis of no correlation can be rejected in favour of the alternative hypothesis of positive correlation. Hence, there seems to be evidence to support the view of a positive correlation between the times obtained in the dry and in the wet on all-weather tracks.

Exploration 13.4 — *Checking for positive correlation*

a) Is there evidence, at the $2\frac{1}{2}$ per cent level, that there is a positive correlation between dry-weather grass track and all-weather track race times?

b) Test the hypothesis, at the ten per cent level, that there is a negative correlation between wet-weather grass track and all-weather track race times.

c) Do the available data support the view, at the ten per cent level, that race times on grass in dry and wet weather are correlated?

Tables of critical values for *r*

So far we have discovered properties of the sample correlation coefficient for eight data pairs. Clearly the distribution of values of *r* will differ with more than eight or fewer than eight data pairs.

If you have a programmable calculator, try the following Calculator activity.

CALCULATOR ACTIVITY

Bivariate data – Sample size 10

How does the program in the previous Calculator activity need to be adjusted to calculate the product moment correlation coefficient for, say, ten data pairs? Try altering your program and obtain a

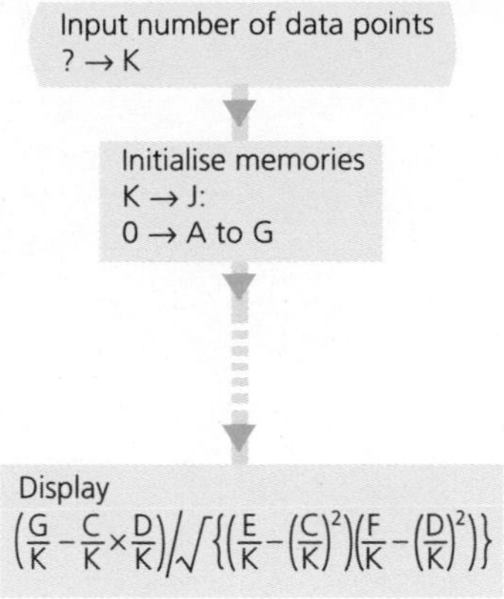

distribution of 100 values of r. Compare the $2\frac{1}{2}$ per cent, five per cent and ten per cent critical values with those found for eight data pairs.

Here is one way of adapting the earlier flowchart. This allows the input of the number of data pairs (it is stored in memory K).

It is also possible to change the underlying bivariate distribution. The one used in the programs is a bivariate rectangular distribution where both X and Y take values between 0 and 1. By changing the content of memories A and B (to represent the values of the random variables X and Y respectively) other bivariate distributions can be explored. It is important that the distributions are symmetric and independent.

Input number of data points
? → K

Initialise memories
K → J:
0 → A to G

Display
$$\left(\frac{G}{K} - \frac{C}{K} \times \frac{D}{K}\right) \bigg/ \sqrt{\left\{\left(\frac{E}{K} - \left(\frac{C}{K}\right)^2\right)\left(\frac{F}{K} - \left(\frac{D}{K}\right)^2\right)\right\}}$$

The following ordered stem-and-leaf displays were obtained for ten and six data pairs using:

Ran #+ Ran #+ Ran #+ Ran #+ Ran #+ Ran #+ Ran #+ Ran #+ Ran # → A

and

Ran # + Ran # + Ran # → B

The stem-and-leaf of product moment correlation coefficient for 10 data pairs $n = 119$ is like this:

Leaf unit = 0.010

depth		
1	−7	3
4	−6	8 4 1
8	−5	8 6 4 3
13	−4	9 7 5 4 0
22	−3	9 8 7 6 6 5 5 5 2
28	−2	8 8 7 3 1 1
43	−1	9 9 9 7 6 5 4 3 3 3 2 1 1 0 0
55	−0	8 7 7 7 6 5 5 5 4 3 3 1
(12)	0	0 1 2 2 3 3 5 7 7 7 8 9
52	1	1 2 2 3 4 5 5 5 7 7 7 8 8
39	2	1 2 3 3 4 4 5 8 8 8 9
28	3	0 0 1 1 2 2 9 9
20	4	1 1 2 3 4 4 5 6 6 6
10	5	0 1 2 3 5 6
4	6	0 3 6 7

The stem-and-leaf of product moment correlation coefficient for six data pairs n = 172 looks like this:

Leaf unit = 0.010

depth		
1	–9	6
4	–8	6 4 2
13	–7	9 9 6 6 5 2 1 0 0
21	–6	9 8 8 4 3 2 2 2
25	–5	6 2 1 0
38	–4	9 8 7 6 6 5 5 4 3 3 3 2 0
48	–3	8 7 6 6 5 5 5 4 4 2
58	–2	9 7 5 5 4 2 2 1 0 0
70	–1	8 7 7 7 7 7 6 4 3 3 2 0
84	–0	9 9 8 6 6 5 5 5 3 3 3 2 1 1
(10)	0	2 2 3 3 4 6 7 7 9 9
78	1	0 0 1 1 1 1 2 2 3 3 4 4 5 5 6 7 7 7 8 8
58	2	2 2 2 2 3 5 5 5 7 8
48	3	0 0 2 4 5 6 6 7 8
39	4	1 1 3 4 4 4 4 5 5 6 7 7
27	5	0 0 0 2 3 4 5 7 8
18	6	0 0 1 4 4 4 4 9
10	7	0 1 4 6 9
5	8	0 2 2 2
1	9	6

The five per cent upper-tail critical values are:

0.55 for ten pairs

between 0.74 and 0.71 for six pairs

Critical values depend on the number of pairs of data values. Tables of critical values are produced for a range of values of n (the number of pairs). These tables indicate the following upper-tail five per cent critical values.

n	Critical value of r
6	0.7293
8	0.6215
10	0.5494

The experimental distributions obtained are in close agreement with these tabulated values.

Example 13.2

a) *Is there evidence to support the view, at the five per cent level, that the following data are positively linearly correlated?*

X	2.71	5.09	4.27	3.36	3.99	4.52
Y	1.50	2.49	1.81	1.01	1.63	1.58

b) *The ten members of a rugby club's squad who consider themselves as forwards took part in out-of-season training. Part of this training involved running 100 metres. In another part, they ran 10 km across country. Their times are shown in the table.*

Distance	Time										
100 m	X (seconds)	22.4	23.5	12.5	22.2	15.5	19.1	23.8	21.9	22.2	19.9
10 km	Y (minutes)	46.9	44.5	49.8	40.3	43.7	52.0	46.9	35.0	41.6	40.8

Do these data support the view that those players who are fast over 100 m tend to be slow over 10 km and vice versa? Conduct your test at the five per cent level.

Solution

a) *Appropriate hypothesis are:*
H_0: *product moment correlation coefficient* $= 0$;
H_1: *product moment correlation coefficient* > 0.
The value of r is 0.7447 which is closer to +1 than the critical value, 0.7293, for six pairs. It is reasonable to conclude that the data are positively linearly correlated.

b) *Appropriate hypothesis are:*
H_0: *product moment correlation coefficient* $= 0$;
H_1: *product moment correlation coefficient* < 0.
The value of r for these ten pairs of data is – 0.3633. The tables do not give 'lower-tail' critical values, but the distribution of r is symmetric about $r = 0$. *This means that the five per cent lower-tail critical value is – 0.5494. We can see that, for this data, r is not closer to –1 than the critical value. Hence there is insufficient evidence to support the negative correlation of the times over 100 m and 10 km. So the data do not support the view that the players who are faster over 100 m tend to be slower over 10 km. The data suggest that there is no linear correlation.*

EXERCISES

13.2 A

1 The maize crop over a six-year period from a farming cooperative is recorded in the table, along with the corresponding rainfall.

Maize crop (tonnes)	69	86	67	80	72	68
Rainfall (mm)	50	65	42	55	60	35

Is there evidence in these data to support the hypothesis, at the ten per cent level, that the maize crop is linearly correlated to rainfall? State your alternative hypothesis.

2 Records of nine randomly-selected births were examined to determine if there is any evidence of a correlation between gestation period and the acidity (measured as pH level) of the mother's blood during labour.

The data are shown in the table.

Gestation	34	42	37	40	34	38	41	39	36
Blood pH	7.36	7.28	7.38	7.34	7.35	7.32	7.30	7.36	7.21

State suitable hypotheses and conduct a test of the correlation coefficient for these data. Make your conclusions clear.

3 The times of the births recorded in question **2** are given below.

Blood pH	7.36	7.28	7.38	7.34	7.35	7.32	7.30	7.36	7.21
Time of birth	20:15	00:30	13:45	18:00	10:15	03:00	15:30	21:15	14:00

The times are given using the 24-hour clock in hours : minutes.

i) a) Plot a scattergraph of these data.

b) Comment on the extent of correlation you feel is evident from your scattergraphs.

c) Conduct a test at the five per cent level of significance to determine if there is a positive correlation between blood pH and time of birth.

ii) Determine if there is any correlation between time of birth and gestation period.

4 Is there evidence in the following car production figures to support the view that car production and steel production are negatively correlated?

Year	1	2	3	4	5	6	7	8	9	10	11
Cars (000s)	9.8	19	27	32.1	43.8	41.1	37.8	43.1	61.1	78	71.8
Steel production	634	690	530	645	730	655	437	380	570	320	300

Justify your response.

5 The table records the average daily hours of sunshine each month for two years.

Month		Jan	Feb	Mar	Apr	May	June	July	Aug	Sept	Oct	Nov	Dec
Average	Y1	2.0	1.63	2.78	5.64	7.32	5.5	5.0	4.77	4.66	3.58	1.92	1.75
Sunshine	Y2	1.64	2.48	2.13	4.37	4.61	5.23	5.0	6.08	5.28	3.71	1.71	1.48

a) Plot a scattergraph of these data. Are the data correlated?

b) Justify your responses in **a)** by carrying out an appropriate hypothesis test.

6 The table contains production figures claimed by industry and the seasonally adjusted figures put out by a government.

Actual	42.1	30.7	37.4	25.8	35.2	35.2	36.2	34.1	36.6	35.6	32.5	31.0	31.2	31.6	21
Government	42.4	37.9	35.6	35.2	35.2	35.2	35.3	35.4	35.1	34.3	33.0	32.0	31.6	31.7	31.8

Test the claim that these data are positively correlated.

EXERCISES

13.2 B

1 The following are the percentage shares of the vote predicted for eight election candidates, x, and the percentage shares actually obtained, y.

x	57	70	46	35	15	44	30	55
y	71	72	49	41	16	39	58	77

Is there evidence to support the hypothesis that the data are positively linearly correlated? Use a five per cent level of significance.

2 Calculate the product moment correlation coefficient for these data:

x	1002	1003	997	995	1000	1006
y	63	66	52	52	57	70

a) between the given values of x and y,
b) between the transformed values of x and y:
$X = x - 1000$ and $Y = y - 50$.

3 An experiment produced the following results.
Product moment correlation coefficient between x and $y = 0.73$
Product moment correlation coefficient between x and $z = -0.64$
Product moment correlation coefficient between y and $z = 0.79$
Explain why there must be an error in these results.

4 **a)** Test the following set of bivariate data to see whether there is a significant negative correlation.

x	4.1	4.5	5.2	5.7	6.3	7.2
y	9.2	7.3	3.1	3.4	2.9	2.1

b) Draw a scattergraph of the data. Does this confirm your calculation in **a)**?

5 The prices of two commodities x and y were recorded each December for a period of twelve years. The results are given in the table below.

Price of x (£)	1.20	1.50	2.00	2.50	3.30	3.00	3.10	3.70	4.00	4.80	5.00	5.50
Price of y (£)	3.20	4.50	2.20	5.80	3.00	4.60	5.70	5.70	3.50	3.60	5.90	4.40

a) Draw a scatter diagram to illustrate the data.
b) State with reasons whether you would believe that the two prices were correlated.
c) Conduct a test at the five per cent level to determine whether there is a significant non-zero correlation between the two variates.

6 Consider the following set of data.

x	−3	−2	−1	0	1	2	3
y	9	4	1	0	1	4	9

a) Calculate the product moment correlation coefficient for these data.
b) Draw a scattergraph of the data.
c) Comment on your results.

LEAST SQUARES REGRESSION MODEL

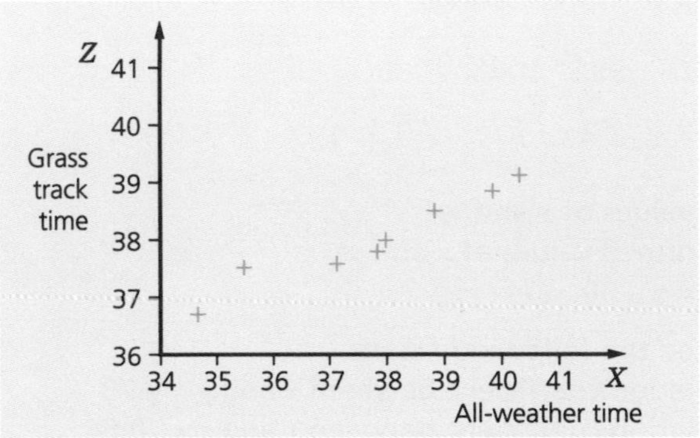

In Exploration 13.2 you probably found that the correlation coefficient between the 'dry' times is 0.967. This is sufficient evidence, at the one per cent level of significance, to support the view that there is positive linear correlation between the times to race 10 km on an all-weather track and on a grass track when the weather conditions are dry. Linear correlation suggests that a straight line is a reasonable model for the relationship between the two variates, Z and X.

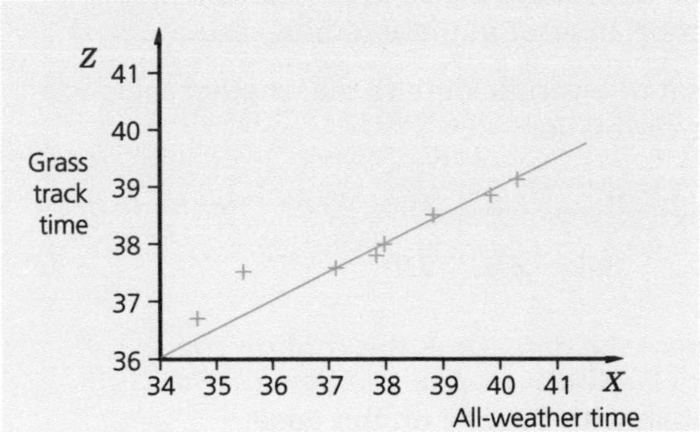

There are several ways of arriving at a linear model for bivariate data. The one we shall study is known as a **least squares regression line** or, simply, a **regression line**. However, before formally considering the derivation of this line, let us look at this diagram.

A line has been drawn representing a model for the relationship. This line passes through the points with coordinates:

(36.0, 37.0) and (40.0, 39.0)

It therefore, has gradient, m, where:

$$m = \frac{39 - 37}{40 - 36} = 0.5$$

Since, it passes through (36, 37) and has a gradient of 0.5, its equation is:

$$\hat{Z} - 37 = 0.5(X - 36)$$

(the hat symbol, $\hat{Z}$, is used to distinguish model from data). This can be rearranged into:

$$\hat{Z} = 0.5X + 19$$

The 'squares' referred to in the 'least squares regression' are the squares of the residuals of the difference between the data and the model i.e. $(Z - \hat{Z})^2$. More precisely, it is the total of all these squared residuals for each of the eight points. For the model:

$$\hat{Z} = 0.5X + 19$$

The calculation of this sum of squared residuals is shown in this table.

Data		Model	Squared difference
X	Z	$\hat{Z} = 0.5X + 19$	$(Z - \hat{Z})^2$
35.5	37.5	36.75	0.5625
40.4	39.1	39.2	0.01
38.8	38.5	38.4	0.01
39.9	38.8	38.95	0.0225
38.0	38.0	38.0	0.0
37.2	37.6	37.6	0.0
34.7	36.7	36.35	0.1225
37.9	37.8	37.95	0.0225
		$\sum\left\{(Z-\hat{Z})^2\right\} =$	0.75

This particular model produces a value of 0.75 for the sum of the squared differences.

Exploration 13.5 *Sum of squared residuals*

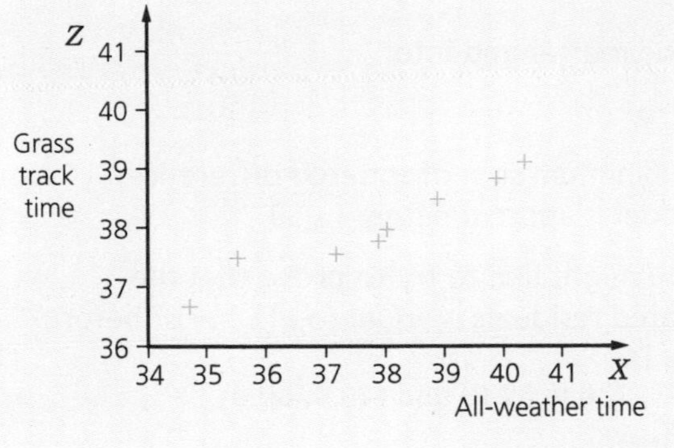

Make a copy on graph paper of this diagram, from the data on pages 354 and 357. On your copy draw what you think is an appropriate straight line to model the relationship. Now *calculate* the equation of the line you have drawn. Use the same process of identifying two points on the line, as above.

Continue with the process and obtain the value of:

$$\sum\left\{(Z-\hat{Z})^2\right\}$$

for your linear model. Compare the value you obtain with the one on this page.

Gradient of the least squares regression line

The 'least squares regression' line is the linear model which produces the smallest possible value for the sum of the squared differences between data and model. The least squares line always passes through the **mean point** which implies that its equation is of the form:

$$\hat{Z} - \bar{Z} = \text{gradient} \times (X - \bar{X})$$

The gradient of the least squares line is the ratio of the covariance of x and z to the variance of x. In other words:

$$\text{gradient} = \frac{s_{xz}}{s_x^2}$$

which leads to the equation for the least squares regression line:

$$\hat{z} - \bar{z} = \frac{s_{xz}}{s_x^2}(x - \bar{x})$$

For these data, $s_{xz} = 1.305$ and $s_x^2 = 3.435$, which leads to the least squares model.

$$\hat{z} - 38.0 = \frac{1.305}{3.435}(x - 37.8)$$

Simplifying this:

$$\hat{z} \approx 23.6393\ldots + 0.379913x$$

The value of the sum of squared differences:

$$\sum\left\{(z - \hat{z})^2\right\} = 0.2737\ldots$$

which is smaller than the total for the model on page 372.

Note

A formal derivation of the regression equation appears in Appendix II.

Exploration 13.6

Calculating the regression equation

a) In Exploration 13.3 you showed that the correlation coefficient between all-weather wet condition times and grass track dry condition times was 0.867. This indicates, at the one per cent level of significance, that there is a positive linear relation between the times. Calculate the regression equation:

$$\hat{z} - \bar{z} = m(y - \bar{y})$$

and show that it can be rearranged into:

$$\hat{z} \approx 22.8076 + 0.3998\,y$$

Hence show that the minimum sum of squared differences between data and model is approximately 1.050.

b) If you like algebra, you might like to try to prove that the minimum sum of squared residuals is equal to $n(1 - r^2)s^2$ before referring to Appendix II.

Factor and response

Each of the regression models developed above has treated the dry grass race time as being *dependent upon* or being a *response to* another variable. In each case, the other variable was a race time on the all-weather track. The all-weather time is given the status of the **factor** producing the response. It is important to distinguish between the factor variable and the response variable before embarking on the calculation of the equation of a regression line. The 'response-factor' issue may be reflected in the description given to a regression. For instance, the regression model:

$$\hat{z} = 23.64 + 0.3799x$$

developed earlier can be described as the equation of the line of regression of z on x to indicate that the response, z, depends on the factor, x.

It may be relevant to calculate the equation of the line of regression of x on z. In which case, the equation would be:

$$\hat{x} - \bar{x} = \frac{s_{xz}}{s_z^{\ 2}}(z - \bar{z})$$

The model would be trying to fit all-weather track times for each given dry grass time. This is different from the process which leads to the $\hat{z}$ model. The $\hat{x}$ model turns out to be:

$$\hat{x} - 37.8 = \frac{1.305}{0.53}(z - 38.0)$$

This x on z model is derived by minimising $\sum\left\{(x - \hat{x})^2\right\}$ not $\sum\left\{(z - \hat{z})^2\right\}$.

The equation simplifies to give:

$$\hat{x} = 2.462\,264\,\hat{z} - 55.766$$

It is not obvious that the two models are not the same since they are not in quite the same form. When this $\hat{x}$ model is rearranged as:

$$z = 22.65 + 0.4061\hat{x}$$

it can be compared directly to the $\hat{z}$ model:

$$\hat{z} = 23.64 + 0.3799x$$

We can now see that the two equations do not represent the same straight line.

Example 13.3

A young sprinter was looking back over the best times she had recorded in 100 m races each year. The times are shown in the table.

Ages, a (years)	12	13	14	15	16	17	18	19
Time, T (seconds)	19.1	15.5	15.1	14.6	12.2	12.1	11.3	10.8

a) *Calculate the product moment correlation coefficient for these data. Comment on the value you obtain.*

b) *Calculate the equation of an appropriate least squares regression line model for these data.*

c) *Provide an interpretation for the gradient of the line.*

d) *Use the equation you have calculated to predict the best time the sprinter will take for 100 m when she is aged:*
i) 20 ii) 30.
Comment on the predictions.

Solution

a) *The correlation coefficient is:*

$$r = \frac{s_{at}}{s_a s_t} = \frac{5.656\,25}{\sqrt{5.25}\sqrt{6.749\,84}} \approx -0.9502$$

This indicates that there is negative correlation between the sprinter's age and her best time for 100 m. The value obtained is highly significant, at less than one per cent level, and would support a linear model.

b) *You need to decide which of the variables is the response and which is the factor. The sprinter's best 100 m time is dependent on her age. Hence the regression equation needed is the 'time on age':*

$$\hat{t} - \bar{t} = \frac{s_{at}}{s_a^{\,2}}(a - \bar{a})$$

i.e.

$$\hat{t} - 13.8375 = \frac{-5.656\,25}{5.25}(a - 15.5)$$

which simplifies to:

$$\hat{t} = 30.5369 - 1.0774a$$

c) *The gradient, – 1.0774, indicates that the sprinter's best time reduces by 1.0774 seconds each year.*

d) *This requires us to use the equation to find the values of* $\hat{t}$ *for* $a = 20$ *and* $a = 30$.

i) $\hat{t}(20) = 30.5369 - 1.0774 \times 20 = 8.99$

ii) $\hat{t}(30) = 30.5369 - 1.0774 \times 30 = -1.78$

Note

The first of the values suggests that this sprinter is likely to be a world record holder by the time she is 20. The second value is utterly ridiculous! The first is unlikely, the second is impossible.

The final part of Example 13.3 raises an important issue regarding the use of regression lines. A regression line provides a linear model which is valid only in the range of the factor variable. Using the model outside the range of the factor is not justified and can, clearly, lead to silly results.

A further issue is that of cause and effect. Clearly the sprinter's times improved as she got older. But was her age the cause of the improvement? It may be a contributory factor but other factors such as training are going to play a role. Try to avoid confusing correlation with causation.

CALCULATOR ACTIVITY

Regression calculations

Calculators which allow bivariate data to be input often offer regression calculations. The model they often use is the equation of the line of regression of y on x and they present this in the simplified form:

$$\hat{y} = a + bx$$

Use the data from Example 13.3 to discover how to extract the regression coefficients, a and b, and model predictions, $\hat{y}$, from your statistical calculator.

EXERCISES

13.3 A

1 A survey of children in a primary school revealed the following data.

Age of child, a	4	5	6	7	8	9	10	11	12
Pocket money, p	25	80	100	130	140	140	200	180	250

a) Plot the data on a scattergraph. Mark the mean point.
b) Draw a line, through the mean point, which you feel is an appropriate model for the relationship between age and pocket money.
c) Calculate the equation of the line you have drawn and use this to determine the residuals. Determine the sum of the squared residuals.
d) Determine the product moment correlation coefficient for the data. What interpretation is it reasonable to put on its value?
e) Calculate the equation of the line of regression of pocket money or age.
f) Use this equation to determine the sum of squared residuals and compare this with the result you obtained in **c)**.

2 **a)** Determine the equation of the line of regression of y and x for the data.

x	3	5	7	9	11	13	15	17	19	21
y	8	13	15	18	20	24	26	30	31	34

b) Use your equation to provide an estimate of y when $x = 10$.
c) Explain briefly why it is inappropriate to use the equation in **a)** to determine what value of x gives rise to $y = 25$.
d) Why is it inappropriate to use the equation in **a)** to provide an estimate of y when $x = 30$?

3 There is a belief that the risk of heart disease is negatively related to the red wine intake.

Wine intake	3	3.5	4	5	6	8	10	25	35	75	80
Heart disease risk	72	65	108	61	53	58	47	43	31	21	18

a) Determine the correlation coefficient for the data in the table. Does this support the belief?
b) Use an appropriate least squares regression line to estimate the risk when the wine intake is 52.
c) Calculate the equation of the least squares regression line which is appropriate to estimate the wine intake for a given heart disease risk. Comment on this process.

4 Drinking water goes through a variety of purification processes before it enters the domestic supply. A new filtration process is tested for its effectiveness in removing suspended matter from flowing water.

Flow of water, x	1	2	3	6	10	20
Measure of removal, y	0.36	0.30	0.27	0.10	0.051	0.022

A) a) Determine the product moment correlation coefficient for these data.

b) Calculate the equation of the line of regression of removal (y) on flow (x).

c) How do you interpret the gradient of this equation?

d) Estimate the removal of suspended matter when the flow is:
i) 5 **ii)** 15 **iii)** 25.

e) Comment on the likely reliability of each of the estimates.

B) An alternative model of the relationship between the flow (x) and the removal (y) is:

$$\frac{1}{y} = a + bx$$

a) Calculate the product moment correlation coefficient between x and $\frac{1}{y}$. Compare this with the value in **A) a)**.

b) Using an appropriate least squares method, determine values of a and b.

c) Use your new equation to estimate the removal when the flow is 15.

5 The growth of a kitten is carefully monitored by a concerned owner. The results are shown in the table.

Age	11	12	13	14	15	16	17	18	20
Weight	357	375	387	423	440	471	541	678	1080

a) Calculate the equation of regression of weight on age and use this to estimate the weight of the kitten at age 19 weeks.

b) Plot a scattergraph of the data and explain whether you feel the estimate in **a)** under – or over – estimates the kitten's weight.

c) Do you feel that the line whose equation you found is a reasonable model for the data?

d) A refinement to the model is suggested in which $(\text{age})^3$ is the independent variable. Obtain a least squares model based on this and use the model to estimate the weight at age 19.

e) Which of the two estimates do you feel is better?

6 An investigation in the development of the ear of a cocker spaniel produced the following data.

Age (weeks)	2	4	8	9	10	12	15
Area	6	18	45	57	68	112	245

a) Calculate the product moment correlation coefficient of age and area. Test the hypothesis that the area of a cocker spaniel's ear is a positive linear function of its age.
b) A model of the development of a dog in its puppy stage is proposed where the relationship between age and ear area is quadratic. Calculate the product moment correlation coefficient of $(\text{age})^2$ and area. Test, at one per cent, the hypothesis that area and $(\text{age})^2$ are positively linearly related.
c) Determine an appropriate model of the relationship between age and ear area, based on these data.

7 a) Plot a scattergraph of these data.
b) Determine the product moment correlation coefficient, r, for these data.

x	2.0	2.5	2.6	3.1	3.8	3.9	4.7	5.0	5.7	5.9	6.5	6.7	7.2
y	9.8	10.1	10.9	11.8	12.2	12.6	13.1	12.8	14.6	15.7	16.2	16.8	17.7

c) Test, at the one per cent level, the hypothesis that there is positive correlation between the variables in the table above.
d) Calculate the value of $13(1 - r^2)s_y^2$ and state what this represents.

EXERCISES

13.3 B

1 The value of two variables, x and y, are recorded on eight separate occasions during the course of an experiment. The values are given in the table below.

x	2	4	6	8	10	12	14	16
y	11.9	11.0	10.3	9.5	9.1	8	7.3	6.6

a) It is suspected that there might be a linear relationship between x and y.
Calculate the product moment correlation coefficient r and state with reasons whether you believe that such a relationship exists.
b) Show that the regression line of y on x is $y = 12.6 - 0.374x$.
c) Calculate the equation of the regression line of x on y.
d) By looking at the table explain why it may not be very useful to obtain the regression line of x on y.

2 Two variables x and y are supposed to be related by a law of the form $y = a + bx$.

An experiment yields the following parts of values for x and y.

x	1	2	3	4	5	6	7
y	1.0	2.5	1.7	2.0	3.0	5.0	4.5

a) Calculate the regression line of y on x.

b) Calculate the product moment correlation coefficient, r.
c) Determine the sum of the squared residuals in this case and verify that it is equal to $n(1 - r^2)s_y^2$.

3 The amount of a chemical, y (in grams), is related to the time, t (in seconds), for which a particular chemical reaction has been taking place. The table below gives the observations taken during the conduct of an experiment involving this reaction.

Time, t (s)	300	430	600	790	900	1200
Quantity, y (g)	7	9	15	22	23	29

a) Calculate the product moment correlation coefficient, r, for the data and state, with reasons, whether you think that the relationship between y and t is a linear one.
b) Which variable is the factor and which is the response?
c) Calculate the equation of the appropriate regression line, bearing in mind your answer to **b)**.
d) Predict the average amount of chemical produced after:
i) 800 seconds,
ii) 2000 seconds.
Comment on the reliability of each prediction.

4 In an experiment, six randomly-selected English A-level students were given a certain number of lines (x) of Shakespearian verse to memorise; each took a number of minutes (y) to do this.

x	10	15	20	25	30	35
y	12	17	26	37	48	58

a) Plot a scatter diagram of this data and calculate the equation of the regression line of y on x.
b) Interpret the gradient of this line.
c) Use your equation to estimate y when $x = 100$. Comment on your result.
d) Suggest an alternative way of modelling the relationship between x and y.

5 The following class test scores were obtained from a randomly-chosen group of four students.

Student	A	B	C	D
Maths (x)	8	11	16	17
Statistics (y)	12	–	15	18

Before the Statistics mark for student B was mislaid, the least squares regression line of y on x was calculated and found to be $y = 7.04 + 0.574\,x$.
Use this equation to recover the missing mark for student B.

6 The data below refer to the mass, m (in grams), of a plant t days after it was first planted.

No of days, t	5	10	16	22	30	34	40
Mass, m (grams)	2.4	3.2	5.0	7.2	11.5	17.8	25.4

a) Plot the data on a scatter diagram and state with reasons whether you think that m and t are connected by a linear relationship.
b) Calculate the product moment correlation coefficient between m and t. Does this confirm your answer to part **a)**? Give reasons.

It is now suggested that $m = ab^t$ is the appropriate form of the relationship between m and t.

c) **i)** Find an expression for $z = \log m$ in terms of t.
ii) Calculate the equation of the regression line of z on t, and use this to estimate a and b.
d) Draw the graph of the curve $m = ab^t$ using your estimates of a and b on your scatter diagram.
e) Calculate the product moment correlation coefficient between z and t and state whether this confirms the view that the relationship between z and t is more likely to be linear than that between m and t.

7 A medical researcher recorded the level of absorption (y) corresponding to given doses (x) of a specific drug.

x	1	2	3	4	5
y	0.95	0.72	0.62	0.57	0.49

a) Plot a scatter diagram of y against x and calculate the equation of the regression line of y on x in the form $y = a + bx$.
b) The researcher believes that x and y may be related by an equation of the form $\frac{1}{x}+\frac{1}{y}=\frac{1}{k}$, where y is a constant. Plot a scatter diagram of $\frac{1}{y}$ against $\frac{1}{x}$ and calculate the equation of the regression of $\frac{1}{y}$ on $\frac{1}{x}$ in the form $\frac{1}{y} = a + \frac{b}{x}$. Compare the gradient, b with that predicted by the researcher's equation.
c) Compare the two models for the data.

MOVING AVERAGES

An important source of bivariate data is where data are collected or reported upon at regular intervals. The world of business and finance is concerned with such periodical data. Consider the case of a company involved in selling houses. The company has a number of sales outlets located in the major districts of a large city. The directors of the company are reviewing the monthly sales over a three-year period. These are shown in the following tables.

Month	No. of sales	Month	No. of sales	Month	No. of sales
1	15	13	16	25	17
2	14	14	13	26	15
3	13	15	14	27	16
4	14	16	14	28	15
5	13	17	14	29	17
6	14	18	17	30	19
7	17	19	19	31	20
8	15	20	16	32	22
9	18	21	18	33	20
10	19	22	19	34	19
11	16	23	18	35	19
12	15	24	17	36	14

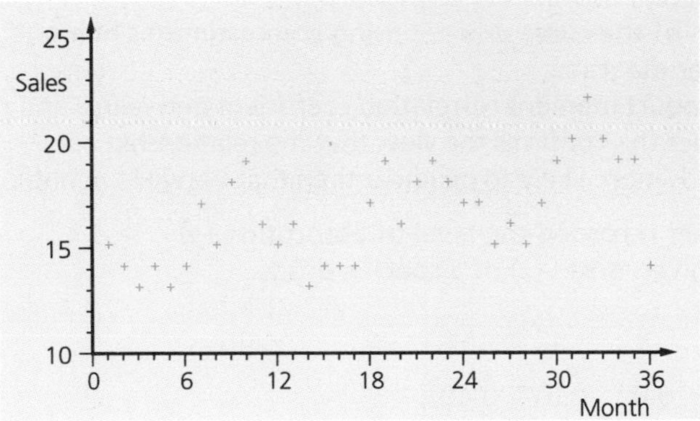

A scattergraph of these data is shown in the diagram.

The company directors want to find out if there is any trend in the data. The scattergraph indicates that there is plenty of variation in the data and hints at a possible cycle of lows and highs. One way of examining this is to use a **moving average**. This involves calculating the average sales for, say, months 1, 2, 3, then calculating the average for months 2, 3, 4, then for months 3, 4, 5, then months 4, 5, 6, and so on. In this way the average moves through all the data until finally the average for months 34, 35, 36 is found. The reason that the average for three months is chosen is simply that this is the span of a 'season' and 'seasonal variation' may well account for the fluctuations in the sales.

Month	3 month MA	Month	3 month MA	Month	3 month MA
1	–	13	14.7	25	16.3
2	14	14	14.3	26	16
3	13.7	15	13.7	27	15.3
4	13.3	16	14	28	16
5	13.7	17	15	29	17
6	14.7	18	16.7	30	18.7
7	15.3	19	17.3	31	20.3
8	16.7	20	17.7	32	20.7
9	17.3	21	17.7	33	20.3
10	17.7	22	18.3	34	19.3
11	16.7	23	18	35	17.3
12	15.7	24	17.3	36	–

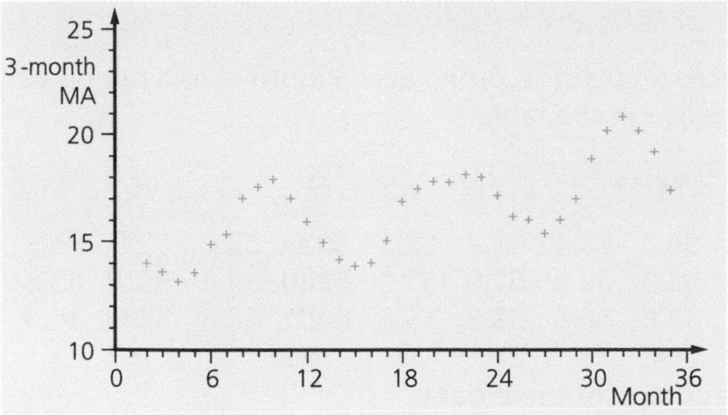

Notice that the first of the moving averages, the one for months 1, 2, 3, is recorded against the middle month. This leads to there being no moving average for month 1 nor indeed for month 36. The resulting scattergraph is shown in this diagram.

This graphical presentation is much smoother than the original data and reveals the highs and lows in the sales.

Companies usually report on an annual basis, so it might be relevant to calculate annual moving averages, i.e. twelve-monthly moving averages. These are done in a similar fashion to the three-monthly averages. The difficulty arises in determining the middle month. The first twelve-month average will be recorded against 'month $6\frac{1}{2}$' since this is the middle of months 1 to 12. The next twelve-month average is then recorded against 'month $7\frac{1}{2}$' and so on.

Month	12 month MA	Month	12 month MA	Month	12 month MA
$6\frac{1}{2}$	15.25	$14\frac{1}{2}$	15.92	$22\frac{1}{2}$	16.75
$7\frac{1}{2}$	15.33	$15\frac{1}{2}$	15.92	$23\frac{1}{2}$	17
$8\frac{1}{2}$	15.25	$16\frac{1}{2}$	15.92	$24\frac{1}{2}$	17.17
$9\frac{1}{2}$	15.33	$17\frac{1}{2}$	16.08	$25\frac{1}{2}$	17.25
$10\frac{1}{2}$	15.33	$18\frac{1}{2}$	16.25	$26\frac{1}{2}$	17.75
$11\frac{1}{2}$	15.42	$19\frac{1}{2}$	16.33	$27\frac{1}{2}$	17.92
$12\frac{1}{2}$	15.67	$20\frac{1}{2}$	16.5	$28\frac{1}{2}$	17.92
$13\frac{1}{2}$	15.83	$21\frac{1}{2}$	16.67	$29\frac{1}{2}$	18
				$30\frac{1}{2}$	17.75

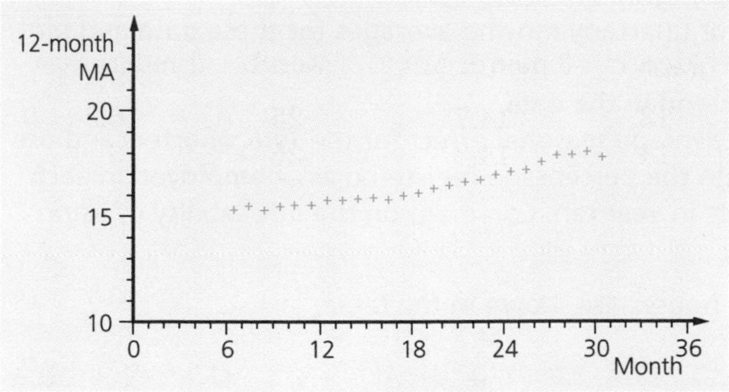

A different picture emerges. Extending the span of the moving average to 12 months has further smoothed the data, to the extent that the highs and lows are no longer apparent. However, there now appears to be a steady increase in the average monthly sales over the three-year period, from 15 sales per month at the beginning to 18 sales per month at the end of the period. If this trend is real, then the next 12-month moving average will also be about 18. This suggests that the total sales for that 12-month period i.e. the months 26 to 37, amount to $12 \times 18 = 216$. However, the total for months 26 to 36 is 196. So, the predicted sales for month 37 is the difference between these i.e. 20 sales. This is a common use of moving averages.

EXERCISES

13.4A

1 The numbers of births (in 000s) recorded each month over a period of three years are recorded in the table.

Month	J	F	M	A	M	J	J	A	S	O	N	D
Year 1	59.9	61.7	59.5	66.4	60.3	63.4	61.3	62.1	60.6	59.6	57.9	55.8
Year 2	56.7	57.6	54.3	63.4	56.7	59.9	57.2	57.2	55.2	53.4	55.2	52.2
Year 3	53.6	55.2	49.4	57.5	53.9	56.6	52.8	55.8	54.7	53.5	53.0	48.5

a) Plot a time series graph of these data.
b) Calculate the values of a 12-month moving average for these data and plot these on your time series graph.
c) What trend is evident in these data? Draw an appropriate trend line through your moving average plot and estimate the number of births in: **i)** January, year 4 **ii)** August, year 4.
d) Calculate quarterly moving averages for the data in the table and plot these on your graph. Describe the pattern of births revealed by the quarterly smoothed data.

2

	Year								
Quarter	1	2	3	4	5	6	7	8	9
1	4.1	4.0	7.0	9.1	10.2	10.5	10.8	11.1	10.9
2	4.1	4.4	7.7	9.3	10.4	10.6	10.9	11.2	10.5
3	4.0	5.0	8.3	9.6	10.6	10.7	10.9	11.2	10.0
4	4.0	5.9	8.8	9.9	10.6	10.9	11.0	11.1	9.5

The table shows the percentage workforce registered as unemployed each quarter for a period of nine years.

a) Plot a time series graph of these data.
b) Determine four quarterly moving averages for these data and plot them on your graph.
c) Describe the trend in the data.
d) Calculate the average seasonal effect for the four quarters and use this to estimate the percentage registered as unemployed in each of the quarters in year ten. Comment on the advisability of using this approach.

3 Monthly sales of houses are shown in the table.

Month	J	F	M	A	M	J	J	A	S	O	N	D
Year 1	118	122	126	121	124	120	114	124	129	102	130	96
Year 2	97	111	112	110	132	115	116	122	105	98	83	77
Year 3	86	89	85	100	111	123						

a) Plot these data on a time series graph.
b) Calculate annual moving average sales figures and plot these on the graph.

c) Describe any trend you observe.
d) Calculate the average annual effect and use this to estimate the monthly home sales for July to December in Year 3.
e) Calculate quarterly moving average sales and plot these on your graph. Describe any seasonal variations revealed.

4 The total annual rainfall over a period of 16 years is shown in the table.

Year	1	2	3	4	5	6	7	8
Rainfall	753	794	925	905	1002	978	999	973
Year	9	10	11	12	13	14	15	16
Rainfall	879	899	885	992	919	923	814	838

a) Plot a graph of these data.
b) Calculate 3-year moving averages for these data and plot these on your graph.
c) Use your moving average data to estimate the rainfall for years 17 and 18.
d) Calculate 7-year moving averages and use these to estimate the rainfall in years 17 and 18.
e) Compare your estimates found in **c)** and **d)**.

EXERCISES

13.4B

1 The following data refer to the cost of a commodity over twelve months.

Month	1	2	3	4	5	6	7	8	9	10	11	12
Cost	6.0	5.9	6.1	6.9	6.5	6.4	6.5	8.1	7.7	7.1	7.2	8.1

a) Plot these data on a scatter diagram.
b) Calculate the 3-monthly moving averages for the given data and plot the moving averages on a separate scattergraph.
c) Draw by eye a line of best fit to represent the trend.
d) Does the price appear to be increasing?
e) Predict the price in month 13.

2 The cost of a commodity in pence over a 30 month period is given in the table below.

Month	Cost(p)	Month	Cost(p)	Month	Cost(p)	Month	Cost(p)	Month	Cost(p)
1	107	7	117	13	123	19	127	25	145
2	105	8	112	14	118	20	134	26	154
3	99	9	114	15	130	21	147	27	150
4	106	10	120	16	141	22	143	28	148
5	108	11	131	17	136	23	139	29	143
6	121	12	125	18	130	24	136	30	157

a) Plot these data on a scattergraph.
b) Calculate 5-monthly moving averages and plot these on the scattergraph.
c) Draw a line of best fit through the 5-monthly moving averages.
d) Predict the cost of the commodity in month 31.

3 Repeat question **2** using 8-monthly averages.

4 The daily humidity over a period of 45 days is shown in the table.

Day	1	2	3	4	5	6	7	8	9	10	11	12	13	14	15
Humidity	64	71	86	78	68	55	59	75	73	86	80	72	66	68	68
Day	16	17	18	19	20	21	22	23	24	25	26	27	28	29	30
Humidity	66	65	73	70	67	62	69	78	75	69	81	74	70	70	71
Day	31	32	33	34	35	36	37	38	39	40	41	42	43	44	45
Humidity	65	72	72	71	73	73	67	70	59	60	59	60	55	79	63

i) **a)** Plot a time series graph of these data.
b) Describe any pattern you are able to detect in the data.
ii) **a)** Calculate 15-day moving averages for the data and plot these on your graph.
b) Describe any trend you perceive in the data.
iii) **a)** Calculate 8-day moving averages and use these to determine if there is a pattern in the data.

CONSOLIDATION EXERCISES FOR CHAPTER 13

1

x	4.3	4.9	9.0	10.2	11.8	17.0	21.2	23.0
y	5.2	5.3	6.8	12.6	15.0	16.5	21.6	27.2

a) Plot a scattergraph of the data.
b) Calculate the value of the product moment correlation coefficient.
c) Explain whether you feel that the variables are linearly related.

2 An experiment is set up which involves suspending various masses from a coiled metal spring. The total length of the spring is measured for each mass. The results are shown in the table.

Mass (x)	5	10	15	20	25	30	40	50
Length (y)	143	150	158	160	163	170	180	193

a) Determine the equation of the line of regression of length on mass in the form:
$y = a + bx$
b) Use your equation to estimate the length of the spring when the mass is 45.
c) What meaning may be attached to the gradient, b?

3

Age	4:5	5:6	6:8	8:7	9:7	10:1	11:1	12:5
Hours	14	14.2	14.4	15.5	15.5	15.8	15.9	18.0

The ages of eight randomly-selected children are recorded (years:months) together with an estimate of the time (in hours) they spent awake the previous day.

a) Calculate the product moment correlation coefficient for these data.
b) Interpret the value you obtain.
c) Plot a graph of the data and review your interpretation.

4

Preservative	0.25	0.35	0.40	0.60	0.65	0.80	0.95	1.00
pH	7.5	7.4	7.3	7.3	7.4	7.6	7.7	7.9

a) Calculate the product moment correlation coefficient for these data and interpret its value.
b) Plot a scattergraph of the data and review your interpretation.
c) Investigate whether (preservative)^2 is a better independent variable.

5

	Year				
Quarter	1	2	3	4	5
1	27	42	36	39	48
2	39	33	42	45	42
3	30	36	36	39	42
4	50	57	66	75	83

The data represents the quarterly income of an expanding company.

a) Plot a time series graph of the data.
b) Calculate 4-quarterly moving averages and plot these on your graph.
c) Describe any trend or variation you perceive in the data.
d) Estimate the income in each quarter in Year 6.

6 The table below shows the scores, x, of ten salespersons who took a test designed to assess their aptitude for the job, and their corresponding sales productivity scores, y.

x	41	35	34	40	33	42	37	42	30	43
y	32	20	35	24	27	28	31	33	26	41

$$\left[\sum x = 377, \sum y = 297, \sum x^2 = 14\,397, \sum y^2 = 9145, \sum xy = 11305\right]$$

a) Plot a scatter diagram of the data, with x on the horizontal axis and y on the vertical axis.
b) Calculate the regression line of y upon x, and use your line to calculate an estimate of y when $x = 31$. Comment on the use of the regression line to calculate this estimate in this case.

c) Draw the regression line on your scatter diagram and give the coordinates of one point which the line must go through. Draw some of the distances which compromise the minimal sum of squares from which the equation of the regression line is derived.
d) Calculate the product moment correlation coefficient for the given data and interpret the result in terms of your scatter diagram.

(UCLES Question 8, Specimen Modular S1, 1992)

7 The accountant of a company monitors the number of items produced per month by the company, together with the total cost of production. The following table shows the data collected for a random sample of twelve months.

Number of items, x (1000s)	21	39	48	24	72	75	15	35	62	81	12	56
Production cost, y (£1000)	40	58	67	45	89	96	37	53	83	102	35	75

a) Plot these data on a scatter diagram. Explain why this diagram would support the fitting of a regression equation of y and x.
b) Find an equation for the regression line of y on x in the form $y = a + bx$.

Use $\sum x^2 = 30\,786$ *and* $\sum xy = 41444$.

The selling price of each item produced is £2.20.
c) Find the level of output at which total income and total costs are equal. Interpret this value.

(ULEAC Question 9, Paper S2, January 1993)

8 A chemist measured the speed, y, of an enzymatic reaction at twelve different concentrations, x, of the substrate and the results are given below.

x	$\frac{1}{2}$	$\frac{1}{3}$	$\frac{1}{4}$	$\frac{1}{6}$	$\frac{1}{7}$	$\frac{1}{8}$	$\frac{1}{9}$	$\frac{1}{10}$	$\frac{1}{11}$	$\frac{1}{12}$	$\frac{1}{13}$
y	0.204	0.218	0.189	0.172	0.142	0.149	0.111	0.125	0.123	0.112	0.096

The chemist thought that the model relating y and x could be of the form

$$y = a + \frac{b}{x}$$

a) Plot a scatter diagram of y against $\frac{1}{x}$.
b) Find the equation of the regression line in the above form, giving coefficients to three significant figures.

Use $\sum\left(\frac{1}{x}\right)^2 = 793$ and $\sum\left(\frac{y}{x}\right) = 11.23$.

c) Find, to two significant figures, the sum of squares of residuals for your equation.

Originally the data included an observation $\left(\frac{1}{5}, 0.269\right)$.

d) Plot this point on your scatter diagram and explain why you think this value has been omitted.

The sum of squares of the residuals of the equation which included the observation $\left(\frac{1}{5}, 0.269\right)$ is 0.0086.

e) Compare this residual sum of squares with the value calculated in **c)** and comment whether the difference is consistent with your answer to **d)**.

(ULEAC Question 8, Specimen paper T2, 1994)

9 A company monitored the number of days (x) of business trips taken by executives of the company and the corresponding claims (£y) they submitted to cover the total expenditure of these trips.

A random sample of ten trips gave the following results.

x (days)	10	3	8	17	5	9	14	16	21	13
y (£)	116	39	85	159	61	94	143	178	225	134

a) Plot these data on a scatter diagram.
Give a reason to support the calculation of a regression line through these points.

b) Find and equation of the regression line of y on x, in the form $y = a + bx$.

Use $\sum x^2 = 1630$ and $\sum xy = 17\,128$.

c) Interpret the slope b and intercept a of your line.

d) Find the expected expenditure of a trip lasting eleven days.

e) State, giving a reason, whether or not you would use the line to find the expected expenditure of a trip lasting two months.

(ULEAC Question 9, Paper S2, January 1993)

10 The yield of a batch process in the chemical industry is known to be approximately linearly related to the temperature, at least over a limited range of temperatures. Two measurements of the yield are made at each of eight temperatures within this range, with the following results.

Temperature, x (°C)	180	190	200	210	220	230	240	250
Yield, y (tonnes)	136.2 136.9	147.5 145.1	153.0 155.9	161.7 167.8	176.6 164.4	194.2 183.0	194.3 175.2	196.5 219.3

Use $\sum x = 1720$ and $\sum x^2 = 374\,000$.

a) Plot all the data on a scatter diagram.

b) For each temperature, calculate the mean of the two yields.
Calculate the equation of the regression line of these mean yields on temperature. Draw this regression line on your scatter diagram.

c) Predict, using your regression line, the yield of a batch at each of the following temperatures.

i) 185°C **ii)** 245°C **iii)** 300°C

Discuss the amount of uncertainty in each of your three predictions.

d) In order to improve predictions of the mean yield at various temperatures in the range 180°C to 250°C, it is decided to take a further eight measurements of yield. Recommend, giving a reason, the temperatures at which these eight measurements should be carried out.

(AEB Question 8, Specimen paper 2, 1994)

11 In an experiment on memory, five groups of people (chosen randomly) were given varying lengths of time to memorise the same list of 40 words. Later, they were asked to recall as many words as they could in t seconds. Show the data in a scatter diagram.

t	20	40	60	80	100
y	12.1	18.5	22.8	24.6	24.0

a) Calculate the equation of the regression line for y on t.
b) Use your regression line to calculate y when $t = 20$. Comment on the result.
c) Use your regression line to calculate y when $t = 160$. Comment on the usefulness or otherwise of this result.
d) Discuss briefly whether the regression line provides a good model or whether there is a better way of modelling the relationship between y and t.

(MEI Question 1 (adapted), Paper S2, June 1993)

12 A random sample of students who are shortly to sit an examination are asked to keep a record of how long they spend revising, in order to investigate whether more revision time is associated with a higher mark. The data are given below, with x hours being the revision time (correct to the nearest $\frac{1}{2}$ hour) and y% being the mark scored in the examination.

x	0	3	4.5	3.5	7	5.5	5	6.5	6	10.5	2
y	36	52	52	57	60	61	63	63	64	70	89

$n = 11 \quad \sum x = 53.5 \quad \sum y = 667 \quad \sum x^2 = 338.25$

$\sum y^2 = 42129 \quad \sum xy = 3366.5$

a) Obtain the value of the product moment correlation coefficient for the data.
b) Specify appropriate null and alternative hypotheses, and carry out a suitable test at the five per cent level of significance.
c) Without further calculation, state the effect of the data $x = 2$, $y = 89$ on the value of the product moment correlation coefficient. Explain whether or not this point should be excluded when carrying out the hypothesis test.

(MEI Question 1, Paper S2, January 1994)

13 The following data relate to trials carried out in a laboratory to examine the relationship between the amount of a certain chemical used in a process and the concentration of the final product.

Amount of chemical, x (g)	22	24	30	32	34	36	40	42	44	46
Concentration, y		1.1	1.6	0.9	1.9	1.5	1.1	1.8	2.4	1.2

$\sum x = 350 \quad \sum y = 15.2 \quad \sum x^2 = 12\,852 \quad \sum xy = 544.8$

a) **i)** Draw a scatter diagram of the data. Your x-axis should run from 0 to 120 and your y-axis from 0 to 8.

ii) Calculate the equation of the regression line of y on x and draw this line on your scatter diagram.

iii) Use the regression equation to estimate the concentration if 80 g of chemical is used.

iv) Give two reasons why the estimate made in **a) iii)** is likely to be unreliable.

A further trial is carried out using 120 g of chemical. The resulting concentration is 7.5.

b) **i)** Add the further information to your scatter diagram.

The regression equation for all eleven points is $y = -0.766 + 0.0662x$.

ii) Draw this line on your scatter diagram.

An estimate is required of the concentration when 80 g of chemical is used.

c) Explain how you would make such an estimate and how confident you would feel about your estimate in each of the following circumstances.

i) There is no time for further trials but it is believed that the relationship over the observed amounts of chemical is approximately linear and the last point was reliable.

ii) There may have been an error in the last observation. There is time to carry out three more trials.

(AEB Question 8E, Paper 9, Winter 1994)

14 A technician monitoring water purity believes that there is a relationship between the hardness of the water and its alkalinity. Over a period of ten days, she recorded the data in this table.

Alkalinity (mg/l)	33.8	29.1	22.8	26.2	31.8	31.9	29.4	26.1	28.0	27.2
Hardness (mg/l)	51.0	45.0	41.3	46.0	48.0	50.0	46.3	45.0	45.3	43.0

a) Plot the data on graph paper with 'Alkalinity' on the horizontal axis. Mark the mean point.

b) The technician decides to calculate the equation for the least squares regression line of hardness on alkalinity. Show that this line has gradient 0.821 and find its equation.

(OUDLE Question 4, Specimen S1, 1994)

15 Explain the use of moving averages when analysing time-series.

The table below shows the number of cubic metres of gas used per quarter in the domestic household of a family with three young children for the years 1990 to 1993.

Year	Gas Used (cubic metres)			
	Jan – March	April – June	July – Sept	Oct – Dec
1990	727	410	105	555
1991	832	348	150	589
1992	854	323	209	696
1993	770	173	289	781

a) Plot these time-series data on a graph.
b) Calculate the values of an appropriate moving average for these data. Plot these on your graph and hence draw in a suitable trend line.
c) Comment on the seasonal variations and the trend displayed in your graph.
d) Estimate the seasonal effect for the quarter Jan – March.
e) Forecast the amount of gas used by this family during Jan – March of 1994.

(NEAB Question 6, AS, June 1994)

16 The quarterly figures for the United Kingdom domestic expenditure on goods and services at market prices from 1989 to mid 1992 are given in the table below.

Year	Quarter	Domestic Expenditure £(millions)	Moving Averages
1989	1	76.7	
	2	80.2	
			82.6
	3	85.1	
			84.1
	4	88.4	
			85.3
1990	1	82.9	
			86.5
	2	85.0	
			87.6
	3	89.6	
			88.3
	4	92.9	
			89.4
1991	1	85.8	
			90.6
	2	89.3	
			92.0
	3	94.5	
			93.2
	4	98.3	
			94.4
1992	1	90.6	
	2	94.2	

Source: Economic Trends, 1993 Annual Supplement

a) Draw, on graph paper, a time-series graph of the quarterly domestic expenditure over this period.

b) Plot the moving average values on your graph and hence draw in a trend line.

c) Describe briefly the trend and seasonal variations shown by your graph.

d) Calculate the average seasonal effect for the fourth quarters and hence obtain an estimate for the domestic expenditure in the fourth quarter in 1992.
Suggest one reason for the seasonal effect apparent in the fourth quarter of each year.

(NEAB Question 10, Specimen paper 8)

Summary

- The covariance of x and y:

$$\text{cov}(x,y) = s_{xy} = \frac{1}{n}\sum xy - \bar{x}\bar{y}$$

- Pearson's product moment correlation coefficient:

$$r = \frac{Cov(x,y)}{\{sd(x)sd(y)\}} = \frac{s_{xy}}{s_x s_y}$$

- The equation of the line of regression of y on x:

$$\hat{y} - \bar{y} = \frac{s_{xy}}{s_x^{\ 2}}(x - \bar{x})$$

- The equation of the line of regression of x on y:

$$\hat{x} - \bar{x} = \frac{s_{xy}}{s_y^{\ 2}}(y - \bar{y})$$

- The minimum sum of squared residuals is:
  $n(1 - r^2)\, s_y^{\ 2}$

- Moving averages are used in:
  exploring seasonal or cyclical variation
  revealing trends in time-series.

STATISTICS

14 The normal probability model

In this chapter:

- *we introduce a widely-discussed probability model – the classical bell-shaped normal distribution,*
- *we discover how all normally-distributed variables are characterised by their mean and variance,*
- *we shall use standardisation and the distribution function for the standardised normal variate,*
- *we discover properties of linear combinations of normal variates,*
- *we use $\frac{(O-E)^2}{E}$ to measure the fit of a normal model for data.*

THE NORMAL PROBABILITY DENSITY FUNCTION

On a field work week, a group of students explored coastal footpaths.

Unit = 0.10 cm

1	1E	8
1	2O	
1	2T	
2	2F	5
6	2S	6 7 7 7
12	2E	8 8 8 8 9 9
22	3O	0 0 0 0 0 0 0 1 1 1
32	3T	2 2 2 2 2 2 2 2 3 3
46	3F	4 4 4 4 4 4 4 5 5 5 5 5 5 5
58	3S	6 6 6 6 6 6 6 6 7 7 7 7
80	3E	8 8 8 8 8 8 8 8 8 8 8 8 9 9 9 9 9 9 9 9 9 9
105	4O	0 0 0 0 0 0 0 0 0 0 0 0 0 0 1 1 1 1 1 1 1 1 1 1 1
133	4T	2 2 2 2 2 2 2 2 2 2 2 2 2 2 2 2 2 2 2 2 3 3 3 3 3 3 3 3
(30)	4F	4 4 4 4 4 4 4 4 4 4 4 4 4 4 4 5 5 5 5 5 5 5 5 5 5 5 5 5 5 5
137	4S	6 6 6 6 6 6 6 6 6 6 6 6 6 6 6 6 7 7 7 7 7 7 7 7 7 7 7 7 7 7 7
107	4E	8 8 8 8 8 8 8 8 8 8 8 8 8 8 8 9 9 9 9 9 9 9 9 9 9 9
81	5O	0 0 0 0 0 0 0 0 0 0 0 1 1 1 1 1 1 1 1 1
61	5T	2 2 2 2 2 2 2 2 2 2 2 2 3 3 3 3 3 3 3
42	5F	4 4 4 4 4 4 4 5 5 5 5 5 5
29	5S	6 6 6 6 7 7 7 7 7
20	5E	8 8 8 9 9 9 9 9
12	6O	0 0 0 1 1
7	6T	3 3 3 3
3	6F	4
2	6S	6
1	6E	8

One activity involved them in measuring random samples of leaves on natural hedgerows. The site had been chosen for the diversity of natural flora. In this diversity were examples of honeysuckle hedging. The group measured a total of 300 honeysuckle leaves. Their results are shown in the ordered five-part stem-and-leaf display on page 394.

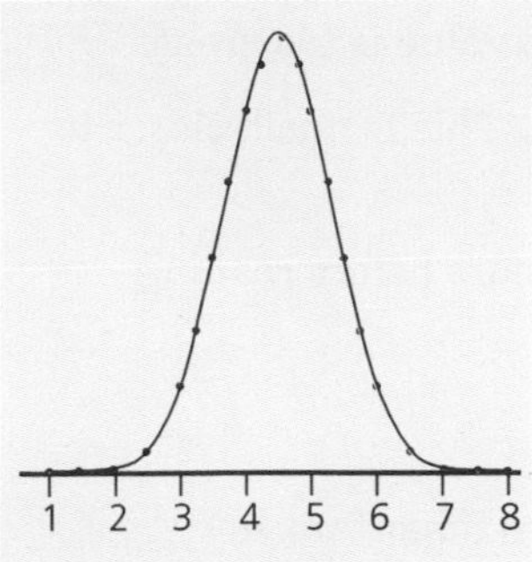

The distribution of lengths is symmetric with a peak in the middle and with tails at both ends. The **normal distribution** is a model with these characteristics. The **density function** for the normal model, with the same mean (4.50) and standard deviation (0.84) as the collection of honeysuckle leaf lengths, is shown in this diagram.

We can compare the data and the model by putting them into similar form. The **relative frequency histogram** from Chapter 1, *Data*, is appropriate. The table below provides the information needed to construct the histogram shown next to it. Remind yourself how relative frequency densities are calculated.

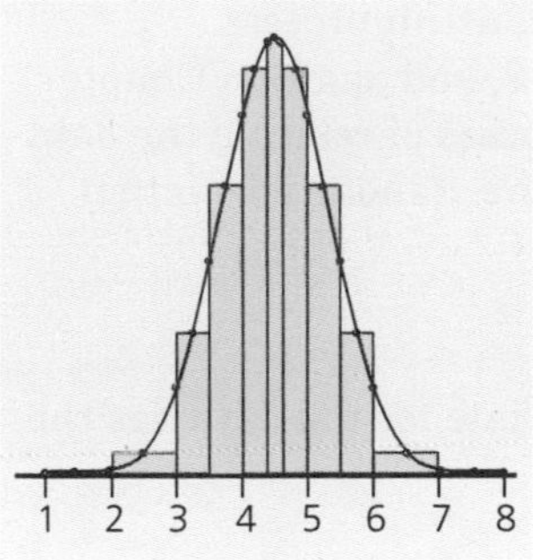

Length, l cm	Class width	Frequency	Relative frequency	Relative frequency density
$1 \le l < 2$	1	1	1/300	0.0033
$2 \le l < 3$	1	11	11/300	0.0367
$3 \le l < 3.5$	0.5	27	27/300	0.18
$3.5 \le l < 4$	0.5	41	41/300	0.2733
$4 \le l < 4.4$	0.4	53	53/300	0.4417
$4.4 \le l < 4.6$	0.2	30	30/300	0.5
$4.6 \le l < 5$	0.4	56	56/300	0.4667
$5 \le l < 5.5$	0.5	46	46/300	0.3067
$5.5 \le l < 6$	0.5	23	23/300	0.1533
$6 \le l < 7$	1	12	12/300	0.04

Is the curve a reasonable model for the distribution of lengths of honeysuckle leaves? Clearly the curve is not in any way an exact fit for the histogram. However, the curve does seem to follow the pattern of the histogram. Try the following Calculator activity, to find out something about the **normal curve**.

CALCULATOR ACTIVITY

The normal curve

The normal curve is symmetric. It peaks on its axis of symmetry and it tails off to zero. The curve is asymptotic to the x-axis in both directions. An example of a function which possesses these properties is:

$f(x) = e^{-x^2}$

a) Try drawing the graph $y = e^{-x^2}$. Describe what you find. You should find that the curve has the desired properties, but its axis of symmetry is $x = 0$.

b) You need to translate the graph so that its axis of symmetry is $x = 4.5$. Try replacing x by $(x - 4.5)$ and drawing a graph of the resulting function. You should now have a curve which begins to resemble the one above, but its maximum value is too great.

c) Try drawing the graph of $y = 0.5e^{-(x-4.5)^2}$. This is much closer to the desired curve.

d) You might like to explore the curve still more before reading further.

Interpreting the results

Both the mean and the standard deviation are important in determining the appropriate normal curve to model the relative frequency density. Remember the **standardisation** process introduced in Chapter 3, *Summary statistics 2*, and again in Chapter 13, *Bivariate data*. Standardisation is the process of relating the data to the mean and expressing this in terms of the standard deviation. The **standardised value** of x is:

$$\left(\frac{x - 4.5}{0.84}\right)$$

The normal curve **density function** appropriate for the data has the form:

$$f(x) = \frac{1}{0.84\sqrt{2\pi}} e^{-0.5\left(\frac{x-4.5}{0.84}\right)^2}$$

Return to your calculator to see what difference this makes to the appearance of the curve.

Comparing data and model

Once the mean and standard deviation have been identified, the normal model is uniquely defined. This model has a number of properties related to these parameters. For instance, approximately five per cent of the normal model's variates are more than two standard deviations away from the mean i.e. approximately five per cent are either more than 6.18 ($4.50 + 2 \times 0.84$) or less than 2.82 ($4.50 - 2 \times 0.84$). What proportion of the actual honeysuckle leaf lengths lies in these regions?

It is possible to obtain an estimate of this proportion from the relative frequency histogram using the technique we met in Chapter 1, *Data*. The proportion of the data in that histogram is shown in the table.

Data	Proportion
more than 6.18	$0.04 \times (7 - 6.18) \approx 0.03$
less than 2.82	$0.0033 + 0.0367\ (2.82 - 2) \approx 0.03$

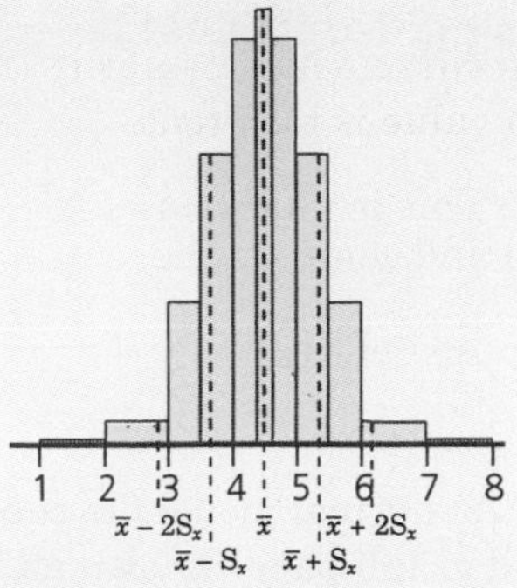

This suggests that about six per cent of the leaf lengths are more than two standard deviations away from the mean. The normal model is quite close to this. The proportion is shown in this relative frequency histogram.

Another property is that about 68 per cent of the normal model's variate is within one standard deviation of the mean. This implies that the proportion of the model's variates lying between 3.66 and 5.34 is about 68 per cent. From the histogram the proportion of the data between these limits is as shown in the table.

Data	Proportion
3.66 to 4.0	$0.2733(4 - 3.66)$
4.0 to 4.4	$0.4417(4.4 - 4)$
4.4 to 4.6	$0.5(4.6 - 4.4)$
4.6 to 5.0	$0.4667(5 - 4.6)$
5.0 to 5.34	$0.3067(5.34 - 5)$
	total 66%

So, again the model is reasonably close to this estimate of the proportion of the data.

Exploration 14.1 *Measuring pebbles*

Another activity during the fieldwork week involved taking samples at various parts of the beach at the coastal location. One of the samples had 168 pebbles and the length of each pebble was measured. They recorded their result in the table shown below.

Length l mm Number of pebbles, f	$l < 4$ 0	$4 \le l < 4.5$ 3	$4.5 \le l < 5.5$ 5	$5.5 \le l < 6$ 14	$6 \le l < 6.5$ 18
l mm f	$6.5 \le l < 7$ 12	$7 \le l < 7.5$ 17	$7.5 \le l < 8$ 19	$8 \le l < 8.5$ 18	
l mm f	$8.5 \le l < 9$ 17	$9 \le l < 10$ 21	$10 \le l < 11$ 12	$11 \le l < 12$ 7	$12 \le l < 13$ 4
l mm f	$13 \le l < 14$ 1	$l \ge 14$ 0			

The mean length of pebble was 8.02 mm and the standard deviation was 1.82 mm.

- Draw a relative frequency density histogram for these data.
- Use the histogram to estimate the proportion of the pebbles with lengths more than two standard deviations from the mean.
- Estimate the proportion with lengths within one standard deviation of the mean.

The appropriate normal probability model has the density function:

$$f(x) = \frac{1}{1.82\sqrt{2\pi}} e^{-0.5\left(\frac{x-8.02}{1.82}\right)^2}$$

Draw this density function on your histogram and compare the model's proportions with the data's.

Normal probability paper

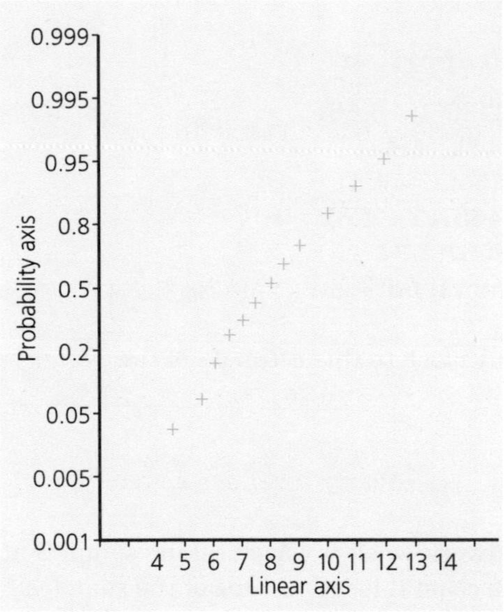

It is possible to judge how well the normal model fits the data by using specially-prepared graph paper known as **normal probability paper**. The x-axis of the paper has a perfectly standard linear scale. The y-axis is anything but linear. It is designed so that a **cumulative proportion** plot of a normal variate produces a straight line graph. This is a plot of the pebble data, using normal probability paper.

We need to take care reading the y-scale because of its non-linear nature. Note that it is not possible to plot the final point. The way to interpret this plot is simple. If the plot seems to be a straight line then there is reason to believe that the normal distribution is a reasonable fit.

Try plotting the honeysuckle data on normal probability graph paper and interpret the outcome.

EXERCISES

14.1 A

1 The data records the speeds, in miles per hour, of 1018 cars entering a speed check area on a main road.

Speed, v mph Number of cars, f	$30 < v \le 40$ 16	$40 < v \le 50$ 106	$50 < v \le 60$ 232
v f	$60 < v \le 65$ 138	$65 < v \le 70$ 152	$70 < v \le 75$ 132
v f	$75 < v \le 80$ 104	$80 < v \le 90$ 98	$90 < v \le 110$ 30

a) Illustrate these data in a histogram.

b) Calculate an estimate of:

i) the mean speed **ii)** the standard deviation of the speed.

c) Mark on your histogram the values corresponding to:
   i) the mean, ii) the mean ± standard deviation,
   iii) the mean ± two standard deviations.

d) From your histogram estimate the proportion of speeds which:
   i) are more than two standard deviations away from the mean,
   ii) lie within one standard deviation of the mean,
   iii) lie between one and two standard deviations from the mean.

e) Compare the proportions in **d)** with those of a normal distribution.

f) Write down the density functions for a normal model with the same mean and standard deviation as yours.

2 a) Using the data in question **1**, construct a cumulative proportion table.

b) Plot the results in **a)** on normal probability paper.

c) Does your plot support the view that the data may be modelled using a normal distribution?

d) Use your plot to estimate:
   i) the mean, ii) the standard deviation of the speeds.

e) Use your plot to estimate the proportion of cars with speed greater more than 77 mph.

f) Use your plot to estimate the speed exceeded by 75 per cent of the cars.

3 a) Construct a relative frequency density histogram for the data in question **1**.

b) Superimpose the density function from your answer to question **1 f)** onto the relative frequency density histogram.

c) Compare the curve and the histogram.

4 The data represent the monthly sales income, in £000s, of a random selection of 150 estate agents.

Sales, x Number, f	$80 \le x < 100$ 6	$100 \le x < 110$ 17	$110 \le x < 120$ 21	$120 \le x < 125$ 12
x f	$125 \le x < 130$ 12	$130 \le x < 140$ 20	$140 \le x < 150$ 20	$150 \le x < 160$ 15
x f	$160 \le x < 180$ 15	$180 \le x < 220$ 12		

a) Represent these data in a histogram.

b) Calculate estimates of the mean and variance of x.

c) Determine the proportion of sales:
   i) more than two standard deviations from the mean,
   ii) within one standard deviation of the mean,
   iii) between one and two standard deviations from the mean.

d) Compare the proportions in **c)** with those of an appropriate normal model.

e) Superimpose on your histogram the frequency density function for the normal model with mean equal to 137.5 and standard deviation equal to 27.5. Compare your model and the data.

5 **a)** Plot the sales data in question **4** on normal probability paper.
b) Use your plot to estimate the mean and standard deviation of the sales.
c) Use your plot to estimate:
i) the proportion of sales more than £175 000,
ii) the monthly sales exceeded by 90 per cent of the estate agents.
d) Do you feel that a normal model is appropriate for these data?

EXERCISES

14.1 B

1 The heights (to the nearest cm) to which a particular species of plant grow are recorded for a sample of 200 such plants. The results are given in the table below.

Height (cm)	Frequency
1	9
2	9
3	13
4	21
5	32
6	36
7	23
8	41
9–15	16

Calculate an estimate of the mean and the standard deviation of the data. Draw a relative frequency histogram of the data and use it to estimate the proportion of heights more than two standard deviations away from the mean. Compare this with what you would expect from a normal distribution.

2 The salaries of a group of 200 graduates were recorded when the graduates first found employment after leaving university. The cumulative frequency distribution below shows the results.

Salary (£)	Cumulative frequency
≤ 12 500	7
≤ 13 000	35
≤ 13 500	102
≤ 14 000	164
≤ 14 500	197
≤ 15 000	200

Plot the data on normal probability paper and show that it is reasonable to assume that the data are normally distributed. Use your graph to estimate the mean and standard deviation of the data.

3 The following data refer to the masses of a group of 200 adult males.

Mass (m) in kg	Frequency
$50 \leq m < 55$	4
$55 \leq m < 60$	28
$60 \leq m < 65$	47
$65 \leq m < 70$	41
$70 \leq m < 75$	46
$75 \leq m < 80$	26
$80 \leq m < 85$	8

a) Estimate the mean and standard deviation of these data and plot a relative frequency density histogram.

b) Plot the data on normal probability paper and use this diagram to get a second estimate of the mean and standard deviation of the data.

4 For normally distributed data, about 16 per cent of the variate is more than one standard deviation below the mean, and about 16 per cent is more than one standard deviation above the mean.

Check this statement with the discussion above, which states that about 68 per cent of the normal model's variate is within one standard deviation of the mean.

Using the graph on page 398, use the 50th percentile on the cumulative proportion scale to estimate the mean length of pebbles. Then use the 84th percentiles to estimate the standard deviation of the lengths, which is given by their difference divided by 2.

Compare these estimates with the actual sample mean length of 8.02 mm and sample standard deviation of 1.82 mm.

5 Repeat question **4** for the honeysuckle data.

CALCULATING NORMAL PROBABILITIES

The normal curve is a model for the density function of a continuous variate. Recall from Chapter 11, *Continuous random variables*, that calculating probabilities for these variates involves integration of the density function. What is required is the value of a definite integral. This can be done using a calculator.

Some calculators have a definite integral facility built in, i.e. the calculator will produce the value of:

$$\int_a^b \mathrm{f}(x)\,\mathrm{d}x$$

The function, $\mathrm{f}(x)$, is the appropriate density function:

$$\mathrm{f}(x) = \frac{1}{\sigma\sqrt{2\pi}}\mathrm{e}^{-\frac{1}{2}\left(\frac{x-\mu}{\sigma}\right)^2}$$

and μ and σ are the mean and standard deviation of the normal variate. The limits a and b are the values of the variate. The definite integral then produces, to a reasonable degree of accuracy, the value of the probability:

$$P(a < x < b)$$

that the variate lies between the limits a and b.

If you have a calculator with this facility, try the next Calculator activity, part **a)**. If, on the other hand, you have a programmable calculator you could have a go at part **b)**.

CALCULATOR ACTIVITY

Finding probabilities

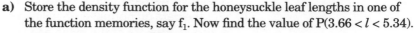

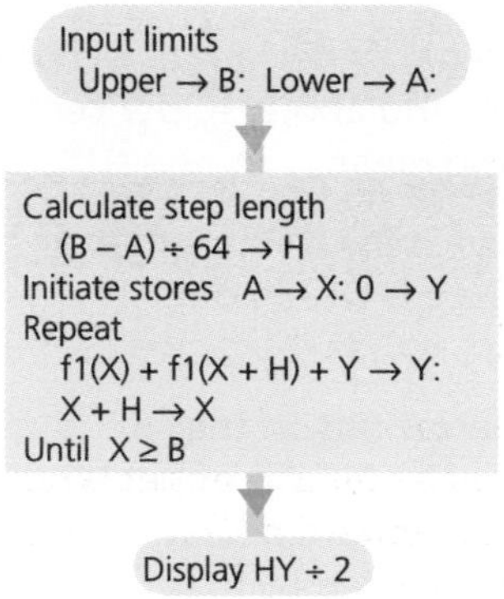
Input limits
Upper → B: Lower → A:
Calculate step length
(B – A) ÷ 64 → H
Initiate stores A → X: 0 → Y
Repeat
f1(X) + f1(X + H) + Y → Y:
X + H → X
Until X ≥ B
Display HY ÷ 2

a) Store the density function for the honeysuckle leaf lengths in one of the function memories, say f_1. Now find the value of $P(3.66 < l < 5.34)$.

$$P(3.66 < l < 5.34) = \int_{3.66}^{5.34} f_1(x)\,dx$$

using 2^6 steps if neccessary. You should find that you get a value of about 0.683. Now have a go at finding other probabilities associated with:

i) honeysuckle lengths, **ii)** pebble lengths.

b) This flowchart represents a program which attempts to duplicate the definite integration process in **a)** above. The flowchart assumes that:

$$f_1(x) = \frac{1}{0.84\sqrt{2\pi}} e^{-\frac{1}{2}\left(\frac{x-4.5}{0.84}\right)^2}$$

Once you have stored the program, try the same activity as in **a)**.

Tables of the distribution function of N(0, 1)

The standardised value of a random variable is important when using the normal probability model. The process of standardisation allows direct comparison of an observation from one distribution with one from another distribution. It also allows probabilities to be estimated for any normal model from the standardised normal variate.

Suppose that X is a random variable which is normally distributed with mean μ and variance σ^2. The notation for this is generally:

$$X \sim N(\mu, \sigma^2)$$

The process of standardising an observation, x, of X produces the value, z, where:

$$z = \frac{x - \mu}{\sigma}$$

The random variable, Z, defined in this way is the standardised normal variate with mean equal to zero and variance equal to unity i.e.

$$Z \sim N(0, 1)$$

The density function for Z is usually given the symbol $\phi(z)$ {ϕ is the lower case Greek f and is pronounced phi}. The form for $\phi(z)$ is:

$$\phi(z) = \frac{1}{\sqrt{2\pi}} e^{-\frac{1}{2}z^2} \quad \text{where } -\infty < z < \infty$$

Knowledge of the probability for a range of values of Z allows us to write down the probability for the corresponding range of another normal variate. For instance:

$$P(-1 < Z < 2) \approx 0.8186$$

implies that, in the normal model:

$$P(6.20 < \text{pebble length} < 11.66) \approx 0.8186$$

This is because the standardised values are:

$$2 = \frac{\text{pebble length} - 8.02}{1.82}$$

and:

$$-1 = \frac{\text{pebble length} - 8.02}{1.82}$$

Rearranging these produces the given values. Have a go at showing that:

$$P(3.66 < \text{honeysuckle length} < 6.18) \approx 0.8186$$

for the model of the distribution of lengths of honeysuckle leaves.

The real value of this correspondence between all normal variates and the Z variate lies in the tabulation of probabilities for Z. The table on page 518 shows the commulative probability function (i.e. the **distribution function**), $\Phi(z)$. Φ is the upper case Greek letter phi.

$$\Phi(z) = P(Z < z)$$

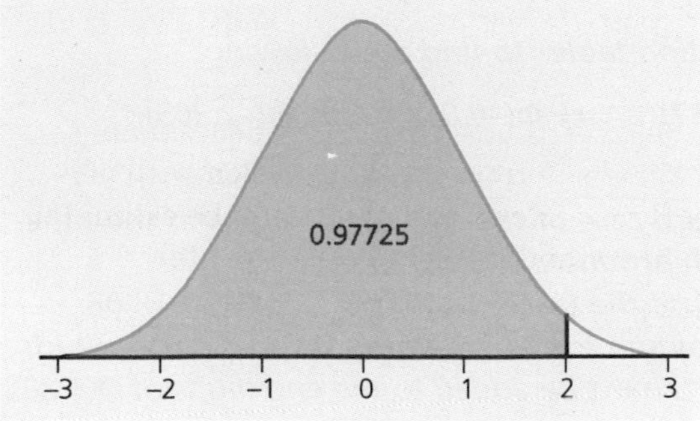

The tables give values for $Z \geq 0$. How does the symmetry of the density function make it possible to find $\Phi(z)$ for $Z < 0$?

Have a look in the tables to find $\Phi(2)$. You should find that:

$$\Phi(2) = P(Z < 2) = 0.977\,25$$

This is the area under the curve for the density function of the standard normal probability model shown here.

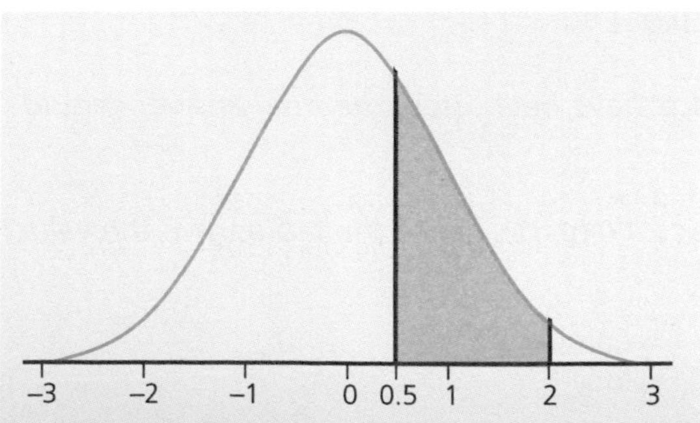

To find the probability:

$$P(0.5 < Z < 2)$$

we need to find the difference:

$$P(Z < 2) - P(Z < 0.5)$$

i.e.

$$\Phi(2) - \Phi(0.5)$$

Reading directly from the tables, this is:

$$0.977\,25 - 0.691\,46 \approx 0.2858$$

However, to find:

$$P(-1 < Z < 2) = \Phi(2) - \Phi(-1)$$

we need to make use of the symmetry of the density function to evaluate $\Phi(-1)$. This is not available directly from the tables but can be found from $\Phi(+1)$.

This diagram helps to illustrate their general relationship.

$$\Phi(-a) = 1 - \Phi(a)$$

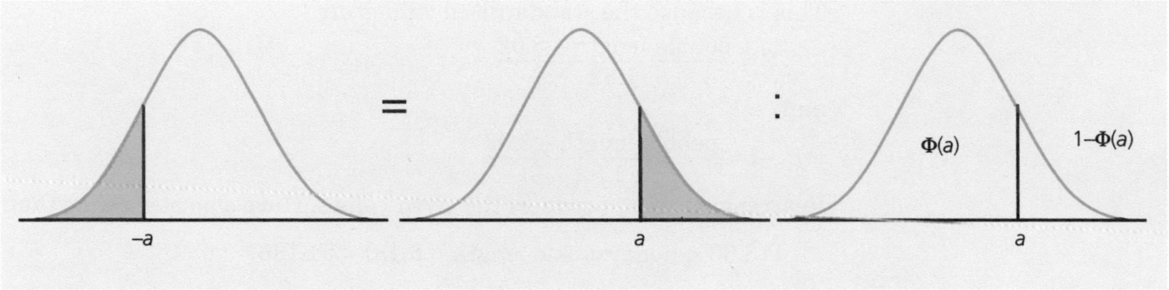

This results in:

$$\begin{aligned} P(-1 < Z < 2) &= \Phi(2) - \Phi(-1) \\ &= 0.977\,25 - (1 - \Phi(1)) \\ &= 0.977\,25 - (1 - 0.841\,34) \\ &\approx 0.8186 \end{aligned}$$

Example 14.1

Use the distribution function tables to find the following.

a) *$\Phi(0.88)$* ***b)*** *$\Phi(1.367)$* ***c)*** *$\Phi(-0.81)$* ***d)*** *$\Phi(-2.305)$*

Solution

a) *This is a straightforward case of the row for* $Z = 0.8$ *and then the column for 0.08 which produces* $\Phi(0.88) = 0.810\,57$.

b) *To find* $\Phi(1.367)$, *we use the tables to find* $\Phi(1.36) = 0.913\,08$. *Then, we go to the column for 7 in the proportional parts, which shows 112. These digits must be added to the end digits of* $\Phi(1.36)$. *This is equivalent to saying:*

$$\begin{aligned} \Phi(1.367) &= 0.913\,08 + 0.001\,12 \\ &= 0.914\,20 \end{aligned}$$

Since proportional parts have been used, the final answer should be given as:

$$\Phi(1.367) \approx 0.9142$$

c) $\Phi(-0.81)$ *is the same as* $1 - \Phi(0.81)$, *hence the tables give the value:*

$$1 - 0.79103 = 0.20897$$

d) $\Phi(-2.305) = 1 - \Phi(2.305)$
$= 1 - (0.989\,28 + 0.000\,13)$ *using the proportional parts '13'*
$= 0.010\,59$
≈ 0.0106

Example 14.2

Calculate the following probabilities, given that Z is a standardised normal variate.

a) $P(Z > 0.85)$ ***b)*** $P(0.083 < Z < 0.945)$ ***c)*** $P(-1.62 < Z < 2.35)$
d) $P(Z > -1.65)$ ***e)*** $P(|Z| < 1.375)$

Solution

a) *The concept of complementary events implies:*
$P(Z > 0.85) = 1 - P(Z \le 0.85)$
Since Z is a continuous variate, there is no distinction between $P(Z \le 0.85)$ *and* $P(Z < 0.85)$. *This suggests that:*
$$P(Z > 0.85) = 1 - P(Z < 0.85) = 1 - \Phi(0.85) = 0.197\,66$$

b) $$P(0.083 < Z < 0.945) = \Phi(0.945) - \Phi(0.083) = (0.826\,39 + 0.001\,27) - (0.531\,88 + 0.001\,19) \approx 0.2946$$

c) $$P(-1.62 < Z < 2.35) = \Phi(2.35) - \Phi(-1.62) = 0.990\,61 - (1 - \Phi(1.62)) = 0.990\,61 - (1 - 0.947\,38) = 0.937\,99 \approx 0.9380$$

d)

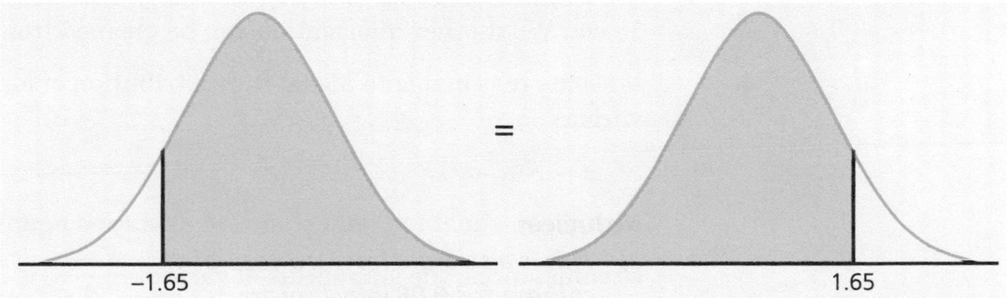

$P(Z > -1.65)$ *has the same value as* $P(Z < +1.65)$ *since the curve is symmetric. This leads to the straightforward result:*
$$P(Z > -1.65) = \Phi(1.65) = 0.950\,53$$

e) *This requires interpretation of* $|Z| < 1.375$. *This modulus in an equality is equivalent to the double inequality:*
$-1.375 < Z < 1.375$
Hence:
$$P(-1.375 < Z < 1.375) = \Phi(1.375) - \Phi(-1.375) = \Phi(1.375) - (1 - \Phi(1.375)) = 2\Phi(1.375) - 1 = 2(0.914\,66 + 0.000\,80) - 1 \approx 0.8309$$

Example 14.3

a) *Given that X is a random variable with distribution modelled by the normal, $N(5, 1.44)$, estimate the probability that an observation of X is less than 6.8.*

b) *If $X \sim N(8, 0.04)$, calculate $P(X > 8.4)$.*

Solution

a) *The notation, $N(5, 1.44)$, says that the mean of X is 5 and that the variance is 1.44. Hence the standard deviation is 1.2. Standardised values of X are distributed as $N(0, 1)$. The standardised value of 6.8 is $\frac{6.8-5}{1.2} = 1.5$, hence the probability:*

$$\begin{aligned} P(X < 6.8) &= P(Z < 1.5) \\ &= \Phi(1.5) = 0.933\ 19 \end{aligned}$$

b) *Since $X \sim N(8, 0.04)$, standardising $\frac{X-8}{\sqrt{0.04}} \sim N(0, 1)$ produces a standard normal variate.*

$$\begin{aligned} P(X > 8.4) &= P\left(Z > \frac{8.4-8}{0.2}\right) \\ &= P(Z > 2) \\ &= 1 - \Phi(2) \\ &= 0.022\ 75 \end{aligned}$$

Tables of percentage points of N(0, 1)

The students enjoying their field work week measured lengths of the coastal flower known as Sea Pinks. Tragically, they lost the data collected apart from two pieces of information. Eighty-five per cent of the lengths were less than 20 cm. Thirty-nine per cent were more than 18 cm. What other information can be gleaned from these two bits?

It seems reasonable to model the distribution of lengths as a normal variate:

$$X \sim N(\mu, \sigma^2)$$

with mean equal to μ and standard deviation equal to σ. Under these circumstances, the standardised variate $\frac{X-\mu}{\sigma}$ is $N(0, 1)$.

Hence, the two pieces of information available are equivalent to:

$$P(X < 20) = P\left(Z < \frac{20-\mu}{\sigma}\right) = 0.85$$

$$P(X > 18) = P\left(Z > \frac{18-\mu}{\sigma}\right) = 0.39$$

The first of these can be written in terms of the distribution function as:

$$\Phi\left(\frac{20-\mu}{\sigma}\right) \doteq 0.85$$

hence

$\frac{20-\mu}{\sigma}$ must be equal to 1.036 since $\Phi(1.036) = 0.85$.

The second relationship, concerning $P\left(Z > \frac{18-\mu}{\sigma}\right)$ can, with care, be written in terms of the distribution function:

$$1 - \Phi\left(\frac{18-\mu}{\sigma}\right) = 0.39$$

hence:

$$\Phi\left(\frac{18-\mu}{\sigma}\right) = 0.61$$

The 'percentage points' tables indicate:

$$\Phi(0.279) = 0.61$$

So this indicates that $\frac{18-\mu}{\sigma}$ is equal to 0.279.

So far the information has yielded two equations involving the unknown mean and standard deviation:

$$\frac{20-\mu}{\sigma} = 1.036 \quad \text{and} \quad \frac{18-\mu}{\sigma} = 0.279$$

These can be rewritten as linear equations:

$$20 - \mu = 1.036\sigma \quad \text{and} \quad 18 - \mu = 0.279\sigma$$

Subtracting the second equation from the first eliminates μ and produces:

$$(20 - \mu) - (18 - \mu) = 1.036\sigma - 0.279\sigma$$
$$\Rightarrow 2 = 0.757\sigma$$

From this the standard deviation can be estimated as:

$$2 \div 0.757 \approx 2.64$$

Similarly, we can find that $\mu \approx 17.3$.

Example 14.4

Information of a normal variate, X, indicates that:
$P(X < 6.3) = 0.995$
and
$P(X < 4.9) = 0.75$
Use this information to obtain estimates of the mean and standard deviation of X.

Solution

Assume that μ and σ are the required mean and standard deviation. Standardising the values 6.3 and 4.9 it is possible to write the information in terms of $\Phi(Z)$:

$$\Phi\left(\frac{6.3-\mu}{\sigma}\right) = 0.995 \quad \textit{and}$$
$$\Phi\left(\frac{4.9-\mu}{\sigma}\right) = 0.75$$

The 'percentage points' table indicates:
$\Phi(2.576) = 0.995$ *and* $\Phi(0.674) = 0.75$

Hence, two equations result:

$$\frac{6.3-\mu}{\sigma}=2.576 \quad \textit{and} \quad \frac{4.9-\mu}{\sigma}=0.674$$

These can be solved to produce:

$\sigma \approx 0.736$ *and* $\mu \approx 4.40$

EXERCISES

14.2 A

1 Use a calculator to determine the following.

a) $P(4 < x < 5)$ when $X \sim N(3, 4)$
b) $P(25.6 < x < 32.8)$ when $X \sim N(28, 25)$
c) $P(18.7 \le x \le 19.6)$ when $X \sim N(20, 0.8^2)$
d) $P(1 < x \le 2.2)$ when $X \sim N(0, 1)$

2 Use tables of the standard normal variate to estimate the following.

a) $P(|Z| > 1.23)$ **b)** $P(0.153 < Z < 1.47)$ **c)** $P(-0.34 < Z < 1.04)$
d) $P(Z < 1.645)$ **e)** $P(|Z-1| < 1.5)$ **f)** $P(Z > 1.854)$

3 Use tables to estimate the value of a standard normal variate when:

a) $\Phi(Z) = 0.945$ **b)** $\Phi(-Z) = 0.372$
c) $\Phi(Z-1) = 0.75$ **d)** $\Phi(1-Z) = 0.05$

4 Use tables to estimate the following.

a) $P(4.8 < x < 5.79)$ given $X \sim N(4, 1.21)$
b) $P(3.65 < x < 6.28)$ given $X \sim N(5, 1.44)$
c) $P(|x-3| < 0.45)$ given $X \sim N(2, 2)$
d) $P(|2-x| < 1.5)$ given $X \sim N(1.75, 0.25)$

5 Assume that $X \sim N(1.52, 0.442)$.

a) Calculate:
i) $P(x > 0.8)$ **ii)** $P(0.75 < x < 2.25)$.
b) Determine the value of a if $P(x < a) = 0.99$.
c) Calculate the quartiles of X, and hence the interquartile range.

6 Calculate the mean and standard deviation of each of the following normally distributed variates given that:

a) $P(X < 2.60) = 0.739$ and $P(X > 1.30) = 0.897$
b) $P(Y < 9.49) = 0.005$ and $P(Y > 9.52) = 0.0435$
c) $P(W > 80.5) = 88\%$ and $P(W < 92.5) = 34\%$

EXERCISES

14.2 B

1 Jars of coffee are stated to contain 200 g. In fact, there is a legal requirement that at least 95 per cent of jars must contain not less than the printed weight.

a) If the weight of coffee is normally distributed, with $\sigma = 0.4$ g, find the minimum setting for the mean weight which will conform with this requirement.

b) When the mean is at this setting, calculate the interquartile range of the weights.

2 Make the assumption that the IQ score of a randomly-chosen person is normally distributed with mean 100 and standard deviation 15.

a) A person is described as 'gifted' if their IQ is over 140. Calculate the percentage of the population who are gifted.
b) What is the IQ score below which 85 per cent of the population would be expected to be?
c) State any properties of the normal distribution which you think may not be realistic assumptions to make about IQ scores.

3 Examination marks are normally distributed with a mean of 45 per cent and a standard deviation of 18 per cent. The pass mark is 40 per cent.

a) If 'A' grades are awarded for marks of 70 per cent and over what proportion of candidates obtain an A grade?
b) What is the probability that a randomly-chosen candidate with an A grade achieved over 75 per cent in the examination?
c) What is the probability that a randomly-chosen candidate who failed the examination achieved less than 30 per cent?
d) What proportion of candidates who pass the examination obtain an A grade?

4 The time taken to complete an application form is a normally distributed random variable with $\mu = 17.2$ minutes and $\sigma = 3.6$ minutes.

a) Find the time within which 90 per cent of applicants will have completed the form.
b) If five applicants are chosen at random, find the probability that exactly three of them take over 20 minutes to complete the form.
c) If six applicants are chosen at random, find the probability that at least half of them take at least 15 minutes to complete the form.

5 Given that $X \sim \mathrm{N}(6, 25)$:

a) use tables to calculate $\mathrm{P}(X > 11)$
b) calculate $\mathrm{P}(5 \leq X \leq 7)$
c) find a value t such that $\mathrm{P}(X < t) = 0.85$
d) find a value s such that $\mathrm{P}(X \leq s) = 0.77$.

6 Jars of marmalade have a stated mass of 454 g. An inspector wishes to investigate this and so he weighs 300 jars and finds 255 of them weigh over 450 g and 200 weigh less than 455 g.

a) Assume that the mass of marmalade may be modelled as a normal variate. Calculate an estimate of the true mean mass and the standard deviation of the mass.
b) What conclusion should the inspector draw from this experiment?

LINEAR COMBINATIONS OF NORMAL VARIATES

The students spend one day of their field work collecting and measuring shells of small molluscs. They focused particularly on empty limpet shells and cockle shells. The data they collected led them to model the diameters of the shells as independent normal variates with the following parameters.

	Mean (mm)	Variance
Limpet, L	22	64
Cockle, C	18	36

How might we estimate the likelihood that the first cockle shell measured had a greater diameter than the first limpet shell?

This requires us to estimate the probability:

$P(C > L)$

The difficulty here is that, we have two random variables. It can be transformed into a problem involving just one random variable:

$P(L - C < 0)$

or $P(D < 0)$

if a new variable, D, is defined as:

$D = L - C$

Then the results concerning linear combinations can be used.

Mean: $\mu_D = \mu_L - \mu_C$

Variance: $\sigma_D{}^2 = \sigma_L{}^2 + \sigma_C{}^2$

Linear combinations of normal variates are themselves normally distributed. This is the case for the difference between the lengths.

$D \sim N(22 - 18, 8^2 + 6^2)$

i.e.

$D \sim N(4, 100)$

So the problem is to find the probability:

$P(D < 0)$

Standardising this produces:

$$P\left(Z < \frac{-0.4}{\sqrt{100}}\right) = 1 - \Phi(0.4)$$

$$= 0.3446$$

It seems reasonable to conclude that there is a chance of between 34 and 35 per cent that the first cockle shell will be longer than the first limpet shell.

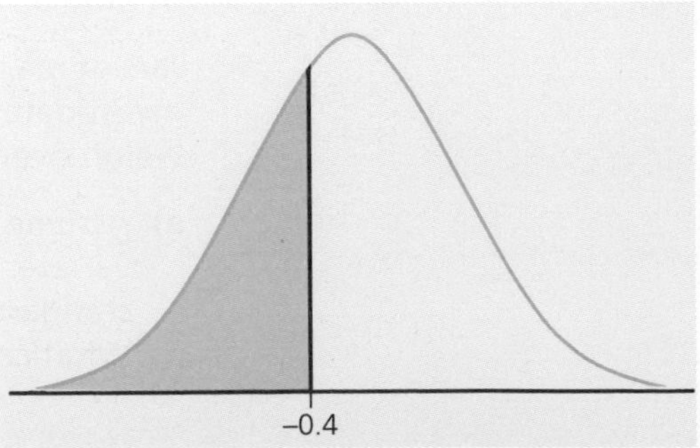

Example 14.5

The random variables X *and* Y *are normally distributed.*
$X \sim N(5, 4)$ *and* $Y \sim N(6, 4.84)$
Calculate these probabilities.
a) $P(Y > X)$ ***b)*** $P(3X - 2Y < 1)$

Solution

a) *We have to convert the problem so that it concerns only one random variable. One approach is to define* W *as* $Y - X$*. Then* W *is a linear combination of normal variates and so it is normally distributed. Hence:*
$W \sim N(6 - 5, 4 + 4.84) = N(1, 8.84)$
and the probability to find is $P(W > 0)$*. Standardising this value of* W*:*

$$P(W > 0) = P\left(Z > \frac{0 - 1}{\sqrt{8.84}}\right) = P(Z > -0.336)$$

Using the symmetry of the Z distribution:
$P(Z > -0.336) = \Phi(0.336) = 0.632$
Hence the required probability, $P(Y > Z)$*, is approximately 0.632.*

b) *Again we need to identify a suitable random variable.*
Let $V = 3X - 2Y$*, so the mean of* V *is* $3 \times 5 - 2 \times 6 = 3$ *and its variance is* $3^2 \times 4 + 2^2 \times 4.84 = 55.36$*. Hence,* $V \sim N(3, 55.36)$ *and the required probability is* $P(V < 1)$*.*

$$P(V < 1) = P\left(Z < \frac{1 - 3}{\sqrt{55.36}}\right) = P(Z < -0.269)$$
$$= 1 - \Phi(0.269)$$
$$\approx 0.394$$

Testing the fit of a normal model

One of the students had collected 127 observations of the difference, $L - C$, between the limpet and cockle shell diameters. She presented the data in the frequency table below. She wanted to test, at the five per cent level, the hypothesis that the data were consistent with the normal distribution N(4, 100). How do you think she could have done this test?

Difference, d mm	Frequency
$d < -10$	7
$-10 \le d < -5$	11
$-5 \le d < 0$	23
$0 \le d < 5$	29
$5 \le d < 10$	25
$10 \le d < 15$	15
$15 \le d < 20$	11
$d \ge 20$	6

She decided to use the $\sum \frac{(O-E)^2}{E}$ measure of fit. She set up her hypotheses:

H_0: data are consistent with N(4, 100); H_1: this model does not fit.

Then she set about the task of calculating the expected frequencies, E. Some of her calculations are:

1 $$P(D < -10) = P\left(Z < \frac{-10-4}{\sqrt{100}}\right) = 1 - \Phi(+1.4) = 0.080\ 76$$

$\Rightarrow$ expected frequency for $d < -10$ is $127 \times 0.080\ 76 \approx 10.26$.

2 $P(-10 < D < -5) = P(D < -5) - P(D < -10)$

$= 1 - \Phi(0.9) - 0.080\ 76 = 0.103\ 30$

$\Rightarrow$ expected frequency for $-10 < d < -5$ is 13.12.

She continued in this way to produce the following table.

O	7	11	23	29	25	15	11	6
E	10.26	13.12	20.39	24.80	23.61	17.60	10.27	6.95

From this she calculated the value of her measure of fit to be 3.071. She realised that the number of degrees of freedom of the appropriate chi-squared statistic was 7 and she found the five per cent upper-tail critical value of 14.067. So, she concluded that the data were consistent with coming from N(4, 100).

EXERCISES

14.3 A

A bar of chocolate is wrapped in foil and paper. Assume that the mass of the bar, the foil and the paper are independent normal variates. The mean and standard deviations of the masses are given in the table.

	Mean (g)	Standard deviation (g)
bar	28.7	0.03
foil	1.8	0.025
paper	2.5	0.08

a) Determine the mean and variance of the mass of a wrapped bar.

b) Estimate the proportion of wrapped bars where the total mass is more than 33.2 g.

c) The manufacturer displays a 'minimum mass' on the outer wrapper. Fewer than 0.1 per cent of wrapped bars should weigh less than this. What minimum do you advise the manufacturer to display?

2 A hoist is capable of lifting a load not exceeding 300 kg. It is required to raise a mixture of boxes and barrels. The weight of a box and the weight of a barrel are assumed to be normally distributed with the parameters shown in the table.

	Mean (kg)	Standard deviation (kg)
box	12	0.5
barrel	18	0.8

a) Calculate the probability that the capacity of the hoist will be exceeded when loaded with:
i) 25 boxes **ii)** 17 barrels **iii)** 5 barrels and 17 boxes.
b) Calculate an estimate of the probability that the total weight of six boxes is at least 1 kg more than the total weight of four barrels.

3 Assume that $X \sim \mathrm{N}(5, 9)$ and $Y \sim \mathrm{N}(8, 13)$.

a) What are the mean and standard deviation of the distributions of:
i) $X + Y$ **ii)** $Y - X$ **iii)** $4X - 3Y$?
b) Calculate the following probabilities.
i) $\mathrm{P}(X + Y > 20)$ **ii)** $\mathrm{P}(Y < X)$
iii) $\mathrm{P}(Y - X > 6)$ **iv)** $\mathrm{P}(4X < 3Y - 4)$

4 Assume that the distribution of the weight of boys may be modelled using N(58, 60) and the weight of girls using N(52, 35). Estimate the probability that a randomly-chosen girl weights less than a randomly-chosen boy.

5 The speeds of 500 vehicles passing through a notorious accident spot on a highway are shown in cumulative form in the table.

Maximum speed of vehicle, V	30	40	50	55	60	65	70	75	80	90	100
Number of vehicles travelling more slowly than this	2	8	60	175	245	320	335	440	490	498	500

a) A model of the distribution of speeds is proposed which suggests that the speeds are normally distributed with a mean of 60 and a standard deviation of 10.
i) Determine the frequencies expected by this model in the classes:
$v < 40,\ 40 \le v < 50,\ 50 \le v < 55,\ 55 \le v < 60,\ 60 \le v < 65,$
$65 \le v < 70,\ 70 \le v < 80,\ v \ge 80.$
ii) Obtain the observed frequencies in each of the classes.
iii) Carry out a chi-squared goodness-of-fit test to determine if this is a suitable model.
b) Plot the data from the table on normal probability paper. Interpret the results of your plot and compare with the results of the chi-squared test.

EXERCISES

14.3 B

1 If $X \sim N(2, 25)$ and $Y \sim N(3, 16)$ and X and Y are independent, calculate the following probabilities.

a) $P(X + Y) < 7)$ b) $P(2X - 3Y > -10)$
c) $P(X > Y)$ d) $P(|X - Y| < 5)$

2 In a penalty shoot-out competition, the time, T, to complete the taking of a penalty is normally distributed with mean 35 seconds and standard deviation four seconds. Ten penalties are taken, one after another, with no pause in between.

a) Calculate the probabilities that:
   i) half the penalties will be completed within three minutes,
   ii) the total time for all ten penalties is more than five minutes and less than six minutes.

b) A TV executive televising this competition needs to know how long the competition will take. Find the total time for the ten penalties that it is 99 per cent certain a given competition will not exceed.

3 A journey to school consists of a walk followed by a bus ride and then another walk. The length of each part of the journey is approximately normally distributed.

	Mean (minutes)	Standard deviation (minutes)
First walk	8	1.2
Bus journey (including wait)	32	8.5
Second walk	4	0.7

a) Find the standard deviation of the total length of the journey to school.
b) Find the probability that the journey on a randomly-chosen school day takes longer than 46 minutes.
c) Given that such a journey to school does take longer than 46 minutes, find the probability it takes longer than 50 minutes.

4 To make a fruit drink, the juice from three oranges is mixed with the juice from one lemon.

The volume of juice from one orange is normally distributed with $\mu = 15$ ml, $\sigma = 1$ ml.

The volume of juice from one lemon is normally distributed with $\mu = 8$ ml, $\sigma = 0.6$ ml.

a) Find the probability that the volume of such a drink exceeds 50 ml.
b) Find the probability that the juice of four randomly-chosen lemons has a larger volume than that of two randomly-chosen oranges.
c) In what proportion of the fruit drinks does lemon juice form at least 15 per cent of the total volume?

5 The heights of 50 people, in cm, is given in the table below.

Height (h) (cm)	Frequency
$150 \le h < 155$	1
$155 \le h < 160$	3
$160 \le h < 165$	8
$165 \le h < 170$	12
$170 \le h < 175$	13
$175 \le h < 180$	7
$180 \le h < 185$	5
$185 \le h < 190$	1

a) Test the hypothesis that the data is normally distributed with mean 170 cm and standard deviation 8.
b) Calculate the actual mean and standard deviation of the data.
c) Test the hypothesis that the data is normally distributed.

CONSOLIDATION EXERCISES FOR CHAPTER 14

1 Oranges are graded according to their weight as in the table.

	Large	Medium	Small
Weight, w	$w > 125$	$100 < w \le 125$	$w \le 100$

Assume that the weight of oranges may be modelled using a normal distribution with mean 112 and standard deviation 10. Estimate the proportion of oranges in each grade.

2 Assume that the life of a watch battery may be modelled using a normal distribution with mean 2 years and standard deviation 0.25 year.

a) The manufacturer quotes life expectancy of 18 months. What proportion of batteries may be expected to expire before the quoted time?
b) Consumer protection legislation is introduced in which manufacturers are required to quote a life expectancy which should be exceeded by 99.5 per cent of the batteries. What should this be?

3 A mechanical digger is used to excavate a trench to take ducting to carry cable. One scoop of the digger removes a quantity of soil and stones which may be modelled as a normal distribution. The mean of this distribution is 55 kg and the standard deviation is 10 kg.

a) Estimate the proportion of scoops which contain more than 60 kg.
b) A dumper truck can hold 500 kg of soil and stones. What is the probability that the truck can hold the contents of ten scoops?

4 Ball bearings emerge from a production line. The diameter of these is modelled as a random variable following a normal distribution. Given that 99 per cent of the production have diameters in excess of 4.90 mm

and that 2.5 per cent are more than 5.1 mm in diameter, determine the mean and standard deviation of the model.

5 **a)** Criticise the statement, 'A battery has a life expectancy which is normally distributed.'
b) What properties would you need to identify in a random variable in order to model its distribution using the normal density function?

6

Class	$x \le 5$	$5 < x \le 7$	$7 < x \le 9$	$9 < x \le 10$
Frequency, f	5	17	38	26
Class	$10 < x \le 11$	$11 < x \le 12$	$12 < x \le 14$	$14 < x \le 16$
f	29	29	49	32
Class	$16 < x \le 18$	$x > 18$		
f	16	9		

a) Plot the data on normal probability paper and interpret your result.
b) Assume that the distribution of the data may be modelled using the random variable $X \sim \mathrm{N}(12, 12)$. Determine the frequencies expected by this model in the same classes as in the table.
c) Use the chi-squared goodness of fit test to determine whether this is an approximate model for these data.

7 The UK operates a 'first past the post' electoral system. A cluster sample of 135 constituencies is selected. A measure of the closeness of the election in each constituencies is given in the table.
a) Calculate an estimate of the mean assuming that the lower bound of the '$x < 0.1$' class is zero.

Measure, x	$x < 0.1$	$x < 0.2$	$x < 0.3$	$x < 0.4$
Cumulative number of constituencies	23	31	50	73
x	$x < 0.5$	$x < 0.6$	$x < 0.7$	$x < 0.8$
Cumulative number of constituencies	94	117	128	132
x	$x < 0.9$	$x < 1.0$		
Cumulative number of constituencies	135	135		

b) Use your mean to obtain an estimate of the standard deviation of the measure.
c) Plot the data on normal probability paper and use the plot to estimate mean and standard deviation. Compare the results with your calculations.
d) To what extent does your plot support the hypothesis that this measure is normally distributed?

e) Use the chi-squared test to judge the hypothesis that N(0.35, 0.04) is an appropriate model.

8 The random variables W, X and Y are distributed independently normally with the same mean, μ, and the same variance, σ^2. Describe as fully as possible the distributions of:

a) $W + X + Y$ b) $2W$ c) $W - X$

d) $2W + X$ e) $W + 2$ f) $X - \mu$

g) $\dfrac{X - \mu}{\sigma}$ h) $\frac{1}{2}Y$ i) $2W - X - Y$.

9 A machine produces rubber balls with diameters that are normally distributed with mean 5.50 cm and standard deviation 0.08 cm. Calculate the proportion of balls which have diameters:

a) less than 5.60 cm

b) between 5.34 and 5.44 cm.

(AEB Question 2, Specimen paper 2, 1994)

10 A machine cuts a very long plastic tube into short tubes. The length of the short tubes is modelled by a normal distribution with mean m cm and standard deviation 0.25 cm. The value of m can be set by adjusting the machine. Find the value of m for which the probability is 0.1 that the length of a short tube, picked at random, is less than 6.50 cm. The machine is adjusted so that $m = 6.40$, the standard deviation remaining unchanged. Find the probability that a tube picked at random is between 6.30 cm and 6.60 cm long.

(UCLES Question 5, Specimen, Module S1, 1994)

11 Eggs are graded according to length L mm. A size 3 egg is one for which $60 < L < 65$. An egg producer finds that 37 per cent of her eggs are larger than size 3 while 21 per cent are smaller. Assume that L is normally distributed.

a) Show the information given on a sketch of the distribution of L.

b) Write down two equations involving the mean and standard deviation and solve them.

Further investigation shows that **i)** $L > 70$ for about seven per cent of eggs, and **ii)** $L < 55$ for less than one per cent of eggs.

c) Determine whether **i)** is consistent with the distribution found in **b)**.

d) Determine whether **ii)** is consistent with the distribution found in **b)**.

(MEI Question 4, Paper S2, January 1994)

12 The speeds of cars passing a certain point on a motorway can be taken to be normally distributed. Observations show that of cars passing the point, 95 per cent are travelling at less than 85 mph and ten per cent are travelling at less than 55 mph.

a) Find the average speed of the cars passing the point.

b) Find the proportion of cars that travel at more than 70 mph.

(ULEAC Question 5, Paper S1, June 93)

13 The maximum load a lift can carry is 450 kg. The weights of men are normally distributed with mean 60 kg and standard deviation 10 kg. The weights of women are normally distributed with mean 55 kg and standard deviation 5 kg. Find the probability that the lift will be

overloaded by five men and two women, if their weights are independent.

(ULEAC Question 3, Specimen, T1, 1994)

14 A certain type of light bulb has a lifetime which is normally distributed with mean 1100 hours and standard deviation 80 hours.

a) Find the probability that a randomly chosen bulb will last at least 1000 hours.

b) Find the lifetime which is exceeded by 95 per cent of bulbs.

A newly-installed light fitting takes six of these bulbs. The lifetimes of the bulbs are independent of one another.

c) Show that the probability that the light fitting can run for 1000 hours without any of the bulbs failing is a little over 0.5.

d) The probability that the light fitting can run for t hours without any of the bulbs failing is 0.95. Find t.

(MEI Question 4, Paper S2, Summer 1993)

15 **a)** The normal random variables X_1, X_2 and X_3 are independently and identically distributed, each with mean μ and variance σ^2. State the distribution of:

i) $X_1 + X_2 + X_3$ **ii)** $X_2 - X_3$ **iii)** $2X_1$.

b) The time, in seconds, taken by an electrician to complete the final wiring of a 13 amp socket is normally distributed with mean 175 seconds and standard deviation 20 seconds. He completes the wiring of three independent sockets in a particular room.

i) Calculate the probability that his total time to complete the wiring of all three sockets is less than ten minutes.

ii) Calculate the probability that he takes more than 30 seconds longer to complete the wiring of the second socket than he does to complete the wiring of the third socket.

iii) What is the probability that, in completing the wiring of the first socket, he takes more than 30 seconds longer than his mean time taken to complete the wiring of all three sockets?

(NEAB Question 9, Specimen paper 9, 1994)

16 **a)** The heights of foxglove plants growing in *woodland* are known to be normally distributed with mean 27.5 cm and standard deviation 3.5 cm. Calculate the probability that a randomly-chosen woodland foxglove is more than 35 cm high.

b) On *riverbanks*, the heights of foxgloves are distributed normally with mean 32.0 cm and standard deviation 4 cm. Calculate the probability that a randomly-chosen riverbank foxglove is less than 35 cm high.

c) Calculate the probability that a randomly-selected woodland foxglove is taller than a randomly selected riverbank foxglove.

(Oxford Question 6, Paper 3, 1995)

17 A food manufacturer delivers tins of baked beans packed in cardboard boxes, each box containing 24 tins. The mass, F, of a full tin of baked beans is a normally distributed random variable having

mean 510 g and standard deviation 2 g.
The mass, B, of an empty cardboard box is a normally distributed random variable having mean 150 g and standard deviation 1.5 g.
Find, to three decimal places:

a) the probability that the mass of a randomly chosen full tin will lie between 505 g and 515 g,

b) the probability that the total mass of a randomly chosen box of tins will exceed 12 400 g.

(NEAB Question 11, Specimen paper 4, 1994)

Summary

The *normal probability distribution* is a distribution used to model many symmetrically distributed continuous variates where a bell-shaped density function is appropriate.

- $X \sim \mathrm{N}(\mu, \sigma^2) \Rightarrow \mathrm{f}(x) = \dfrac{1}{\sigma\sqrt{2\pi}} \mathrm{e}^{-\frac{1}{2}\left(\frac{x-\mu}{\sigma}\right)^2}$
- $Z \sim \mathrm{N}(0, 1) \Rightarrow \phi(z) = \dfrac{1}{\sqrt{2\pi}} \mathrm{e}^{-\frac{1}{2}z^2}$
- X may be standardised as a Z-variate using the linear transformation:
  $$Z = \frac{X - \mu}{\sigma}$$
- The distribution function for Z is known as $\Phi(z)$.
- A linear graph on **normal probability paper** of the cumulative proportion of a random variable indicates the appropriateness of a normal model.
- Linear combinations of normal variates are normally distributed:

  if $X \sim N\left(\mu_x, \sigma_x^{\,2}\right)$ *and* $Y \sim \mathrm{N}\left(\mu_y, \sigma_y^{\,2}\right)$,

  then $aX + bY \sim \mathrm{N}\left(a\mu_x + b\mu_y, a^2\sigma_x^{\,2} + b^2\sigma_y^{\,2}\right)$

STATISTICS

15

Approximating distributions

In this chapter:

- *we discover how certain well-tabulated probability models may be used to approximate the probabilities of other models.*

POISSON APPROXIMATION TO BINOMIAL

The **binomial distribution** provides an appropriate model when we are counting the number of successes in a fixed number of Bernoulli trials. In Chapter 7, *The binomial probability model*, we discovered the **probability function** for $X \sim \mathrm{B}(n, p)$:

$$\mathrm{P}(X = r) = {}_n\mathrm{C}_r \, p^r (1-p)^{n-r}$$

and the recurrence relation:

$$\mathrm{P}(X = r) = \frac{n+r-1}{r} \times \frac{p}{1-p} \times \mathrm{P}(X = r-1)$$

Recall also the **Poisson distribution** we discovered in Chapter 9, *The Poisson distribution*. This is an appropriate model we can use when we are counting the number of random events in a fixed interval. The probability for $\mathrm{X} \sim \mathrm{Poi}(\lambda)$ is:

$$\mathrm{P}(X = r) = \frac{e^{-\lambda}\lambda^r}{r!}$$

and the recurrence formula is:

$$\mathrm{P}(X = r) = \frac{\lambda}{r} \times \mathrm{P}(X = r-1)$$

Consider the binomial distribution: $X \sim \mathrm{B}(50, 0.03)$.

What is the mean of X? The probabilities, P(0), P(1), P(2) are shown below.

	B(50, 0.03)	Poi(1.5)
X	P(X)	P(X)
0	0.218 065	0.223 130
1	0.337 214	0.334 695
2	0.255 518	0.251 021

Try calculating $\mathrm{P}(X = r)$ for $r = 3, 4, \ldots, 9$.

The mean of X is $50 \times 0.03 = 1.5$. The probabilities for a Poisson variable with mean equal to 1.5 are also shown in the table above.

Try calculating $\mathrm{P}(X = r)$ for $r = 3, 4, \ldots, 9$ for the Poisson variable.

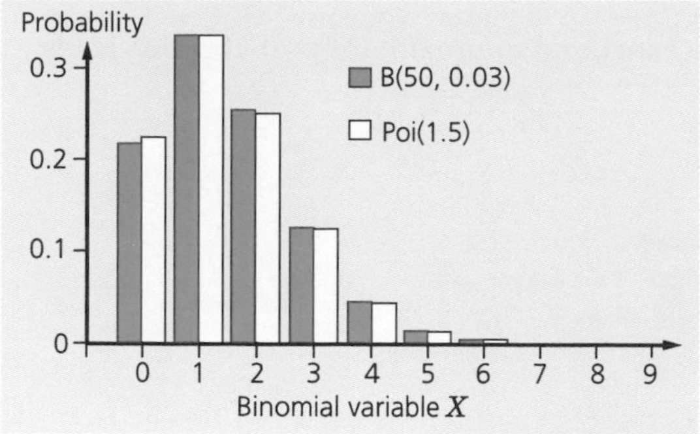

A graphical comparison of the two sets of probabilities is shown in this diagram.

Do you feel that the Poisson probabilities are approximately the same as the binomial probabilities? A numerical comparison of the probabilities displayed in the diagram is shown in the following table, where the **absolute difference** is the numerical difference between the values.

	B(50, 0.03)	Poi(1.5)	
X	P(X)	P(X)	Absolute difference
0	0.218 065	0.223 130	0.005 065
1	0.337 214	0.334 695	0.002 519
2	0.255 518	0.251 021	0.004 497
3	0.126 442	0.125 511	0.000 931
4	0.045 949	0.047 067	0.001 117
5	0.013 074	0.014 12	0.001 046
6	0.003 033	0.003 530	0.000 497
7	0.000 590	0.000 756	0.000 167
8	0.000 098	0.000 142	0.000 044
9	0.000 014	0.000 024	0.000 009

One way of evaluating the quality of the approximation is to find the individual absolute differences in the binomial and Poisson probabilities, then to sum these absolute differences. If this is done when using Poi(1.5) to approximate B(50, 0.03), then the total obtained is 0.0159 (to four decimal places). This is a small value but remember that we are dealing with probabilities. Since absolute differences have been used, this total is larger than any error in the calculation of B(50, 0.03) probabilities using Poi(1.5).

The maximum error in approximating an individual probability is less than 0.005. This occurs in using $e^{-1.5}$ in place of $(0.97)^{50}$ for P($X = 0$).

Clearly, it is possible to approximate binomial probabilities using the Poisson distribution – but which binomial distributions?

If you have access to a programmable calculator you might like to try the following calculator activity. If you have a spreadsheet available, you could carry out a similar comparison between the B(n, p) and Poi(np) probabilities.

CALCULATOR ACTIVITY

Poi(1.5) $\approx$ B(n, $\frac{1.5}{n}$)

The purpose of this activity is to give you a feeling for the binomial distributions which may reasonably be approximated by Poi(1.5). The program calculates P($X = r$) for both B(n, p) and Poi(1.5) until the absolute difference in the exact and approximate probabilities is less than 10^{-6}.

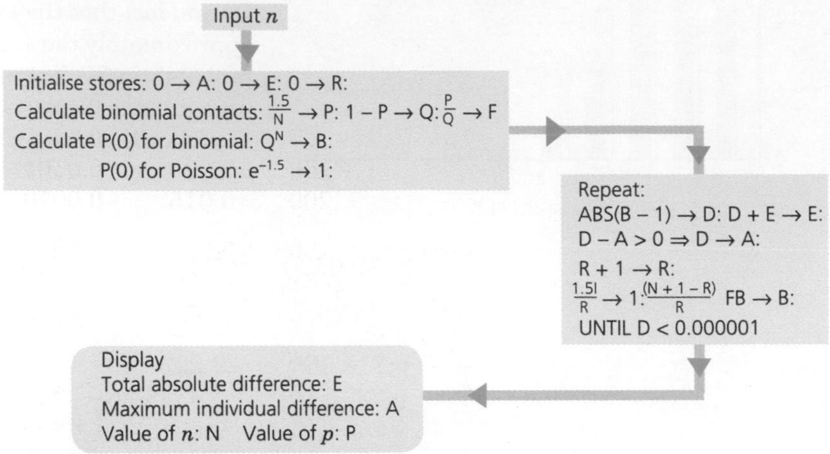

Input n

Initialise stores: $0 \to A$: $0 \to E$: $0 \to R$:
Calculate binomial contacts: $\frac{1.5}{N} \to P$: $1 - P \to Q$: $\frac{P}{Q} \to F$
Calculate P(0) for binomial: $Q^N \to B$:
P(0) for Poisson: $e^{-1.5} \to 1$:

Repeat:
ABS(B – 1) \to D: D + E \to E:
D – A > 0 \Rightarrow D \to A:
R + 1 \to R:
$\frac{1.5I}{R} \to 1 \cdot \frac{(N+1-R)}{R}$ FB \to B:
UNTIL D < 0.000001

Display
Total absolute difference: E
Maximum individual difference: A
Value of n: N Value of p: P

Interpreting the results

The sort of results that you might obtain are shown in the following table.

n	p	Total absolute difference	Maximum difference
10	0.15	0.0839	0.0262
20	0.075	0.0405	0.0128
75	0.02	0.0106	0.0034
100	0.015	0.0079	0.0025
500	0.003	0.0016	0.0005

These results suggest that the larger the value of n and, consequently, the smaller the value of p, the closer the Poisson probabilities are.

How could you adapt the program shown above to find out if other Poisson distributions can be used to approximate binomial probabilities? One approach would be to input the mean and then replace 1.5 by the input mean.

CALCULATOR ACTIVITY

$\text{Poi}(\lambda) \approx \text{B}(n, \frac{\lambda}{n})$

Now try to find out how well appropriate Poisson distributions can be used to approximate binomial distributions.

Interpreting the results

The sort of results that you might obtain are summarised in the following table.

	Binomial		Total ABS difference	Maximum difference
	n	p		
Poi(3)	10	0.3	0.1728	0.0428
	50	0.06	0.0308	0.0076
	100	0.03	0.0152	0.0034
	150	0.02	0.0101	0.0023
	200	0.015	0.0076	0.0017
Poi(5)	100	0.04	0.0205	0.0040
	250	0.02	0.0099	0.0018
	1000	0.005	0.0025	0.0004
Poi(10)	100	0.1	0.0517	0.0068
	250	0.04	0.0200	0.0026
	500	0.02	0.0099	0.0013
	1000	0.01	0.0049	0.0006

It is increasingly evident that the value of p plays a crucial part in determining if Poi(np) probabilities are going to be a close approximation to B(n, p) probabilities. The binomial parameter, p, must be very small. How small it is depends to some extent on whether we want to approximate individual probabilities, e.g. $P(X = 4)$, or whether we want to approximate an accumulation of probabilities, e.g. $P(X > 4)$ or $P(2 < X \leq 8)$.

When $p \leq 0.02$ the approximations are generally very good, regardless of the event. When $0.02 < p \leq 0.05$, approximations for individual probabilities are generally reliable but accumulations of probabilities need to be viewed with caution. We should avoid approximating binomial probabilities, other than for individual events when $p > 0.05$.

Rare events

When we discussed the Poisson distribution in Chapter 9, *The Poisson distribution*, it was used as a model for the number of random events occurring in a continuous medium. We now know that it can be used to model binomial variables when p is very small. Such situations are often called **rare events**. This leads to the application of the Poisson distribution in modelling the occurrence of rare events.

Example 15.1

A school has 1200 students. On average, three students per year receive national recognition for their sporting achievements. Let X be the number of students receiving some recognition in one particular year.

a) *Obtain estimates for:*
i) $P(X = 0)$ *ii)* $P(X \leq 5)$.

b) *One year there were seven students given national recognition, is there evidence that the average number of nationally-recognised students is on the increase?*

Solution

a) *It may be reasonable to model the distribution of X as Poi(3) since this appears to be a 'rare event'. Under this assumption:*

i) $P(X = 0) = e^{-3} \approx 0.0498$

ii) $P(X \le 5) = 0.9161$ *(from tables)*

b) *Appropriate hypotheses are:* H_0: mean = 3; H_1: mean > 3.

Then from tables for Poi(3):

$P(X \ge 7) = 1 - P(X \le 6) = 0.0335$

which is smaller than a five per cent level of significance.

Hence, there is reason to reject H_0 *in favour of* H_1.

EXERCISES

15.1 A

1 A telesales representative makes 40 calls per evening and on average makes two sales. Let X be the number of sales per evening.

a) Obtain estimates for:

i) $P(X > 4)$ **ii)** $P(X < 6)$.

Let Y be the number of sales in a two-day period.

b) Obtain estimates for:

i) $P(Y < 1)$ **ii)** $P(3 \le Y \le 4)$.

2 On average two applicants each year wanting to train as pilots are colour-blind. Let X be the number of colour-blind applicants in one year.

a) Obtain estimates for:

i) $P(X = 3)$ **ii)** $P(X > 3)$ **iii)** $P(X \ge 4)$.

b) Find the probability of fewer than two colour-blind applicants in:

i) each of four successive years,

ii) just one of four successive years.

3 A football team has 34 players including reserves. On average only one player fails to turn up for a match.

a) If three fail to turn up for a match is there evidence that the number of players failing to turn up has increased?

b) In a series of four matches there were only two absences in total. Is there evidence that the number of players failing to turn up for matches is decreasing?

4 State which approach to use to calculate the following probabilities and then use it to find the required probability.

a) $P(X \ge 3)$ when $n = 200$ and $p = 0.006$.

b) $P(X \le 2)$ when $n = 5$ and $p = 0.24$.

c) The probability of more than one faulty floppy disc in a box of 40 discs, if on average each box contains only one faulty disc.

EXERCISES

15.1 B

1 A literary editor reads 60 manuscripts a month and, on average, recommends one for future publication. Let X be the number of manuscripts recommended on one month.

a) Obtain estimates for:
  i) $P(X > 2)$ **ii)** $P(X < 4)$.

Let Y be the number of manuscripts recommended in six months.

b) Obtain estimates for:
  i) $P(Y < 8)$ **ii)** $P(11 \leq Y \leq 13)$.

2 On average, five out of the 300 new students at a college each year are left-handed. Let X be the number of new left-handed students in one year.

a) Obtain estimates for:
  i) $P(X = 5)$ **ii)** $P(X > 5)$ **iii)** $P(X \geq 6)$.

b) Find the probability of fewer than four new left-handed students:
  i) in each of three successive years,
  ii) in just one of three successive years.

3 An orchestra has 80 members. On average only one fails to attend a concert.

a) If four members fail to attend a concert, is there evidence that absentee numbers have increased?

b) In a special series of six concerts, there were only three absences in total. Is there evidence that the number of absences significantly declined for this series?

4 Which model would you use to evaluate the following probabilities?

a) $P(X \leq 4)$ where $n = 100$ and $p = 0.01$

b) $P(X \leq 4)$ where $n = 10$ and $p = 0.1$

c) The probability of there being not more than four broken eggs in a crate if, on average, a crate contains only one broken egg.

NORMAL APPROXIMATION TO POISSON

Consider this diagram, which shows the distribution $X \sim \text{Poi}(9)$. How could we describe the shape of the graph?

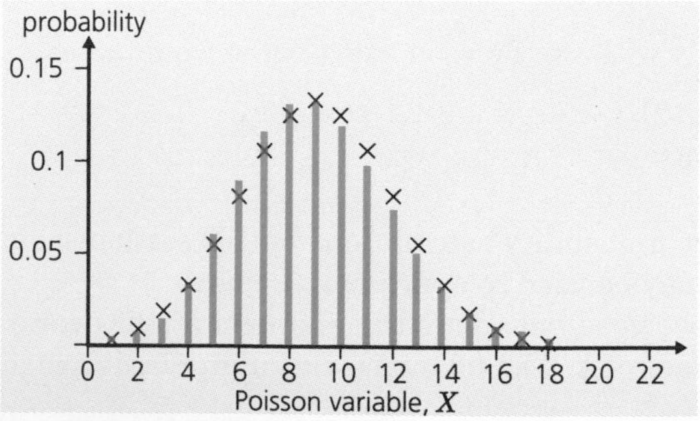

It is symmetric. It is bell-shaped. It tails off in both directions.

This description is very like the one used in Chapter 14, *The normal probability model*, to describe the normal distribution's **density curve**.

Notice in this graph that the points marked with a cross do not quite coincide with the ends of the vertical lines representing the Poisson probabilities. These points correspond to the normal distribution $N(9, 3^2)$ with the same mean and variance as the Poisson variable. It appears that the normal distribution provides reasonably accurate approximations to Poisson probabilities – but under what circumstances? And how are the approximations obtained?

The circumstances are easily identified – it occurs when the Poisson distribution has the symmetric bell-shape of the normal distribution. This is when the mean of the Poisson is large. To determine what is meant by 'large' we need to address the second question – how the approximations are made.

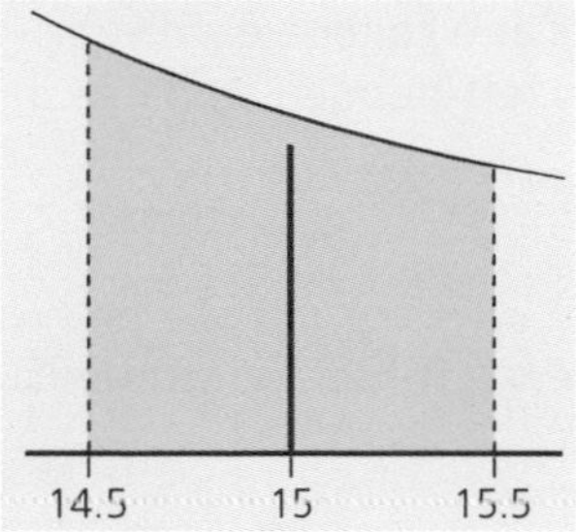

Continuity correction

The difficulty with this is in the fundamental difference between the variates. A Poisson variate is discrete and a normal variate is continuous. If $X \sim \text{Poi}(9)$, then the probability of X being equal to 15 is readily obtained.

$$P(X = 15) = \frac{e^{-9} \times 9^{15}}{15!} = 0.019\,43$$

However, if $X \sim N(9, 9)$, then the probability that X is 15 has to be found as the area under the density curve between $X = 15.5$ and $X = 14.5$. The diagram illustrates this. Try finding the value of this area.

The process of finding the value is:

$$\begin{aligned} P(X = 15) &= P(X \le 15.5) - P(X \le 14.5) \\ &= \Phi\left(\frac{15.5 - 9}{\sqrt{9}}\right) - \Phi\left(\frac{14.5 - 9}{\sqrt{9}}\right) \\ &= 0.984\,89 - 0.966\,60 \\ &= 0.0183 \end{aligned}$$

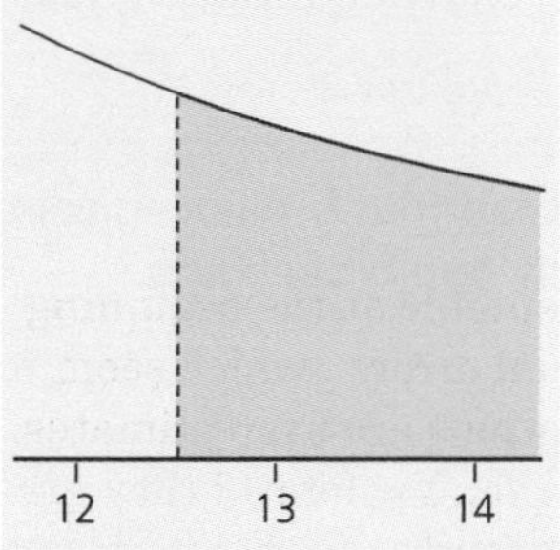

There is quite a lot of work involved in calculating an individual probability using the normal distribution. Calculating cumulative probabilities is often less work! For instance, to evaluate $P(X > 12)$ using the normal approximation, we need firstly to notice that we are interested in the event 'greater than 12' which does not include 12, hence the region under the normal curve is the area from 12.5 onwards.

So we have to apply a continuity correction to the discrete event being approximated as a continuous event.

$$\begin{aligned} P(x > 12) &\approx 1 - P(x \le 12) \\ &= 1 - 0.878\,42 \\ &= 0.1216 \end{aligned}$$

This compares favourably with the Poisson exact value from tables.

$$\begin{aligned} P(x > 12) &= 1 - P(x \le 12) \\ &= 1 - 0.8758 \\ &= 0.1242 \end{aligned}$$

The diagram and the two probability calculations in this section seem to indicate that N(9, 9) may be used to approximate Poi(9) probabilities. However, we need to explore the approximations over a much wider range of values of X. The following Calculator activity is designed for this.

CALCULATOR ACTIVITY

N(9, 9) ≈ Poi(9)

The format of the program for the previous Calculator activity can be followed here. The difference between the two situations lies in evaluating the normal probabilities. The subroutine evaluates $\Phi(z)$ by adapting the programs developed in Chapter 14, *The normal probability model*.

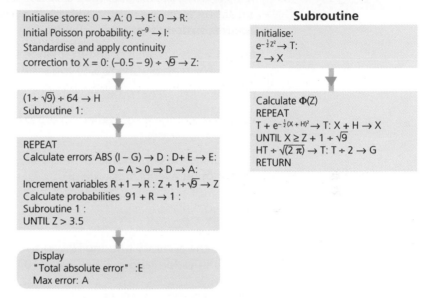

Initialise stores: $0 \to A$: $0 \to E$: $0 \to R$:
Initial Poisson probability: $e^{-9} \to I$:
Standardise and apply continuity correction to X = 0: $(-0.5 - 9) \div \sqrt{9} \to Z$:

$(1 \div \sqrt{9}) \div 64 \to H$
Subroutine 1:

REPEAT
Calculate errors ABS $(I - G) \to D$: $D + E \to E$:
$D - A > 0 \Rightarrow D \to A$:
Increment variables $R + 1 \to R$: $Z + 1 \div \sqrt{9} \to Z$
Calculate probabilities $91 + R \to 1$:
Subroutine 1 :
UNTIL Z > 3.5

Display
"Total absolute error" :E
Max error: A

Subroutine

Initialise:
$e^{-\frac{1}{2}Z^2} \to T$:
$Z \to X$

Calculate $\Phi(Z)$
REPEAT
$T + e^{-\frac{1}{2}(X+H)^2} \to T$: $X + H \to X$
UNTIL $X \geq Z + 1 \div \sqrt{9}$
$HT \div \sqrt{(2\pi)} \to T$: $T \div 2 \to G$
RETURN

Discover how well N(9, 9) approximates Poi(9) probabilities.

Interpreting the results

You should discover that the maximum difference between a Poi(9) probability and the N(9, 9) approximation is 0.01. The total absolute difference is about 0.08.

These errors are rather larger than were acceptable at the beginning of this chapter. It is useful to explore the actual errors., which seem to show a pattern. For $X = 0$ to 3, the approximations are overestimates. For $X = 4$ to 8 they are underestimates. Then, for $X = 9$ to 13 they are overestimates and for all $X \geq 14$ they are underestimates. You may be able to see this in the graph on page 425.

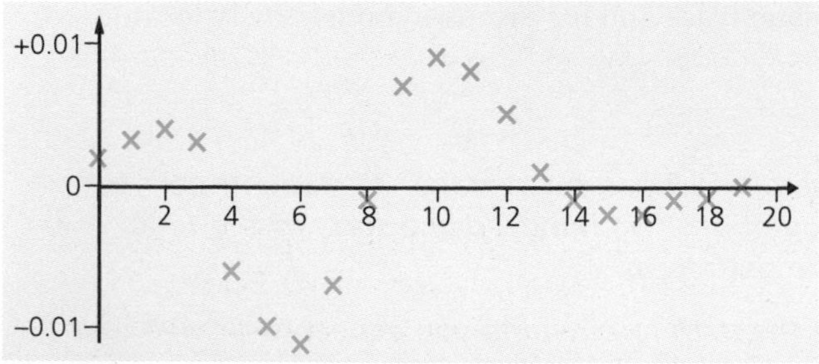

Alternatively, you might like to adapt your calculator program to display the actual error for each probability. This diagram is a graph of the actual errors.

Interpreting these errors in approximating Poisson probabilities is fairly straightforward. If the Poisson event is reasonably symmetric about the mean, e.g. $P(4 < X \leq 12)$, the positive and negative differences tend to cancel out, leading to reasonable accuracy. If the event is largely in the upper tail e.g. $P(X > 14)$, then the approximation is an underestimate. The reverse is true in a lower tail.

Exploration 15.1

Poi(9)

- Demonstrate that for $X \sim \text{Poi}(9)$:

a) $P(4 < X \leq 12) = 0.821$ and that when this probability is approximated using N(9, 9) the result is 0.812,

b) $P(X > 14) = 0.0415$ and that the normal approximation is 0.0334.

What percentage error is there in using the normal approximation in each of these cases?

- Explore the percentage error in using the normal approximation for other events of $X \sim \text{Poi}(9)$.

Exploration 15.2

$N(\lambda, \lambda) \approx Poi(\lambda)$

Explore the accuracy of using $N(\lambda, \lambda)$ to approximate $Poi(\lambda)$ probabilities for various λ, e.g. $\lambda = 5, 10, 15, \ldots$.

Interpreting the results

You might expect results such as those in the following table.

	Absolute difference	Maximum difference
5	0.11	0.015
10	0.08	0.010
15	0.06	0.005
25	0.05	0.003
50	0.04	0.002

Clearly, the total difference and maximum individual difference decrease as the mean of the Poisson distribution becomes larger. A pattern of errors emerges that is similar to that for Poi(9) for upper and lower tail probabilities and for events symmetrically located around the mean.

Example 15.2

The school in Example 15.1 has, on average, 75 students per year given county recognition. Two hundred and forty of the 1200 students are in the sixth form.

a) *How many of the sixth form would you expect to receive county recognition each year?*

b) *One year there were 24 sixth-formers selected for county awards. Does this indicate that the sixth form were particularly merit-worthy that year?*

Solution

a) *If it is reasonable to assume that the number of students recognised at county level is a Poisson variate, then there would be, on average, 75 ÷ 5 = 15 sixth-formers per year.*

b) *Appropriate hypotheses for the rest are:*
H_0: = 15; H_1: > 15
i.e. one-tailed.
The probability of there being at least 24 students under the hypothesis of an average of 15 is:

$$\begin{aligned} P(X \geq 24) &= P(X = 24) + P(X = 25) + \ldots \\ &= P\left(z \geq \frac{23.5 - 15}{\sqrt{15}}\right) \\ &= 1 - \Phi(2.195) \\ &\approx 0.014 \end{aligned}$$

This suggests that it is highly unlikely to find 24 or more sixth-formers receiving county honours. It is reasonable to conclude that the average number has increased.

EXERCISES

15.2 A

1 On average, an estate agent sells 30 houses per month. In the last month 25 were sold. Was this significantly fewer than usual?

2 The number of responses to an advertisement in a local newspaper follows a Poisson distribution with a mean of 8.

a) Find the probability that there are fewer than five responses.
b) Find the probability that there are more than 50 responses to five consecutive daily advertisements.

3 A fisherman catches, on average, four fish, when he fishes at a certain lake.

a) On one visit to the lake he caught only two fish. Does this indicate that it has become more difficult to catch fish at this lake?
b) If he caught four fish from two visits, does this suggest that it has become more difficult to catch fish at this lake?
c) How would you respond if he caught eight fish in four visits?

4 A computer crashes, on average, twelve times a month.

a) Find the probability of more than 15 crashes in a month.
b) Find the probability of less than 30 crashes in a three-month period.
c) Find the probability of exactly twelve crashes in a month.
d) Find the probability of exactly 36 crashes in a three-month period.
e) Compare your answers to **d)** and **c)**.

EXERCISES

15.2 B

1 On average, a company appoints 20 new trainees each year. Last year, 15 were appointed. Was this significantly fewer than usual?

2 The number of full-page advertisements placed in a monthly magazine follows a Poisson distribution with a mean of 5.

a) Find the probability that there are fewer than four full-page advertisements in an issue.
b) Find the probability of at least 50 full-page advertisements in a year's run of the magazine.

3 **a)** A salesman wins, on average, 25 new contracts each month. One month he won only 20 new contracts. Does this indicate a deterioration in his efficiency?
b) If the salesman won only 40 new contracts in two months, would this indicate that his performance had deteriorated?
c) If the salesman won only 80 new contracts in four months, would this indicate a deterioration?

4 The number of train cancellations per week on a certain line has an average of 10.

a) Find the probability of at least twelve cancellations in a week.
b) Find the probability of fewer than 50 cancellations in a four-week period.
c) Find the probability of exactly ten cancellations in a week.
d) Find the probability of exactly 40 cancellations in four weeks.
e) Compare your solutions to **c)** and **d)**.

NORMAL APPROXIMATION TO THE BINOMIAL

Consider this diagram, which shows the binomial distributions B(20, 0.5) and B(50, 0.2). How could we describe the similarities between the distributions?

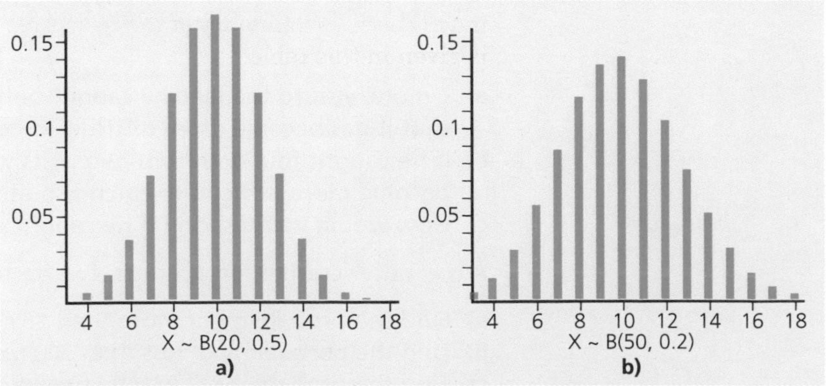

Clearly the distributions appear symmetric, although there is an element of asymmetry in B(50, 0.2). This distribution is more spread than B(20, 0.5). Their variances are 5 and 8 using the result $\sigma^2 = np(1-p)$

that we discovered in Chapter 8, *Sums and differences of distributions*. They look as if they could be approximated by a normal distribution. For which do you think will the normal approximation be better?

Recall that the normal distribution is symmetric. This implies that if the actual distribution is also symmetric then the approximation is likely to be better than if it were not symmetric. The appropriate normal distribution to choose to approximate binomial probabilities must have the correct mean and variance. So N(10, 5) is appropriate to approximate B(20, 0.5).

Consider the probabilities $P(X = 8)$, $P(X = 9)$, …, $P(X = 12)$. The normal approximations to these can be evaluated using the continuity correction, as in the case of approximating Poisson probabilities. Below is an illustration of the calculation of $P(X = 12)$.

B(20, 0.5)	N(10, 5)
0.120	$\Phi\left(\frac{12.5-10}{\sqrt{5}}\right)-\Phi\left(\frac{11.5-10}{\sqrt{5}}\right)\approx\Phi(1.118)-\Phi(0.671)$ $\approx 0.868\,15-0.748\,89$ ≈ 0.119
B(50,0.2)	**N(10,8)**
0.103	$\Phi\left(\frac{12.5-10}{\sqrt{8}}\right)-\Phi\left(\frac{11.5-10}{\sqrt{8}}\right)\approx\Phi(0.884)-\Phi(0.530)$ $\approx 0.811\,68-0.701\,94$ ≈ 0.110

The normal approximation to the perfectly symmetric binomial distribution B(20, 0.5), is much closer than to the less symmetric B(50, 0.2). Compare B(20, 0.5) probabilities and their approximations using N(10, 5).

You should have found that the comparison between the symmetric binomial distribution B(20,0.5) and its approximating distribution is much closer. A summary of the measures of closeness for the two cases is given in this table.

Binomial		Normal approximation			
n	p	μ	σ^2	Σ (absolute difference)	Maximum difference
20	0.5	10	5	0.008	0.001
50	0.2	10	8	0.053	0.007

You can see that the approximation to the symmetric binomial distribution is definitely the better of the two.

Exploration 15.3

Normal and binomial distributions with mean 10

- Find out how well appropriate normal distributions approximate binomial distributions where the mean is equal to 10. You might like to develop a calculator activity or to use a spreadsheet for this.
- Now repeat the investigation when the mean is larger than 10.

Interpreting the results

Some results you might have discovered are given in the table below.

	Binomial		Normal approximation	
Mean	n	p	Σ (absolute difference)	Maximum difference
10	30	0.333	0.033	0.005
	40	0.25	0.046	0.007
15	30	0.5	0.005	0.001
	45	0.333	0.027	0.003
	60	0.25	0.038	0.004
	75	0.2	0.043	0.005
20	40	0.5	0.004	0.000
	50	0.4	0.015	0.002
	80	0.25	0.033	0.003
	100	0.2	0.037	0.004

It is clear that symmetry is also the crucial factor in determining the closeness of the normal approximation to binomial probabilities. A binomial distribution is perfectly symmetric when $p = 0.5$. So if p is close to 0.5, then the normal distribution with mean np and variance $= np(1-p)$ will provide a reasonable approximation to $B(n, p)$ probabilities.

Example 15.3

An observation of a binomial variate, X, is made. It is found that $X = 5$. Test, at five per cent, the hypothesis that $X \sim B(50, 0.2)$ using:

a) *actual binomial probabilities,*
b) *appropriate normal probabilities.*
against the alternative hypothesis that $p < 0.2$. Comment on the results.

Solution

a) *The hypotheses are:*
H_0: $X \sim B(50, 0.2)$; H_1: $X \sim B(50, p)$ *where* $p < 0.2$.
Since the alternative hypothesis is directional, we are required to use a one-tailed test. The observed value of X is 5 so we need to determine whether an observation as small as this is unusual at five per cent. So:

$$P(X \leq 5) = 0.8^{50} + 50 \times 0.2 \times 0.8^{49} + \ldots$$
$$\approx 0.0480$$

This is less than five per cent, hence, the ***observation is unusual.*** *We conclude that p is less than 0.2.*

b) *The test is the same, but approximations to the probabilities are going to be made using N(10, 8).*

$$\begin{aligned} P(X \le 5) &= \Phi\left(\frac{5.5 - 10}{\sqrt{8}}\right) \\ &\approx 1 - \Phi(1.591) \\ &\approx 1 - 0.9442 \\ &= 0.0558 \end{aligned}$$

This is greater than five per cent, hence, the ***observation is not unusual.*** *We conclude that p is 0.2.*
The conclusions contrast with one another. The one reached using the actual binomial probabilities is more reliable since the second one was based on an approximation to the model.

A cautionary note

Example 15.4 is an illustration of the potential dangers of using approximations. Clearly, the work involved in using the normal approximation was considerably less than using the exact distribution. However, the conclusion reached using the approximation was wrong! If p is not close to 0.5, then use of the normal approximation can lead to wrong conclusions.

B(n, 0.2)		Normal approximation	
Mean	n	Σ (absolute difference)	Maximum difference
10	50	0.053	0.007
15	75	0.043	0.005
20	100	0.037	0.004
25	125	0.034	0.003
30	150	0.031	0.002
40	200	0.027	0.002
50	250	0.024	0.001

Consider the information shown in the table. It was compiled by considering the normal approximation for a binomial distribution where $p = 0.2$. We can see that the measure of error reduces as the value of n increases. Clearly, since p is fixed in this compilation, increasing n has the effect of increasing the mean.

The message is that if p is not close to 0.5, then n needs to be large to reduce the potential for error in using the normal approximation.

Example 15.4

Obtain an estimate of the probability that there will be more than 60 left-handed students among the 240 sixth-formers at the school in Example 15.1, if it is reasonable to assume that 23 per cent of people are left-handed.

Solution

The number of left-handed students may be modelled by the binomial B(240, 0.23). The value of p *is not close to 0.5 but 240 is a fairly large value and the mean, 55.2, is large.*
Under these circumstances it is reasonable to approximate the required probability using the normal distribution as N(55.2, 42.504).
Hence:

$$\begin{aligned}\mathrm{P}(X > 60) &\approx \mathrm{P}\left(z \le \frac{60.5 - 55.2}{\sqrt{42.504}}\right) \\ &= 1 - \Phi(0.813) \\ &= 0.208\end{aligned}$$

EXERCISES

15.3 A

1 A politician visits houses during the run-up to an election. She finds that at 40 per cent of homes the occupants agree with her party's policies.

a) Find the probability that if five houses are visited all will have occupants that agree with her.
b) Estimate the probability that at least half the occupants of 20 houses will agree with the politician.
c) Repeat **b)** for 50 homes and comment on the two probabilities.

2 It is estimated that nationally 35 per cent of households have a computer. In a certain town 45 households out of a sample of 100 had a computer. Is there evidence that the level of ownership of computers is above average for this town?

3 If 70 per cent of university students take out student loans, find the probabilities that:

a) at least 50 in a group of 90 students have taken out student loans,
b) more than 60 in a group of 100 students have taken out student loans.

Two groups of 100 students are following different courses. Find the probability that:
c) at least 75 students in both groups have taken out student loans.

4 It is claimed that only 20 per cent of the longterm unemployed remain unemployed after attending a job preparation scheme.

a) In a sample of 400, 100 remained unemployed. Test the claim at the five per cent significance level.
b) In a sample of 100, 25 remained unemployed. Test the claim at the five per cent significance level.
c) Compare your answers to **a)** and **b)**.

EXERCISES

15.3 B

1 A charity usually receives donations from about 60 per cent of households in a door-to-door collection.

a) Find the probability that every household in a street of five makes a donation.
b) Estimate the probability that at least half the households in a street of 20 make a donation.
c) Estimate the probability that a least half the households in a street of 50 make a donation.

2 Nationally, 28 per cent of households own a pet. In a certain region, 35 households in a random sample of 100 own a pet. Is there evidence that pet ownership is significantly above average in this region?

3 If 65 per cent of students play sport at least once a week, estimate the probability that:

a) at least 45 in a group of 60 play sport in a given week,
b) more than 50 in a group of 80 play sport in a given week.
A college has five houses, each with 120 students. Estimate the probability that:
c) at least 80 in each house play sport in a given week,
d) only two of the houses have at least 80 students who play sport that week.

4 It is claimed that, after vaccination, only 25 per cent of adults will catch 'flu during the winter.

a) In a random sample of 200, 72 caught 'flu. Test the claim at the five per cent significance level.
b) In a random sample of 50, 18 caught 'flu. Test the claim at the five per cent significance level.
c) Compare your answers to **a)** and **b)**.

OTHER DISTRIBUTIONS

Distributions of continuous or discrete random variables which have the characteristic bell-shape of the normal distribution may be approximated using $N(\mu, \sigma^2)$. You have discovered this in the case of two discrete random variables. Consider the continuous variate, X, which is the sum of three independent observations of a rectangularly distributed variate: Rect(0, 1). The mean of the rectangular distribution is 0.5 and its variance is $\frac{1}{12}$. Hence:

$$\mu = 0.5 + 0.5 + 0.5 = 1.5 \text{ and } \sigma^2 = \tfrac{1}{12} + \tfrac{1}{12} + \tfrac{1}{12} = \tfrac{1}{4}$$

Recall that the sum of two independent observations from Rect(0, 1) produces a symmetric triangular distribution. It is reasonable to suppose that X will be symmetrically distributed . You could obtain a sample of observations on X using your calculator repeatedly to produce and add up three random decimals. The following is a five-part stem-and-leaf of 300 observations of X.

```
Leaf    Unit = 0.10
depth

   4 | 0T | 2233
  16 | 0F | 444555555555
  28 | 0S | 677777777777
  48 | 0E | 88888888888889999999
  81 | 1O | 000000000000000111111111111111111
 124 | 1T | 2222222222222222222223333333333333333333333
(48) | 1F | 444444444444444444444444444555555555555555555555
 128 | 1S | 66666666666666666666666666677777777777777777777777
  78 | 1E | 888888888999999999999999999999
  48 | 2O | 0000000000001111111
  29 | 2T | 222222222233333
  14 | 2F | 4444445555
   4 | 2S | 67
   2 | 2E | 88
```

How well might a normal distribution provide a fit for these observed frequencies? This can be addressed using the familiar measure of fit.

$$\sum \frac{(O-E)^2}{E}$$

The expected frequencies are calculated using the normal distribution with the same mean and variances as X i.e. N(1.5, 0.25). Since X is a continuous variate, there is no need (or justification) to use a continuity correction.

The normal approximation to $P(X < 0.4)$ is:

$$\Phi\left(\frac{0.4-1.5}{\sqrt{0.25}}\right) = \Phi(-2.2)$$
$$= 1 - \Phi(2.2)$$
$$= 0.013\,90$$

Hence, the expected frequency is 4.17. The normal approximation to $P(0.4 \le X < 0.6)$ is:

$$\Phi\left(\frac{0.6-1.5}{\sqrt{0.25}}\right) - \Phi(-2.2) = \Phi(2.2) - \Phi(1.8)$$
$$= 0.022\,03$$

Hence the expected frequency is 6.609. Continue this process of obtaining expected frequencies corresponding to the branches of the stem-and-leaf diagram above. You will need to make $X \ge 2.6$ the final class to avoid too small an expected frequency.

O	4	12	12	20	33	43	48	50	30	19	15	10	4
E	4.17	6.609	13.449	23.37	34.677	43.947	47.556	43.947	34.677	23.37	13.449	6.609	4.17

This table represents a summary of the calculations. The value of $\sum \frac{(O-E)^2}{E}$ amounts to 14.192. The significance of this can be judged using the five per cent upper-tail chi-squared statistic with $(13-1)$ degrees of freedom, 21.026. Clearly the results is not significant, hence it suggests that N(1.5, 0.25) is a reasonable fit for these data. However, the data were obtained by summing random decimals. This suggests that the distribution of the sum of three random decimals may be approximated by a normal distribution.

This is an example of a more general result leading to the use of the normal distribution in approximating the distribution obtained by summing repeated observations of a random variable.

Example 15.5

Describe the distribution which could be used to model the total time for the swings of a pendulum.

Solution

The time for one swing may be assumed to be a random variable, T, with mean, μ_T and variance σ_T^2. The time for ten swings is equivalent to the sum of ten observations on T. Hence, the time for ten swings may be modelled by a normal distribution $N(10\mu_T, 10\sigma_T^2)$.

EXERCISES

15.4 A

A company executive travels on the underground to his office in central London. The journey has a mean length of 42 minutes and a standard deviation of twelve minutes.

a) State the distribution that would model the total time spent travelling to and from work in a five-day week.
b) Estimate the probability that he spends less than six hours per week travelling.
c) Estimate the probability that he spends between 6.5 and 7.5 hours per week travelling.

The mass of a good-quality banana is a random variable with mean 200 grams and standard deviation 30 grams.

a) What would be the distribution of the mass of a bunch of twelve bananas?
b) Estimate the probability that a bunch of twelve bananas has a mass between 2.2 and 2.5 kilograms.

The quantity of apples put in a plastic bags by customers at a supermarket has a mean mass of 1.5 kg with a standard deviation of 0.4 kg.

a) Estimate the probability that the next 20 customers buy more than 32 kg of apples.
b) Estimate the probability that the next 200 customers buy more than 320 kg of apples.

4 The four continuous random variables X_1, X_2, X_3 and X_4 are rectangularly distributed on the interval [10, 20]. Let X be the sum of these four random variables.

a) Estimate the probability that $X > 70$.
b) Find the value of k if the probability that $X > k$ is 0.90.

EXERCISES

15.4 B

1 A student's journey time to and from school has a mean of 23 minutes and standard deviation of three minutes.

a) Describe the distribution which could be used to model the total time spent travelling in a school week.
b) Estimate the probability that he spends over 4 hours travelling in a given week.
c) Estimate the probability that he spends between 3.5 hours and 4 hours travelling in a given week.

2 The mass of a grade A apple is a random variable with a mean of 120 g and a standard deviation of 8 g.

a) Describe the distribution which could be used to model the total mass of a bag of twelve grade A apples.
b) Estimate the probability that this total mass lies between 1.42 kg and 1.45 kg.

3 The amount spent in a supermarket may be assumed to be a random variable with a mean of £10.52 and a standard deviation of £4.20.

a) Estimate the probability that the next 20 customers spend over £200 in total.
b) Estimate the probability that the next 100 customers spend over £1000 in total.

4 The continuous variate X is the sum of five independent observations of a rectangularly distributed variate Rect (0, 10).

a) Estimate the probability that $X > 30$.
b) Find the value k such that the probability $(X > k) = 0.95$.

CONSOLIDATION EXERCISES FOR CHAPTER 15

1 **a)** A magazine sells, on average, 300 subscriptions each month. After an advertising campaign, 320 subscriptions were sold in a month. Does this indicate a significant increase in sales?
b) To what level would subscriptions need to rise to indicate a significant improvement?

2 Market research suggests that 40 per cent of a newspaper's readers look at the arts pages, but only two per cent of these read the dance reviews. If 1000 readers are selected at random, estimate the probability that:

a) at least 380 read the arts pages,
b) at least ten read the dance reviews.

3 At a certain university, 15 per cent of the new intake of students are mature students. In previous years, ten per cent of mature students have been over 50. If a random sample of 80 new students is taken:

a) find the probability that at least ten are mature students,
b) find the most likely number of students over 50.

In fact, four members of the sample are over 50.

c) Does this provide evidence that the proportion of mature students over 50 is still ten per cent?

4 It is claimed that equal numbers of those watching the main evening news choose channel A and channel B. Using appropriate methods, test this claim if:

a) in a random sample of 15, five choose channel A,
b) in a random sample of 150, 50 choose channel A.

5 State the conditions under which a binomial distribution may be approximated by a Poisson distribution.

The probability that an adult suffers an allergic reaction to a particular inoculation is 0.0018. If 5000 adults are given the inoculation, estimate the probability that:

a) at most ten suffer an allergic reaction,
b) from five to 15, inclusive, suffer an allergic reaction.

(NEAB Question 3, Specimen paper 9, 1996)

6 A factory mass-produces certain items, four per cent of which are defective. A random sample of n items is chosen from a day's total output.

a) Calculate the smallest value of n for which the probability that the sample contains at least one defective item is greater than 0.99.
b) When $n = 100$, find an approximate value for the probability that the sample contains five or more defective items.

7 The number of plants found in squares of equal area on a large moor follows a Poisson distribution with mean 27. Using a suitable approximation, find the probability that a particular square contains 30 or more plants.

(AEB Question 5 (part), Paper 9, June 1994)

8 Among the blood cells of a certain animal species, the proportion of cells which are of type A is 0.037 and the proportion of cells which are of type B is 0.004.

a) Find the probability that in a random sample of eight blood cells, at least two will be of type A.

Using suitable approximations, find the probability that:

b) in a random sample of 200 blood cells the total number of type A and type B is at least 81,

c) in a random sample of 300 blood cells there will be at least four cells of type B.

(ULEAC Question 8, Specimen, T1, 1994)

9 A Statistics student studied young children's ability to estimate. Each child was presented with a straight line AB of length 12 cm drawn on a piece of paper. The child was required to put a mark X on the line where he or she believed the midpoint of AB was.

As a preliminary model of this experiment the student assumed that the child would mark the point X in such a way that the distance AX had a uniform distribution over the interval [3, 9].

a) Calculate the mean and variance of the distance AX using this model.

The student tested 100 children and the results obtained can be summarised as follows: $\Sigma x = 590$ and $\Sigma x^2 = 3571$, where x is the distance AX, in centimetres.

b) Calculate the mean and variance of AX for the student's data.

c) Explain briefly why the student may wish to alter the model in the light of these findings.

The student studied the results for the 100 children again and noticed that five children had AX smaller than 4.3 cm and 17 had AX more than 6.9 cm.

d) Assuming that AX has a normal distribution with mean μ and variance σ^2, use this new information above to estimate the values of μ and σ^2.

e) Compare your estimates in **d)** with the values found in **b)** and comment on the suitability of the normal distribution as a model for this experiment.

(ULEAC Question 9, Specimen, T1, 1994)

10 The quantity of milk in bottles from a dairy is normally distributed with mean 1.036 pints and standard deviation 0.014 pints.

Show that the probability of a randomly-chosen bottle containing less than a pint is very nearly 0.5 per cent.

For the rest of the question take the answer of 0.5 per cent to be exact. A crate contains 24 bottles. Find the probability that:

a) no bottles contain less than a pint of milk,

b) at most one bottle contains less than a pint of milk.

A milk float is loaded with 150 crates (3600 bottles) of milk. State the expected number of bottles containing less than a pint of milk.

Give a suitable approximating distribution for the number of bottles containing less than a pint of milk. Use this distribution to find the probability that more than 20 bottles contain less than a pint of milk.

(MEI Question 4, Specimen paper S2, 1994)

11 The probability that I see a shooting star on any given night is 0.01. If I look for shooting stars on 80 nights, and observations on any one night are independent of observations on any other night, use a suitable model to find the probability that, in the 80-night period:

a) only one shooting star is observed,
b) more than two but fewer than five shooting stars are observed.

(UCLES Question 5, Specimen paper S2, 1994)

12 The number of accidents per week at a certain intersection has a Poisson distribution with parameter 2.5. Find the probability that:

a) exactly five accidents will occur in a week,
b) more than 14 accidents will occur in four weeks.

13 In parts **a)**, **b)** and **c)** of this question use the binomial, Poisson and normal distributions, according to which you think is the most appropriate. In each case, draw attention to any feature of the data which supports or casts doubt on the suitability of the model you have chosen. Indicate, where appropriate, that you are using one distribution as an approximation to another.

a) The annual income of all employees of a large firm has a mean of £14 500 with a standard deviation of £2200. What is the probability that a mean income of 100 employees selected at random is between £14 000 and £15 000?
b) A technician looks after a large number of machines on a night shift. She has to make frequent minor adjustments. The necessity for these occurs at random at a constant average rate of eight per hour. What is the probability that:
  i) in a particular hour she will have to make five or fewer adjustments,
  ii) in an eight hour shift she will have to make 70 or more adjustments?
c) A number of neighbouring allotment tenants bought a large quantity of courgette seeds which they shared among them. Overall 15 per cent failed to germinate. What is the probability that a tenant who planted 20 seeds would have:
  i) five or more failing to germinate,
  ii) at least 17 germinating?

(AEB Question 7, Specimen paper 2, 1996)

14 Evidence accumulated in a large college suggests that the time taken by A-Level History students to complete their essays, after initial preparation, is a normally distributed random variable with mean 86 minutes and standard deviation 8 minutes. Find, to three decimal places, the probability that a randomly-chosen essay has taken between 70 and 90 minutes to complete.

Five essays are chosen at random. Find, to three decimal places, the probability that exactly one of them has taken more than 90 minutes to complete.

In a given year 200 essays were submitted. Use a suitable method of approximation to estimate, to three decimal places, the probability that over 70 of these took more than 90 minutes to complete.

(NEAB Question 15, Specimen paper 2, 1996)

15 On average my train is late on 45 journeys out of 100. Next week I shall be making five train journeys. Let X denote the number of times my train will be late.

a) State one assumption which must be made for X to be modelled by a binomial distribution.
b) Find the probability that my train will be late on all of the five journeys.
c) Find the probability that my train will be late on two or more out of five journeys.
Approximate your binomial model by a suitable normal model to estimate the probability that my train is late on 20 or more out of 50 journeys.

(UCLES Question 7, Specimen paper S1, 1994)

Summary

It is possible, under appropriate circumstances, to approximate probabilities of one model by those of another.

Description	**Approximation**	**Condition**
$B(n, p)$	$Poi(np)$	large n, small p $p < 0.02$ usually good
$Poi(\lambda)$	$N(\lambda, \lambda)$	large λ, at least 10 with care $\lambda \geq 25$ reasonable
$B(n, p)$	$N(np, np(1-p))$	symmetry is key issue ideally $p \approx 0.5$ otherwise n must be large
ΣXi	$N(n\mu_x, n\sigma^2)$	sums of repeated observations

STATISTICS

16

The central limit theorem

In this chapter we:

- *meet one of the most widely accepted uses of the normal distribution,*
- *use it to provide a reasonable model for the distribution of means of samples,*
- *discover how it is possible to build limits for a population mean based on a sample mean.*

MEANS OF SAMPLES FOR A NORMAL DISTRIBUTION

By the end of Chapter 15, *Approximating distributions*, we were using the normal distribution in connection with sums of observations of a random variable. It is a simple step from summing observations to finding their average. The normal distribution plays an important role in modelling the distribution of the mean of a sample of data.

We found in Chapter 14, *The normal probability model*, that if the independent random variables, X and Y, are normally distributed, then linear combinations such as $X + Y$, $\frac{1}{2}X + \frac{1}{2}Y$, $\frac{1}{2}(X + Y)$, etc. are also normally distributed. In particular, consider the case where:

$$X \sim \mathrm{N}(\mu_X, \sigma_X^2)$$

and

$$Y \sim \mathrm{N}(\mu_Y, \sigma_Y^2)$$

What is the mean of the random variable, R, where:

$$R = \tfrac{1}{2}(X + Y)?$$

R can be considered as a linear combination:

$$\tfrac{1}{2}X + \tfrac{1}{2}Y$$

of X and Y. Hence, the mean of R is the same linear combination of the means of X and Y:

$$\mu_R = \tfrac{1}{2}\mu_X + \tfrac{1}{2}\mu_Y = \tfrac{1}{2}\left(\mu_X + \mu_Y\right)$$

i.e. the average of the mean of X and Y. But, what about the variance of R?

Recall that variance is a squared measure of spread and this prompts the familiar results:

$$\sigma_R{}^2 = \left(\tfrac{1}{2}\right)^2 \sigma_X{}^2 + \left(\tfrac{1}{2}\right)^2 \sigma_Y{}^2$$
$$= \frac{\sigma_X{}^2 + \sigma_Y{}^2}{2^2}$$

This can be thought of as half the average of the variances.

If X and Y have the same normal distribution, then $\frac{1}{2}(X + Y)$ is the distribution of the mean of samples of size 2 taken randomly from that normal distribution. What do our discoveries imply about the distribution of the mean of samples of size 2?

Under these circumstances, X and Y have the same distribution, say $X \sim \mathrm{N}(\mu, \sigma^2)$ (dropping the distinguishing subscripts). Then the distribution of $\frac{1}{2}(X + Y)$ can conveniently be written as the random variable, $\overline{X}_2$ representing the mean of a sample of size 2. The results suggest that the mean of the distribution of $\overline{X}_2$ is the average of the means of X and Y. So in other words the mean of $\overline{X}_2$ is $\frac{1}{2}(\mu + \mu) = \mu$. The variance of $\overline{X}_2$ is half the average of the variances, i.e:

$$\frac{1}{2}\left(\frac{\sigma^2 + \sigma^2}{2}\right) = \frac{\sigma^2}{2}$$

Hence, the distribution of the mean of a sample of size 2 is:

$$\overline{X} \sim \mathrm{N}\left(\mu, \tfrac{1}{2}\sigma^2\right)$$

Exploration 16.1

Distribution of the mean of a sample

- Assume that the random variables, U, V, W are normally distributed as:
  $U \sim \mathrm{N}(\mu_U, \sigma_U{}^2)$, $V \sim \mathrm{N}(\mu_V, \sigma_V{}^2)$, $W \sim \mathrm{N}(\mu_W, \sigma_W{}^2)$.

  Consider the random variable, R, where
  $R = \frac{1}{3}(U + V + W)$

  Describe its distribution and use this to explore the distribution of the mean of a sample of size 3 drawn from a normal distribution, $\mathrm{N}(\mu, \sigma^2)$.

- Explore the distribution of the mean of a sample of size 4 drawn from $\mathrm{N}(\mu, \sigma^2)$. Make a conjecture about the distribution of means of samples of size n drawn from a normal distribution.

Interpreting the results

Your exploration may have revealed that the mean of the distribution of sample means, or the mean of $\overline{X}_n$, is the same as the mean of X. You may have found that the variance of the distribution of $\overline{X}_n$ gets smaller as the sample size increases. The variance of $\overline{X}_n$ is inversely proportional to the size of the sample.

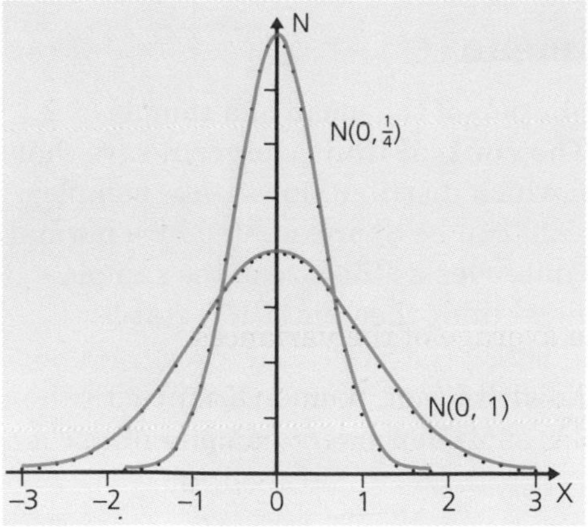

This is summarised in the diagram, which shows the distribution of means of samples of size 4 drawn from the standard normal distribution N(0, 1). The distribution of sample means is clearly considerably less spread than the standard distribution.

So if $X \sim N(\mu, \sigma^2)$ then means of samples of size n are:

$$\overline{X}_n \sim N\left(\mu, \frac{\sigma^2}{n}\right)$$

Standard error

The variance of the sample means is $\frac{\sigma^2}{n}$. Hence, the standard deviation of the sample means is:

$$\sqrt{\frac{\sigma^2}{n}} = \frac{\sigma}{\sqrt{n}}$$

The standard deviation of the distribution of means is known as the **standard error of the mean**. The term 'standard error' is given to the standard deviation of the distribution of a sample statistic.

Example 16.1

The energy in one portion of muesli served at breakfast in a large hotel may be modelled as a random variable with a mean of 125 calories. The variance may be assumed to be 100.

a) *Calculate the probability that a serving of muesli will contain more than 140 calories.*

b) *Each of a family of five has a serving of muesli. Calculate the probability that the mean energy that each of them has is more than 140 calories. Comment on the results.*

Solution

a) *Suppose X is the random variable representing the energy in a serving of muesli. Then $X \sim N(125, 100)$. The question is seeking:*

$$P(X > 140) = 1 - \Phi\left(\frac{140 - 125}{10}\right) = 0.0668$$

b) *This part of the example concerns the distribution of the mean energy in a sample of five servings. The distribution of the mean of a sample of size 5 is:*

$$\overline{X}_5 \sim N\left(125, \frac{10^2}{5}\right) = N(125, 20)$$

Hence: $P(\overline{X}_5 > 140) = 1 - \Phi\left(\frac{140 - 125}{\sqrt{20}}\right) = 0.000\,39$

Clearly there is considerably less chance of the mean serving having over 140 calories than an individual serving.

Distribution of sample mean

We have now seen that the distribution of the mean of a sample of a normal variate is itself normal. The **central limit theorem** says that it does not generally matter from which distribution we take samples. The distribution of the sample mean can be approximated by a normal distribution. The approximation improves as the size of the sample gets larger. More strictly, the central limit theorem (CLT) states:

> Given the random variable, X, with mean, μ, and variance σ^2, then the distribution of the mean, $\overline{X}_n$, of independent samples of size n approaches $\mathrm{N}\left(\mu, \dfrac{\sigma^2}{n}\right)$.

Notice that the central limit theorem requires that μ and σ^2 merely exist. It does not demand that X is symmetric nor that it is continuous. This implies that the distribution of the mean of a sample drawn from, say a Poisson distribution, Poi(λ), can be approximated by $\mathrm{N}\left(\lambda, \dfrac{\lambda}{n}\right)$ where n is the size of the sample. This approximation is valid irrespective of the value of λ, in contrast with the sort of approximations considered in Chapter 15, *Approximating distributions*.

The question which so far remains unanswered is, 'What size of sample is sufficient?' There is no exact answer. Let us consider the following illustration. Suppose random samples of size 5 are drawn from the binomial distribution, $X \sim \mathrm{B}(4, 0.2)$. This binomial variate has mean, $\mu = 0.8$, and variance, $\sigma^2 = 0.64$. The means $\overline{X}_5$, of samples of size 5, should be distributed with a mean also equal to 0.8, and variance $\frac{1}{5}(0.64) = 0.128$. This table contains the results of obtaining 200 samples of size 5 and records the means of those samples.

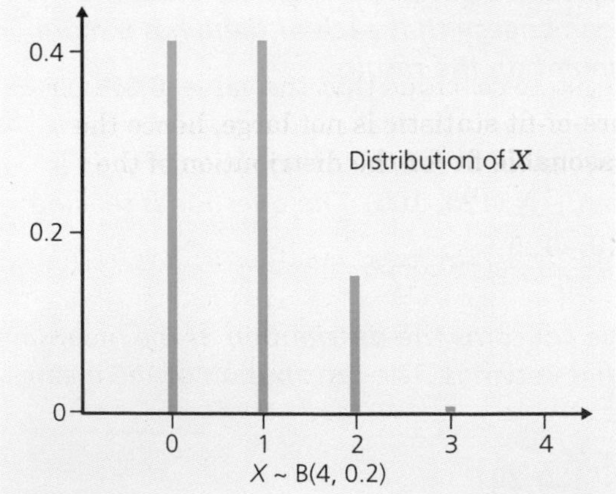

Mean of sample	Frequency
0.0	2
0.2	9
0.4	30
0.6	34
0.8	43
1.0	36
1.2	26
1.4	12
1.6	5
1.8	3
2.0	0

Does the normal distribution: $\mathrm{N}\left(0.8, \dfrac{0.64}{5}\right)$ fit these data?

Since X is discrete, then a **continuity correction** should be applied in considering the means of samples of observations of X.

Recall that the continuity correction, −0.5, is used in approximating exact probabilities associated with a random variable which takes integer values. In this case, the sample mean is the variate and so the continuity correction becomes $\pm\frac{0.5}{n}$, where n is the sample size, i.e. 0.1 here. Hence, the probability $P(\overline{X}_5 = 0.2)$ is approximated by the standardised variates probability:

$$P\left(z \le \frac{0.3-0.8}{\sqrt{0.128}}\right) - P\left(z \le \frac{0.1-0.8}{\sqrt{0.128}}\right) \approx 0.0558$$

Try calculating some of the approximate probabilities for the values of $\overline{X}_5$ in the table above.

The table below shows the probabilities and expected frequencies.

$\overline{X}_5$	Observed	Approximate probability	Expected
0.0	2	0.0252	5.04
0.2	9	0.0558	11.16
0.4	30	0.1198	23.96
0.6	34	0.1890	37.80
0.8	43	0.2204	44.08
1.0	36	0.1890	37.80
1.2	26	0.1198	23.96
1.4	12	0.0558	11.16
≥ 1.6	8	0.0252	5.04

If we conduct the goodness-of-fit test, H_0: $\overline{X}_5 \sim N(0.8, 0.128)$; H_1: the normal is not an appropriate model at the five per cent level, the number of degrees of freedom is 8. The critical value of the appropriate chi-squared statistic is 15.507.

Now try calculating the measure of fit $\sum \frac{(O-E)^2}{E}$ for these data.

We should now be in a position to conclude that the value, 6.244 (or thereabouts), of the measure-of-fit statistic is not large, hence the normal distribution is a reasonable fit for the distribution of the sample means.

We have seen that the sample size may be as small as 5, even if the variate is discrete. But we should use the appropriate continuity correction matched to the size of the samples when the sample size is small. When n is large, e.g. in excess of 10, and X is a discrete variate there is no need to use the continuity correction.

Exploration 16.2

A collection of activities on the central limit theorem

- Return to the data in the table on page 446. Calculate:

a) the mean, **b)** the variance,
of the means of the samples of size 5. Compare the calculated values with the values suggested by the central limit theorem.

- Remind yourself about the distribution of random decimals. What is the mean of the distribution? What is the variance of the distribution? This table represents the results of obtaining a sample of four random decimals from a computer and calculating its mean. There are results for 300 samples.

$\overline{X}_4$	Frequency
$\overline{X}_4 < 0.25$	17
$0.25 \le \overline{X}_4 < 0.3$	12
$0.3 \le \overline{X}_4 < 0.35$	19
$0.35 \le \overline{X}_4 < 0.4$	26
$0.4 \le \overline{X}_4 < 0.45$	34
$0.45 \le \overline{X}_4 < 0.55$	89
$0.55 \le \overline{X}_4 < 0.6$	36
$0.6 \le \overline{X}_4 < 0.65$	27
$0.65 \le \overline{X}_4 < 0.7$	15
$0.7 \le \overline{X}_4 < 0.75$	14
$\overline{X}_4 \ge 0.75$	17

a) Calculate estimates of the mean and variance of the empirical distribution of sample means and compare that with the values anticipated by the central limit theorem.

b) Conduct a goodness-of-fit test based on the hypothesis:

$\overline{X}_4 \sim \mathrm{N}\left(0.5, \frac{1}{48}\right)$

- Devise a program for a calculator or a spreadsheet which obtains samples of size 5 from a Poisson distribution, say Poi (1.0) and calculates their means. Carry out a test to determine if the normal distribution suggested by the central limit theorem provides a reasonable fit for the distribution of means.

Example 16.2

The life of a cut-price light bulb is a random variable which may be modelled by an exponential distribution with mean 300 hours. Hence, its variance is 300². These cut-price bulbs are sold in special packs of nine. Use the central limit theorem to estimate the probability that the average life of the bulbs in a special pack is in excess of 500 hours.

Solution

The central limit theorem suggests that the mean life of a pack of nine bulbs is distributed:

$$\mathrm{N}\left(300, \frac{300^2}{9}\right) = \mathrm{N}\left(300, 100^2\right)$$

Hence: $P\left(\overline{X}_9 > 500\right) = 1 - \Phi\left(\frac{500 - 300}{100}\right)$

$= 0.022\,75$

EXERCISES

16.1 A

1 The daily takings from a corner shop is a normal variate with a mean of £800 and a variance of 2500.

a) Find the probability that the takings for one day exceed £900.
b) Find the probability that the mean daily takings for a five-day period are greater than £850.
c) Find the probability that the mean daily takings for a six-day period lie between £780 and £820.
d) Find the interquartile range for the mean daily sales over five days.

2 The time taken to serve customers at a post office can be modelled as a random variable with a mean of 3 minutes and a variance of 4.

a) Find the probability that a transaction takes less than two minutes.
b) Find the probability that the mean time for ten transactions is greater than 3.5 minutes.
c) When Rachel joins the queue one person is about to be served and three others are standing in front of her in the queue. Find the probability that she will reach the counter in less than ten minutes.
d) Emma is standing behind Rachel in the queue. Find the probability that Emma is served within 14 minutes.

3 The length of straws is a random variable that is uniformly distributed between 20 and 21 cm.

a) Find the mean and variance of this random variable.
b) Estimate the probability that the average length of 40 straws exceeds 20.6 cm.
c) Estimate the probability that the total length of ten straws is less than 202 cm.

4 An organic food store buys apples that are described as 'Top grade'. The mass of these apples is a random variable with mean 200 grams and variance 100. Apples from other grades are lighter.

a) The mean mass of apples from a box of 20 is 195 grams. Carry out a test to investigate whether these are 'Top grade' apples or some other grade.
b) Repeat the test for another box of 20 apples with a mean mass of 198 grams.
c) What is the minimum mass for a box of 20 apples which could satisfy an inspector that they are 'Top grade'?

EXERCISES

16.1 B

1 The quantity of cheese sold each week in a grocery shop is a normal variate with a mean of 60 kg and a variance of 225.

a) Find the probability that less than 40 kg will be sold in a week.
b) Find the probability that the mean weekly sales in a five-week period are less than 55 kg.

c) Find the probability that mean weekly sales over eight weeks lie between 58 kg and 62 kg.
d) Find the interquartile range for mean weekly sales over eight weeks.

2 The time taken to conduct a foreign exchange transaction may be modelled as a random variable with a mean of 8 minutes. The variance may be assumed to be 4.

a) Find the probability that a transaction will take longer than ten minutes.
b) Find the probability that the mean time taken for the next four transactions is over 10 minutes.
c) Jean is standing fourth in the queue. Find the probability that she will reach the counter in less than 20 minutes.
d) Jane is standing seventh in the queue. Find the probability that she will reach the counter in less than an hour.

3 The length of wood off-cuts is a random variable, uniformly distributed between 0 and 0.5 m.

a) Estimate the probability that the average length of 20 off-cuts is over 0.3 m.
b) Estimate the probability that the combined length of ten off-cuts exceeds 2 m.
c) Miscellaneous lots of 100 off-cuts are sold. Find the interquartile range for the average length of off-cut in each lot.

4 The weight of 'A' grade oranges is a random variable with a mean of 150 g and a variance of 25. Other grades of oranges weigh less.

a) The mean weight of oranges from a bag of twelve is 140 g. Carry out a test to investigate whether these oranges are grade 'A' or another grade of orange.
b) Repeat the test for a bag of twelve oranges with a mean weight of 145 g.
c) What is the minimum weight for a bag of twelve oranges which would reasonably satisfy an investigator that the bag does consist of grade 'A' oranges?

CONFIDENCE INTERVALS

In Chapter 14, *The normal probability model*, we discovered that about five per cent of observations of a normal variate are more than two standard deviations away from the mean. The converse of this is that 95 per cent of observations of a normal variate are within two standard deviations of the mean. In terms of the standard normal variate, this is equivalent to saying:

$$P(|z| \le 2) \approx 95\%$$

A closer examination of the standard normal tables will reveal that $P(|z| \le 2)$ is more accurately given as 95.5 per cent. Strictly, 95 per cent of normal variates are within 1.96 standard deviations of the mean. So a more accurate statement is:

$$P(|z| \le 1.96) = 95\%$$

Recall that when X is a normal variate, with mean μ and variance σ^2, then the sample mean, $\overline{X}_n$, is normally distributed with mean μ and variance $\frac{\sigma^2}{n}$. This implies that:

$$P\left(\left|\frac{\overline{x}_n - \mu}{\sigma/\sqrt{n}}\right| \le 1.96\right) = 95\%$$

which is to say that 95 per cent of all sample means are within 1.96 standard errors of the underlying population mean, μ.

This result has useful implications when sampling is being used to estimate an otherwise unknown population mean, μ. It is possible to set up an interval for μ by algebraically rearranging this result.

We know that the modulus function, $|x|$, has the property that:

$$|x| = |-x|$$

This implies that:

$$\left|\frac{\overline{x}_n - \mu}{\sigma/\sqrt{n}}\right| = \left|\frac{\mu - \overline{x}_n}{\sigma/\sqrt{n}}\right|$$

Also, if $|x| < a$, then, this is equivalent to saying that x lies between $-a$ and a. In other words, the 95 per cent statement is the same as:

$$-1.96 \le \frac{\mu - \overline{x}_n}{\sigma/\sqrt{n}} \le 1.96$$

Firstly multiplying by $\frac{\sigma}{\sqrt{n}}$ gives:

$$-1.96 \times \frac{\sigma}{\sqrt{n}} \le \mu - \overline{x}_n \le 1.96 \times \frac{\sigma}{\sqrt{n}}$$

and then adding $\overline{x}_n$ gives:

$$\overline{x}_n - 1.96 \times \frac{\sigma}{\sqrt{n}} \le \mu \le \overline{x}_n + 1.96 \times \frac{\sigma}{\sqrt{n}}$$

so we discover an interval for the population mean, μ, based on a sample mean, $\overline{x}_n$.

Interpretation

The best way to interpret this interval is to imagine taking a sample of size n from a population, calculating its mean, constructing the interval then repeating the whole process many, many times. We can have hundreds and hundreds of intervals. This result says that 95 per cent of the intervals will contain the population mean and five per cent will not.

Such intervals are known as **95 per cent confidence intervals**, or more accurately, **symmetric 95 per cent confidence intervals** for the population mean, μ:

$$\bar{x}_n \pm 1.96\frac{\sigma}{\sqrt{n}}$$

Example 16.3

A sample of five servings of muesli had an average energy content of 120 calories. Assume that the energy content of a serving of muesli is normally distributed and has a variance of 100. Obtain a symmetric 95 per cent confidence interval for the population mean energy content of a serving.

Solution

A symmetric 95 per cent confidence interval is $\bar{x}_n \pm 1.96\frac{\sigma}{\sqrt{n}}$ *which in this case is:*

$$120 \pm 1.96 \times \frac{10}{\sqrt{5}} = 120 \pm 8.77$$

This is equivalent to saying that the 95 per cent confidence interval for μ *is:*

$$111.2 \le \mu \le 128.8$$

Other confidence intervals

The argument which led to the concept of a 95 per cent confidence interval was based on:

$$P(|z| \le 1.96) = 95\%$$

of the standard normal variate. If we explore the normal tables we find that:

$$P(|z| \le 2.326) = 98\%$$

This leads to a symmetric 98 per cent confidence interval for μ:

$$\bar{x}_n \pm 2.326\frac{\sigma}{\sqrt{n}}$$

Similar investigation leads to the results shown in this table.

	Confidence interval
Level	Symmetric limits
90%	$\bar{x}_n \pm 1.645\frac{\sigma}{\sqrt{n}}$
95%	$\bar{x}_n \pm 1.96\frac{\sigma}{\sqrt{n}}$
98%	$\bar{x}_n \pm 2.326\frac{\sigma}{\sqrt{n}}$
99%	$\bar{x}_n \pm 2.576\frac{\sigma}{\sqrt{n}}$

Example 16.4

Nine students bought new calculator batteries from their college shop. They agreed to keep an accurate record of the life-span, in hours, of the batteries, with the following results.

71.7, 80.0, 130.7, 129, 85.8, 103.4, 101.7, 91.0, 93.7

a) *Calculate the mean life-span of this collection.*
b) *Calculate a symmetric 99 per cent confidence interval for the mean life span of this make of calculator battery assuming that the standard deviation of the life span is 12 hours.*

Solution

a) *The total of the life span is 887 hours, hence the mean is $887 \div 9 = 98.556$ hours.*
b) *We are seeking a 99 per cent confidence interval for the population mean, μ, which leads to:*

$$98.556 \pm 2.576 \times \frac{12}{\sqrt{9}} \quad \text{i.e.} \quad 98.556 \pm 10.304$$
$$\Rightarrow 88.25 \le \mu \le 108.86$$

Example 16.5

Last year there were 17 reported losses of fishing boats in coastal waters. Estimate a 90 per cent confidence interval for the mean number of fishing vessels lost annually. State any assumptions you feel are relevant.

Solution

If we assume that losses of fishing vessels are random events in time, then a Poisson distribution may be appropriate to model the number of vessels lost in a year. The evidence suggests that the mean is quite large, hence, it may be possible to approximate the probability distribution using N(17, 17). Under these circumstances, we have information from one year's observations and hence an approximate 90 per cent confidence interval is:

$$17 \pm 1.645 \times \frac{\sqrt{17}}{\sqrt{1}}$$
$$\Rightarrow 10.2 \le \mu \le 23.8$$

Large samples

We now think about the information required to produce a confidence interval: we need to be able to establish that the sample mean is normally distributed, and we need to know the actual population variance. The first of the conditions is determined by the nature of the random variable under observation. The second is somewhat perplexing.

In constructing a confidence interval, we are trying to produce limits for the unknown population mean. Yet, in order to do this we are required to know the population variance. This is somewhat implausible!

The issue at stake is the use of normal probabilities to model the distribution of sample means. This is justifiable provided the population variance is known or may be assumed. Otherwise, the justification is not valid – unless the sample size is large. In this case, 'large' may be taken as greater than 30 for most populations. Under these circumstances it is reasonable to estimate the population variance by the unbiased variance, s_{n-1}^2, which you may recall from Chapter 3, *Summary statistics 2*, or you may have seen on your calculator.

$$s_{n-1}^2 = \frac{n}{n-1}s^2$$
$$= \frac{n}{n-1}\left(\frac{\sum x^2}{n} - \bar{x}^2\right)$$

Then we use the standard deviation associated with this unbiased variance in place of σ. The confidence intervals then take the form:

$$\bar{x} \pm z_c \times \frac{s_{n-1}}{\sqrt{n}}$$

where z_c takes the value appropriate to the confidence level required.

Example 16.6

A survey of 150 households asked how many people regularly eat muesli for breakfast. The results of the survey are summarised as follows.
$n = 150$, $\Sigma x = 173$, $\Sigma x^2 = 355$

a) *Obtain an unbiased estimate of the population variance.*
b) *Obtain approximate 95 per cent confidence interval for the mean number of people who regularly eat muesli for breakfast.*

Solution

a) s_{n-1}^2 *is an unbiased estimate of population variance, and hence:*

$$s_{n-1}^2 = \frac{150}{149}\left\{\frac{355}{150} - \left(\frac{173}{150}\right)^2\right\}$$
$$= 1.0434$$

b) *The sample size, 150, is large. We are reasonably justified to use the normal distribution, even though the population variance is unknown, to construct a 95 per cent confidence interval. This gives the confidence interval as:*

$$\frac{173}{150} \pm 1.96 \times \frac{\sqrt{1.0434}}{\sqrt{150}} \quad \text{i.e.} \quad 1.18 \pm 0.163$$

Sample size estimation

Statisticians are often asked, 'How large a sample should I take?' The simple answer is, 'Well, it depends …' which is about as accurate an answer as can be given without more information. However, our discoveries in this chapter put us in a position to seek the further information and thus to provide a more satisfactory answer.

Suppose we have discovered that the sampling is from a normal distribution with known variance, σ^2. Our enquirer wants to use the sampling to be 90 per cent certain that the sample mean is within, say five units, of the population mean. We can use the symmetric 90 per cent confidence interval:

$$\bar{x}_n \pm 1.645\frac{\sigma}{\sqrt{n}}$$

and compare this with the enquirer's requirements:

$$\bar{x}_n \pm 5$$

This implies that the enquirer's error, 5, satisfies the inequality:

$$5 \geq 1.645\frac{\sigma}{\sqrt{n}}$$

Multiplying by $\sqrt{n}$ produces:

$$5\sqrt{n} \geq 1.645\sigma$$

Squaring both sides and dividing through by 5^2 suggests:

$$n \geq 0.108\,241\sigma^2$$

With the variance at our finger tips we should be able to complete the calculation and advise the enquirer how large a sample should be.

Example 16.7

A student wants to determine the mass of the contents of crisp packets. How many packets should be included in the student's sample if he wants to be 95 per cent certain that the sample mean is within 0.1 g of the population mean? (Assume that the variance of the mass of crisps in a packet is 0.0625.)

Solution

It is reasonable to assume that the mass of crisps in a packet is normally distributed: $N(\mu, 0.0625)$. *A symmetric 95 per cent confidence interval for* μ *is:*

$$\bar{x}_n \pm 1.96 \times \frac{\sqrt{0.0625}}{\sqrt{n}}$$

if $\bar{x}_n$ *is to be within 0.1 g of* μ *then:*

$$0.1 \geq 1.96 \times \frac{0.25}{\sqrt{n}}$$

Hence:

$$n \geq \left(\frac{1.96 \times 0.25}{0.1}\right)^2 = 24.01$$

which suggests that he should take a sample of 25 crisp packets.

EXERCISES

16.2 A

1 In a survey, 200 people were asked the length of time that they spent in the shower, the last time that they took one. The results were as follows.
$\Sigma x = 909 \qquad \Sigma x^2 = 4555$

a) Find an unbiased estimate of the population variance.
b) Find a 95 per cent confidence interval for the mean time spent in the shower.

c) Estimate the percentage of the sample that spent more than six minutes in the shower.

2 A transport company records the number of lorry breakdowns per quarter. For a certain year the number in the four quarters were: 6, 8, 10, 7.

a) Find an approximate 90 per cent confidence interval for the mean number of breakdowns per quarter. State any assumptions you make.
b) Combine the figures and find an approximate 90 per cent confidence interval for the mean number of breakdowns per year.
c) Compare your answers to **a)** and **b)**.

3 A promoter wants to find the mean number of people who buy tickets for concerts, but who do not attend the concert.

a) The promoter obtained information from 50 concerts. His results for the number of non-attendances were as follows.
$\Sigma x = 214 \qquad \Sigma x^2 = 1144$
Find, stating any assumptions you make, a 98 per cent confidence interval for the mean number of non-attendances.
b) The promoter makes a policy of selling five more tickets than there are seats available. Estimate the proportion of concerts with more people than seats. State your assumptions.

4 A random sample of 40 boxes of apples was weighed and gave the following results.
$\Sigma x = 442.9 \qquad \Sigma x^2 = 4906.05$

a) Find unbiased estimates of the variance and mean of the mass of a box of apples.
b) Find a 95 per cent confidence interval for the mean mass of a box of apples.
c) If a 95 per cent confidence interval was found using 80 boxes instead of 40, how would you expect the width of the interval to compare with the interval obtained in **b)**?
d) How many boxes of apples should be weighed to find a 95 per cent confidence interval that has a width that is approximately 20 per cent of the width of the interval obtained in **b)**?

EXERCISES

16.2 B

1 In a survey, the time spent by each visitor to an art exhibition was recorded. The results, for 300 visitors, were as follows.
$\Sigma x = 12\ 381$ minutes $\qquad \Sigma x^2 = 571\ 016$

a) Obtain an unbiased estimate of the population variance.
b) Find an approximate 90 per cent confidence interval for the mean time spent at the exhibition.
c) Estimate the percentage of visitors who spend longer than half an hour at the exhibition.

2 At a public library, the numbers of books reported lost in each quarter of a certain year were:
15, 20, 24, 19.

a) Stating any assumptions you make, obtain an approximate 95 per cent confidence interval for the mean number of books reported lost each quarter.
b) Combine the four quarters figures together and estimate an approximate 95 per cent confidence interval for the mean number of books reported lost each year.
c) Compare your answers to **a)** and **b)**.

3 An airline manager wants to determine the mean number of people who fail to appear for a certain regular flight.

a) How many flights should be in his sample if he wants to be 98 per cent certain that the sample mean is no more than 0.2 different from the population mean? (Assume that the variance of the number of non-travellers is 2.25.)
b) The manager decided to obtain data from 400 flights. His results were as follows.
$\Sigma x = 2107$ non-travellers $\quad \Sigma x^2 = 12\,138$
Use these results to calculate an approximate 98 per cent confidence interval for the mean number of non-travellers.
c) The airline makes a practice of selling three more tickets for each flight than the number of seats available. Estimate the proportion of flights with more prospective passengers than seats.

4 A random sample of 36 cartons of soup gave the following results.
Total volume of contents, $\Sigma x = 20.54$, $\Sigma x^2 = 15.02$.
a) Calculate unbiased estimates of the mean and variance of the volume of soup contained in a carton.
b) Calculate a 95 per cent confidence interval for the mean volume of soup in a carton, stating clearly any assumptions you make.
c) If a random sample of 144 cartons of soup was taken, and a 95 per cent confidence interval for the mean volume was calculated, how (approximately) would you expect the width of such an interval to differ from the width of the interval found in **b)**?
d) In order to obtain a 95 per cent confidence interval for the mean volume of soup with a width just one-tenth of that found in **b)**, approximately what size of sample should be taken?

HYPOTHESIS TESTING

There is a direct link between symmetric confidence intervals and two-tailed hypothesis tests about the mean of a normal variate. The following are typical hypotheses:

H_0: population mean $= \mu$; $\quad H_1$: population mean $\neq \mu$

If the test is conducted at the five per cent level of significance the process is analogous with constructing a 95 per cent confidence interval. If the hypothesised mean is within the confidence limits, then the test result is not significant. Equally, if the population mean is outside the limits the test result is significant.

Clearly, a two-tailed test at α per cent may be compared with a $(100 - \alpha)$ per cent symmetric confidence interval.

We can use symmetric confidence intervals in connection with one-tailed hypothesis testing. We need to exercise caution in linking the significance level and the confidence level. An α per cent one-tailed level must be linked with a $(100 - 2\alpha)$ per cent confidence interval. It is not sufficient for the population mean to be outside the confidence limits. A one-tailed test is directional, so the direction of the alternative hypothesis dictates which of the limits of the confidence interval should be used.

Upper limit for H_1: population mean $< \mu$

Lower limit for H_1: population mean $> \mu$

Example 16.8

A sample of ten jars of home-made marmalade were weighed. The total mass of the marmalade was 4663 g. Assume that the standard deviation of the jar of marmalade in a jar is 12 g.

a) *Construct symmetric 90 per cent confidence limits for the population mean of marmalade per jar.*
b) *Test at the five per cent level of significance if there is evidence to support the view that the mean mass of marmalade per jar is less than 470 g.*

Solution

a) *The confidence limits are:*

$$\frac{4663}{10} \pm 1.645 \times \frac{12}{\sqrt{10}} \quad \text{i.e.} \quad 466.3 \pm 6.24$$

Hence the 90 per cent limits on α are [460.06, 472.54].

b) *The alternative hypothesis for this test is:*

H_1: population mean < 470

This is a five per cent one-tailed test and the hypothesised mean is below the upper limit. Hence, there is no reason to support the view that the mean is less than 470.

An alternative approach to using the confidence interval would be to standardise the sample mean, 466.3, using the hypothesised population mean, 470. This results is the test statistic:

$$\frac{466.3 - 470}{12/\sqrt{10}} = -0.975$$

which is clearly not less than the z-value, −1.645, which corresponds to a five per cent level of significance. Hence, the same conclusion is reached.

CONSOLIDATION EXERCISES FOR CHAPTER 16

1 The total number of train cancellations in a certain region, in a 30-day month was 47.

a) Construct a 98 per cent confidence interval for the mean daily number of cancellations. Assume that, from previous studies, the variance of the number of cancellations is 1.4.
b) Test, at the one per cent level of significance, if the evidence is consistent with the claim that, on average, fewer than one cancellation per day occurs.
c) Estimate the probability of at least 60 cancellations in a 30-day month.

2 The number of accidents on a certain stretch of road in each of the last five years was:
19, 22, 23, 14, 22.

a) Stating any assumptions you make, construct a 90 per cent confidence interval for the mean number of accidents per year.
b) Test, at the five per cent significance level, the claim that on average there are fewer than 24 accidents per year.
c) Test this claim at the one per cent significance level.
d) Compare and comment on your answers to **b)** and **c)**.

3 Forty randomly-chosen groups of five students picked mushrooms and the masses, in kg, picked by each group were recorded. Unbiased estimates of the population mean and variance of the mass picked per group were found. A 95 per cent confidence interval for the population mean mass picked per group was calculated as [18.5 kg, 20.9 kg].

a) Find the unbiased estimate of the variance of the mass picked per group. Hence find an estimate of the variance of the mass picked per student.
b) Calculate a 95 per cent confidence interval for the population mean mass picked per student.
c) Find the probability that a group chosen at random picks over 20 kg.
d) Find the probability that a student chosen at random picks over 4 kg.
e) Find the probability that at least three students in a randomly chosen group pick over 4 kg.

4 In a store, the time taken to serve a customer may be taken as normally distributed with a mean of seven minutes and a standard deviation of three minutes.

a) Find the probability that it takes longer than 15 minutes to serve a customer.
b) What percentage of customers who take at least ten minutes to serve finally take longer than 15 minutes?

A group of three customers enter together.

c) What is the probability that the mean time taken to serve them is over ten minutes?

d) If the mean time taken to serve them is over ten minutes, what is the probability that it is over twelve minutes?

5 The safety notice in a lift states that the lift is designed to carry not more than 20 persons or a maximum of 1600 kg. The mass of a person is normally distributed with mean 72 kg and standard deviation 9 kg. Use the central limit theorem to find the probability that the mean mass of 20 persons selected at random will exceed 80 kg. Comment on this result with respect to what the safety notice says.

(UCLES Question 6, Specimen paper S2, 1994)

6 A company employs a large number of clerical staff. Applicants for employment are timed carrying out a standard task which involves entering data into a computer. It is observed that the time each applicant takes is normally distributed with mean 340 seconds and standard deviation 80 seconds.

a) **i)** What proportion of the applicants take longer than 420 seconds?

ii) What proportion of the applicants take between 240 and 420 seconds?

iii) What time is exceeded by five per cent of the candidates?

Applicants who take longer than 420 seconds are automatically rejected and those who take less than 240 seconds are automatically accepted. The remainder are interviewed before a decision is taken.

b) What is the median time taken by those applicants who are interviewed?

Six applicants take the test at the same time.

c) Assuming that they may be regarded as a random sample of all applicants, what is the probability that:

i) their mean time is less than 360 seconds,

ii) they each take less than 360 seconds?

(AEB Question 6, Paper 9, November 1994)

7 Describe what you understand by a '95 per cent confidence interval for μ', illustrating your description by making reference to any experiment you may have conducted.

A random sample of 36 girls in Year 9 had their heights recorded. The resulting data can be summarised as follows.
Total height, Σx, is 57.74 m.

Write down an estimate for the mean height of the population from which the sample was drawn. Calculate a 95 per cent confidence interval for the population mean assuming that the population variance is 0.0049 m^2.

(Oxford Question 8, Specimen paper S2, 1994)

8 A machine is regulated to dispense liquid into cartons in such a way that the amount of liquid dispensed on each occasion is normally distributed with a standard deviation of 20 ml.

Find 99 per cent confidence limits for the mean amount of liquid dispensed if a random sample of 40 cartons has an average content of 266 ml.

(ULEAC Question 1, Specimen paper T2, 1994)

9 A plant produces steel sheets of which the masses are know to be normally distributed with a standard deviation of 2.4 kg. A random sample of 36 sheets had a mean mass of 31.4 kg. Find 99 per cent confidence limits for the population mean.

(ULEAC Question 2, Paper S2, January 1993)

10 Why is the normal distribution an important part of statistics?

A manufacturer of potato crisps produces packets of crisps with a stated mass of 25 g. The masses of the packets of crisps are normally distributed with mean 25.5 g and standard deviation 0.35 g. Calculate, correct to two decimal places, the proportion of the packets of crisps that will be below the stated mass.

A Trading Standards Inspector informs the manufacturer that no more than one packet in 1000 should weigh less than the stated 25 g. To satisfy the inspector's specification, calculate:

a) the minimum mean mass of the packets of crisps produced if the standard deviation remains at 0.35 g,
b) the largest standard deviation permitted if the mean mass of the packets of crisps produced is 26 g.

The manufacturer takes a sample of ten packets of crisps from his current production line and their masses, in grams, are as follows.

25.88	25.51	26.01	25.25	25.40
26.15	25.70	25.82	26.45	25.63

Investigate at the five per cent significance level, whether there is sufficient evidence to conclude that the true mean mass of the packets of crisps is less than 26 g. Assume that this sample comes from a normal distribution with a known standard deviation of 0.35 g.

Based upon your conclusion what action could the manufacturer take regarding his current production process?

(NEAB Question 8, Paper A5, June 1994)

11 A politician, speaking to a journalist claims that school-leavers in his constituency have, on average, six GCSEs. The journalist checks the claim by interviewing a random sample of 100 school-leavers. The data he obtains are summarised below; x denotes the number of GCSEs per person.

$n = 100 \quad \Sigma x = 431 \quad \Sigma x^2 = 2578$

a) Obtain the mean and standard deviation of the data.
b) Construct a 95 per cent confidence interval for the mean number of GCSEs per person.
c) Explain, without further calculation, whether or not the politician's claim is consistent with the journalist's findings.

d) By considering again the mean and standard deviation of the data as calculated in **a)**, explain why the number of GCSEs per person seems to be normally distributed. Show in a sketch a possible shape for the distribution.

e) Explain whether lack of normality does or does not invalidate the confidence interval found in **b)**.

12 A food processor produces large batches of jars of jam. In each batch the gross mass of a jar is known to be normally distributed with standard deviation 7.5 g.

The gross masses, in grams, of a random sample from a particular batch were 517, 481,504, 482, 503, 497, 512, 487, 497, 503, 509.

a) Calculate a 90 per cent confidence interval for the mean gross mass of this batch.

b) The manufacturer claims that the mean gross mass of a jar in a batch is at least 502 g. Test this claim at the five per cent significance level.

c) Explain why, if the manufacturer had claimed that the mean gross mass was at least 496 g, no further calculations would be necessary to test this claim.

(AEB Question 7, Paper 9, June 1994)

13 In a marketing campaign, a free sample of Bianco soap powder is delivered to each of a large number of households in one region of the country. Six weeks later an interviewer calls at a random sample of 200 of these households and asks how many packets of Bianco the householders have bought since they received the free sample. The data are as follows, with x denoting the number of packets bought.

$n = 200 \qquad \Sigma x = 362 \qquad \Sigma x^2 = 1233$

a) Calculate the mean and variance of the data.

b) Construct a 95 per cent confidence interval for the mean number of packets bought in all households receiving the free sample.

Throughout the country as a whole, 16 per cent of households use Bianco, and buy it at an average rate of 0.8 packets per week.

c) Throughout the country as a whole, how many packets of Bianco per household are bought in six weeks? What does this suggest about the effectiveness of the marketing campaign, and why?

d) To what extent, if any, does the method of acquiring the data affect your conclusions?

(MEI Question 4, Paper S3, January 1994)

14 The masses of Granny Smith apples are taken as normally distributed with mean 110 grams and standard deviation 8 grams. The apples are sent to the wholesaler in bags which must contain at least 10 kilograms of apples.

a) Write down the mean, the standard deviation, and the distribution of the total mass of 90 randomly-chosen apples. Show that the probability that this total mass is at least 10 kilograms is approximately 9.4 per cent.

Now suppose that n randomly chosen apples are put into a bag, and that their total mass is T grams.

b) Write down the mean and variance of T in terms of n. Show that the condition for $P(T > 10\ 000)$ to be at least 99 per cent is: $10\ 000 - 110n \leq -18.608\sqrt{n}$

c) Hence find the smallest value of n which gives a probability of at least 99 per cent that the total mass of apples exceeds 10 kilograms.

(MEI Question 2, Specimen paper S3, 1994)

15 A food processor produces large batches of jars of jam. In each batch, the gross mass of a jar is known to be normally distributed with standard deviation 7.5 g. (The gross mass is the mass of the jar plus the mass of the jam.)

The gross masses, in grams, of a random sample from a particular batch were:

514, 485, 501, 486, 502, 496, 509, 491, 497, 501, 506, 486, 498, 490, 484, 494, 501, 506, 490, 487, 507, 496, 505, 498, 499

a) Estimate the proportion of this batch with gross mass over 500 g. Calculate an approximate 95 per cent confidence interval for this proportion.

b) Calculate a 90 per cent confidence interval for the mean gross mass of this batch.

The mass of an empty jar is known to be normally distributed with mean 40 g and standard deviation 4.5 g. It is independent of the mass of the jam.

c) **i)** What is the standard deviation of the mass of the jam in a batch of jars?

ii) Assuming that the mean gross mass is at the upper limit of the confidence interval calculated in **b)**, calculate limits within which 99 per cent of the masses of the contents would lie.

d) The jars are claimed to contain 454 g of jam. Comment on this claim as it relates to this batch of jars.

(AEB Question 5, Paper 9, June 1994)

16 The random variable, X, is normally distributed with mean μ and variance σ^2.

a) Write down the distribution of the sample mean, $\overline{X}$, of a random sample of size n.

An efficiency expert wishes to determine the mean time taken to drill a fixed number of holes in a metal sheet.

b) Determine how large a random sample is needed so that the expert can be 95 per cent certain that the mean time will differ from the true mean by less than 15 seconds.

Assume that it is known from previous studies that $\sigma = 40$ seconds.

(ULEAC Question 5, Paper S1, June 1992)

17 A machine is supposed to produce keys to a nominal length of 5.00 cm. A random sample of 50 keys produced by the machine was such that $\Sigma x = 250.50$ cm and $\Sigma x^2 = 1255.0290$ cm^2, where x denotes the length of a randomly-chosen key produced by the machine.

a) Calculate unbiased estimates of the mean and variance of the length of keys produced by the machine, giving your answers correct to an appropriate degree of accuracy.

b) Calculate a 95 per cent symmetric confidence interval for the mean length of keys produced by the machine, giving your answers correct to an appropriate degree of accuracy.

c) Carry out a hypothesis test at the five per cent significance level to test the assertion that the machine is producing keys which are too long, giving suitable null and alternative hypotheses and stating your conclusion clearly.

Summary

- The distribution of sample mean may be modelled as: $N\left(\mu, \frac{\sigma^2}{n}\right)$
- A confidence interval

$$\bar{x} \pm z_c \frac{\sigma}{\sqrt{n}}$$

provides limits within which the population mean, μ, may lie with a degree of certainty dependent on the value, z_c.

z_c	Confidence level
1.645	90%
1.960	95%
2.326	98%
2.576	99%

- If the population variance is unknown then the unbiased variance, s_{n-1}^2, should be used and the sample size should be large.

STATISTICS

17 *Non-parametric tests*

It is often the case that little is known about the parameters of a population.

In this chapter we introduce statistical tests which:

- *do not require us to know much of the characteristics of the population from which the sample is drawn,*
- *are based simply on relative sizes, on ranks, or a combination of these.*

RANK CORRELATION

In Chapter 13, *Bivariate data*, we discussed the **product moment correlation coefficient** (PMCC) and the way in which it is used to assess evidence of linear association. The value of the sample product moment correlation coefficient, r, for a collection of data pairs is not a reliable guide for assessing evidence of non-linear association. Neither is r reliable when one or both of the original variables has been replaced by positions or ranks (1st, 2nd, 3rd, ... etc.).

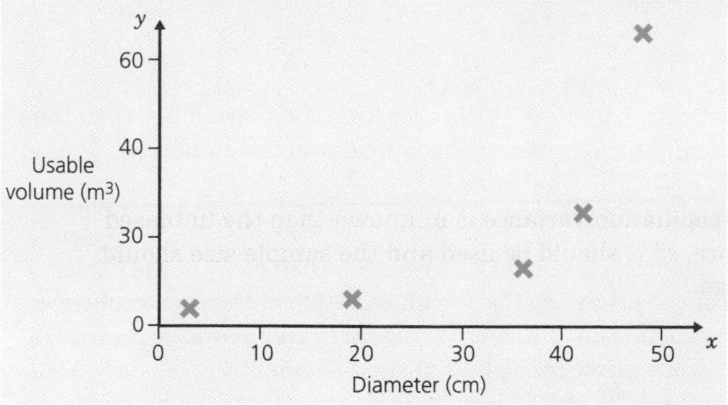

x	y
Diameter (cm)	Usable wood (m^3)
3	5.78
19	8.98
36	19.56
42	38.65
48	69.20

Study the scattergraph above. Do the data appear to be linearly related or might there be some other sort of relationship? Calculate the product moment correlation coefficient for these data.

You should find that $r = 0.8464$. A test, conducted at the five per cent level, of the hypothesis that the population correlation coefficient, ρ, is positive would suggest that this value is significant. The five per cent one-tailed critical value is 0.8054. This supports the view that there is a positive linear correlation in these data. This contrasts with the appearance of a 'curved' relationship.

Non-linear relationship

The data in the diagram and table on page 465 represent the base diameter of a tree and the usable wood in the tree, for five trees of different ages. It is reasonable to expect a positive association between the diameter of a tree and the wood in the tree. But it is not plausible for the relationship to be linear. There is a need for a measure of correlation that supports a positive (or negative) association, where appropriate, and that can be used for non-linear relationships.

Product moment correlation coefficient of the ranks

The simplest answer is to replace the data by their individual positions or ranks. For example, the smallest value of x is 3, so this is replaced by the rank 1. The next smallest x-value is 19, so this is replaced by the rank 2 and so on. The smallest y-value is 5.78 and this is replaced by the rank 1 and so on. The result of this is shown below.

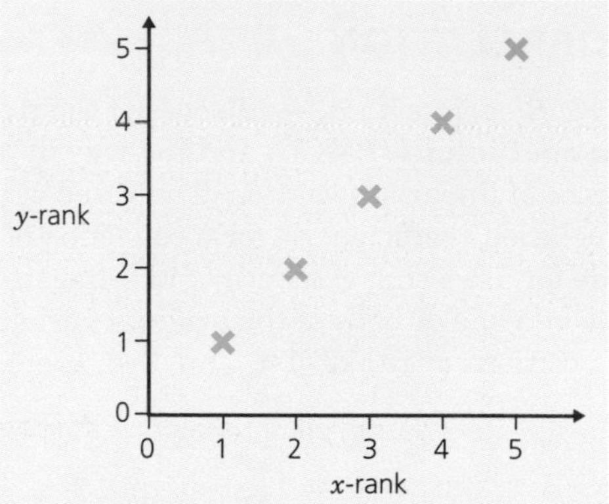

x-rank	y-rank
1	1
2	2
3	3
4	4
5	5

It is quite clear that there is perfect positive correlation between the ranks. What might the correlation coefficient of the ranks be?

Spearman's coefficient

Calculating the product moment correlation coefficient of the ranks rather than of the original values is an example of **rank correlation**. The resulting coefficient is known as **Spearman's rank correlation coefficient**. Notice that the values of the coefficient for the data and for the ranks are different. This is an indication that the distribution of the product moment correlation coefficient and Spearman's correlation coefficient are different. Try Exploration 17.1 to discover something of the distribution of r_S (r_S is the notation used for Spearman's coefficient).

Exploration 17.1

Distribution of r_S

The value of r_S for the ranks in the diagram and table above is exactly 1. Is this sufficient evidence at the five per cent level to support the assertion that the ranks are perfectly positively correlated?

- Show that there are 5! possible distinct arrangement of the five ranks, 1, 2, 3, 4, 5.
- Under the hypothesis that all arrangements of ranks are equally likely show:
  a) $P(r_S = 1) \approx 0.083$ **b)** $P(r_S \leq 0.9) \approx 0.042$ **c)** $P(r_S \leq 0.8) \approx 0.067$.
- Show that the five per cent critical value is 0.9.
- Conduct the hypothesis test and consider your interpretation.

Critical values

Tables of critical values of r_S may be found after the tables for the product moment correlation coefficient. The five per cent critical value for five pairs of ranks is 0.9. The value, 1, for the ranks in the diagram and data on page 466 is greater than this and so it is reasonable to conclude that there is a positive correlation between tree diameter and usable wood in the tree.

These tables may also be used to test two-sided hypotheses and to test for negative correlation.

Example 17.1

Forestry officials claim that, on mountains, trees grow at a slower rate, the higher up the mountain they are planted. The evidence they put forward is in the table below. They measured the growth rate of trees in six plantations at different altitudes. A growth rate of 1 represents a faster growing tree than rate 2 etc. Test, at five per cent, the claims of the officials.

	Lowest altitude			Highest altitude		
Plantation	A	B	C	D	E	F
Growth rate	1	4	2	3	5	6

Solution

The six plantations are at different heights and so can be positioned. We can give the highest plantation the rank 1 and so on. This produces the following pairs of ranks.

x	6	5	4	3	2	1
y	1	4	2	3	5	6

The hypotheses for the test are:

H_0: no correlation; H_1: negative correlation.

The value of r_S (i.e. the product moment correlation coefficient for the ranks) is:

$$r_S = \left(\frac{\sum xy}{6} - 3.5 \times 3.5\right) \div \left(\sqrt{\frac{35}{12}}\sqrt{\frac{35}{12}}\right)$$

$$\approx -0.8286$$

The tables of critical values for Spearman's coefficient indicate that 0.8286 is the upper five per cent one-sided critical value. This has to be interpreted as –0.8286 for the lower five per cent critical value. Hence, the result is significant indicating that there is evidence against the null hypothesis. The forestry officials seem to be justified in making their claim.

An alternative method of calculation

In calculating the product moment correlation coefficient for n pairs, (x, y), of ranks, the means and the variances can be pre-determined. The n ranks are going to be 1, 2, 3, ..., n in some order. Hence:

the mean of ranks = $\frac{n+1}{2}$

and the variance of ranks = $\frac{n^2-1}{12}$

This can be put to good use in calculating the product moment correlation coefficient of ranks.

$$r_S = s_{xy} \div s_x s_y$$

But $s_{xy} = \frac{1}{n}\sum xy - \left(\frac{n+1}{2}\right) \times \left(\frac{n+1}{2}\right)$

$$= \sum \frac{xy}{n} - \left(\frac{n+1}{2}\right)^2$$

and $s_x s_y = \sqrt{\frac{n^2-1}{12}}\sqrt{\frac{n^2-1}{12}} = \frac{n^2-1}{12}$

Hence the only variable in this is xy, the sum of the products of the ranks. This results in:

$$r_S = \frac{12\sum xy - 3n(n+1)^2}{n(n^2-1)}$$

which does not seem to be much of an improvement over using the familiar approach. However, Σxy can be expressed fairly simply in terms of the sum of the squares of the differences between the individual ranks. Consider defining the difference:

$$d = x - y$$

for each pair of ranks. Then:

$$\Sigma d^2 = \Sigma(x - y)^2 = \Sigma x^2 + \Sigma y^2 - 2\Sigma xy.$$

But $\Sigma x^2 = \Sigma y^2 = \frac{1}{6}n(n + 1)(2n + 1)$

Hence $2\Sigma xy = \frac{1}{3}n(n + 1)(2n + 1) - 2\Sigma d^2$

Then:

$$12\Sigma xy - 3n(n + 1)^2 = 2n(n + 1)(2n + 1) - 3n(n + 1)^2 - 6\Sigma d^2$$
$$= n(n^2 - 1) - 6\Sigma d^2$$

Hence:

$$r_S = \frac{n(n^2-1) - 6\Sigma d^2}{n(n^2-1)}$$

$$= 1 - \frac{6\Sigma d^2}{n^3 - n}$$

This alternative way of calculating Spearman's coefficient is preferred by some statisticians. However, you need to consider the facilities you may have available on your calculator to determine if indeed this is a preferred method for you. Notice, also the cautionary point about 'ties' following Example 17.2.

Example 17.2

Two judges at a horticultural show placed entries in the 'miniature tree' class in the order shown in the table. Test at the five per cent level, if there is evidence of agreement between the judges.

Entry	A	B	C	D	E	F	G
Judge X	3	5	1	2	7	4	6
Judge Y	2	7	3	1	5	4	6

Solution

Appropriate hypotheses for the test are:
H_0: no correlation; H_1: positive correlation.
The differences, d, in the ranks are shown below.

	A	B	C	D	E	F	G
d	+1	−2	−2	+1	+2	0	0

Hence, d^2 = 14 and Spearman's rank correlation coefficient is:

$$r_S = 1 - \frac{6 \times 14}{343 - 7} = 0.75$$

The five per cent critical value for seven pairs of ranks is 0.7143, the calculated value for the judges is closer to +1, hence the judges are in agreement.

A cautionary note – what do you do when there are ties?

Suppose that judge X could not decide between her top two miniature trees and so made them 'equal first'. To make the average of her ranks come to 4 as it ought to be, you need to give each of the 'equal first ' the *average*, $1\frac{1}{2}$, *of the two ranks*, 1 and 2, which would otherwise have been the case.

Suppose that judge Y could not decide between the last three and so made them all 'equal fifth'. The average of his ranks must also be kept at 4, and so his final three trees would get the *average*, 6, *of the three ranks*, 5, 6 and 7.

The new table would now look like this.

Entry	A	B	C	D	E	F	G
Judge X	3	5	$1\frac{1}{2}$	$1\frac{1}{2}$	7	4	6
Judge Y	2	6	3	1	6	4	6

The value of Σd^2 is 5.5, hence the value of $1-\frac{6\Sigma d^2}{n^3-n}$ is 0.9018.

However, calculation of r_S using the product moment correlation coefficient of the ranks produces 0.8976. The correct value is the latter. So there is not much difference between the values and it is acceptable to use the Σd^2 result provided there are not too many ties. However, it is far better to use the product moment correlation coefficient of the ranks at all times.

EXERCISES

17.1 A

1 A group of students gained the following marks in their Pure Mathematics and Statistics Modular examinations.

Student	A	B	C	D	E	F	G	H
Pure Mathematics	62	52	48	79	36	47	44	42
Statistics	74	66	52	71	56	73	40	57

a) Rank each set of marks.
b) Calculate Spearman's rank correlation coefficient.
c) Test at five per cent the hypothesis that there is a positive correlation between relative performance in the two examinations.

2 Five 200-metre runners kept a record of the times achieved in training in dry and in wet conditions. The table records the average times.

Competition	A	B	C	D	E
Dry	25.5	25.7	26.2	25.8	23.4
Wet	24.4	25.3	24.8	25.4	26.3

a) Rank these times.
b) Calculate the value of r_S for the ranks.
c) Is there evidence at the ten per cent level that there is a correlation between their relative performance in different conditions?

3 The mass and length of eight pine cones were recorded. The table shows the rank of the data. Low ranks correspond to high mass and great length.

a) Calculate the value of r_S.

Cone	A	B	C	D	E	F	G	H
Mass	2	6	1	5	3	4	7	8
Length	1	5	2	8	3	4	6	7

b) Is there evidence of positive correlation between the mass and length of pine cones?

4 In a ski-jumping contest each competitor made two jumps. The orders of merit for the ten competitors who completed both jumps are shown below.

Ski-jumper	A	B	C	D	E	F	G	H	I	J
First jump	2	9	7	4	10	8	6	5	1	3
Second jump	4	10	5	1	8	9	2	7	3	6

a) Calculate Spearman's rank correlation coefficient for the performances of the ski-jumpers in the two jumps.

b) Using a five per cent level of significance, interpret your result.

(ULEAC Question 4, Specimen paper T2, 1994)

5 A building society analyst used a combination of twelve ratios to measure the overall performance during 1992 of each of Britain's top ten building societies. The table below ranks these societies according to their performance rating. The table also indicates the size of each society as measured by its assets.

Building society	Performance rating	Assets (£million)
Cheltenham & Gloucester	1	14 789
Halifax	2	58 710
Leeds Permanent	3	16 631
National & Provincial	4	10 708
Bradford & Bingley	5	11 910
Britannia	6	8 524
Woolwich	7	20 165
Nationwide	8	34 119
Alliance & Leicester	9	20 479
Bristol & West	10	7 141

Sources: Wrigglesworth, J., 'Major Players' (UBS) 1993
Building Societies Yearbook 1993

Calculate an appropriate measure of correlation for these two sets of data, justifying your choice of measure. Interpret the value you obtain.

(NEAB Question 6, Specimen paper 8, 1996)

6 At the end of a word-processing course the trainees are given a document to type. They are assessed on the time taken and on the quality of their work. For a random sample of twelve trainees the following results were obtained.

Trainee	A	B	C	D	E	F	G	H	I	J	K	L
Quality (%)	97	96	94	91	90	87	86	83	82	80	77	71
Time (seconds)	210	230	198	204	213	206	200	186	192	202	191	199

a) Calculate Spearman's coefficient of rank correlation for the data. Explain what the sign of your correlation coefficient indicates about the data.

b) Carry out a test, at the five per cent level of significance, of whether or not there is any correlation between time taken and quality of work for trainees who have attended this word-processing course. State clearly the null and alternative hypotheses under test and the conclusion reached.

(MEI Question 1, Specimen paper S2, 1994)

7 The marks awarded by two judges to ten contestants are given in the table.

Contestant	1	2	3	4	5	6	7	8	9	10
Judge X	5	6	8	4	7	5	9	4	6	5
Judge Y	8	7	8	6	9	8	9	7	8	7

a) Rank the marks given by each judge taking care over the tied ranks.

b) Obtain the value of Spearman's rank correlation coefficient.

c) Test, at five per cent, the hypothesis that there is positive agreement between the judges.

8 Eight children took a mathematics test and an English test. The marks they each got are shown in the table.

Child	A	B	C	D	E	F	G	H
Mathematics	9	6	9	5	9	5	8	0
English	4	6	9	2	1	9	4	1

a) Carefully rank the marks in each test.

b) Obtain the value of r_S.

c) Test the hypothesis that there is positive correlation between their relative performance in the subjects.

EXERCISES

17.1 B

1 Two judges independently judge the exhibits of five contestants in a flower show. The judges rank them as follows.

Contestant	A	B	C	D	E
Judge X	4	3	1	2	5
Judge Y	4	1	2	3	5

Calculate Spearman's rank correlation coefficient for these data and state, in general terms, how your answers confirms that the two judges agree reasonably closely.

(UCLES Question 3, Specimen paper S1, 1992)

2 In an investigation into the relationship between the speed of athletes and their IQ, nine athletes were timed over a measured distance, and their times and their IQs were recorded, as shown below.

Athlete	A	B	C	D	E	F	G	H	I
Time	12.1	12.5	14.2	13.5	12.1	14.1	14.6	13.4	12.6
IQ	114	110	120	117	102	121	115	109	91

a) Rank the times and the IQs.
b) Calculate the value of Spearman's rank correlation coefficient.
c) Do these data support the view that IQ and speed of an athlete are correlated?

3 Top-Exec is a recruitment agency which specialises in recruiting executives for top management positions. One management vacancy that was advertised attracted 94 applications. The ten best applicants were invited to attend for a day's activities, and were each subsequently given scores according to their performance in interview (I), past experience (P) and performance in a short examination (E) designed to test their professional knowledge. The scores in the interview (I) and the examination (E) were as follows.

Applicant	1	2	3	4	5	6	7	8	9	10
Interview, I	6	4	5	8	9	3	4	9	10	2
Examination, E	3	5	7	7	6	8	7	10	9	4

Calculate the value of Spearman's rank correlation coefficient for these data and comment on your result.

The rank order correlation coefficient obtained between interview (I) and past experience (P) was 0.91 and between past experience (P) and examination (E) was 0.32.

Top-Exec was contemplating dropping the examination in future. Considering the three rank correlation coefficient values, comment on the wisdom of taking this action.

(NEAB Question 3, Paper AS, June 1994)

4 Three judges independently adjudicate the eight finalists in a piano competition. The orders of merit awarded to the pianists by each judge are given on the following page.

Pianist	A	B	C	D	E	F	G	H
Judge X	7	5	3	6	1	2	4	8
Judge Y	2	5	3	6	4	1	8	7
Judge Z	5	2	3	1	6	4	7	8

a) Calculate Spearman's coefficient of rank correlation between:
i) X and Y, **ii)** Y and Z, **iii)** Z and X.
b) Comment on the results you find in **a)**.

5 In a national survey into whether low death rates are associated with greater prosperity, a random sample of 14 areas was taken. The areas, arranged in order of their prosperity, are shown in the table below, together with their death rates. The death rates are on a scale for which 100 is the national average.

Area	A	B	C	D	E	F	G	H	I	J	K	L	M	N
Death rate	66	76	84	83	102	78	100	110	105	112	122	131	165	138

a) Calculate an appropriate correlation coefficient and use it to test, at the five per cent level of significance, whether or not there is such an association. State your hypotheses and your conclusion carefully.
b) A newspaper carried this story under the headline, 'Poverty causes increased deaths'. Explain carefully whether or not the data justify this headline.
c) The data include no information of the age distributions in the different areas. Why would such additional information be relevant?

(MEI Question 1, Paper S2, January 1995)

6 An experiment was conducted to see if there is a correlation between the colour of paint and its drying time. The results, for the 'seven colours of the rainbow' are shown in the table.

Colour	R	O	Y	G	B	I	V
Drying time	10.6	7.3	9.9	9.6	8.8	8.2	9.2

a) Rank the colours and their drying times.
b) Test the data to determine if there is correlation.

7 Nine voters in Scotland and nine in England were asked to give the government a score, out of ten, on each of nine issues. The results are shown in the table.

Issue	A	B	C	D	E	F	G	H	I
Scotland	62	54	46	34	54	46	36	29	14
England	76	59	46	37	35	27	46	17	17

Is there evidence of positive correlation between the opinions of voters in Scotland and England?

8 As part of a survey, two students were selected, randomly, from each of Years 7 to 12 in a large school. One of the questions asked concerned the pocket money they received. The answers were ranked and the results are given in the table.

Child	A	B	C	D	E	F	G	H	I	J	K	L
Year	7	7	8	8	9	9	10	10	11	11	12	12
Pocket money	10	9	8	7	2	6	5	1	4	3	11.5	11.5

a) Rank the years of the students.
b) Calculate the value of r_S.
c) Test the hypothesis that there is a negative correlation between school year and pocket money.
d) Repeat the above test with students K and L excluded. Comment.

SIGN TEST (BINOMIAL)

Single sample case

A forester has decided to fell all trees when their diameter reaches 30 cm. He claims that they then have, on average, 15 m^3 of usable wood. A random sample of 13 felled trees produce the following quantities of usable wood.

11.3, 8.8, 6.4, 8.7, 14.6, 16.2, 12.4, 7.1, 15.2, 10.3, 9.5, 17.2, 7.8

A researcher decides to test the forester's claim against the alternative that the average is less than 15 m^3.

A simple test based on the binomial distribution may be devised on the assumption that the average referred to is the **median**. Recall that the median is the middle value. The hypotheses for the test are:

H_0: median = 15;H_1: median < 15.

The next step in the test involves noting the number of data items in the random sample which are above and below the hypothesised median. A convenient way of identifying this is to code the data using:

\+ if the item of data is above the median
− if the item of data is below the median.

The result of the coding in this case is:

−, −, −, −, −, +, −, −, +, −, −, +, −

These are three '+' signs and ten '−' signs. Recall that the median is the value which divides the population so that half is above it and half below. This implies that the number, X, of '+' (or '−') signs in a random sample of 13 signs may be modelled as:

$X \sim B(13, 0.5)$

If the hypothesised median is too high, there are likely to be fewer '+' signs in the coded sample. So what we need to do to conduct this test is to determine the probability of three or fewer:

$P(X \leq 3) = 0.0461$

(from binomial distribution tables).

This probability is smaller that a five per cent level of significance, so this suggests that the observed number of '+' signs is unusually small. Hence the test result is significant. It is reasonable to reject the H_0 hypothesis in favour of H_1 and conclude that the forester's claim is not supported by these data.

This single sample sign test is a common procedure to test an hypothesised median by using the binomial model:

$X \sim B(n, 0.5)$

where n is the number of 'signs' which result from the coding process. Note that n is sometimes less than the size of the original sample. This happens when there are data that happen to have the same value as the hypothesised median.

Paired sample case

A similar use can be made of the binomial probability model:

$B(n, 0.5)$

when we have paired data. Imagine an extensive estate which has mixed terrain and conditions – high and low ground, steep and shallow slopes, exposed and sheltered, sunny and shaded aspects, acid and alkaline soils etc. The estate manager is proposing a major tree-planting exercise but does not know which of two species of tree to plant.

She decides to conduct an experiment. She identifies twelve distinct growing sites in the estate. In each she plants one of each of the two tree species and then monitors the growth over an extensive period. The trees in any given site are matched at the time of planting. The height of the trees at the end of the period are recorded in this table.

Site	A	B	C	D	E	F	G	H	I	J	K	L
Species X	23	18	26	27	34	28	32	18	14	19	24	16
Species Y	28	24	34	24	34	23	37	28	16	20	24	22

Appropriate hypotheses to test if there is a difference between the growth of the species are:

H_0: median of $(Y - X) = 0$; H_1: median of $(Y - X) \neq 0$.

These hypotheses refer to the single random variable $(Y - X)$. A test similar to the single sample sign test can be conducted by recording the sign of the difference, $Y - X$, for the data:

+, +, +, −, 0, −, +, +, +, +, 0, +

Whilst there are twelve differences, there are only ten signs. The number, N, of '−' signs (or '+') may be modelled as:

$N \sim B(10, 0.5)$

In this case there are two '−' signs and the probability:

$P(N \leq 2) = 0.0547$

The alternative hypothesis indicates a two-tailed test, so we need to consider:

$P(N \geq 8) = 0.0547$

in addition. Clearly the total of these is greater than a five per cent level of significance. (Indeed, the total is greater than ten per cent.) It would be reasonable to conclude that this is not an unusual result under the null hypothesis. Hence, there is insufficient evidence to reject the hypothesis of no average difference in growth between the species.

Large samples

Recall that $B(n, 0.5)$ is a symmetric distribution. If, in conducting a binomial sign test, the number of signs is large, we may be justified in approximating the probabilities by using the normal distribution:

$N(\frac{1}{2}n, \frac{1}{4}n)$

incorporating the continuity correction.

EXERCISES

17.2

1 Applicants to university may have to wait before being invited to interviews. One university claims that the median waiting time for its applicants is 31 days. A random sample of 18 successful applicants, asked how long they waited before being interviewed, gave these times.

33	32	35	30	33	37	30	31	28
29	31	31	32	34	30	31	33	30

Test, at the ten per cent level, the claim that the median is 31 days against the alternative that it is longer.

2 Airlines occasionally run courses for people who have a fear of flying. Ten people who attended such a course were asked to rate the course on a scale of 1 (very dissatisfied) to 10 (very satisfied). The results were as follows.

1	2	4	8	8	8	9	9	10	10

Use a single-sample sign test to test, at the ten per cent significance level, the null hypothesis $M = 6$ against the alternative hypothesis $M > 6$, where M is the median rating given for such courses. Give full details of your method.

(UCLES Question 3, Specimen paper S3, 1994)

3 The times, in minutes, spent travelling to school by the 15 members of a statistics class are as follows.

13 23 15 21 22 18 30 25 45 12 17 24 32 28 29

The school prospectus claims that 'half of the pupils live less than 20 minutes from the school'.

a) Use a one-tailed sign test, at the five per cent level, to test this claim.

b) Find the critical region of this test.

(ULEAC Question 2, Specimen paper T2, 1994)

4 At the beginning of 1990, the southern part of Britain experienced winds of a ferocity more commonly associated with more northerly parts. A television presenter commented that the winds were predominantly southern in direction and cited, in support of his assertion, a period of 20 days spanning the end of January and beginning of February where the records showed the following.

Predominant wind direction on 20 consecutive days
S N S S S N N S S S S S S S N S S N N S

Superficially the data appear to support the presenter's point of view. A statistically-minded viewer decides to see to what extent this stands up to closer examination.

a) If the viewer makes the assumption that there is no preferred direction, what is the probability that the wind direction is south (S) rather than north (N)? Write down suitable full null and alternative hypotheses to test the claim of the presenter.

b) Briefly explain why it is appropriate to use a binomial distribution, stating its parameters, and carry out the test to judge the claim of the presenter. Conduct your test at the five per cent level of significance and make your conclusion clear.

c) Explain briefly what is meant by the phrase 'five per cent level of significance'.

(Oxford Question 3, Specimen paper S2, 1994)

5 An enlightened college canteen manager introduced a new range of foods described as appropriate for the 'health-conscious'. As an inducement, she offered this range of foods at a discounted price. The average cost of a meal had been £1.12.

a) Twelve students were selected randomly and asked how much they had spent on this meal. The results were:

£0.97 £1.15 £1.10 £1.02 £1.07 £1.13
£0.98 £1.12 £1.09 £1.10 £1.15 £1.11

Test, at the ten per cent level, the hypothesis that students spend less, on average, after the introduction of the 'healthy' food.

b) One week after the 'healthy' campaign, 500 students were asked if they had spent more, less, or the same on food during that week than normally. The results are summarised in the following table.

	More	The same	Less
Number of students	164	135	201

Test at the five per cent level, the hypothesis that students spend less on food during the 'healthy' week. Explain the procedure you adopt.

6 A petrol company claims that cars will achieve a higher mileage per litre if they use their 'Super unleaded' petrol as opposed to the 'Ordinary unleaded' petrol. Twelve cars of the same model were tested, using 25 litres of each type of petrol and the miles they travelled are shown in the table below.

	Miles completed	
Car	Ordinary	Super
1	165	168
2	164	164
3	172	176
4	160	163
5	178	175
6	159	166
7	162	162
8	169	167
9	173	178
10	166	174
11	164	168
12	170	172

Perform a sign test, at the five per cent significance level, to investigate the hypothesis that the 'Super unleaded' petrol gives a higher mileage than the 'Ordinary unleaded' petrol.

(NEAB Question 3, Paper AS, June 1994)

7 Ten people volunteered to carry out a task before and after consuming two measures of alcohol. The results are shown below. Test, at the five per cent level, if there is evidence that the times after the consumption of alcohol are longer, on average, than before.

	Volunteer	A	B	C	D	E	F	G	H	I	J
Time	Before (s)	42	25	50	43	43	63	61	58	63	29
	After (s)	41	24	49	44	42	62	62	59	62	28

8 Twenty students on a mathematics course are paired on the basis of their perceived ability at the start of the course. One student in each pair is taught the course by a lecturer who uses modern technology extensively. The other students are taught by a lecturer using traditional approaches. Their end-of-course results are shown in the table below.

Pair	1	2	3	4	5	6	7	8	9	10
Traditional	70	74	37	60	68	61	44	72	61	51
Modern	72	69	47	64	70	55	50	63	58	54

Is there evidence at the ten per cent level, that the modern methods are more effective than traditional approaches?

9 Twelve A-Level students attended an intensive revision course. Their performance before and after the course was measured. The results are shown in the table below.

Student	A	B	C	D	E	F	G	H	I	J	K	L
Before	38	48	31	63	36	72	59	58	31	37	60	62
After	45	54	30	72	48	63	74	69	29	44	78	74

Test, at the ten per cent level, the hypothesis that the revision course improves performance.

10 **a)** Fifteen people attended a fitness course. One of the vital statistics recorded is the weight loss. This is shown in the table. A negative loss is equivalent to a gain!

Person	A	B	C	D	E	F	G	H	I	J	L	M	N	O	P
Loss	9	8	–3	5	–5	–2	2	1	2	2	3	–2	1	5	6

Test the hypothesis that the fitness course results in a weight loss for the participants. Conduct your test at:

i) the five per cent level

ii) the ten per cent level.

b) A National Fitness Week took place. Twelve thousand people took part, and 5826 of these reported a weight loss. There were 610 who reported no change and the remainder gained weight. Test the hypothesis that participation in the Fitness Week resulted in a weight loss. Conduct your test at the one per cent level of significance.

SIGN TEST (SIGNED RANK)

Paired data

Although the binomial sign test is simple to apply, it does not always take into account all the available information. In the **paired-data case**, there is often a lot of information which is neglected.

Consider, again, the estate manager's experimental results, on page 476. The binomial sign test takes into account the *sign of the difference* in growth but not the *magnitude* of the difference. It is possible for the positive differences to be considerably larger than the negative differences but the binomial sign test ignores this additional information. The table below shows the actual differences of the growth of the two species at the twelve sites, distinguished into positive and negative differences.

	Site	A	B	C	D	E	F	G	H	I	J	K	L
Difference	+	5	6	8				$5\frac{1}{2}$	$9\frac{1}{2}$	2	$1\frac{1}{2}$		6
$Y - X$	−				$3\frac{1}{2}$		5						

When the resulting differences are ranked ignoring the sign of the differences (but still keeping track of whether it is positive or negative), the following table is produced.

	Site	A	B	C	D	E	F	G	H	I	J	K	L
Rank	+	$4\frac{1}{2}$	$7\frac{1}{2}$	9				6	10	2	1		$7\frac{1}{2}$
	−				3		$4\frac{1}{2}$						

Notice that the differences in sites E and K are zero, hence they are not included. Where the ranks are tied the average rank is given to each difference. The reason for keeping track of the signs is that the totals of the 'positive ranks' and 'negative ranks' are now calculated:

$t_+ = 47\frac{1}{2}$ $\qquad t_- = 7\frac{1}{2}$

Critical values

Tables of critical values of the *smaller* of these totals are the tables of the **Wilcoxon signed rank statistic**, T. The assumption that there is no average difference in the growth rates is equivalent to the hypothesis that the ranks are equally likely to be positive or negative. This in turn, implies that the T_+ and T_- totals are the same. Hence, if the smaller total is unusually small there would be reason to conclude that the null hypothesis of no difference was not valid.

In the case of the estate manager, she wanted to know if there was a difference i.e. the test to be carried out is two-tailed. The hypotheses are as before:

H_0: median difference = 0; $\qquad H_1$: median difference $\neq 0$.

The tables indicate that the two-sided, five per cent critical value for ten pairs of data is 8. The test statistic is t_-, in this case:

$t_- = 7\frac{1}{2} < 8$, the critical value.

This is smaller than the critical value hence the result is unusual under the null hypothesis. This suggests that there is a difference in growth between the two species of tree.

The result using this test in this instance is in contrast to the previous test. The **Wilcoxon signed rank test** is a more powerful test.

The conclusions we reach are more reliable because they are based on more of the information contained in the data than is used in the binomial sign test.

Example 17.3

An experiment was set up to establish if a high-fibre diet was beneficial in a fitness programme. Sixteen athletes took part in the experiment, with the results as shown in the table. Is there evidence, at one per cent, to support the view that the diet produces an improvement in performance?

	Athlete	A	B	C	D	E	F	G	H	I	J	K	L	M	N	O	P
Time	After	47	43	46	42	41	45	46	42	44	49	45	46	46	42	45	43
	Before	46	48	49	49	49	41	48	47	43	48	48	50	45	49	45	48

Solution

A large positive value of the measure, (time before – time after), indicates an improvement in performance. The signed differences are as shown below.

+		5	3	7	8		2	5			3	8		4		5,
–	1					4			1	1			1			

The ranks of these differences, taking into account ties are as shown below.

+		10	$6\frac{1}{2}$	$12\frac{1}{2}$	$14\frac{1}{2}$		5	10			$6\frac{1}{2}$	$14\frac{1}{2}$		$12\frac{1}{2}$	.	10
–	$2\frac{1}{2}$					8			$2\frac{1}{2}$	$2\frac{1}{2}$			$2\frac{1}{2}$		.	

The totals of the positive and the negative ranks are:

$t_+ = 102 \qquad t_- = 18$

The hypotheses for the test are:
H_0: median difference = 0; H_1: median difference < 0.
There are 15 signed ranks since there was no change for one athlete, and the one-sided, one per cent level, critical value is 19. The test statistic in this case is t_-:

$t_- = 18 < 19$

Hence this is unusually small.
It is reasonable to conclude that the performance is improved following a high-fibre diet.

Distribution of *T*

The assumption underlying the Wilcoxon signed rank statistic is that the ranks are equally likely to be positive or negative. Consider the case where there are six signed ranks. Since there are two signs possible for each of the six ranks, the total number of distinct and equally likely arrangements of signs is 2^6. The probability of any given arrangement is therefore:

$$\frac{1}{2^6} = 0.015\,625$$

The following table shows ten of the 64 possible arrangements. It starts with the arrangement where $t_- = 0$.

Rank	1	2	3	4	5	6	t_-	Cumulative probability
	+	+	+	+	+	+	0	$\frac{1}{2^6} \approx 0.0156$
	−	+	+	+	+	+	1	$\frac{2}{2^6} \approx 0.0313$
	+	−	+	+	+	+	2	$\frac{3}{2^6} \approx 0.0469$
	+	+	−	+	+	+	3	$\frac{5}{2^6} \approx 0.0781$
	−	−	+	+	+	+		
	+	+	+	−	+	+	4	$\frac{7}{2^6} \approx 0.1094$
	−	+	−	+	+	+		
	+	+	+	+	−	+	5	$\frac{10}{2^6} \approx 0.015\,63$
	−	+	+	−	+	+		
	+	−	−	+	+	+		

We can determine critical values of T from the part of the distribution for six-signed ranks shown in the table. The ten per cent level critical value for a one-sided test is 3 since:

$$P(T \le 3) = 0.078 < 10\%$$

$$P(T \le 4) = 0.109 > 10\%$$

and it is fair to comment that a test statistic equal to 4 should give cause for consideration of rejection of H_0 since 10.9 per cent is rather close to the significance level.

What would be an appropriate critical value for a one-sided test conducted at five per cent? The table indicates:

$$P(T \le 2) = 0.047 < 5\%$$

$$P(T \le 3) = 0.078 > 5\%$$

and hence, the critical value is 2. Try determining the two per cent one-sided critical value. Check your answer in the tables. Note that there is no one per cent critical value.

Exploration 17.2

Five signed ranks

- Construct a table, similar to the one above, for five signed ranks and show that the one-sided critical value:

a) for a five per cent significance level is 0,
b) for a ten per cent significance level is 2.

- Construct a similar table for seven signed ranks and check your agreement of the critical values identified in the tables.
- Repeat the last part for eight signed ranks.

EXERCISES

17.3

1 To investigate the effect of alcohol on reaction time, ten subjects were each given a stimulus and their reaction time to that stimulus was recorded. On one occasion the stimulus followed the consumption of alcohol and on another the stimulus was given when alcohol had not previously been consumed. The reaction times, in seconds, were as follows.

Subject	No alcohol	After alcohol
1	5.1	4.7
2	3.1	3.6
3	3.0	2.8
4	3.3	3.0
5	3.6	4.3
6	3.8	4.7
7	5.3	5.9
8	4.9	6.1
9	6.4	6.5
10	1.7	2.5

Stating clearly your hypotheses, carry out:

a) a sign test,
b) a Wilcoxon matched-pairs, signed-ranks test
to ascertain whether or not alcohol increases reaction time. In both cases use a five per cent level significance.
Comment on your results.

(ULEAC Question 6, Specimen paper T3, 1994)

2 In order to assess the difference in wear on the two tyres, six motorcycles, each initially with new tyres, were ridden for 1000 miles. After this time, the depth of tread on each tyre was measured and recorded. The results are shown in the table below.

Motorcycle	A	B	C	D	E	F
Depth of tread on front tyre (mm)	2.4	1.9	2.3	1.9	2.4	2.5
Depth of tread on rear tyre (mm)	2.0	2.1	2.0	2.0	1.8	2.0

Use a paired-sample Wilcoxon signed rank test to test the assertion, at the ten per cent significance level, that there is no difference in wear for the front and rear tyres. You need not state null and alternative hypotheses but should give a clear conclusion.

(UCLES Question 6 (part 1), Specimen paper S3, 1994)

3 Eight swimmers decided to experiment with their method of starting. Their times for one length of a pool using different starting techniques are shown in the table.

Swimmer	A	B	C	D	E	F	G	H
Technique 1	24.07	24.24	24.35	24.33	24.21	24.11	24.03	24.42
Technique 2	24.13	24.29	24.41	24.34	24.18	24.08	24.11	24.44

Test at the ten per cent level, whether there is a difference in the times using the two techniques.

4 Eight students studying History are assessed in their understanding of Contemporary Issues and Ancient Issues. The results are shown in the table below.

Student	A	B	C	D	E	F	G	H
Ancient	65	83	76	61	32	27	43	37
Contemporary	63	64	43	76	38	64	57	51

Is there evidence to support the hypothesis that their understanding of Ancient Issues is less than that of Contemporary Issues? Conduct your test at the five per cent level of significance.

5 Candidates for clerical positions in a large bank are given two tests, an oral test and a written test. Is there evidence in the data to support the view that there is a difference in the outcomes of the two tests? Conduct the test at the ten per cent level.

Candidate	1	2	3	4	5	6	7	8	9	10
Oral	80	53	89	75	50	73	81	79	65	64
Written	79	49	80	65	52	65	84	82	60	55

6 A large food-processing firm is considering introducing a new recipe for its ice-cream. In a preliminary trial a panel of eleven tasters were asked to score ice-cream, made from both the existing and the new recipe, for sweetness. The results, on a scale of 0 to 100 with the sweeter ice-cream being given the higher scores, were as shown in this table.

Taster	A	B	C	D	E	F	G	H	I	J	K
Old	88	35	67	17	24	32	8	44	73	47	25
New	94	49	66	82	25	96	14	56	27	44	79

Test whether the new recipe is sweeter than the old one using a five per cent level of significance and:

a) the binomial sign tests,
b) the Wilcoxon signed rank test.

7 It is believed that alpine flower seeds have a better germination rate when refrigerated before planting. Consider the following data which records the germination rate for nine varieties of alpine flowers.

Variety	A	B	C	D	E	F	G	H	I
Refrigerated	88.6	75.2	82.3	85.2	76.9	97.1	63.1	68.5	99.0
Unrefrigerated	40.3	38.1	53.0	85.4	84.5	54.3	50.2	31.2	75.3

a) Test the belief at the five per cent level of significance:
 i) using the binomial sign test,
 ii) using the signed rank test.
b) Comment on the results.
c) Explain the expression 'five per cent level of significance'.

8 In a training session, nine athletes run 100 m, then rest 15 minutes and run another 100 m. The times are recorded below.

Athlete	A	B	C	D	E	F	G	H	I
First 100 m	10.4	11.2	10.1	10.7	10.6	10.9	10.9	10.6	10.7
Second 100 m	10.6	11.3	11.9	10.6	10.7	10.6	12.3	10.4	11.7

Is there sufficient evidence to suggest that the times for the second 100 m are slower than for the first?

9 A solicitor estimates how long each job will take and charges clients on the basis of these estimates. In practice, a job may take more time or less time than estimated, but then there is neither an additional charge nor a refund.

A regular client complains that jobs rarely take as long as estimated, and that he is being overcharged. The solicitor decides to investigate the client's claim. She consults the records on a random sample of recent jobs and determines for each the number of hours by which the estimated time exceeded the actual time taken. The figures obtained are as listed below.

6.25, –2.00, –1.50, 2.75, 3.50, 2.50.

a) Write down the null and alternative hypotheses for an appropriate significance test. Define any symbols you use, and explain why the alternative hypothesis takes the form it does.
b) Carry out the hypothesis test, using a five per cent significance level. State any assumption(s) required for the test to be valid.
c) What objection(s) might the client reasonably make regarding the solicitor's investigation?

(MEI Question 3, Paper S3, January 1995)

10 i) Conduct Wilcoxon signed rank tests on the data in the following questions of Exercises 17.2.
a) question **6** b) question **7** c) question **8**
d) question **9** e) question **10 a)**
ii) In which situations do you draw different conclusions?
iii) Which of the conclusions are the more reliable?

TWO SAMPLE RANK TEST

Whether we regard it as environmental or financial, forestry is a long-term investment. It takes a minimum of ten years before even considering harvesting and it is likely to take 20, 30 or more years for trees to reach a reasonable state of maturity. This may have implications on the sort of experimentation the estate manager is able to carry out. If she has to wait ten years or more to see any significant growth, it may be better for her to obtain data from other sources.

A neighbouring estate is planted with the Stardust variety of the evergreen specie she wants to use. A more distant estate has the Blue Gown variety. She contacts their managers and asks them to give her some data on ten-year-old trees. The data they supplied is shown in this table, which shows the heights of the trees, in metres.

Stardust	1.9	1.5	1.7	2.4	2.3	2.0	3.4
Blue Gown	3.7	2.6	2.1	3.6			

Clearly, the data are in no sense paired. However, it is possible to order the data and consequently to assign ranks, as shown in this table.

Ordered	Stardust			3.4		2.4	2.3		2.0	1.9	1.7	1.5
	Blue Gown	3.7	3.6		2.6			2.1				
Ranked	S			3		5	6		8	9	10	11
	B	1	2		4			7				

Wilcoxon rank sum: *R*

Critical values of the sum of ranks of the smaller collection are available, as the **Wilcoxon rank sum statistic, *R***. Assuming that the *average height of the two varieties* is the same is equivalent to assuming that the *average of the ranks given to each collection* is the same. This, in turn, has implications on the totals of the ranks for each collection. An unusually small total is sufficient evidence to reject the hypothesis of equality of averages.

In this case, there were fewer observations for the Blue Gown variety and the total ranks are:

$$\Sigma r_B = 14$$

Suppose the manager wants to determine if there is any difference in the heights reached after ten years of growth. The hypotheses for the test are:

H_0: no difference in average height; H_1: average heights are different.

This is a two-tailed test and the tables for the Wilcoxon rank sum statistic, R, where $m = 4$ and $n = 7$ (i.e. the numbers of items of data in each collection) indicate that the five per cent critical value is:

$R = 13$

Hence, in this case:

$\Sigma r_B = 14 > 13$

which is the critical value. This suggests that the sum of ranks is not unusually small. Hence, there is insufficient evidence to suggest a difference in heights of the two varieties.

Note:
R provides critical values for the sum of ranks of the smaller collection. The ordering procedure can affect this total. It would have been possible to give the tree with height 3.7 m the highest rank of 11. So care is needed in the assignment of ranks.

For a two-tailed test, assume that the smaller collection is assigned ranks so the smaller possible total results.

For a one-tailed test, we must assign ranks so that it reflects the alternative hypothesis. Example 17.4 should clarify this point.

Example 17.4

*The heights of an evergreen specie known as **Plicata Zebrina** were obtained from two different plantations. The trees had been planted in the same year but the conditions in the plantation were different. The data are given in this table.*

Plantation A	7.26	7.16	7.11	7.07	5.21		
Plantation B	6.33	6.16	4.98	4.40	4.20	4.14	2.86

a) *Test, at the five per cent level, the hypothesis that the conditions in plantation A are more favourable to growth of Zebrina than those in plantation B.*

b) *Test the hypothesis that the average height of **Plicata Zebrina** in plantation A is less than in plantation B.*

Solution

a) *The hypotheses are:*
H_0: average height in A = average height in B;
H_1: average height in A > average height in B.
The results are:

A	1	2	3	4	7		
B	5	6	8	9	10	11	12

and the sum of the ranks of the smaller collection is:

$\Sigma r_A = 17$

This one-sided critical value for R *when* $m = 5$ *and* $n = 7$ *is:*

$R = 21$

Hence:

$\Sigma r_A = 17 \le 21$

which suggests that this rank sum is unusually small under the null hypothesis. So there is evidence to support the view that A *has the more favourable conditions.*

b) *The alternative hypothesis for this test is:*

H_1: *average height in* A < *average in* B.

It is inappropriate to rank the data in the identical fashion to that used in ***a)****. The appropriate ranking is shown below.*

A	12	11	10	9	6		
B	8	7	5	4	3	2	1

This produces $\Sigma r_A = 48 > 21$*, hence the result is not significant. There is insufficient evidence in support of the view that* ***Plicata Zebrina*** *grows to greater average height in plantation* B.

Exploration 17.3

Mann-Whitney U statistic

- Recall some work from Chapter 4, *Probability*, involving an application of arrangements and probability. Read through pages 122–126. What conclusions could you reach regarding the questions in Example 4.7 if the tests were conducted at the five per cent level of significance?

- There is another statistic known as the **Mann-Whitney *U*-statistic** which is equivalent to the **Wilcoxon *R*-statistic**. The *U*-statistic is calculated by considering the possible arrangements of the ranked coding of the data. Suppose that the coding used is *A* and *B*.

  The number, U_{AB}, of times *A* occurs before *B* in the arrangement is calculated.
  The number, U_{BA}, of times *B* occurs before *A* is calculated.

  The smaller of these is the *U*-statistic. Tables of critical values are of the form:

  $U_{\text{observed}} \le U_{\text{critical}}$

  Review the Stardust and Blue Gown data on page 487 and calculate the Mann-Whitney *U*-statistic for these data. Conduct the same hypothesis test as set up for the *R*-statistic.

Example 17.5

Refer to the data in Example 17.4. Conduct a Mann-Whitney U-statistic test for the hypotheses in ***a)****.*

Solution

It is convenient to record the coded ranks as in this table.

1	2	3	4	5	6	7	8	9	10	11	12
A	A	A	A	B	B	A	B	B	B	B	B

$U_{AB} = 7 + 7 + 7 + 7 + 5 = 33$
$U_{BA} = 1 + 1 + 0 + 0 + 0 + 0 + 0 = 2$
The smaller of these is 2. Tables indicate that the five per cent critical value for U is 6.
Since the test statistic, 2, is less that the critical value, 6, the result is unusual under the null hypothesis.
The conclusion is the same as reached using the R-statistic.

Exploration 17.4

Distribution of R and U

- Try developing the R-statistic for the case $m = 5$, $n = 7$. The table below shows how the process begins.

Ranks of A								$\sum R_A$	Cumulative probability
1	2	3	4	5				15	$\frac{1}{692} \approx 0.0013$
1	2	3	4		6			16	$\frac{2}{692} \approx 0.0025$
1	2	3	4			7		17	$\frac{4}{692} \approx 0.0051$
1	2	3		5	6				
1	2	3	4				8	18	$\frac{7}{692} \approx 0.0084$
1	2	3		5		7			
1	2		4	5	6				

a) Show that the one-sided one per cent critical value of R is 18.

b) Determine the 2.5 per cent critical value.

- Now try developing the U-statistic for the case $m = 5$, $n = 7$. The table below shows how the process begins.

1	2	3	4	5	6	7	8	9	10	11	12	*UBA*	Cumulative probability
A	A	A	A	A	B	B	B	B	B	B	B	0	0.0013
A	A	A	A	B	A	B	B	B	B	B	B	1	0.0025
A	A	A	B	A	A	B	B	B	B	B	B	2	0.0051
A	A	A	A	B	B	A	B	B	B	B	B		
A	A	A	A	B	B	B	A	B	B	B	B	3	
A	A	A	B	A	B	A	B	B	B	B	B		
A	A	B	A	A	A	B	B	B	B	B	B		

a) Show that the one-sided one per cent critical value is 3.
b) Determine the 2.5 per cent critical value.
c) Explain why the R and U statistics are equivalent. Determine the numerical relationship between the critical values of R and of U.

EXERCISES

17.4

Use either the R or the U statistic in these exercises. Questions **1** to **5** refer to the same numbered questions in Exercises 4.3A.

1 Is there evidence at the five per cent level?

2 Is there evidence at the five per cent level?

3 Is there evidence at the 2.5 per cent level?

4 Is there evidence at the one per cent level?

5 Is there evidence at the five per cent level?

6 Random samples $x_1, x_2, \ldots, x_m$ and $y_1, y_2, \ldots, y_n$ are taken from two independent populations. The Wilcoxon rank sum test is to be used to test:
H_0: the two populations have identical distributions;
against:
H_1: the two populations have identical distributions except that their location-parameters differ.

Accordingly, the complete set of $m + n$ observations is ranked in ascending order (it may be assumed that no two observations are exactly equal), and S denotes the sum of the ranks corresponding to $x_1, x_2, \ldots, x_m$.

a) Consider the case $m = 4$, $n = 5$.
  i) Show that, if all the xs are less than all the ys, the value of S is 10.
  ii) List all possible sets of ranks of the xs that give rise to a value of S such that $S \leq 12$.
  iii) What is the total number of ways of assigning four ranks from the available nine to the xs?
  iv) Deduce from your answers to **ii)** and **iii)** that the probability that $S \leq 12$ if H_0 is true is $\frac{4}{126}$.
  v) Compute the value of this probability as given by the normal approximation $N(\frac{1}{2} m(m + n + 1), \frac{1}{12}mn(m + n + 1))$ to the distribution of S if H_0 is true.

b) The following are the numerical values of the data for a case with $m = 6$, $n = 8$.

Sample 1 ($x_1, x_2, \ldots, x_m$):	4.6	6.6	6.0	5.2	8.1	9.5		
Sample 2 ($y_1, y_2, \ldots, y_m$):	5.5	7.9	7.1	8.4	6.3	6.8	10.2	9.0

Test H_0 against H_1 at the five per cent level of significance, using the normal approximation given in part **a), v)** above, for the distribution of S under H_0 or otherwise.

(MEI Question 3, Specimen paper S4, 1994)

7 A farmer decides to test a new diet for increasing egg production. A group of twelve hens is randomly divided into two groups, each of six hens. One group is fed the new diet and the other the old one. The following egg yields were recorded in the first year of the test.

New diet	248	255	260	258	245	259
Old diet	227	261	230	250	244	242

Stating clearly your hypotheses, use an appropriate non-parametric test to examine whether or not the new diet leads to the same average egg yield as the old diet. Use a five per cent level of significance.

(ULEAC Question 3, Specimen paper T3, 1994)

8 In an experiment, subjects were asked to judge a short video film for 'gratuitous violence'. The film was divided into twelve scenes, and subjects indicated which scenes they thought contained gratuitous violence. The experimenter predicted that females would judge more scenes to be gratuitously violent than males would.

Males	5	6	4	3	7	4	4	3	1	3
Females	7	9	8	2	5	10	7	7	6	8

a) State the null and alternative hypotheses.
b) Which is the most appropriate test for the experimenter to use to test the prediction? Give reasons for your answer.
c) Carry out the test, and present your conclusion clearly.

(Nuffield Question 4, Specimen paper 4, 1994)

9 A study was undertaken to investigate the effect of vitamin C on the common cold. Fifteen students, each of whom had recently developed a cold, were randomly assigned to two groups. Group A acted as the control group and received unknowingly only a daily sugar tablet, whereas Group B received one gram of vitamin C per day. The table below shows the duration, in days, of cold symptoms for each student.

Group A	13	11	12	9	18	7	12	
Group B	8	14	7	10	9	12	13	6

Using the Mann-Whitney test, investigate the claim that vitamin C shortens the duration of common cold symptoms.

(AEB Question 3 (part), Paper AS, June 1994)

10 There were ten women and ten men attending an Access course. Their results in the first of their module assessments are shown in the tables that follows. Is there evidence to suggest that women do better than men on Access courses? (Conduct your test at the five per cent level).

Men	21	29	38	39	42	43	50	60	62	72
Women	32	36	37	46	49	51	52	64	65	75

11 A group of five boys and three girls were each set the same task to perform. The times (in seconds) taken to complete the task were recorded.

Boys	144	86	91	103	97
Girls	74	57	95		

a) Using B or G as appropriate, code and rank these times, giving the time 57 the rank 1. Hence, show that the sum of the ranks of the girls is 8.

b) How many distinct arrangements of five Bs and three Gs are possible? Assuming that each of these distinct arrangements is equally likely, evaluate the probability that the sum of the ranks of the three Gs is no more than 8.

c) Conduct an hypothesis test to determine whether there is sufficient evidence at:

i) the five per cent level of significance,

ii) the ten per cent level of significance,

to support the view that girls are, on average, faster than boys in performing this task. Make your conclusions clear.

d) Explain carefully what is meant by the phrase, 'at the ten per cent level of significance'.

(Oxford Question 2, Paper 3, June 1994)

12 To compare a new treatment for a certain disease with a standard treatment, a group of twelve patients was divided at random into a group of seven who received the new treatment and a group of five who received the standard treatment. The improvements shown by the patients after a given period are represented by the following measurements.

New treatment	3.2	5.5	3.4	4.9	3.0	2.4	5.1
Old treatment	2.3	3.8	2.2	1.9	2.6		

Use an appropriate non-parametric test to decide whether there is significant evidence that the new treatment is better than the old one.

13 Derive the distribution of:

a) the Wilcoxon rank sum distribution,

b) the Mann-Whitney distribution on the hypothesis of no difference for the following cases.

i) $m = 4, n = 3$ **ii)** $m = 5, n = 3$

14 Determine the ten per cent one-sided critical values of:

a) R **b)** U
in the case $m = 2$, $n = 5$.

15 Prove that there is no one per cent one-sided critical value for R when $m = n = 4$ and define the ten per cent one-sided critical value.

SAMPLE PROPORTION

Point estimate

Let us return to the data about the usable wood in 13 felled trees, from page 475. Three of the trees had more than 15 m^3 of usable wood. What does this tell us about the proportion of such useful trees? The most immediate response is quite simply 'three out of 13', which works out as about 23 per cent. We know from the outcome of the sign test (binomial) we conducted that the proportion is not likely to be as high as a half. It would be helpful to the forestry manager if she could identify a range of values likely to contain the proportion of equally valuable trees of that particular species.

Interval estimate

An estimation of the range sought can be provided by a **confidence interval** similar to those we met in Chapter 16, *The central limit theorem*. There, interval estimates for the **mean** of a population based on a sample mean were found, but we want an interval estimate for a **population proportion**. In order to do this we need to determine an appropriate model for the distribution of proportions found in samples of some fixed size, n, say.

Distribution of sample proportion

Assume that the proportion of trees of this species which are this productive is p. The number, X, of productive trees in a sample of size n may then be modelled by a binomial distribution.

$$X \sim \mathrm{B}(n, p)$$

Under the circumstances discussed in Chapter 15, *Approximating distributions*, binomial probabilities may be approximated using the normal distribution with mean equal to np and variance equal to $np(1-p)$. It is a short step from knowing the distribution of the number of productive trees in a sample of given size, to working out a reasonable model for the distribution of the proportion. Recall that the proportion, $\frac{3}{13}$, of productive trees was found by dividing the number, 3, by the sample size, 13. Thus in a general case, sample proportion is found in the same way, i.e. by dividing the number observed, X, by the size of the sample, n. Since, $X \approx \mathrm{N}(np, np(1-p))$, it means that sample proportion, $\frac{x}{n}$, is distributed approximately normally.

$$\frac{x}{n} \approx \mathrm{N}\left(\frac{np}{n}, \frac{np(1-p)}{n^2}\right)$$
$$= \mathrm{N}\left(p, \frac{p(1-p)}{n}\right)$$

This indicates that the distribution of the sample proportion is approximately normal, and has mean equal to the population proportion, p, and variance equal to $\frac{p(1-p)}{n}$, which depends on the size of the sample and on p.

Confidence interval

We are now in a position to identify a confidence interval for p based on the proportion obtained in a sample. Recall the argument presented in Chapter 16, *The central limit theorem*, which led from the distribution of sample mean, $\overline{X}_n \sim \mathrm{N}\left(\mu, \frac{\sigma^2}{n}\right)$, to the 95 per cent symmetric confidence interval based on $\bar{x}$:

$$\bar{x} \pm 1.96\frac{\sigma}{\sqrt{n}}$$

By a similar argument, the analysis presented here indicates that:

$$\hat{p} \pm 1.96\sqrt{\frac{p(1-p)}{n}}$$

provides 95 per cent symmetric confidence interval for population proportion, p.

Note two points:

- $\hat{p}$ (called 'p hat') is the sample proportion,
- the confidence interval is dependent on p but this is unknown!

This situation is similar to the perplexing problem posed on page 454. A simple solution here is to use the sample proportion $\hat{p}$. This gives rise to the approximate 95 per cent confidence interval for p:

$$\hat{p} \pm 1.96\sqrt{\frac{\hat{p}(1-\hat{p})}{n}}$$

To return to the productive trees: $\hat{p} = \frac{3}{13}$, hence the 95 per cent confidence interval for the species' proportion is approximately:

$$\frac{3}{13} \pm 1.96\sqrt{\frac{\frac{3}{13}\left(1-\frac{3}{13}\right)}{13}}$$

i.e. 0.2308 – 0.2290 ≈ 0.002 to 0.460.

This suggests that the proportion of this specie of tree which produce at least 15 m^3 of usable wood lies between 0.2 per cent and 46 per cent. Note that this confirms the rejection of the 50 per cent hypothesis.

Example 17.6

A sample of 50 conifers is chosen at random from a plantation. Twelve of those are found to be over 10 m in height. Obtain a 95 per cent confidence interval for the proportion of these conifers in the plantation which are at least 10 m in height.

Solution

A point estimate, $\hat{p}$, for the sample proportion is $\frac{12}{50}$, i.e. 0.24. Hence, an approximate 95 per cent confidence interval for p is given by:

$$0.24 \pm 1.96\sqrt{\frac{0.24(1-0.24)}{50}}$$

This provides the interval: $0.122 \leq p \leq 0.358$.

Note:

Notice that in Example 17.6, the point estimate, 0.24, is very close to the 0.23 for the productive trees. But the confidence intervals are of widely differing widths. Which of the two intervals is wider, and what has brought this about? The real difference in the two examples lies in the sizes of the samples, 13 and 50. There is a factor 4 difference in sample size. There is a factor of approximately 2 in the widths of the intervals. The smaller sample produces the wider interval.

CALCULATOR ACTIVITY

Confidence intervals and sample sizes

- Use your calculator to help you discover how the width of a 95 per cent confidence interval varies with changes in sample size. You might like to stick with samples where $\hat{p} \approx 0.25$ to be in a position to compare with those found already.
- Use your calculator to explore how the widths of 95 per cent confidence intervals change with differing sample proportion. Choose a convenient sample size, such as $n = 100$, and explore for $\hat{p} = 0.01$ to 0.99. Try to determine where the maximum width occurs. Will this be the case for other sample sizes?

Upper bound for interval estimate

You may have discovered that the width of a confidence interval has an inverse relationship with the sample size. More precisely, the width of a given confidence interval is *proportional to* $\sqrt{\frac{1}{n}}$. This implies that to *halve the width* of the interval, we need to *quadruple the sample size*.

You may also have discovered that for any given confidence level, the interval has its maximum width when $p = 0.5$. This is useful in providing an upper bound to the width of a confidence interval. In particular, in the case of a 95 per cent confidence interval for p, by using 2 in place of 1.96 and putting $p = 0.5$, the interval becomes:

$$\hat{p} \pm 2\sqrt{\frac{0.5(1-0.5)}{n}} \quad \text{i.e.} \quad \hat{p} \pm \sqrt{\frac{1}{n}} \quad \text{or} \quad \hat{p} \pm n^{-\frac{1}{2}}$$

Sample size determination

Under these circumstances, the term, $n^{-\frac{1}{2}}$, may be considered as the maximum error in using $\hat{p}$ as an estimate for p. For example, if $n = 25$, then the maximum error associated with using $\hat{p}$ is roughly $25^{-\frac{1}{2}}$ i.e. 0.2 or 20 per cent.

This may be used in determining the minimum sample size required for a specified degree of accuracy. For instance, suppose that the forestry manager wanted to know the proportion of productive trees to within eight per cent. This means that she would be prepared for an interval of the form $\hat{p}$ – 0.08. The above analysis suggests that:

$$0.08 \geq \sqrt{\frac{1}{n}}$$
$$\Rightarrow (0.08)^2 \geq \frac{1}{n}$$
$$\Rightarrow n \geq \frac{1}{0.0064} = 156.25$$

Thus she would need a sample of size 157.

Example 17.7

a) *A forester wants to be 95 per cent sure that he knows the proportion of conifers over 10 m high, to within 0.1 m. He decides to select n conifers randomly and measure their height. How large should n be?*

b) *Suppose the forester in **a)** above wanted to be 99 per cent sure. How would you advise him?*

Solution

a) *The 95 per cent confidence interval for population proportion of maximum width is:*

$$\hat{p} \pm n^{-\frac{1}{2}}$$

To estimate p within 0.1 implies that:

$$0.1 \geq n^{-\frac{1}{2}}$$
$$\Rightarrow 0.01 \geq n^{-1}$$
$$\Rightarrow n \geq \frac{1}{0.01}$$
$$\Rightarrow n \geq 100$$

i.e. n is at least 100.

b) *A 99 per cent confidence interval for p is:*

$$\hat{p} \pm 2.576\sqrt{\frac{p(1-p)}{n}}$$

This interval is of maximum width when p is 0.5, and so to be 99 per cent sure of being within 0.1:

$$0.1 \geq 2.576\sqrt{\frac{0.5(1.05)}{n}}$$
$$\Rightarrow 0.1 \geq 1.288n^{-\frac{1}{2}}$$
$$\Rightarrow n \geq 165.9$$

Thus the sample should be at least 166.

Difference in proportion

Example 17.8

A random sample of 100 conifers in a plantation consisted of 36 Blue Gown and 64 Stardust. The trees were measured. It was found that 22 of the Blue Gown and 38 of the Stardust were under 8 m high. Obtain a 95 per cent confidence interval for the difference between the proportions of these varieties of conifer under 8 m in height.

Solution

Point estimates for the proportions are:

$$\hat{p}_B = \frac{22}{36} \qquad \hat{p}_A = \frac{38}{64}$$

We need a confidence interval for the difference in proportions. A point estimate for this difference is:

$$\frac{22}{36} - \frac{38}{64} = \frac{6}{288} \approx 0.0174$$

In order to find a confidence interval, we need to know the standard deviation of the distribution of the difference. Estimates for the variances of the distributions of the separate proportions are:

$$\frac{\frac{22}{36}\left(1-\frac{22}{36}\right)}{36} \approx 0.006\,60 \qquad \frac{\frac{38}{64}\left(1-\frac{38}{64}\right)}{64} \approx 0.003\,77$$

Hence, the variance of the distribution of the difference is the sum of these separate variances i.e. 0.010 37. Then the estimate of the 95 per cent confidence interval for the difference in proportions is:

$$0.0174 \pm 1.96\sqrt{0.010\,37} \approx 0.0174 \pm 0.1996$$

This gives the interval –0.1822 to 0.217. Clearly the negative proportion is unrealistic and so the confidence interval is 0 to 0.217.

EXERCISES

17.5

1 A random sample of 100 apples on sale in supermarkets is inspected and 43 apples are found to be bruised. Calculate a 95 per cent confidence interval for the proportion of bruised apples offered for sale.

2 A sample of 200 nails produced by a machine is carefully measured, and 43 of the nails are found to be undersized. Calculate a 90 per cent confidence interval for the proportion of undersized nails produced.

3 A survey of 1225 people showed that 283 had changed their car in the previous twelve months. Calculate a 95 per cent confidence interval for the proportion of people who change their car annually.

4 At a previous election, 30 per cent of the electorate voted for the MHT party. Prior to the next election a telephone survey was conducted to attempt to predict the forthcoming result. Out of 1000 people telephoned, 330 said that they intended to vote for the MHT party.

a) The number of people who intend to vote for the MHT party is modelled by a binomial distribution. Use a normal approximation to this binomial distribution to carry out a hypothesis test at the five per cent significance level to test the claim that support for the MHT party has changed. Give suitable null and alternative hypotheses and state your conclusion clearly.

b) State briefly why this survey might not give an accurate prediction of the forthcoming result.

(UCLES Question 9, Specimen paper S2, 1994)

5 Out of a random sample of 200 people asked, 54 said that they travelled by train regularly. Calculate a symmetric 95 per cent confidence interval for the proportion of the population who travel by train regularly.

(UCLES Question 3, Specimen paper S2, 1994)

6 In an election there are two candidates, A and B. A random sample of 1000 voters is taken and, of these, 545 say they will vote for A. State a point estimate for the fraction of A-voters in the whole population. Explain carefully what is meant by a 95 per cent confidence interval for the fraction of A-voters in the whole population and obtain such an interval from the sample given.

7 After a young child was knocked down outside her home by a speeding stolen car, a group of neighbours conducted a vigorous campaign to have speed bumps introduced on all roads in their large estate. The local evening newspaper sampled local opinion and, from a random sample of 90 adult residents of the estate, found that 34 out of the 50 women and 16 out of the 40 men in the sample supported the introduction of speed bumps. Calculate an approximate, symmetric 95 per cent confidence level for the difference between the proportions of women and men living on the estate who support the idea of introducing speed bumps on their roads.

(NEAB Question 5, Specimen paper 8, 1994)

8 A new drug-cure for the common cold is being trialled. The pharmaceutical company want to be 90 per cent sure that they know the true proportion, with a maximum error of 0.05, of patients who recover using the drug. How large should the trial be?

9 How many voters should be interviewed if you want to be 95 per cent of knowing the proportion, within two per cent, of voters who intend supporting a particular party?

10 An observer at a railway station records whether trains arrive on time (which he defines as within two minutes of the scheduled arrival time). The results from a random sample of 25 trains follow (L indicates late and O indicates on time).

O, O, L, O, O, L, L, O, O, L, O, L, O, O, O, O, O, O, L, O, O, L, O, L, L

Assume that the probability, p, of a train arriving late is a constant and that the arrival times of all trains are independent.

a) **i)** Estimate p.

ii) Calculate an approximate 95 per cent confidence interval for p.

b) Each weekday 150 trains arrive at a station. Using your estimate of p from **a), i)** evaluate the mean and variance of X, the number of trains arriving late of a weekday.

The timetable is revised in an attempt to improve the situation. Following this revision 28 out of a random sample of 200 trains are observed to arrive late.

c) **i)** Estimate the value of p after the timetable revision.

ii) Using the value of p estimated in **c), i)**, evaluate the mean and variance of Y, the number of trains arriving late on a weekday after the timetable revision. The number of weekday arrivals is unchanged.

The variable Z is defined by $Z = X - Y$, where X and Y are independent.

iii) Estimate the mean and variance of Z.

iv) Calculate an interval containing approximately 95 per cent of the values of Z.

d) Comment on the effect of the timetable change and also the assumption that arrival times are independent.

(AEB Question 7, Paper 9, November 1994)

Summary

Spearman's rank correlation coefficient

- r_S = product moment correlation coefficient (PMCC) of ranks
  $= 1 - \dfrac{6\sum d^2}{n^3 - n}$, except when there are ties.

Sign tests

- Binomial based on B(n, 0.5)
- Wilcoxon signed rank, T, for paired data

Two sample tests

- Wilcoxon rank sum, R
- Mann-Whitney, U-statistic

Distribution of sample proportion

- $\hat{p} \sim \mathrm{N}\left(p, \dfrac{p(1-p)}{n}\right)$
- The 95 per cent confidence interval for p is $\hat{p} \pm \sqrt{\dfrac{\hat{p}(1-\hat{p})}{n}}$
- The upper bound is $\hat{p} \pm n^{-\frac{1}{2}}$

STATISTICS

18

Coursework in statistics

Throughout your course you should be developing an appreciation of Statistics and how it can be applied to help you understand and interpret real-world situations and problems. In many courses you will have to demonstrate this by doing a piece of coursework, or a project. Examiners will want to see your ability to apply the correct statistical techniques in a practical situation.

BEFORE YOU START

There are several things that you need to consider carefully before starting your coursework.

- Make sure that you understand the **purpose** of the coursework, and any **particular requirements** that have to be met.
- Choose a topic that interests you and meets the needs of the coursework. You will want to do something that is **practical** and **planned** so that you can demonstrate your understanding of statistics.
- Have a clear idea of the **length** of the project, the **time** that you should spend on it and what needs to be included, and the **format** of your final report. The examination syllabus gives details of the marks available and an indication of the amount of time that you should spend on your coursework. You may be required to spend no more than six to eight hours, or as much as 20 hours.
- Know what use you are going to make of **IT** – computers and calculators – before you start, and how you are going to demonstrate this in your final report.
- Find out how much **learning support** is available for your particular project.
- Become familiar with your Examination Board **syllabus** and what it expects from coursework.

THE PURPOSE OF COURSEWORK

Some coursework tasks are purely for your own 'internal consumption', to consolidate your work in Statistics, and help you apply your knowledge to real situations. Coursework can also contribute to your examination award. These separate roles of coursework may have different requirements: make sure you know what is required of you in each case. An 'investigation of single variable data' is going to have to meet different criteria from 'the use of statistical methods to investigate a subject'. Are you going to be 'testing assertions' or 'estimating parameters'? The full requirements of an 'in-depth investigation into correlation and regression' may well be a little more demanding than an 'investigation of bivariate data'. You *must* know and understand the requirements and regulations.

CHOOSING YOUR PROJECT

Wherever possible, avoid using secondary source material, i.e. data that have been collected by someone else. You are likely to have to justify the integrity of the data and to demonstrate that you have taken any possible bias into consideration. Suppose that you belong to an Athletics Club and you decide to collect data from a local half-marathon. It would be perfectly acceptable for you to use the complete set of times for all of the competitors. Suppose, however, that you decided to use the times from a national event. It is unlikely that you would have personally collected the times for all the competitors, so you will almost certainly be expected to discuss their accuracy, even if you decide to sample from the entire collection rather than use all of the data.

Make sure that the data that you collect are suitable for the purpose. If you are 'investigating single variable data' it is perfectly legitimate and appropriate to split the data into the times for men and for women, or for runners in particular age categories. However, if you are 'investigating bivariate data' you may need to obtain times for the same people in two different races, or you may need to identify the ages of the competitors a little more precisely.

Suppose that you want to 'test an assertion' about the degree of correlation between two variables. You will need to identify the nature of your data carefully: do not draw inferences using the wrong test statistic. The conditions under which Pearson's coefficient is appropriate are different from those for Spearman's, for example.

LEARNING SUPPORT

It can be very useful to discuss your work and ideas with other students. You may even be able to work collectively with others, but you should check first to see if this is acceptable. You may want to seek help or guidance from your teacher, lecturer or tutor. Do so; the worst they can say is 'No'. There are clear regulations about the type and extent of the support that they can offer you. You are supposed to be learning about your subject, and tutors are there to help you learn, as well as to find out what you have learned.

You may have access to powerful computer-based software. Make sure that you edit any output so that you present only what is relevant, and make sure that you understand the results that the computer produces. Computers, like calculators, are tools to free you from tedious calculation and analysis, and give you more time to think about the task, methods and outcome.

THE REPORT: PRESENTING YOUR COURSEWORK

When you present your coursework you need to have in mind a clear plan of what you are going to include. Here is a sample, skeleton structure that you may like to start from.

1 Title

Give your report a concise, appropriate and relevant title so that someone picking up your work for the first time can get a good idea of what you have investigated.

2 Extract or summary

This should be a composed of a few sentences that summarise your main findings. Do not go into great detail in this section. You should choose your language very carefully. Use phrases such as 'My findings seem to indicate that ...' rather than 'I proved that ...'. Between 50 and 200 words should be enough, but check what is required.

3 Aims

Describe the area that you are studying and state what your intentions or aims are. Again, be careful with your wording: 'I intend to investigate if men are faster than women in half-marathon races' is more appropriate than 'I am going to prove that men are faster than women'. In your introduction, make sure you justify the study you undertook. It is usually good practice to write your introduction *before* you carry out the task. It can be valuable to reflect on it and to refine it after completing the coursework.

4 Data collection

In this section you should describe what data you collect and the method(s) used to collect that data. Explain *why* you use your chosen approach and what alternatives are open to you. Discuss the difficulties that you experience and what precautions you take regarding bias (think carefully about the sample size; watch out for experimental error, etc.). You may consider carrying out a pilot study to assess the suitability of the method of data collection that you have chosen.

Avoid presenting huge arrays or tables of data within the main text of your report. Use an appendix for this and make appropriate reference to it in this section. Also, remember that 'a picture speaks a thousand words' and consider presenting simple yet informative diagrammatic representations of your data here.

Data can be gathered from a designed experiment, direct observation, questionnaires and surveys or by using secondary sources or simulation (but see the warning about secondary sources of data earlier in this chapter).

5 Modelling and analysis

Discuss the statistical models that you use and the assumptions that you make about the data to justify the use of these models. Describe the analysis of the data and do not forget that diagrams and graphs are a part of the analysis. When you are comparing, for example, men's race times with women's, you might like to use a back-to-back stem-and-leaf display. Alternatively you could use a schematic plot. Be selective in your choice of diagrams and avoid overdoing things. You will not get the best marks for simply creating as many different sorts of graph or diagram as you can. The diagrams are there to illustrate your text.

Avoid presenting lengthy calculations; instead, give some outcome of your calculation and use an appendix for the details of the number crunching. If you are testing hypotheses or other assertions, it is important to clarify the significance level of the test, whether it is one-tailed or two-tailed; you could give a thumb-nail sketch of the model distribution with the critical region highlighted. And just as for diagrams, make sure that everything that you include is relevant.

It makes life easier for the examiner if you can keep comparisons on the same page.

6 Interpretation and refinement

This is where you turn a statement like 'The result was significant' into everyday language such as: 'The result suggested that men are faster than women by 6 minutes on average.' Discuss the limitations of your analysis. Point out any doubts or concerns that you might have. Suggest clearly and concisely, how you might refine the approach that you adopted if you were to investigate this topic again.

It may be easier – for you and the reader – to include this as part of the *Modelling and analysis* section.

7 Conclusions

Try to draw together all that you have done and include any criticisms of your approach that you want to emphasise (e.g. the sample may not have been large enough or the sample was biased because ...).

A project may be inconclusive, in which case you should state precisely what further work could be done, to reach a conclusion.

QUESTIONNAIRES AND SURVEYS

If you are collecting your own data, it is quite likely that you will need to use some sort of survey. Questionnaires are very difficult to write. Far too many projects are based on surveys that have no real relevance or purpose, or are badly conducted and so provide doubtful data.

Think very carefully about the design of any questionnaires. Here are some things to bear in mind.

- Use simple, unambiguous language.
- Use a small number of questions.
- Avoid long, over-complicated questions.
- Avoid using leading questions.
- Avoid questions that are outside people's direct experience.
- Avoid embarrassing questions – you may not get truthful answers.
- Be very careful with lists of alternative answers, as people often base their answer on what they can remember (such as the first or last response in the list).
- Do not rely on people's memories.
- Make time for a pilot study to test your questionnaire.

ASSESSMENT

You must have a clear idea of the criteria against which your work is to be judged. Make sure that you know what you have to do in order to reach particular levels. For instance, you may need only to 'give an explanation of how you collected your data'; on the other hand, you may be expected to 'show that you can identify a variety of sources, refine and process data and discuss their reliability'. You may be expected to 'communicate clearly, concisely and with precision and to organise results systematically' or merely to 'communicate some work clearly'. Spend time discovering what you need to do – and do it!

Summary

- Know what your Examination Board wants from your coursework.
- Choose something that interests you, that has a clear purpose and that allows you to demonstrate a wide range of skills.
- Use IT appropriately; understand how calculators and computers produce various results and what these results mean.
- Record data carefully and write up your work as you go along, using a clear structure for your final report.
- Include relevant and accurate diagrams and calculations.
- Use the Report checklist in this chapter.

Answers

CHAPTER 1
Data

Exercises 1.1A *(p.5)*

3

0	V	G	A	P
12	20	15	10	3

4

1	2	3	4	5	6
15	20	22	23	11	4

Exercises 1.1B *(p.6)*

1 c, d, d, c, c a), d), e) are continuous; b), c) are discrete

2 **a)** qualitative, discrete
b) quantitative, continuous
c) qualitative, discrete
d) quantitative, discrete
e) quantitative, discrete
f) qualitative, discrete
g) quantitative, continuous

3

B	F	G	I
14	7	6	5

Exercises 1.2A *(p.15)*

1 a)

Class	*f*
$300 \le x < 400$	1
$400 \le x < 500$	3
$500 \le x < 600$	10
$600 \le x < 700$	4
$700 \le x < 800$	2

b)

Class	*f*
$0 \le x < 1000$	4
$1000 \le x < 2000$	8
$2000 \le x < 3000$	7
$3000 \le x < 4000$	1

c)

Class	*f*
$0.6 \le x < 0.7$	2
$0.7 \le x < 0.8$	4
$0.8 \le x < 0.9$	1
$0.9 \le x < 1.0$	2

2 a) *Unit = 1*

1F	5
1S	7 7 6 6 7
1E	9 8 8 9
2O	0 1
2T	2

b) *Unit = 100*

1E	8 8 9
2O	1 0
2T	3
2F	5

c) *Unit = 0.001*

0–	3
0+	8 8 8 9
1–	0 1

3 23, 20, 24, …, 50

4 4, 5, 5, …, 15

5 *Unit = 1*

4+	9 9
5–	0 1 1 3 3 2 0 2 1 3 2 1 2 3 1 3 2 0

6 *Unit = 1*

0	0 1 0 0 0 0 1 0 1 1 1 1
T	2 3 2 2
F	5 4
S	
E	9 8
1O	1 0 0 0
1T	2 2 2 2 2

7 **a)** 9.5, 14.5 **b)** 5, 5, 5, 5, 10
c) frequency density ∝ 1.2, 2, 2.2, 1.6, 1

8 **a)** 25, 30, 35, 45, 55
b) 3, 5, 5, 10, 10; frequency density ∝ 20, 4, 2.4, 0.5, 0.1

9 **a)** $0 \le t < 9.5$, $9.5 \le t < 19.5$, $19.5 \le t < 29.5$, $29.5 \le t < 59.5$, $59.5 \le t < 119.5$; 9.5, 10, 10, 30, 60
b) frequency density ∝ 2, 3.8, 2, 0.6, 0.083

Exercise 1.2B *(p.17)*

1 **a)** 10, 10, 20, …, 490
b) 4, 2, 0, …, 26
c) 0.00, 0.01, 0.02, …, 0.14

2 *Unit = 1*

3	7
4	8 3
5	9 0 2 6
6	4 8 5 0 5 3
7	5 9 6 2 9
8	3 8

3 *Unit = 10*

7+	8 8 9 9 9
8–	0 0 0 0 1 1 1 1 2 2 3 3 3 4 4

4 *Unit = 0.1*

1O	1
1T	
1F	4 4 4 4
1S	6 6 6 7 7 6 7 7 7
1E	9 9 8 8 8 9 8
2O	0 1 1 0
2T	2 3 3
2F	4 5

5 *Unit = 1*

13+	8 6 9
14–	3 1 4 2 2 3 4 2 1 4 4
14+	6 6 7 7 9 6 7 9
15–	1 3 3 2 3 2 2 1 2 3 3 0 0 0 4 1
15+	5 7 7 5 6 6 7 8 7
16–	3 2
16+	9

6

Class	*f*
$0 \le v < 10$	1
$10 \le v < 20$	9
$20 \le v < 30$	9
$30 \le v < 40$	16
$40 \le v < 50$	10
$50 \le v < 60$	4
$60 \le v < 70$	1

7 *Unit = 0.1*

2	4
3	2 6
4	5 3 4 7 8 4
5	8 6 9 3 2 8 4 6 9
6	2 6 3 4 5 8
7	6 2 1 5 4
8	5 7 2 1 9
9	6 3 4
10	2 4
11	8

8 **a)** 10, 10, 10, 20, 6

Exercises 1.3A *(p.22)*

2 a) 0.16, 0.38, 0.30, 0.08, 0.08
b) 16, 13 **c)** 3, 3, 3, 3, 6
d) 0.053, 0.127, 0.1, 0.027, 0.013

3 b) 34%, 21%

4 a) frequency density ∝ 2, 4, 2, 1.4, 0.8, 1.3
scaled: 0.04, 0.08, 0.04, 0.028, 0.016, 0.026
b) 29%

Exercises 1.3B *(p.23)*

1 c) 14%, 63%

2 b) 53%

Consolidation Exercises for Chapter 1 *(p.25)*

1 a) continuous **b)** discrete
c) continuous **d)** discrete
e) discrete **f)** discrete

2 a) qualitative, discrete
b) quantitative, continuous
c) qualitative, discrete
d) quantitative, discrete
e) qualitative, discrete
f) qualitative, discrete

9 a) Use a five-part stem with unit = 10

12 a) 19.5, 24.5

13 a) 12.5, 15.5 **b)** 9.5, 3, 3, 3, 3
c) frequency density ∝ 3.16, 6, 5, 2.33, 1.67

14 *Unit = 1*

0+	5 8
1–	4 2 4
1+	7 6 7 9
2–	2 0
2+	6
3–	1

15 a) i) discrete

CHAPTER 2
Summary statistics 1

Exercises 2.1A *(p.33)*

3 b) 2.63

4 b) 25.2

5 a) 24.5 **b)** 25.5

Exercises 2.1B *(p.34)*

3 b) 0.76

4 b) 7

5 b) 3.54

Exercises 2.2A *(p.45)*

1 3

2 44

3 c) 2.7 **d)** 2.71

4 a) 18.56 **c)** 18.8 **d)** 18.84

7 b) 587.3

Exercises 2.2B *(p. 47)*

2 523

3 c) 3.88

4 78.6

5 a) *Unit = 0.1*

depth		
1	14	7
6	15	2 3 6 6 6
12	16	7 7 7 8 8 9
(5)	17	3 3 4 8 9
13	18	2 2 4 4 7 7 8
6	19	3 6
4	20	2 4 5
1	21	2

b) 17.6

6 a)

3	45	2 3 5
8	46	1 2 3 6 7
14	47	1 1 2 5 7 8
21	48	0 1 3 5 6 6 9
(7)	49	2 2 3 5 6 7 8
27	50	0 0 1 2 3 5 5 6 7 8 9
16	51	0 1 2 3 4 5 6 7 7
7	52	0 2 3 7
3	53	2 5 6

b) 49.8

7 c) 24 **d)** 2.65

Exercises 2.3A *(p.53)*

1 a) 40.7 **b)** 0.667

2 a) 2.25 **b)** 2.25
c) 2

3 a) 0.755 **b)** 368.875

4 2.76

5 a) 8, 11, 13, 14.5, 15.5, 17, 24
b) 17.0

6 a) 25.07

7 41

Exercises 2.3B *(p.55)*

1 a) 52.9 **b)** 55.7

2 a) 1.98 **b)** 1.98 **c)** 2

3 a) 0.877 **b)** 163.8

4 1.53

5 8.86

6 4.86

7 65.55

Exercises 2.4A *(p.59)*

1 a) 54
b) 8, 16, 21, 18, 8, 9; 57.8

3 b) 366.2

4 b) 3.29, 2.91
d) positively skewed

5 positively skewed since mean > median

Exercises 2.4B *(p.61)*

1 a) 0.517 **b)** 0.58
c) negatively skewed

3 b) 162.9

4 b) 12.10, 9.63
c) positively skewed

5 a) 14.09, 10
c) positively skewed

Exercises 2.5A *(p.65)*

1 a) 185.4, 208.2 **b)** 112

2 94

3 a) 4.5, 4 **b)** 15.4, 15.25
c) 127, 125.24, geometric mean < arithmetic mean

4 £6.25

5 54.0

6 8.5%

Exercises 2.5B *(p.66)*

1 b) 41.75

2 13.4

3 a) 20 **b)** 12 **c)** 26

5 a) 123.2%, 92.5%, 120.9%
b) 11%

Consolidation Exercises for Chapter 2 *(p.67)*

1 b) 37.5

2 a) 19.4 **b)** 9.3

3 a) 12.9 **b)** 9.9

5 a) 7, 7 **b)** 7, 7

6 £7.77

7 105

8 24.9

9 97

10 b) 50.5, 50.43

11 a) 4 **b)** 3.89 **c)** 4

CHAPTER 3
Summary statistics 2

Exercises 3.1A *(p.77)*

1 a) 5, 4 **b)** 0.6, 0.3
c) 26, 17.5 **d)** 0.8, 0.325

2 a) 9 **b)** 0.4
c) 320 **d)** 4.6

3 **a)** 17.3, 10.1, 27.9 **b)** 17.8
4 **a) i)** 39.5 **ii)** 35.7, 44.0
5 **a)** 2.8; 0.95, 5.35
6 **b)** 0.3, 1.4, 2.7, 4.4, 8.9
7 **a) i)** 157 **ii)** 151.7 **iii)** 161.7
b) i) 162.5 **ii)** 158.1 **iii)** 167.5
8 **a)** 23.75, 35.83, 51.67; 30, 41, 48; 20, 30.83, 38.33; 33.33, 44, 53.33

Exercises 3.1B *(p.80)*

1 **a)** 10, 7 **b)** 39, 23
c) 1.3, 0.7 **d)** 17, 8
2 **a)** 6 **b)** 60.5
c) 101.8 **d)** 510
3 22.7, 26.4, 39.4
4 **a)** 32.2; 27.5, 36.4
b) 9.5, 27.5, 32.2, 36.4, 49.5
5 **a)** 26; 17, 36
6 **b)** 78, 96, 111, 126, 155
7 **b)** 0, 7.6, 11.5, 13.9, 40; 0, 11.7, 13.4, 15.5, 40

Exercises 3.2A *(p.82)*

1 **a)** 40.0; 37.25, 43.05
b) 31.45, 48.85; 29.8, 53.03
2 **c)** 358
3 **a)** –3.45, 9.75
b) 10.1, 11.3, 12.3, 12.3
4 8.9
5 **a)** 4.35; 2.36, 7.71
b) –2.99, 13.06
6 **a)** 3.35 **b)** 0.04

Exercises 3.2B *(p.83)*

1 **a)** 68; 54, 78 **b)** 30, 102
c) 12, 15, 20
2 **d)** 526
3 **a)** –2, 54 **b)** 58; 73 is extreme
4 58, 158
5 **a)** 9.35; 8.0, 10.625
b) 5.2, 13.8, 15.6; 15.9 is extreme
6 **a)** 30.1; 33.5, 28 **b)** 22.5, 39.0
d) 20.0; 44.5 far out

Exercises 3.3A *(p.92)*

1 4.071…, 3.638
2 2.143
3 24.7, 108.91, 10.44
4 16.58
5 **a)** 999 **b)** 12.22

Exercises 3.3B *(p.93)*

1 17.8; 0.81
2 1.0
3 **a)** 0.644, 2.074, 1.440
b) 3.144, 2.074, 1.440
4 2.58
5 £2.60, £1.77

Consolidation exercises for Chapter 3 *(p.99)*

1 **a)** 0.0; –0.7, 0.6 **b)** 1.3
2 **a)** 0.88 **b)** 2.28
3 **b)** 46.94; 40.68, 55.45; positive
4 **b)** 20.31; 18.15, 23.46
5 **b)** 29; 26, 35
6 68.75, 62
7 **c)** 252, 6.52
8 **a)** £208.42, £18.30
b) £4644.40
c) £246.39, £38.35
9 **a)** 6,5 **d)** 6.5
e) reduce both **f)** positive
10 **a) i)** 17.9 **ii)** 12.3, 20.4
11 **a)** frequency density ∝ 21, 64, 66, 48, 36, 21, 16, 12, 5.5, 2.4
b) 3.13, 2.29
12 **a)** 13.6, 1.37 **b)** 13.2 **c)** 13.7

CHAPTER 4
Probability

Exercises 4.1A *(p.115)*

1 $\frac{11}{36}$
2 **a)** $\frac{3}{8}$ **b)** $\frac{7}{8}$
3 **a)** $\frac{1}{4}$, $\frac{1}{52}$, $\frac{1}{26}$
b) i) C, K **ii)** none **iii)** the others
4 **b) i)** $\frac{1}{6}$ **ii)** $\frac{3}{4}$ **iii)** $\frac{2}{3}$ **iv)** $\frac{1}{6}$
5 **a)** $\frac{132}{380}$ **b)** $\frac{152}{380}$
6 **a)** $\frac{5}{8}$ **b)** $\frac{3}{5}$

Exercises 4.1B *(p.116)*

1 $\frac{8}{64}$, $\frac{30}{64}$
2 **a)** $\frac{7}{16}$ **b)** $\frac{15}{16}$
3 **b)** $\frac{11}{28}$
4 **a) i)** $\frac{6}{13}$ **ii)** $\frac{6}{13}$ **iii)** $\frac{1}{6}$ **iv)** $\frac{1}{6}$
b) i) $\frac{10}{13}$ **ii)** $\frac{4}{13}$ **iii)** $\frac{1}{4}$ **iv)** $\frac{1}{10}$
5 **b) i)** $\frac{6}{64}$ **ii)** $\frac{49}{64}$ **iii)** $\frac{8}{64}$ **iv)** $\frac{40}{64}$ **v)** $\frac{11}{64}$
6 **a)** $\frac{9}{16}$ **b)** $\frac{5}{9}$

Exercises 4.2A *(p.121)*

1 **a)** 12 **b)** $\frac{1}{3}$
2 **a)** 5! **b)** 96
3 15 890 700
4 **a)** $\frac{1}{5040}$ **b)** $\frac{1}{210}$
5 **a)** 12 376 **b)** 1200

Exercises 4.2B *(p 122)*

1 240
2 **a)** 8! **b)** 2 × 6!
3 **a)** 5! **b)** 5^5
4 **a)** 406 192 **b) i)** 250 800
ii) 38 760 **iii)** 319 770
5 **a)** 792 **b)** 200

Exercises 4.3A *(p.126)*

1 **b)** A A B A B B A B
d) i) 1 **ii)** 1 **iii)** 2 **iv)** 3
e) yes, 10% chance of being wrong
2 **b)** T R R R T T T T T **c)** 84
d) yes, 8.3% chance of being wrong
3 yes, 1.6% chance of non-smokers not having faster reactions
4 yes, with 3.3% chance of error
5 no

Exercises 4.3B *(p.127)*

1 **b)** N L N N L L L
d) i) 1 **ii)** 1 **iii)** 2
e) no, because 11.4%
2 **b)** C B C B B C B B B **c)** 84
d) No, because 13.1%
3 yes, less than 6% chance of error
4 yes, less than 5% chance of error
5 yes

Consolidation Exercises for Chapter 4 *(p.130)*

1 **a)** $\frac{1}{2}$ **b)** $\frac{1}{4}$ **c)** $\frac{5}{12}$
2 **a)** $\frac{1}{2}$ **b)** $\frac{3}{8}$ **c)** $\frac{1}{4}$
3 **a)** $\frac{1}{16}$ **b)** $\frac{1}{4}$ **c)** $\frac{1}{4}$
4 $\frac{7}{15}$
5 **b) i)** $\frac{1}{3}$ **ii)** $\frac{1}{6}$ **iii)** $\frac{1}{4}$ **iv)** $\frac{1}{6}$
6 **a)** $\frac{1}{4}$ **b)** $\frac{1}{2}$ **c)** $\frac{16}{169}$
d) $\frac{88}{169}$ **e)** $\frac{1}{26}$
7 **a)** $\frac{t(t-1)}{(t+s)(t+s-1)}$
b) $\frac{s(s-1)}{(t+s)(t+s-1)}$
c) $\frac{2st}{(t+s)(t+s-1)}$
8 24
9 **a)** 6! **b)** 3 × 5!
c) 3 × 5! **d)** 192
10 **a)** 180 **b) i)** 60 **ii)** 120
11 **a)** 5005 **b) i)** 540 **ii)** 4921
12 **a)** 56 **b)** 0.218 75

13 a) 13 860 **b)** 210 **c)** 4620
14 a) 2 598 960 **b)** 1287 **c)** 0.0020
15 a) i) $\frac{3}{4}$ **ii)** $\frac{9}{16}$ **iii)** $(\frac{3}{4})^5$ **iv)** $(\frac{3}{4})^N$ **b)** 33
16 0.6
17 b) i) $\frac{1}{5}$ **ii)** $\frac{2}{5}$ **iii)** 0 **c)** $\frac{1}{2}$
18 a) $\frac{2}{3}$ **c)** $\frac{1}{4}$ **d)** $\frac{23}{60}$ **e)** $\frac{15}{23}$
19 a) 0.62 **b)** $\frac{1}{31}$
20 a) $\frac{17}{35}$ **b)** $\frac{12}{25}$ **c)** $\frac{97}{175}$
21 b) $2.7l$ **c)** $\frac{1}{45}$
22 a) $\frac{7}{12}$ **b) i)** $\frac{5}{8}$ **ii)** $\frac{3}{8}$
23 a) 126 **b)** No, greater than 10% chance of error
24 a) 120 **b)** Yes, if less than 6% error is acceptable
25 a) 65 780 **b)** 0.000 015 2 **c)** 0.966 464 0
26 a) 13 983 816 **b)** 0.000 000 071 5 **c) i)** 0.435 965 **ii)** 0.413 019 **iii)** 0.132 378 **iv)** 0.018 622
27 a) i) 0.667 **ii)** 0.224 **iii)** 0.944 **iv)** 0.35 **c) i)** 0.716 **ii)** 0.117
28 a) 0.05 **b)** 0.5
29 12
30 a) 0.2 **b)** 0.03 **c)** 0.32
31 a) 0.125 **b)** 0.273
33 0.125
34 a) 0.04 **b)** 0.12 **c)** 0.16
35 a) 0.25 **b)** 0.48 **c)** 0.26
36 a) (A) 0.006 67 **(B)** $\frac{1}{2}$ **(C)** $\frac{2}{3}$ **b)** $\frac{2}{9}$ **c)** $\frac{1}{2}$
37 a) 10 000 **b)** 5040 **c)** 24 **d)** $\frac{1}{24}$ **e)** $\frac{1}{24}$ **f)** $\frac{7}{8}$
38 a) 0.07 **b)** 0.33 **c)** 0.93 **d)** 0.323
39 a) independent **b)** neither **c)** mutually exclusive
40 a) i) 0.0429 **ii)** 0.142 **iii)** 0.1215 **iv)** 0.189 **v)** 0.334 **b)** 0.642
41 a) 0.545 **b)** 0.455 **c)** 0.3

CHAPTER 5
Expectation

Exercises 5.1A (p.144)

1 ii) a) $\frac{1}{32}$ **b)** –21.6p **c)** no
2 a) no **b)** 5.3p
3 a) 0.002 **b)** 0.23; 3.9p
4 a) 282 500 **b)** no
5 a) no **b)** £126.67
6 a) 0.015 625, 0.5625, 0.421 875 **b)** 0.1875 clockwise

Exercises 5.1B (p.146)

1 5.6p
2 a) Church gains £180 **b)** no
3 b) no **c)** loss of 16.7p
4 loss of £60 000
5 a) i) £105 **ii)** £110 **b) i)** £205, £82, £82, £41 **ii)** £178.22, £71.29, £71.29 £35.64

Exercises 5.2A (p.151)

1 3.5, $\frac{35}{12}$
2 2.5, 7.5; 1.25
3 2, 1
4 $\frac{8}{27}, \frac{4}{9}, \frac{2}{9}, \frac{1}{27}$; 1, $\frac{2}{3}$
5 $\frac{1}{36}, \frac{3}{36}, \frac{5}{36}, \frac{7}{36}, \frac{9}{36}, \frac{11}{36}$; 4.47, 1.97
6 a) i) $\frac{1}{6}$ **ii)** $\frac{1}{2}$ **iii)** $\frac{1}{3}$ **b)** 1.17, 0.472
7 0, $\frac{35}{6}$

Exercises 5.2B (p.152)

1 4.5, $\frac{20}{3}$
2 6.54, 9.94
3 2.25, 1.59
4 0, 2.5
5 a) 2.5, 1.25 **b)** $\frac{5}{3}, \frac{10}{9}$
6 5.22, 10.27
7 8.5, $\frac{95}{12}$

Exercises 5.3A (p.154)

1 a) 3 **b)** 6 **c)** 9 **d)** 12 **e)** 10 **f)** 20 **g)** 9 **h)** 12
2 275.50, 330.60
3 a) 7 **b)** 12.69 **c)** 1.41
4 9.5; £1500, 0.6
5 £79.50, 337.25
6 a) £11.875, 8.297 **b)** £28.325, 46.51

Exercises 5.3B (p.156)

1 a) 2.1 **b)** –3.2 **c)** 4.2 **d)** –3.2 **e)** 4.9 **f)** 7
2 £250, 36 000
3 a) 3.8 **b)** 3.36 **c)** 0.84
4 a) $C = 10 + 24X$ **b)** £78.40, 361.44
5 a) 4.974, 1.1313 **b)** 1288, 175 056
6 187.5; 0.4

Consolidation Exercises for Chapter 5 (p.158)

1 0.2p
2 loss of 2.4p
3 $\frac{1}{2}, \frac{2}{7}, \frac{1}{7}, \frac{2}{35}, \frac{1}{70}$; 1.8
4 1
5 1.39p
7 1.875, 0.8594
8 a) $\frac{1}{14}$ **b)** 2.571 **c)** 0.3878
9 a) 0.8 **b)** 3.2
10 a) $\frac{1}{56}$ **b)** $\frac{10}{56}$ **c)** 1.125
11 1.33, 0.556
12 2.75, 2.5625
13 £1595
14 £6375
15 88.17; £30.82
16 14.62, 17.04; yes, probability > 0.5
17 a) 10.4 **b)** £4.85 **c)** £4.16
18 b) 1.92, 0.9736 **c)** 0.18 **d)** 0.431
19 a) i) 3.5, 3.5 **ii)** 1.45, 1.45 **c)** 0, –0.3; Y is negatively skewed
20 a) means are 4, variances are 2 **c)** 0.65, 0.8; Y is more peaked

CHAPTER 6
Probability distribution

Exercises 6.1A (p.170)

1 a) 1 **b)** 0.0965 **c)** 0.579
2 a) 0.6 **b)** 0.24 **c)** 0.096 **d)** 0.0384 **e)** 0.0256
3 a) 0.25 **b)** 0.7344 **c)** 0.4219
4 a) 0.444 **b)** 0.556 **c)** 0.444
5 0.8062
6 30

Exercises 6.1B (p.171)

1 a) 0.1667 **b)** 0.1157 **c)** 0.5787 **d)** 1
2 a) i) 0.25 **ii)** 0.1875 **iii)** 0.1406 **iv)** 0.4219 **b)** 1 **d) i)** 0.4219 **ii)** 0.0751
3 a) 0.4225 **b)** 0.0150 **c)** 0.001 84 **d)** 0.000 03
4 a) 0.237 **b)** 0.763 **c)** 0.316
5 0.969
6 14

Exercises 6.2A *(p.177)*

1 **a)** 1 **b)** 1.429 **c)** 0.6122
3 1.111, 0.1235
4 1.8
5 0.5041

Exercises 6.2B *(p.177)*

1 **a)** 0.128 **b)** 5 **c)** 20
3 4, 28
4 0.7806
5 2.5

Consolidation Exercises for Chapter 6 *(p.181)*

1 **a)** 54 **b)** 0.0545 **c)** 0.378 **d)** 1
2 **a)** 2 **b)** 1
3 **a)** 1 **b)** 8 **c)** 56 **d)** 0.30
4 **a)** mode
5 **a)** no **b)** the pork **c)** 16.6p
6 **a)** 0.049 **b)** 0.607 **c)** 0.344
7 **a)** 0.079 **b)** 0.925 **c)** 0.422
8 8
9 **a)** 0.0387 **b)** 0.735
10 **a) i)** 0.004 12 **ii)** 0.008 23 **iii)** 0.996 **b) i)** 1 **ii)** 0.031 25
11 **a)** 0.1 **b)** 0.0424 **c)** 0.01 **d) i)** 0.740 **ii)** 0.049
12 **a)** 0.5 **b)** 0.125
13 **a)** 0.75 **b)** 0.25 **c)** loss of £1
14 **a) i)** 0.167 **ii)** 0.139 **iii)** 0.579 **b)** £28.94
15 **a)** 0.625 **b)** 0.375 **c)** 1.25
16 0.0477
17 **a)** $\frac{4}{27}$ **b)** $\frac{1}{3}\left(\frac{2}{3}\right)^{n-1}$ **c)** 3
18 **a)** 120 **b) i)** 0.125 **ii)** 0.008 33 **c)** 0.008 33 **d)** 0.008 33 **e)** 0.0583 **f)** 0.233
19 **a)** 0.25 **b)** 0.316 **c)** 0.0791
20 **b) i)** 1 **ii)** 0.760
21 **a)** 0.128 **b)** $0.2 \times 0.8^{(x-1)}$, geometric
22 0.973

CHAPTER 7
The binomial probability model

Exercises 7.1A *(p.191)*

1 **a)** 0.294 **b)** 0.496 **c)** 0.945
2 **a)** 0.192 **b)** 0.233 **c)** 0.745
3 **a)** 0.651 **b)** 0.851 **c)** 0.954
4 **a)** 0.045 75 **b)** 0.3138 **c)** 0.004 430

Exerciese 7.1B *(p.192)*

1 **a)** 0.951 **b)** 0.0751; 2
3 **a)** 0.336 **b)** 0.294 **c)** 0.797 **d)** 0.203
5 **a)** 0.784 **b)** 0.870 **c)** 0.953 **d)** 6 **e)** 10

Exercises 7.2A *(p.198)*

1 0.741
3 **a)** 0.2825 **b)** 0.018 **c)** 0.253
4 **a)** 0.523 **b)** 0.895 **c)** 0.174
5 **a)** 0.507 **b)** 0.253 **c)** 0.158 **d)** 0.038

Exercises 7.2B *(p.199)*

1 4
3 **a)** 0.986 **b)** 0.02 **c)** 0.101 **d)** 0.665
4 **a)** 0.0040 **b)** 0.6171 **c)** 0.2023
5 **a)** 0.412 **b)** 0.994 **c)** 0.994

Exercise 7.3A *(p.205)*

1 **a)** no **b)** yes **c)** no
2 **a)** yes **b)** no
3 **a)** yes **b)** no **c)** yes
4 **b)** 6% or more **c)** $x \geq 16$ **d)** 0.608
5 **c)** 4% or more **d)** 0.405

Exercise 7.3B *(p.207)*

1 **a)** no **b)** yes **c)** no
3 **a)** no **b)** yes
4 **b)** $x < 4$, 2% or more **c)** 0.4114
5 **b)** 13% **c)** 15 **d)** 89.8%

Consolidation Exercises for Chapter 7 *(p.208)*

1 **a)** 0.115 **b)** 22
2 **a)** 0.255 **b)** 0.472 **c)** 0.760
3 **a)** 0.493 **b)** 0.0144 **c)** 0.0886 **d)** 3
4 **a)** no **b)** yes
5 **a)** no **b)** yes **c)** yes
6 **a)** 0.512 **b)** 0.104 **c)** 0.669 **d)** 0.056 25 **e)** 0.238 **f)** 0.299
7 **a)** 0.107 **b)** 0.302 **c)** 0.678 **d)** 0.624
8 no
9 **a)** 0.818 **b)** 0.0159 **c)** 0.001
10 **b)** 0.350 **c)** 0.0106
11 **b)** not bias **c)** $X = 0$
12 improved
13 0.0913 **a)** $X \geq 18$ **b)** $X \geq 19$ **c)** $X = 20$
14 **b) i)** 0.0130 **ii)** 0.0176 **c) i)** $X \leq 9$ or $X \geq 18$ **ii)** $X \leq 9$ or $X \geq 17$ **iii)** $X \leq 7$ or $X \geq 19$ **d)** 9.5% or more
15 **a)** 0.0861, 0.0159 **b)** 0.624, 0.589
16 **a)** 4.6% or more **b)** 0.473
17 **a)** 0.0991, 0.139 **c)** $X \geq 17$ **d)** 0.0991, 0.861
18 **a)** 0.0416 **b)** 0.0592 **c)** 0.0832 **d)** 0.1184 **e)** not significant **f)** $4 \leq w \geq 11$
19 **a)** 0.2024 **b)** 0.0419 **d)** ≥ 11 **e)** 3
20 **a)** no **b)** no **c)** $B(20, \frac{1}{36})$
21 **a)** 0.352 **b)** 0.896 **c)** 0.1905
23 lower
24 0.7515
25 **a)** 0.117 **b)** 0.1275

CHAPTER 8
Sums and differences of distributions

Exercises 8.1A *(p.218)*

1 **a)** 7, 8; 3, 12 **b)** 7.5, 3.75
2 **a)** 1000 **b)** 13.5, 24.75 **c)** 18, 33
3 **a)** 2, 2 **b)** 7, 2 **c)** 9, 4
4 **a)** 7.5, 32.25 **b)** 15, 64.5

Exercises 8.1B *(p.219)*

1 **b)** 5.5, 4.375 **c)** 8.25, 6.5625
2 0.5, 1.25; 1.5, 3.75
3 2.5, 1.917
4 **a)** 7, 6.5 **b)** not unique

Exercises 8.2A *(p.223)*

1 5, 4; 11, 10
2 11.5, 8.917
3 **a)** 15, 18 **b)** 20, 19 **c)** 10, 19 **d)** –4, 11 **e)** –6, 6
4 **a)** £0.24, 1009.4424 **b)** £1.20, 5047.212 **c)** £1.44, 6056.6544 **d)** £0.96, 6056.6544

Exercises 8.2B *(p.224)*

1 **a)** 5, 5; 3.5, 2.917 **b)** –1, 4.167
2 £117; 58
3 **a)** 5, 2 **b)** 8, 1.5 **c)** 2, 0.5
4 20.4, 7.92

Exercises 8.3A *(p.228)*

1 **a)** 0.1493 **b)** 0.2668 **c)** 0.1503
2 **a)** 4, 2 **b)** 6.4,1.28
3 **a)** 0.180 **b)** 0.4044 **c)** 4.8
4 **a)** 4.5, 2.475 **b)** 2, 1.6 **c)** 0.0949
5 **a)** $n = 12, p = 0.85\ q = 0.15,$ **b)** 0.4435

Exercises 8.3B *(p.228)*

1 **a)** 0.9983 **b)** 0.251 **c)** 0.8327
2 0.258
3 **a)** 0.4073 **b)** 0.0222
4 **a)** 4, 0.8; 0.4096 **b)** 11.1, 2.886; 0.2101
5 **a)** B(3, 0.5), B(4, 0.5), B(7, 0.5)

Consolidation Exercises for Chapter 8 *(p.229)*

1 **a)** 0.9722, 10.5826 **b)** 3.889, 42.3302
3 **a)** 3, 0.2 **b)** 15, 5
4 1, 6
5 **a)** 0.6 **b)** 1.04 **c)** 4.16 **d)** 37.44 **e)** 9.36 **f)** 0.26
6 **a)** $5\mu, 25\sigma^2$ **b)** $0, \sigma^2$ **c)** 1, 1 **d)** $10 - 3\mu, 9\sigma^2$
7 4.333, 2.222; 8.667, 8.889
8 4.083
9 **a)** 5, 0.2 **b)** 10, 0.1 **c)** 20, 0.05 **d)** 200, 0.005
10 **a)** 2 **b)** 5, 0.4 **c)** 11.7, 38.9, 51.8, 34.6, 11.5, 1.5
11 **a)** 1.25 **b)** 6.25, 31.25, 62.5, 62.5, 31.25, 6,25
12 **a)** 0.0353 **b)** 49.32, 6.906
13 **a)** 86p, 14 **b)** 86p, 21
14 **a)** 0, 0.5 **b)** 0.625
15 **a)** 0.25; 1.25

CHAPTER 9
The Poisson distribution

Exercises 9.1A *(p.242)*

1 **a)** 0.1641 **b)** 0.0050 **c)** 0.9950 **d)** 0.1016
2 **a)** 0.2975 **b)** 0.7306 **c)** 0.8347
3 **a)** 0.1494 **b)** 0.2240
4 **a)** 0.3679 **b)** 0.2642
5 **a)** 0.1353 **b)** 0.0972 **c)** 0.2707 **d)** 0.2707 **e)** 0.9197 **f)** 0.8458
6 **a)** 0.4036 **b)** 0.2033 **c)** 0.9162 **d)** 0.3730
7 **a)** 0.2217 **b)** 0.8365 **c)** 0.2302

Exercises 9.1B *(p.243)*

1 **a)** 0.2700 **b)** 0.6204 **c)** 0.8386 **d)** 0.2882
2 **a)** 0.2385 **b)** 0.3978 **c)** 0.8407 **d)** 0.3474
3 **a)** 0.1933 **b)** 0.3772 **c)** 0.9281 **d)** 0.6584
4 **a)** 0.3753 **b)** 0.9233
5 **a)** 0.0902 **b)** 0.5940 **c)** 0.9473
6 **a)** 0.0067 **b)** 0.9596 **c)** 0.2246
7 **a)** 0.2306 **b)** 0.4685 **c)** 0.7993

Exercises 9.2A *(p.249)*

1 **a)** 0.4335 **b)** 0.5665 **c)** 0.1465 **d)** 0.4558 **e)** 0.7405
2 **a)** 0.3528 **b)** 0.8153 **c)** 0.7674
3 **a)** 0.4405 **b)** 0.3840 **c)** 0.5595
4 4, 2
5 0.1563
6 **a)** 0.2792 **b)** 0.3424 **c)** 0.8506

Exercises 9.2B *(p.250)*

1 **a)** 0.6792 **b)** 0.3208 **c)** 0.9347
2 **a)** 0.5608 **b)** 0.7283 **c)** 0.5372 **d)** 0.2992
3 1.8, 1.8
4 8, 0.0003
5 0.1606
6 **a)** 0.8148 **b)** 0.7399 **c)** 0.4103 **d)** 0.6771
7 **a)** 0.1967 **b)** 0.1189 **c)** 0.7035 **d)** 0.641; 0.0614 **e)** 10 **f)** 11 **g)** 13

Exercises 9.3A *(p.254)*

1 **a)** 0.5488 **b)** 0.0231 **c)** 0.1606 **d)** 0.9161 **e)** 0.8298
2 **a)** 0.6063 **b)** 0.2560 **c)** 0.4142 **d)** 0.1378 **e)** 0.6335 **f)** 0.7189
3 **a)** 0.1614 **b)** 0.0729 **c)** 0.1944 **d)** 0.4631
4 **a)** 0.1912, 0.9814 **b)** 0.0498, 0.1493, 0.2241
5 **a)** 0.1708 **b)** 0.1157 **c)** 0.4495 **d)** 0.8843
6 **a)** 0.2231 **b)** 0.0111 **c)** 0.0500 **d)** 0.1125 **e)** 0.0023
7 **a)** 0.1353 **b)** 0.0166 **c)** 0.1954 **d)** 5
8 no
9 no
10 satisfactory

Exercises 9.3B *(p.255)*

1 **a)** 0.3712 **b)** 0.1953 **c)** 0.3134 **d)** 0.0382; 9
2 **a)** 0.4416 **b)** 0.5423 **c)** 0.6797 **d)** 0.9450 **e)** 0.2187
3 **a)** 0.1054 **b)** 0.8858 **c)** 0.9260 **d)** 0.0111
4 **a)** 0.6065 **b)** 0.3679 **c)** 0.6065 **d)** 0.6065
5 **a)** 0.0261 **b)** 0.1559 **c)** 10
6 **a)** 4, 2 **b)** 1.4, 2 **c)** 2.6, 2.280 **d)** 6.8, 5.0596
7 yes
8 yes
9 no difference

Consolidation exercises for Chapter 9 *(p.257)*

1 **a)** 0.0408 **b)** 0.3223 **c)** 0.2087 **d)** 0.2226 **e)** 0.1443
2 **a)** 0.0656, 0.8088 **b)** 0.2241, 0.654
3 **a)** 0.0608 **b)** 0.0079 **c)** 0.0001
4 **a)** 0.4232 **b)** 0.3528 **c)** 0.1791 **d)** 0.1606
5 0.0613
6 **a)** 0.2240 **b)** 0.3347 **c)** 0.5276 **d)** 0.0620
7 **a)** 0.0404 **b)** 0.6925 **c)** 0.2650 **d)** 0.0067
8 **a)** 0.1801 **b)** 0.6204 **c)** 0.0470
9 **a)** 0.0067 **b)** 0.2138 **c)** 0.0472
10 **b)** less than 3%
12 5
13 **a)** –2, 7 **b)** 10, 10
14 **b)** at least 7.5% **c)** 0.1847
15 **a)** 0.3027, 0.6665, 0.0273 **b)** 0.0290
16 **a)** 0.7408, 0.0369 **c)** 100
17 6.2%
18 **a)** 0.908 **b)** 9
19 **a)** 0.0288 **b)** 0.744
20 **b)** 1.11, 1.10 **c)** no, time is continuous
21 significant
22 0.3374
23 **a)** 0.0672 **b)** 0.2859 **c)** 0.74 **d)** 0.4652

CHAPTER 10
Samples and populations

Exercises 10.1A *(p.268)*

1 a) 276, 29, 86, 225, 72
b) 3440, 355, 1065, 2812, 897
c) 102, 11, 32, 83, 27
d) 41 269, 4249, 12 779, 33 738, 10 764
e) 137 561, 14 161, 42 595, 112 460, 35 880

Exercises 10.1B *(p.269)*

5 b) 76%

Exercises 10.2A *(p.272)*

6 9.5

Exercises 10.2B *(p.273)*

5 b) 4.33

Consolidation Exercises for Chapter 10 *(p.279)*

3 c) 0.0420
4 a) 0.2×0.8^x, $x = 0, 1, 2, \ldots$
8 a) 5, 3, 1, 7, 8
b) 5, 2, 0, 7, 9 **c)** 1, 0, 0, 1, 1
9 9, 2, 1, 19, 29
19 a) 2, 1.414
22 6

CHAPTER 11
Continuous random variables

Exercises 11.1A *(p.292)*

1 a) 10%
b) $f(x) = 4.58 \times 10^{-8} \times (425 - x) \times (x + 100)$
2 a) 2 **b)** 0; $2x - x^2$; 1
c) 0.293 **d)** 0.4
3 a) 0.002
c) 0; $0.002(x - 0.0005x^2)$; 1
i) 0.36 **ii)** 0.36 **iii)** 292.89
4 a) 12 **b)** 0.667
c) 0; $x^3(4 - 3x)$; 1 **d)** 0.614
5 b) 0.025
c) 0; $0.0125t^2$; $0.2t - 0.8$; 1
d) 6.32 **e)** 0.3875
6 b) 0.25 **c)** 0; $0.125x^2$; $0.125(8x - 8 - x^2)$; 1 **d)** 0.75
7 b) 35% **c)** $0.003x^2$
d) 0.316

Exercise 11.1B *(p.294)*

1 b) 52% **c)** 0.05 **d)** 0.5
2 b) 25%
c) $0.000\,732\,4(x - 16)^3$ **d)** 0.27
3 b) 0.167
c) 0; $0.0833(t^2 + 2t - 3)$; 1
d) 0.583
4 b) 0.0333 **c)** 15
5 a) 0.004
c) 0; $0.004(x - 0.001x^2)$; 1
d) i) 0.36 **ii)** 0.28 **e)** 146.45
6 a) 78.82 **b)** 0.25
c) 61.6%; 77.8%
7 a) 4 **b)** 0; $2 - 4x^{-1}$; 1
c) i) 0.143 **ii)** 0.4
d) 2.67; 2.29, 3.2; 0.914
e) 3.64

Exercises 11.2A *(p.302)*

1 a) 24 **b)** 4, 0.636
c) 3.328 **d)** 0.0215 **e)** 0.304
2 a) 7.5, 2.083 **b)** 0.16
3 a) 0 **b)** 0.822
c) 0; $0.5 + 0.3183x$; 1
4 a) 0.0833 **b)** 1
c) 1.125, 0.6094 **d)** 0.935
5 a) 0.5 **b)** 0, 0.333
c) triangular
d) $0.25y + 0.5$; $0.5 - 0.25y$
e) 0.281 25
6 a) 0.1875 **c)** 0.2375 **d)** 61.4%

Exercises 11.2B *(p.303)*

1 a) k **b)** $0.333k^2$
2 a) $1 \div (b - a)$
b) 0; $(x - a) \div (b - a)$; 1
3 a) 0. 093 75 **b)** 2
c) 0.8944 **d)** 0.156 25
4 a) $1 \div (3a^2)$ **b)** $2.5a^2$
c) 0.417
5 a) 0.006 **b)** 15
c) 2.236 **d)** 0.3778
6 a) 0.167 **c)** 2.78; 1.62
d) 0.056 **e)** 0.0208

Consolidation Exercises for Chapter 11 *(p.308)*

1 a) $1 - e^{-x}$ **b) i)** 0.693
ii) 0.132
2 a) 6.93 **b)** 0.239
c) $0.1e^{-0.1x}$ **d)** 10; 100
3 a) 0.313 **b)** 0.249
4 a) 0.025 **b) i)** 0.487
ii) 0.250 **iii)** 0.128
c) 0.487
5 a) 100, 10 000 **b)** 0.607
6 a) $a^{-1}e^{-\frac{t}{a}}$ **c)** 0.135
7 a) 4 **b)** 1.33, 0.889
d) 1.17; 0.536, 2
8 a) 0.75
b) 0; $0.5 + 0.75t - 0.25t^3$; 1
c) i) 0.156 **ii)** 0.688
d) 0, 0.2
9 a) 7.5, 2.083 **b)** 15, 4.167
c) 0.92
10 a) i) 0.125 **ii)** 0.02
b) 0,0.167; $1 + x$, $1 - x$
11 17.5, 52.083; 0.3456
12 a) 0.777 **b)** 0.204
13 b) $x = 0.75$ **c)** 40, 10.7
d) 0; $x^4(5 - 4x)$; 1
14 a) $0.01e^{-0.01x}$; $1 - e^{-0.01x}$
b) 0.3935
15 b) $e - 1$, $0.5(e - 1)(3 - e)$
16 a) 7.67 **b)** 0, 1 **c)** 7.83
d) 0; $0.25(4.5 - y)(y - 0.5)$; 1
e) $0.5(2.5 - y)$ on $0.5 \le y \le 2.5$
17 a) 0.25 **b)** 0; $0.25x$; 1
c) $0.5y$ on $0 \le y \le 2$
18 a) i) 2.67 **ii)** 0.943
iii) 0.4375 **iv)** 0.2650
b) i) 6p, 0.0625; 10p, 0.1875; 14p, 0.3125, 18p, 0.4375;
ii) 14.5p, 3.71p
19 0.6; 0.5, 0.03; 0.222
20 a) 0 **b)** 0.625; 0.206
21 b) 0; $(x + \sin x) \div \pi$; 1 **d)** 1.03

CHAPTER 12
Validating models – the chi-squared distribution

Exercises 12.1A *(p.327)*

1 a) 48.0, 82.3, 52.9, 15.1, 1.7
c) 18.75 > 9.488
d) model does not fit
2 a) 150, 60, 6, **b)** 11.29 > 5.99, model does not fit
3 b) 5.1, 38.4, 115.2, 172.8, 129.6, 38.9 **c)** 10.09 > 9.236, model does not fit
d) 10.09 < 11.070, model fits
4 b) 36.3, 96.8, 96.8, 43.0, 7.1
c) 18.98 > 9.488, poor fit
d) 9.452 < 9.488, model fits (just)
5 b) 15.4, 9.2, 9.1, 7.2, 9.1
c) 10.13 < 11.668 fits

Exercises 12.1B *(p.329)*

1 **b)** 6.17 < 9.488, good fit
2 **a)** 5.1, 30.7, 69.1, 69.1, 26.0
b) 7.48 < 13.277, fits
3 **a)** B(2, 0.5) **b)** 12.5, 25, 12.5; 7.08 < 7.824, fits
c) i) does not fit **ii)** fits
d) yes at 5%, no at 2%
4 **a)** 33.5, 67.1, 58.7, 29.4, 11.3
b) 0.624 < 7.779, fits
c) very good
5 **a)** 7.3, 85.1, 396.9, 926.1, 1080.5, 504.1
b) 11.02 < 11.07, model fits (just)
d) 22.07 > 11.07, very poor fit

Exercises 12.2A *(p.336)*

1 **b)** 2.01; 13.4, 26.9, 27.1, 18.1, 9.1, 5.4
c) 24.63 > 9.488, very poor fit
2 **a)** 33.3, 36.6, 20.1, 7.4, 2.6
b) 1.70 < 11.668
c) good fit
3 **a)** 250
b) 33.99 > 13.388, very poor fit
4 5.75 < 7.815, good fit
5 **b)** 184, 512, 104
c) 1.36 < 4.605, manuscript may have been written by the novelist
6 **a)** Geo (0.7)
b) 168, 50.4, 15.1, 6.5
c) 23.71 > 11.345, geometric distribution with $p = 0.7$ is a very poor fit
d) Geo (0.609); 146.2, 57.1, 22.3, 8.7, 5.7; 4.22 < 11.345, geometric distribution with $p = 0.609$ fits

Exercises 12.2B *(p.337)*

1 **a)** 106.36 > 15.507, unsuitable
b) 29.82, unsuitable
2 Poi (1.465), 9.454 < 9,488, just about fits
3 20.52 > 9.488, model does not fit
4 9.62 > 7.815, distribution is different
5 **a)** 5.9, 14.8, 38.5, 20.8
b) 0.105 < 6.251, exceptionally good fit
6 **a)** Geo (0.50)
b) i) 8.8 > 6.251, inappropriate
ii) 8.8 > 7.815, inappropriate
iii) 8.8 < 9.837, appropriate

Exercises 12.3A *(p.345)*

1 **a)** 3410 **b)** 2040: 1370; 1287: 1243: 557: 323
c) 769.9, 743.6, 333.2
d) three **e)** strongly associated
2 4.68 < 9.488, independent
3 **a)** yes
b) Support staff decidedly against; few academic staff are against the designated smoking area
4 **b)** 17%; no
c) 12.14 > 10.645, associated
d) The 1200 spin are rarely of reject quality. There is a tendency for the 800 spin machines to need repairing and the 600 spin machines to be reject quality.
5 **a)** 1.462 **b)** 1.639, no evidence of association
6 17.94 > 7.879, strong evidence of association, private patients tend to have short stays

Exercises 12.3B *(p.347)*

1 **a) i)** 40 **ii)** 207.5
iii) 142.5 **b)** three
c) 82.56 > 12.838
d) equal to, males positively associated whereas females negatively associated
2 no association
3 yes, very strongly; first class passengers survived whereas third class did not
4 **b)** 20% **c)** 4.736 < 9.488
d) factors are independent
5 evidence of association; people living 'elsewhere' favour tunnel option, those who live in the town do not want any change
6 not independent; female employees tend to be self–funded

Consolidation Exercises for Chapter 12 *(p.348)*

1 **a)** 246.6, 345.2, 241.7, 112.8, 39.5, 14.2
b) 32.15 > 11.07, do not accept the Poi(1.4) model
2 **b)** 72.5, 95, 67.5, 15
c) 3.00 < 7.815; insufficient evidence to suggest that the new treatment has changed the proportions of apples in the various grades
3 1.038 < 5.991 hence same proportions in the schools
4 **b)** 18, 12, 8, 5.3, 3.6, 7.1; model is a poor fit
5 11.102 > 5.991, evidence of association
6 language and gender are independent
7 **a)** 6.74 < 7.815, independent
b) independent
c) 14.56 > 7.815, not independent
8 **b)** observed: 20, 10, 9, 7, 14; expected: 13.206, 15.414, 15.756, 10.314, 5.31; 23.6 > 7.815; poor fit
9 **a)** 23.5
c) 17.52 > 14.067; theory not supported
10 **b)** 7.3, 12.4, 10.6, 9.7
c) 1.77 < 4.605; reasonable claim
11 521 > 9.21; highly significant evidence of association
12 **a)** consistent
b) exponential fits
13 2.244 < 3.84; not associated

CHAPTER 13
Bivariate data

Exercises 13.1A *(p.360)*

1 **a)** – 0.173 **b)** – 0.2226
c) – 17.3 **d)** – 0.2226; r is unchanged by the linear transformations
2 **b)** yes **c) i)** 14.5 **ii)** 1278.2
iii) 18 697.5 **d)** 0.863
3 **a)** negative **b)** –0.718
d) –0.301
4 **a)** (1, 63.5) **b) i)** –0.909
ii) –0.747
5 **a)** (4,10.3), (19, 4.9)
b) i) – 5.5525 **ii)** –3.9077
c) negative linear
6 0.9298, positive linear correlation
8 –0.0109, not linearly correlated

Exercises 13.1B *(p.362)*

1 **b)** 3.364 **c)** 0.553
2 **b)** 892.83 **c)** 0.945 **d)** yes
3 **b)** 0.1143 **c)** 0.575
d) yes but outlier
4 **a)** E **b) i)** 0.698 **ii)** 0.913
6 no change

Exercises 13.2A *(p.369)*

1 0.7865 > 0.5494 yes
2 –0.2118, compared with –0, 5822, not linearly correlated
3 **c)** 0.3367 < 0.5822, no
d) –0.2697, no
4 –0. 6602 compared with –0.6021, yes at 2.5%
5 **b)** 0.8335 compared with 0.6581, yes at 2%
6 0.6057 compared with 0.5823, yes at 1%

Exercises 13.2B *(p.371)*

1 yes, 0.8355
2 **a)** 0.9762 **b)** 0.9762
4 **a)** yes, –0.8798 compared with –0.8114 at 2.5%
5 **c)** no, 0.2709 compared with 0.4973
6 **a)** 0 **c)** clearly correlated but not linearly

Exercises 13.3A *(p.377)*

1 **d)** 0.962, positive linear correlation
e) $p = 23.5a - 49.7$ **f)** 2665
2 **a)** $y = 5.06 + 1.40x$ **b)** 19.1
3 **a)** –0.7851, compared with –0.6851, yes at 1% **b)** 32.4
c) $x = 69.6 - 0.885y$
4 **A a)** –0.8578
b) $y = 0.304 - 0.0172x$
d) i) 0.218 **ii)** 0.046 **iii)** –0.126
e) estimate for 25 is unreliable
B a) 0.9944, closer to unity
b) $(2.31x - 2.03)y = 1$ **c)** 0.031
5 **a)** $w = 68.4a - 505$; 794
b) under **c)** no
d) $w = 148 + 0.100a^3$; 834
e) cubic
6 **a)** 0.8895 > 0.8329, significant at 1%
b) 0.9694 highly significant
c) $A = 1.03a^2 - 14.7$
7 **b)** 0.9770 **c)** highly significant
d) 3.638, least value of sum of squared deviations

Exercises 13.3B *(p.379)*

1 **a)** –0. 9979 **c)** $x = 33.5 - 2.66y$
d) x controlled
2 **a)** $y = 0.6x + 0.414$
b) 0.8823 **c)** 2.869
3 **a)** 0.9876, significant at 1%
b) time is factor
c) $y = 0.0259t - 0.703$
d) i) 20.0
ii) 51.1; second one is unreliable
4 **a)** $y = 1.91x - 9.94$
b) time required to learn an additional line of text
c) 181, unreliable
5 13
6 **a)** curved, increasing
b) 0.9427, straight line should be acceptable
c) $z = \log a + t \log b$;
$z = 0.2196 + 0.029\,43t$;
$a = 1.658$, $b = 1.070$
e) 0.9982, closer to unity
7 **a)** $y = 0.991 - 0.107x$
b) $y^{-1} = 2.062 - 1.077x^{-1}$

Exercises 13.4A *(p.384)*

1 **c) i)** 50.8 **ii)** 53:0
2 **d)** –0.2325; 10.0, 9.6, 9.1, 8.6
3 **d)** 97, 103, 86, 79, 64, 58
4 **c)** 848, 739 **d)** 815, 801
e) 7-year may be more reliable

Exercises 13.4B *(p.385)*

1 **d)** yes **e)** 8.1
2 **d)** 162.5
3 **d)** 169

Consolidation Exercises for Chapter 13 *(p.386)*

1 **b)** 0.9692
c) yes, significant at 1%
2 **a)** $y = 139 + 1.05x$ **b)** 186
c) increase in length for each unit increase in mass
3 **a)** 0.9281
b) suggests a linear model may be appropriate
4 **a)** 0.7496, evidence supports linear correlation
b) appears non-linear
c) 0.8381, squared is better
5 **d)** 52.6, 46.6, 46.6, 87.6
6 **b)** $y = 7.6 + 0.59x$; 26; poor
c) (38, 30)
d) 0.44, not linearly correlated
7 **b)** $y = 21.0 + 0.978x$
c) 17 000; break even production
8 **b)** $y = 0.231 - 0.011\,x^{-1}$
c) 0.0014 **d)** outlier
e) model fits data without (0.2, 0.269) better
9 **b)** $y = 8{,}64 + 9.89x$
c) b represents cost of each additional day, a represents a fixed cost **d)** £117
e) no, out of range of data
10 **b)** $y = 0.952x - 35.5$
c) i) 141, in range, little variability, should be accurate
ii) 198, in range but much variability, unlikely to be accurate
iii) 250, out of range, unreliable
d) bias readings to upper end, say 215 up to 250
11 **a)** $y = 11.4 + 0.150t$
b) 14.4, reliable
c) 35.4, unreliable
12 **a)** 0.3377 **b)** not significant
c) reduces it, exclude the point
13 **a) ii)** $y = 0.776 + 0.0213x$
iii) 2.5; out of range, data not linearly related
c) use new line for reliable estimate; carry out trials around 80 g
14 **b)** $h = 22.6 + 0.821a$
15 **iii)** high seasonal variation in first and fourth quarters, low in others **e)** 813
16 **c)** steadily rising expenditure; sharp fall in first quarter, rising steadily to peak in fourth quarter
d) £102 million; Christmas

CHAPTER 14
The normal probability model

Exercises 14.1A *(p398)*

1 **b) i)** 65.4 **ii)** 13.4
c) i) 65.4 **ii)** 52.0, 78.8
iii) 38.6, 92.2
d) 4%, 67%, 29%
e) 5%, 68%, 27%
2 **e)** 19% **f)** 74 mph
4 **b)** 137.4, 763.9
c) i) 5.9% **ii)** 69.4% **iii)** 24.7%
5 **c) i)** 8.5% **ii)** £182 000

Exercises 14.1B *(p.400)*

1 6.0, 2.6; 7.6% compared with 5%
2 13 500, 540
3 a) 67.7, 7.22

Exercises 14.2A *(p.408)*

1 a) 0.150 **b)** 0.516 **c)** 0.256 **d)** 0.145
2 a) 0.219 **b)** 0.368 **c)** 0.484 **d)** 0.950 **e)** 0.685 **f)** 0.032
3 a) 1.598 **b)** 0.327 **c)** 1.674 **d)** 2.645
4 a) 0.182 **b)** 0.727 **c)** 0.196 **d)** 0.9936
5 a) i) 0.861 **ii)** 0.741 **b)** 3.07 **c)** 1.07, 1.97; 0.90
6 a) 2.16, 0.682 **b)** 9.508, 0.00700 **c)** 96.82, 10.48

Exercises 14.2B *(p.408)*

1 a) 200.658 **b)** 0.539
2 a) 0.38% **b)** 115.5
3 a) 8.2% **b)** 0.580 **c)** 0.518 **d)** 13.5%
4 a) 21.82 **b)** 0.0036 **c)** 0.950
5 a) 0.159 **b)** 0.159 **c)** 11.18 **d)** 9.70
6 a) 453.5; 3.41 **b)** less than half meet the stated mass

Exercises 14.3A *(p.412)*

1 a) 33, 0.007 925 **b)** 0.0123 **c)** 32.7
2 a) 0.5, 0.9656, 0.0139 **b)** 0.310
3 a) i) 13, $\sqrt{22}$ **ii)** 3, $\sqrt{22}$ **iii)** –4, $\sqrt{261}$ **b)** 0.0678, 0.261, 0.261, 0.5
4 0.731
5 a) i) 11.375, 67.95, 74.945, 95.73, 95.73, 74.945, 67.95, 11.375 **ii)** 8, 52, 115, 70, 75, 15, 155, 10 **iii)** significant difference at 0.5% **b)** plot not linear

Exercises 14.3B *(p.414)*

1 a) 0.623 **b)** 0.626 **c)** 0.438 **d)** 0.560
2 a) 0.712 **ii)** 0.785 **b)** 6 minutes 19 seconds
3 a) 8.61 **b)** 0.408 **c)** 0.595
4 a) 0.9492 **b)** 0.860 **c)** 0.5347
5 a) E: 5.28, 3.76, 8.02, 11.70, 11.70, 8.02, 5.28; 0.69 < 11.07, model fits data very well **b)** 170.4, 7.56 **c)** 4.22, 7.65, 12.07, 12.48, 8.47, 5.11; 0.48 < 7.815, normal fits data well

Consolidation Exercises for Chapter 14 *(p.415)*

1 9.7%, 78.8%, 11.5%
2 a) 2.3% **b)** 1 year 4 months
3 a) 30.9% **b)** 0.0569
4 5.01, 0.0467
5 a) normal distribution is a model **b)** symmetry, bell-shape, asymptotic, continuity
6 a) reasonably linear, normal model may fit **b)** 5.4, 13.2, 20.7, 22.2, 28.1, 28.4, 54.5, 39.4, 20.6, 13.5 **c)** 13.47 < 16.919 difference not significant at 5% level
7 a) 0.37 **b)** 0.211 **c)** 0.38, 0.22 **d)** curved at both ends hence normal may not fit **e)** E: 14.3, 16.3, 23.6, 26.7, 23.6, 16.3, 8.9, 5.3 (amalgamated); 15.01 > 14.067 hence difference significant at 5% level
8 a) $N(3\mu, 3\sigma^2)$ **b)** $N(2\mu, 4\sigma^2)$ **c)** $N(0, 2\sigma^2)$ **d)** $N(3\mu, 5\sigma^2)$ **e)** $N(\mu+2, \sigma^2)$ **f)** $N(0, \sigma^2)$ **g)** $N(0, 1)$ **h)** $N(\frac{1}{2}\mu, \frac{1}{4}\sigma^2)$ **i)** $N(0, 6\sigma^2)$
9 a) 0.894 **b)** 0.204
10 6.91; 0.444
11 b) 63.54, 4.394 **c)** yes **d)** no
12 a) 68.14 **b)** 0.428
13 0.044
14 a) 0.894 **b)** 968 **d)** 909
15 a) i) $N(3\mu, 3\sigma^2)$ **ii)** $N(0, 2\sigma^2)$ **iii)** $N(2\mu, 4\sigma^2)$ **b)** 0.985; 0.145; 0.033
16 a) 0.016 **b)** 0.773 **c)** 0.199
17 a) 0.954 **b)** 0.209

CHAPTER 15 Approximating distributions

Exercises 15.1A *(p.424)*

1 a) i) 0.0527 **ii)** 0.9834 **b) i)** 0.0183 **ii)** 0.1953
2 a) i) 0.180 **ii)** 0.143 **iii)** 0.143 **b) i)** 0.027 **ii)** 0.340
3 a) not at 5% level **b)** no
4 a) Poisson, 0.1205 **b)** binomial, 0.6528 **c)** Poisson, 0.632

Exercises 15.1B *(p.425)*

1 a) i) 0.080 **ii)** 0.981 **b) i)** 0.744 **ii)** 0.039
2 a) i) 0.176 **ii)** 0.384 **iii)** 0.384 **b) i)** 0.019 **ii)** 0.429
3 a) yes at 2% **b)** no

Exercises 15.2A *(p.429)*

1 no
2 a) 0.0996 **b)** 0.0485
3 a) no **b)** no **c)** more difficult
4 a) 0.156 **b)** 0.139 **c)** 0.115 **d)** 0.0664

Exercises 15.2B *(p.430)*

1 no
2 a) 0.2650 **b)** 0.912
3 a) no **b)** not at 5% **c)** yes
4 a) 0.318 **b)** 0.067 **c)** 0.125 **d)** 0.0630

Exercises 15.3A *(p.434)*

1 a) 0.010 24 **b)** 0.247 **c)** 0.136
2 yes at 2.5% level
3 a) 0.999 **b)** 0.989 **c)** 0.0266
4 a) reject claim **b)** accept claim

Exercises 15.3B *(p.435)*

1 a) 0.077 76 **b)** 0.873 **c)** 0.944
2 not at 5%
3 a) 0.068 **b)** 0.721 **c)** 0.008 68 **d)** 0.345
4 a) denies the claim **b)** supports claim

Exercises 15.4A *(p.437)*

1 **a)** N(420, 1440) **b)** 0.0569 **c)** 0.56
2 **a)** N(2400, 10 800) **b)** 0.805
3 **a)** 0.132 **b)** 0.000 21
4 **a)** 0.042 **b)** 50.5

Exercises 15.4B *(p.438)*

1 **a)** N(230, 90) **b)** 0.146 **c)** 0.837
2 **a)** N(1440, 768) **b)** 0.406
3 **a)** 0.710 **b)** 0.892
4 **a)** 0.222 **b)** 12.35

Consolidation Exercises for Chapter 15 *(p.438)*

1 **a)** no **b)** 329
2 **a)** 0.907 **b)** 0.283
3 **a)** 0.783 **b)** 1 **c)** no, more
4 **a)** upheld **b)** rejected
5 **a)** 0.706 **b)** 0.923
6 **a)** 113 **b)** 0.3712
7 0.315
8 **a)** 0.859 **b)** 0.202 **c)** 0.034
9 **a)** 6, 3 **b)** 5.90, 0.90 **d)** 5.95, 1.00
10 **a)** 0.887 **b)** 0.994; 18, N(18, 17.91), 0.723
11 **a)** 0.359 **b)** 0.046
12 **a)** 0.067 **b)** 0.923
13 **a)** normal, 0.977 **b) i)** Poisson, 0.1912 **ii)** normal, 0.246 **c) i)** binomial, 0.1702 **ii)** 0.6477
14 0.669, 0.353, 0.089
15 **b)** 0.018 **c)** 0.744; 0.803

CHAPTER 16 The central limit theorem

Exercises 16.1A *(p.449)*

1 **a)** 0.023 **b)** 0.0127 **c)** 0.673 **d)** £30
2 **a)** 0.3085 **b)** 0.2146 **c)** 0.3085 **d)** 0.4115
3 **a)** 20.5, 0.0833 **b)** 0.0142 **c)** 0.0005
4 **a)** other **b)** top grade **c)** 196.3

Exercises 16.1B *(p.449)*

1 **a)** 0.091 **b)** 0.228 **c)** 0.294 **d)** 7.2 kg
2 **a)** 0.159 **b)** 0.023 **c)** 0.124 **d)** 0.993
3 **a)** 0.061 **b)** 0.863 **c)** 0.02 m
4 **a)** other **b)** other **c)** 147.6 g

Exercises 16.2A *(p.455)*

1 **a)** 2.129 **b)** [4.343, 4.747] **c)** 16%
2 **a)** assume number is a Poisson variate, [5.46, 10.04] **b)** [21.8, 40.2]
3 **a)** [4.124, 4.436] **b)** 68%, assume X is normal
4 **a)** 20.11, 11.07 **b)** [9.68, 12.46] **c)** less **d)** 1000

Exercises 16.2B *(p.456)*

1 **a)** 200.84 **b)** [39.9, 42.6] **c)** 79%
2 **a)** assume number is a Poisson variate, [15.17, 23.83] **b)** [60.7, 95.3]
3 **a)** 305 **b)** [5.08, 5.46] **c)** 6%
4 **a)** 0.571, 0.094 **b)** [0.47, 0.67] **c)** halved **d)** 3600

Consolidation exercises for chapter 16 *(p.459)*

1 **a)** [1.064, 2.069] **b)** no **c)** 0.027
2 **a)** [16.7, 23.3] **b)** yes **c)** no
3 **a)** 15.0, 3.0 **b)** [0.55, 7.34] **c)** 0.479 **d)** 0.486 **e)** 0.474
4 **a)** 0.003 83
5 0.0000
6 **a) i)** 0.1587 **ii)** 0.7357 **iii)** 472 s **b)** 335 s **c) i)** 0.5407 **ii)** 0.046
7 1.60, [1.58, 1.63]
8 [257.9, 274.1]
9 [30.4, 32.4]
10 7.66% **a)** 26.08 **b)** 0.324; mean is less than 26
11 **a)** 4.31, 2.68 **b)** [3.78, 5.05] **c)** not consistent **e)** central limit theorem validates the process
12 **a)** [495.6, 503.0] **b)** data support the claim **c)** within confidence interval
13 **a)** 1.81, 2.89 **b)** [1.574, 2.046] **c)** 0.768, campaign is effective
14 **a)** 9.9 kg, 75.9 g, normal **b)** 110n g, 64n g^2 **c)** 9.3
15 **a)** 0.4, [0.208, 0.592] **b)** [494.7, 499.6] **c) i)** 6 g **ii)** [457.3, 461.9] **d)** claim seems valid
16 **a)** N(μ, $\sigma^2 \div n$) **b)** 28
17 **a)** 5.01, 0.000 489 8 **b)** [5.004, 5.016] **c)** data support the assertion

CHAPTER 17 Non-parametric tests

Exercises 17.1A *(p.470)*

1 **b)** 0.5714 **c)** < 0.6429, not positively correlated
2 **b)** –0.3 **c)** no
3 **a)** 0.8333 **b)** yes at 1%
4 **a)** 0.66 **b)** evidence of correlation
5 0.152, no evidence of correlation
6 **a)** 0.6653 **b)** evidence of correlation
7 **b)** 0.7608 **c)** agreement
8 **b)** 0.1687 **c)** no evidence of correlation

Exercises 17.1B *(p.472)*

1 0.7
2 b) 0.6611 **c)** yes at 10%
3 0.4198; seems a wise decision
4 **a) i)** 0.3810 **ii)** 0.3095 **iii)** 0.0952 **b)** no evidence of agreement
5 **a)** 0.9516, significant evidence **b)** cause is not established
6 **a)** –0.3571, no evidence
7 0.7415, yes
8 **b)** –0.0425 **c)** no evidence **d)** –0.8124, significant evidence at 1%

Exercises 17.2 *(p.477)*

1 not significant
2 not significant
3 **a)** not significant **b)** $T \leq 3$
4 **b)** direction not predominantly southerly **c)** 5% chance of unusual result when null hypothesis is true
5 **a)** not significant

b) N (182.5, 91.25), 1.88 > 1.645, hence evidence to support view that students spent less

6 0.0547, hence not significant at 5%

7 not significant

8 no

9 significant improvement

10 **a)** no loss at 5%; evidence of loss at 10%
b) 2.446 > 2.326, evidence of weight loss

Exercises 17.3 *(p.484)*

1 **a)** not significant
b) $t = 9 < 10$, alcohol increases reaction time. The decreases were small, some increases were considerable.

2 no evidence of difference at 10%

3 no evidence of difference

4 no

5 significant evidence of difference

6 **a)** no difference
b) evidence that the new recipe is sweeter

7 **a)** **i)** no difference
ii) refrigerated seeds have better germination rate
c) 5% chance of getting unusual result when null hypothesis is true

8 insufficient evidence

9 **b)** insufficient evidence

10 **i)** **a)** highly significant improvement **b)** no evidence
c) no evidence
d) highly significant improvement
e) evidence of weight loss at both levels

Exercises 17.4 *(p.491)*

1 no

2 no

3 yes

4 no

5 no

6 **a)** **ii)** 1,2,3,4; 1,2,3,5; 1,2,3,6; 1,2,4,5 **iii)** 126 **v)** 0.0331
b) no difference in location

7 $R = 29$ ($U = 8$), not significant

8 $R = 69.5$ ($U = 14.5$), highly significant

9 insufficient evidence

10 no

11 **b)** 56; 0.0714 **c)** **i)** no **ii)** yes
d) 10% chance of rejecting null hypothesis when it is true

12 evidence that new treatment is better

Exercises 17.5 *(p.498)*

1 0.333 to 0.527

2 0.372 to 0.488

3 0.207 to 0.255

4 **a)** evidence of change
b) telephone bias

5 0.208 to 0.332

6 0.545; 0.514 to 0.576

7 0.08 to 0.48

8 271

9 2500

10 **a)** **i)** 0.36 **ii)** 0.172 to 0.548
b) 54, 34.56 **c)** **i)** 0.14
ii) 21, 18.06 **iii)** 33, 52.62
iv) 18.8 to 47.2

Tables of statistics

Standard normal probability distribution function

Proportional parts (columns 1–9)

z	0.00	0.01	0.02	0.03	0.04	0.05	0.06	0.07	0.08	0.09	1	2	3	4	5	6	7	8	9
0.0	0.50000	0.50399	0.50798	0.51197	0.51595	0.51994	0.52392	0.52790	0.53188	0.53586	40	80	119	159	199	239	279	319	358
0.1	0.53983	0.54380	0.54776	0.55172	0.55567	0.55962	0.56356	0.56749	0.57142	0.57535	39	79	118	158	197	237	276	315	355
0.2	0.57926	0.58317	0.58706	0.59095	0.59483	0.59871	0.60257	0.60642	0.61026	0.61409	39	77	116	155	193	232	271	309	348
0.3	0.61791	0.62172	0.62552	0.62930	0.63307	0.63683	0.64058	0.64431	0.64803	0.65173	38	75	113	150	188	225	263	300	338
0.4	0.65542	0.65910	0.66276	0.66640	0.67003	0.67364	0.67724	0.68082	0.68439	0.68793	36	72	108	144	180	216	252	288	324
0.5	0.69146	0.69497	0.69847	0.70194	0.70540	0.70884	0.71226	0.71566	0.71904	0.72240	34	69	103	137	171	206	240	274	309
0.6	0.72575	0.72907	0.73237	0.73565	0.73891	0.74215	0.74537	0.74857	0.75175	0.75490	32	65	97	129	161	194	226	258	291
0.7	0.75804	0.76115	0.76424	0.76730	0.77035	0.77337	0.77637	0.77935	0.78230	0.78524	30	60	90	120	151	181	211	241	271
0.8	0.78814	0.79103	0.79389	0.79673	0.79955	0.80234	0.80511	0.80785	0.81057	0.81327	28	56	83	111	139	167	195	222	250
0.9	0.81594	0.81859	0.82121	0.82381	0.82639	0.82894	0.83147	0.83398	0.83646	0.83891	25	51	76	102	127	152	178	203	229
1.0	0.84134	0.84375	0.84614	0.84849	0.85083	0.85314	0.85543	0.85769	0.85993	0.86214	23	46	69	92	115	138	161	184	207
1.1	0.86433	0.86650	0.86864	0.87076	0.87286	0.87493	0.87698	0.87900	0.88100	0.88298	21	41	62	82	103	124	144	165	185
1.2	0.88493	0.88686	0.88877	0.89065	0.89251	0.89435	0.89617	0.89796	0.89973	0.90147	18	37	55	73	91	110	128	146	164
1.3	0.90320	0.90490	0.90658	0.90824	0.90988	0.91149	0.91308	0.91466	0.91621	0.91774	16	32	48	64	80	96	112	128	144
1.4	0.91924	0.92073	0.92220	0.92364	0.92507	0.92647	0.92785	0.92922	0.93056	0.93189	14	28	42	56	70	84	98	112	126
1.5	0.93319	0.93448	0.93574	0.93699	0.93822	0.93943	0.94062	0.94179	0.94295	0.94408	12	24	36	48	60	72	84	96	108
1.6	0.94520	0.94630	0.94738	0.94845	0.94950	0.95053	0.95154	0.95254	0.95352	0.95449	10	20	31	41	51	61	72	82	92
1.7	0.95543	0.95637	0.95728	0.95818	0.95907	0.95994	0.96080	0.96164	0.96246	0.96327	9	17	26	35	43	52	60	69	78
1.8	0.96407	0.96485	0.96562	0.96638	0.96712	0.96784	0.96856	0.96926	0.96995	0.97062	7	14	22	29	36	43	50	58	65
1.9	0.97128	0.97193	0.97257	0.97320	0.97381	0.97441	0.97500	0.97558	0.97615	0.97670	6	12	18	24	30	36	42	48	54
2.0	0.97725	0.97778	0.97831	0.97882	0.97932	0.97982	0.98030	0.98077	0.98124	0.98169	5	10	15	20	24	29	34	39	44
2.1	0.98214	0.98257	0.98300	0.98341	0.98382	0.98422	0.98461	0.98500	0.98537	0.98574	4	8	12	16	20	24	28	32	36
2.2	0.98610	0.98645	0.98679	0.98713	0.98745	0.98778	0.98809	0.98840	0.98870	0.98899	3	6	10	13	16	19	22	25	29
2.3	0.98928	0.98956	0.98983	0.99010	0.99036	0.99061	0.99086	0.99111	0.99134	0.99158	3	5	8	10	13	15	18	20	23
2.4	0.99180	0.99202	0.99224	0.99245	0.99266	0.99286	0.99305	0.99324	0.99343	0.99361	2	4	6	8	10	12	14	16	18
2.5	0.99379	0.99396	0.99413	0.99430	0.99446	0.99461	0.99477	0.99492	0.99506	0.99520	2	3	5	6	8	9	11	12	14
2.6	0.99534	0.99547	0.99560	0.99573	0.99585	0.99598	0.99609	0.99621	0.99632	0.99643	1	2	4	5	6	7	8	10	11
2.7	0.99653	0.99664	0.99674	0.99683	0.99693	0.99702	0.99711	0.99720	0.99728	0.99736	1	2	3	4	5	5	6	7	8
2.8	0.99744	0.99752	0.99760	0.99767	0.99774	0.99781	0.99788	0.99795	0.99801	0.99807	1	1	2	3	3	4	5	6	6
2.9	0.99813	0.99819	0.99825	0.99831	0.99836	0.99841	0.99846	0.99851	0.99856	0.99861	1	1	2	2	3	3	4	4	5
3.0	0.99865	0.99869	0.99874	0.99878	0.99882	0.99886	0.99889	0.99893	0.99896	0.99900	0	1	1	2	2	2	3	3	3
3.1	0.99903	0.99906	0.99910	0.99913	0.99916	0.99918	0.99921	0.99924	0.99926	0.99929	0	1	1	1	1	2	2	2	3
3.2	0.99931	0.99934	0.99936	0.99938	0.99940	0.99942	0.99944	0.99946	0.99948	0.99950	0	0	1	1	1	1	1	2	2
3.3	0.99952	0.99953	0.99955	0.99957	0.99958	0.99960	0.99961	0.99962	0.99964	0.99965	0	0	0	1	1	1	1	1	1
3.4	0.99966	0.99968	0.99969	0.99970	0.99971	0.99972	0.99973	0.99974	0.99975	0.99976	0	0	0	0	1	1	1	1	1
3.5	0.99977	0.99978	0.99978	0.99979	0.99980	0.99981	0.99981	0.99982	0.99983	0.99983	0	0	0	0	0	0	0	0	0

Percentage points of standard normal

Φ	0.00	0.01	0.02	0.03	0.04	0.05	0.06	0.07	0.08	0.09
0.5	0.000	0.025	0.050	0.075	0.100	0.126	0.151	0.176	0.202	0.228
0.6	0.253	0.279	0.305	0.332	0.358	0.385	0.412	0.440	0.468	0.496
0.7	0.524	0.553	0.583	0.613	0.643	0.674	0.706	0.739	0.772	0.806
0.8	0.842	0.878	0.915	0.954	0.994	1.036	1.080	1.126	1.175	1.227
0.9	1.282	1.341	1.405	1.476	1.555	1.645	1.751	1.881	2.054	2.326

Φ	0	0.001	0.002	0.003	0.004	0.005	0.006	0.007	0.008	0.009
0.95	1.645	1.655	1.665	1.675	1.685	1.695	1.706	1.717	1.728	1.739
0.96	1.751	1.762	1.774	1.787	1.799	1.812	1.825	1.838	1.852	1.866
0.97	1.881	1.896	1.911	1.927	1.943	1.960	1.977	1.995	2.014	2.034
0.98	2.054	2.075	2.097	2.120	2.144	2.170	2.197	2.226	2.257	2.290
0.99	2.326	2.366	2.409	2.457	2.512	2.576	2.652	2.748	2.878	3.090

Critical values of standard normal variate

Φ	0.5%	1.0%	2.0%	2.5%	5.0%	10.0%						
1 − Φ							10.0%	5.0%	2.5%	2.0%	1.0%	0.5%
z	−2.576	−2.326	−2.054	−1.960	−1.645	−1.282	1.282	1.645	1.960	2.054	2.326	2.576

Chi-squared variate: ν degrees of freedom

ν	0.5%	1.0%	2.0%	2.5%	5.0%	10.0%	50.0%	90.0%	95.0%	97.5%	98.0%	99.0%	99.5%
1	0.00004	0.0002	0.0006	0.0010	0.0039	0.0158	0.455	2.706	3.841	5.024	5.412	6.635	7.879
2	0.0100	0.0201	0.0404	0.0506	0.1026	0.211	1.386	4.605	5.991	7.378	7.824	9.210	10.597
3	0.0717	0.115	0.185	0.216	0.352	0.584	2.366	6.251	7.815	9.348	9.837	11.345	12.838
4	0.207	0.297	0.429	0.484	0.711	1.064	3.357	7.779	9.488	11.143	11.668	13.277	14.860
5	0.412	0.554	0.752	0.831	1.145	1.610	4.351	9.236	11.070	12.832	13.388	15.086	16.750
6	0.676	0.872	1.134	1.237	1.635	2.204	5.348	10.645	12.592	14.449	15.033	16.812	18.548
7	0.989	1.239	1.564	1.690	2.167	2.833	6.346	12.017	14.067	16.013	16.622	18.475	20.278
8	1.344	1.647	2.032	2.180	2.733	3.490	7.344	13.362	15.507	17.535	18.168	20.090	21.955
9	1.735	2.088	2.532	2.700	3.325	4.168	8.343	14.684	16.919	19.023	19.679	21.666	23.589
10	2.156	2.558	3.059	3.247	3.940	4.865	9.342	15.987	18.307	20.483	21.161	23.209	25.188
11	2.603	3.053	3.609	3.816	4.575	5.578	10.341	17.275	19.675	21.920	22.618	24.725	26.757
12	3.074	3.571	4.178	4.404	5.226	6.304	11.340	18.549	21.026	23.337	24.054	26.217	28.300
13	3.565	4.107	4.765	5.009	5.892	7.041	12.340	19.812	22.362	24.736	25.471	27.688	29.819
14	4.075	4.660	5.368	5.629	6.571	7.790	13.339	21.064	23.685	26.119	26.873	29.141	31.319
15	4.601	5.229	5.985	6.262	7.261	8.547	14.339	22.307	24.996	27.488	28.259	30.578	32.801
16	5.142	5.812	6.614	6.908	7.962	9.312	15.338	23.542	26.296	28.845	29.633	32.000	34.267
17	5.697	6.408	7.255	7.564	8.672	10.085	16.338	24.769	27.587	30.191	30.995	33.409	35.718
18	6.265	7.015	7.906	8.231	9.390	10.865	17.338	25.989	28.869	31.526	32.346	34.805	37.156
19	6.844	7.633	8.567	8.907	10.117	11.651	18.338	27.204	30.144	32.852	33.687	36.191	38.582
20	7.434	8.260	9.237	9.591	10.851	12.443	19.337	28.412	31.410	34.170	35.020	37.566	39.997
25	10.52	11.52	12.70	13.12	14.61	16.47	24.34	34.38	37.65	40.65	41.57	44.31	46.93
30	13.79	14.95	16.31	16.79	18.49	20.60	29.34	40.26	43.77	46.98	47.96	50.89	53.67
35	17.19	18.51	20.03	20.57	22.47	24.80	34.34	46.06	49.80	53.20	54.24	57.34	60.27
40	20.71	22.16	23.84	24.43	26.51	29.05	39.34	51.81	55.76	59.34	60.44	63.69	66.77
45	24.31	25.90	27.72	28.37	30.61	33.35	44.34	57.51	61.66	65.41	66.56	69.96	73.17
50	27.99	29.71	31.66	32.36	34.76	37.69	49.33	63.17	67.50	71.42	72.61	76.15	79.49
60	35.53	37.48	39.70	40.48	43.19	46.46	59.33	74.40	79.08	83.30	84.58	88.38	91.95
70	43.28	45.44	47.89	48.76	51.74	55.33	69.33	85.53	90.53	95.02	96.39	100.43	104.21
80	51.17	53.54	56.21	57.15	60.39	64.28	79.33	96.58	101.88	106.63	108.07	112.33	116.32
90	59.20	61.75	64.63	65.65	69.13	73.29	89.33	107.57	113.15	118.14	119.65	124.12	128.30
100	67.33	70.06	73.14	74.22	77.93	82.36	99.33	118.50	124.34	129.56	131.14	135.81	140.17
200	152.24	156.43	161.10	162.73	168.28	174.84	199.33	226.02	233.99	241.06	243.19	249.45	255.26

Critical values of Pearson's correlation coefficient, *r*

1-sided	5%	2.5%	1%
2-sided	10%	5%	2%

n pairs of observations

3	0.9877	0.9969	0.9995
4	0.9000	0.9500	0.9800
5	0.8054	0.8783	0.9343
6	0.7293	0.8114	0.8822
7	0.6694	0.7545	0.8329
8	0.6215	0.7067	0.7887
9	0.5822	0.6664	0.7498
10	0.5494	0.6319	0.7155
11	0.5214	0.6021	0.6851
12	0.4973	0.5760	0.6581
13	0.4762	0.5529	0.6339
14	0.4575	0.5324	0.6120
15	0.4409	0.5140	0.5923
16	0.4259	0.4973	0.5742
17	0.4124	0.4821	0.5577
18	0.4000	0.4683	0.5425
19	0.3887	0.4555	0.5285
20	0.3783	0.4438	0.5155
25	0.3365	0.3961	0.4622
30	0.3061	0.3610	0.4226
35	0.2826	0.3338	0.3916
40	0.2638	0.3120	0.3665
45	0.2483	0.2940	0.3458
50	0.2353	0.2787	0.3281
60	0.2144	0.2542	0.2997
70	0.1982	0.2352	0.2776
80	0.1852	0.2199	0.2597
90	0.1745	0.2072	0.2449
100	0.1654	0.1966	0.2324
200	0.1166	0.1388	0.1644

Critical values of Spearman's rank correlation coefficient

1-sided	5%	2.5%	1%
2-sided	10%	5%	2%

n pairs of observations

4	1.0000		
5	0.9000	1.0000	1.0000
6	0.8286	0.8857	0.9429
7	0.7143	0.7857	0.8929
8	0.6429	0.7381	0.8333
9	0.6000	0.7000	0.7833
10	0.5636	0.6485	0.7455
11	0.5364	0.6182	0.7091
12	0.5035	0.5874	0.6783
13	0.4835	0.5604	0.6484
14	0.4637	0.5385	0.6264
15	0.4464	0.5214	0.6036
16	0.4294	0.5029	0.5824
17	0.4142	0.4877	0.5662
18	0.4014	0.4716	0.5501
19	0.3912	0.4596	0.5351
20	0.3805	0.4466	0.5218
25	0.3369	0.3977	0.4662
30	0.3063	0.3624	0.4251
35	0.2829	0.3347	0.3936
40	0.2640	0.3128	0.3681

Critical values of the Wilcoxon rank-sum R statistic for a sample of size m from population A and n from population B

1-sided	5%	2.5%	1%	5%	2.5%	1%	5%	2.5%	1%	5%	2.5%	1%	5%	2.5%	1%	5%	2.5%	1%	5%	2.5%	1%	5%	2.5%	1%
2-sided	10%	5%	2%	10%	5%	2%	10%	5%	2%	10%	5%	2%	10%	5%	2%	10%	5%	2%	10%	5%	2%	10%	5%	2%
m		n=3			n=4			n=5			n=6			n=7			n=8			n=9			n=10	
3	6	*	*	6	*	*	7	6	*	8	7	*	8	7	6	9	8	6	10	8	7	10	9	7
4	6	*	*	11	10	*	12	11	10	13	12	11	14	13	11	15	14	12	16	14	13	17	15	13
5	7	6	*	12	11	10	18	17	16	20	18	17	21	20	18	23	21	19	24	22	20	26	23	21
6	8	7	*	13	12	11	19	18	16	28	26	24	29	27	25	31	29	27	33	31	28	35	32	29
7	8	7	6	14	13	11	21	20	18	29	27	25	39	36	34	41	38	35	43	40	37	45	42	39
8	9	8	6	15	14	12	23	21	19	31	29	27	41	38	35	51	49	45	54	51	47	56	53	49
9	10	8	7	16	14	13	24	22	20	33	31	28	43	40	37	54	51	47	66	62	59	69	65	61
10	10	9	7	17	15	13	26	23	21	35	32	29	45	42	39	56	53	49	69	65	61	82	78	74

Critical values of the Mann-Whitney U statistic for a sample of size m from population A and n from population B

1-sided	5%	2.5%	1%	5%	2.5%	1%	5%	2.5%	1%	5%	2.5%	1%	5%	2.5%	1%	5%	2.5%	1%	5%	2.5%	1%	5%	2.5%	1%
2-sided	10%	5%	2%	10%	5%	2%	10%	5%	2%	10%	5%	2%	10%	5%	2%	10%	5%	2%	10%	5%	2%	10%	5%	2%
m		n=3			n=4			n=5			n=6			n=7			n=8			n=9			n=10	
3	0	*	*	0	*	*	1	0	*	2	1	*	2	1	0	3	2	0	4	2	1	4	3	1
4	0	*	*	1	0	*	2	1	0	3	2	1	4	3	1	5	4	2	6	4	3	7	5	3
5	1	0	*	2	1	0	3	2	1	5	3	2	6	5	3	8	6	4	9	7	5	11	8	6
6	2	1	*	3	2	1	4	3	1	7	5	3	8	6	4	10	8	6	12	10	7	14	11	8
7	2	1	0	4	3	1	6	5	3	8	6	4	11	8	6	13	10	7	15	12	9	17	14	11
8	3	2	0	5	4	2	8	6	4	10	8	6	13	10	7	15	13	9	18	15	11	20	17	13
9	4	2	1	6	4	3	9	7	5	12	10	7	15	12	9	18	15	11	21	17	14	24	20	16
10	4	3	1	7	5	3	11	8	6	14	11	8	17	14	11	20	17	13	24	20	16	27	23	19

Critical values of the Wilcoxon signed-rank sum T statistic for a sample with n signed ranks

1-sided	2-sided	n	6	7	8	9	10	11	12	13	14	15	16	17	18	19	20	21	22	23	24	25
5%	10%		2	3	5	8	10	13	17	21	25	30	35	41	47	53	60	67	75	83	91	100
2.5%	5%		*	2	3	5	8	10	13	17	21	25	29	34	40	46	52	58	65	73	81	89
1%	2%		*		1	3	5	7	9	12	15	19	23	27	32	37	43	49	55	62	69	76

Poisson distribution function

r \ μ	0.05	0.1	0.2	0.3	0.4	0.5	0.6	0.7	0.8	0.9	1	1.1	1.2	1.3	1.4	1.5	1.6	1.7	1.8	1.9	r
0	0.9512	0.9048	0.8187	0.7408	0.6703	0.6065	0.5488	0.4966	0.4493	0.4066	0.3679	0.3329	0.3012	0.2725	0.2466	0.2231	0.2019	0.1827	0.1653	0.1496	0
1	0.9988	0.9953	0.9825	0.9631	0.9384	0.9098	0.8781	0.8442	0.8088	0.7725	0.7358	0.6990	0.6626	0.6268	0.5918	0.5578	0.5249	0.4932	0.4628	0.4337	1
2	1.0000	0.9998	0.9989	0.9964	0.9921	0.9856	0.9769	0.9659	0.9526	0.9371	0.9197	0.9004	0.8795	0.8571	0.8335	0.8088	0.7834	0.7572	0.7306	0.7037	2
3	1.0000	1.0000	0.9999	0.9997	0.9992	0.9982	0.9966	0.9942	0.9909	0.9865	0.9810	0.9743	0.9662	0.9569	0.9463	0.9344	0.9212	0.9068	0.8913	0.8747	3
4	1.0000	1.0000	1.0000	1.0000	0.9999	0.9998	0.9996	0.9992	0.9986	0.9977	0.9963	0.9946	0.9923	0.9893	0.9857	0.9814	0.9763	0.9704	0.9636	0.9559	4
5	1.0000	1.0000	1.0000	1.0000	1.0000	1.0000	1.0000	0.9999	0.9998	0.9997	0.9994	0.9990	0.9985	0.9978	0.9968	0.9955	0.9940	0.9920	0.9896	0.9868	5
6	1.0000	1.0000	1.0000	1.0000	1.0000	1.0000	1.0000	1.0000	1.0000	1.0000	0.9999	0.9999	0.9997	0.9996	0.9994	0.9991	0.9987	0.9981	0.9974	0.9966	6
7	1.0000	1.0000	1.0000	1.0000	1.0000	1.0000	1.0000	1.0000	1.0000	1.0000	1.0000	1.0000	1.0000	0.9999	0.9999	0.9998	0.9997	0.9996	0.9994	0.9992	7
8	1.0000	1.0000	1.0000	1.0000	1.0000	1.0000	1.0000	1.0000	1.0000	1.0000	1.0000	1.0000	1.0000	1.0000	1.0000	1.0000	1.0000	0.9999	0.9999	0.9998	8
9	1.0000	1.0000	1.0000	1.0000	1.0000	1.0000	1.0000	1.0000	1.0000	1.0000	1.0000	1.0000	1.0000	1.0000	1.0000	1.0000	1.0000	1.0000	1.0000	1.0000	9

r \ μ	2.0	2.1	2.2	2.3	2.4	2.5	2.6	2.7	2.8	2.9	3	3.1	3.2	3.3	3.4	3.5	3.6	3.7	3.8	3.9	ρ
0	0.1353	0.1225	0.1108	0.1003	0.0907	0.0821	0.0743	0.0672	0.0608	0.0550	0.0498	0.0450	0.0408	0.0369	0.0334	0.0302	0.0273	0.0247	0.0224	0.0202	0
1	0.4060	0.3796	0.3546	0.3309	0.3084	0.2873	0.2674	0.2487	0.2311	0.2146	0.1991	0.1847	0.1712	0.1586	0.1468	0.1359	0.1257	0.1162	0.1074	0.0992	1
2	0.6767	0.6496	0.6227	0.5960	0.5697	0.5438	0.5184	0.4936	0.4695	0.4460	0.4232	0.4012	0.3799	0.3594	0.3397	0.3208	0.3027	0.2854	0.2689	0.2531	2
3	0.8571	0.8386	0.8194	0.7993	0.7787	0.7576	0.7360	0.7141	0.6919	0.6696	0.6472	0.6248	0.6025	0.5803	0.5584	0.5366	0.5152	0.4942	0.4735	0.4532	3
4	0.9473	0.9379	0.9275	0.9162	0.9041	0.8912	0.8774	0.8629	0.8477	0.8318	0.8153	0.7982	0.7806	0.7626	0.7442	0.7254	0.7064	0.6872	0.6678	0.6484	4
5	0.9834	0.9796	0.9751	0.9700	0.9643	0.9580	0.9510	0.9433	0.9349	0.9258	0.9161	0.9057	0.8946	0.8829	0.8705	0.8576	0.8441	0.8301	0.8156	0.8006	5
6	0.9955	0.9941	0.9925	0.9906	0.9884	0.9858	0.9828	0.9794	0.9756	0.9713	0.9665	0.9612	0.9554	0.9490	0.9421	0.9347	0.9267	0.9182	0.9091	0.8995	6
7	0.9989	0.9985	0.9980	0.9974	0.9967	0.9958	0.9947	0.9934	0.9919	0.9901	0.9881	0.9858	0.9832	0.9802	0.9769	0.9733	0.9692	0.9648	0.9599	0.9546	7
8	0.9998	0.9997	0.9995	0.9994	0.9991	0.9989	0.9985	0.9981	0.9976	0.9969	0.9962	0.9953	0.9943	0.9931	0.9917	0.9901	0.9883	0.9863	0.9840	0.9815	8
9	1.0000	0.9999	0.9999	0.9999	0.9998	0.9997	0.9996	0.9995	0.9993	0.9991	0.9989	0.9986	0.9982	0.9978	0.9973	0.9967	0.9960	0.9952	0.9942	0.9931	9
10	1.0000	1.0000	1.0000	1.0000	1.0000	0.9999	0.9999	0.9999	0.9998	0.9998	0.9997	0.9996	0.9995	0.9994	0.9992	0.9990	0.9987	0.9984	0.9981	0.9977	10
11	1.0000	1.0000	1.0000	1.0000	1.0000	1.0000	1.0000	1.0000	1.0000	0.9999	0.9999	0.9999	0.9999	0.9998	0.9998	0.9997	0.9996	0.9995	0.9994	0.9993	11
12	1.0000	1.0000	1.0000	1.0000	1.0000	1.0000	1.0000	1.0000	1.0000	1.0000	1.0000	1.0000	1.0000	1.0000	0.9999	0.9999	0.9999	0.9999	0.9998	0.9998	12
13	1.0000	1.0000	1.0000	1.0000	1.0000	1.0000	1.0000	1.0000	1.0000	1.0000	1.0000	1.0000	1.0000	1.0000	1.0000	1.0000	1.0000	1.0000	1.0000	0.9999	13

r \ μ	4.0	4.1	4.2	4.3	4.4	4.5	4.6	4.7	4.8	4.9	5	5.1	5.2	5.3	5.4	5.5	5.6	5.7	5.8	5.9	ρ
0	0.0183	0.0166	0.0150	0.0136	0.0123	0.0111	0.0101	0.0091	0.0082	0.0074	0.0067	0.0061	0.0055	0.0050	0.0045	0.0041	0.0037	0.0033	0.0030	0.0027	0
1	0.0916	0.0845	0.0780	0.0719	0.0663	0.0611	0.0563	0.0518	0.0477	0.0439	0.0404	0.0372	0.0342	0.0314	0.0289	0.0266	0.0244	0.0224	0.0206	0.0189	1
2	0.2381	0.2238	0.2102	0.1974	0.1851	0.1736	0.1626	0.1523	0.1425	0.1333	0.1247	0.1165	0.1088	0.1016	0.0948	0.0884	0.0824	0.0768	0.0715	0.0666	2
3	0.4335	0.4142	0.3954	0.3772	0.3594	0.3423	0.3257	0.3097	0.2942	0.2793	0.2650	0.2513	0.2381	0.2254	0.2133	0.2017	0.1906	0.1800	0.1700	0.1604	3
4	0.6288	0.6093	0.5898	0.5704	0.5512	0.5321	0.5132	0.4946	0.4763	0.4582	0.4405	0.4231	0.4061	0.3895	0.3733	0.3575	0.3422	0.3272	0.3127	0.2987	4
5	0.7851	0.7693	0.7531	0.7367	0.7199	0.7029	0.6858	0.6684	0.6510	0.6335	0.6160	0.5984	0.5809	0.5635	0.5461	0.5289	0.5119	0.4950	0.4783	0.4619	5
6	0.8893	0.8786	0.8675	0.8558	0.8436	0.8311	0.8180	0.8046	0.7908	0.7767	0.7622	0.7474	0.7324	0.7171	0.7017	0.6860	0.6703	0.6544	0.6384	0.6224	6
7	0.9489	0.9427	0.9361	0.9290	0.9214	0.9134	0.9049	0.8960	0.8867	0.8769	0.8666	0.8560	0.8449	0.8335	0.8217	0.8095	0.7970	0.7841	0.7710	0.7576	7
8	0.9786	0.9755	0.9721	0.9683	0.9642	0.9597	0.9549	0.9497	0.9442	0.9382	0.9319	0.9252	0.9181	0.9106	0.9027	0.8944	0.8857	0.8766	0.8672	0.8574	8
9	0.9919	0.9905	0.9889	0.9871	0.9851	0.9829	0.9805	0.9778	0.9749	0.9717	0.9682	0.9644	0.9603	0.9559	0.9512	0.9462	0.9409	0.9352	0.9292	0.9228	9
10	0.9972	0.9966	0.9959	0.9952	0.9943	0.9933	0.9922	0.9910	0.9896	0.9880	0.9863	0.9844	0.9823	0.9800	0.9775	0.9747	0.9718	0.9686	0.9651	0.9614	10
11	0.9991	0.9989	0.9986	0.9983	0.9980	0.9976	0.9971	0.9966	0.9960	0.9953	0.9943	0.9937	0.9927	0.9916	0.9904	0.9890	0.9875	0.9859	0.9841	0.9821	11
12	0.9997	0.9997	0.9996	0.9995	0.9993	0.9992	0.9990	0.9988	0.9986	0.9983	0.9980	0.9976	0.9972	0.9967	0.9962	0.9955	0.9949	0.9941	0.9932	0.9922	12
13	0.9999	0.9999	0.9999	0.9998	0.9998	0.9997	0.9997	0.9996	0.9995	0.9994	0.9993	0.9992	0.9990	0.9988	0.9986	0.9983	0.9980	0.9977	0.9973	0.9969	13
14	1.0000	1.0000	1.0000	1.0000	0.9999	0.9999	0.9999	0.9999	0.9999	0.9998	0.9998	0.9997	0.9997	0.9996	0.9995	0.9994	0.9993	0.9991	0.9990	0.9988	14
15	1.0000	1.0000	1.0000	1.0000	1.0000	1.0000	1.0000	1.0000	1.0000	0.9999	0.9999	0.9999	0.9999	0.9999	0.9998	0.9998	0.9998	0.9997	0.9996	0.9996	15
16	1.0000	1.0000	1.0000	1.0000	1.0000	1.0000	1.0000	1.0000	1.0000	1.0000	1.0000	1.0000	1.0000	1.0000	0.9999	0.9999	0.9999	0.9999	0.9999	0.9999	16
17	1.0000	1.0000	1.0000	1.0000	1.0000	1.0000	1.0000	1.0000	1.0000	1.0000	1.0000	1.0000	1.0000	1.0000	1.0000	1.0000	1.0000	1.0000	1.0000	1.0000	17

r	μ 6.0	6.1	6.2	6.3	6.4	6.5	6.6	6.7	6.8	6.9	7	7.1	7.2	7.3	7.4	7.5	7.6	7.7	7.8	7.9	ρ
0	0.0025	0.0022	0.0020	0.0018	0.0017	0.0015	0.0014	0.0012	0.0011	0.0010	0.0009	0.0008	0.0007	0.0007	0.0006	0.0006	0.0005	0.0005	0.0004	0.0004	0
1	0.0174	0.0159	0.0146	0.0134	0.0123	0.0113	0.0103	0.0095	0.0087	0.0080	0.0073	0.0067	0.0061	0.0056	0.0051	0.0047	0.0043	0.0039	0.0036	0.0033	1
2	0.0620	0.0577	0.0536	0.0498	0.0463	0.0430	0.0400	0.0371	0.0344	0.0320	0.0296	0.0275	0.0255	0.0236	0.0219	0.0203	0.0188	0.0174	0.0161	0.0149	2
3	0.1512	0.1425	0.1342	0.1264	0.1189	0.1118	0.1052	0.0988	0.0928	0.0871	0.0818	0.0767	0.0719	0.0674	0.0632	0.0591	0.0554	0.0518	0.0485	0.0453	3
4	0.2851	0.2719	0.2592	0.2469	0.2351	0.2237	0.2127	0.2022	0.1920	0.1823	0.1730	0.1641	0.1555	0.1473	0.1395	0.1321	0.1249	0.1181	0.1117	0.1055	4
5	0.4457	0.4298	0.4141	0.3988	0.3837	0.3690	0.3547	0.3406	0.3270	0.3137	0.3007	0.2881	0.2759	0.2640	0.2526	0.2414	0.2307	0.2203	0.2103	0.2006	5
6	0.6063	0.5902	0.5742	0.5582	0.5423	0.5265	0.5108	0.4953	0.4799	0.4647	0.4497	0.4349	0.4204	0.4060	0.3920	0.3782	0.3646	0.3514	0.3384	0.3257	6
7	0.7440	0.7301	0.7160	0.7017	0.6873	0.6728	0.6581	0.6433	0.6285	0.6136	0.5987	0.5838	0.5689	0.5541	0.5393	0.5246	0.5100	0.4956	0.4812	0.4670	7
8	0.8472	0.8367	0.8259	0.8148	0.8033	0.7916	0.7796	0.7673	0.7548	0.7420	0.7291	0.7160	0.7027	0.6892	0.6757	0.6620	0.6482	0.6343	0.6204	0.6065	8
9	0.9161	0.9090	0.9016	0.8939	0.8858	0.8774	0.8686	0.8596	0.8502	0.8405	0.8305	0.8202	0.8096	0.7988	0.7877	0.7764	0.7649	0.7531	0.7411	0.7290	9
10	0.9574	0.9531	0.9486	0.9437	0.9386	0.9332	0.9274	0.9214	0.9151	0.9084	0.9015	0.8942	0.8867	0.8788	0.8707	0.8622	0.8535	0.8445	0.8352	0.8257	10
11	0.9799	0.9776	0.9750	0.9723	0.9693	0.9661	0.9627	0.9591	0.9552	0.9510	0.9467	0.9420	0.9371	0.9319	0.9265	0.9208	0.9148	0.9085	0.9020	0.8952	11
12	0.9912	0.9900	0.9887	0.9873	0.9857	0.9840	0.9821	0.9801	0.9779	0.9755	0.9730	0.9703	0.9673	0.9642	0.9609	0.9573	0.9536	0.9496	0.9454	0.9409	12
13	0.9964	0.9958	0.9952	0.9945	0.9937	0.9929	0.9920	0.9909	0.9898	0.9885	0.9872	0.9857	0.9841	0.9824	0.9805	0.9784	0.9762	0.9739	0.9714	0.9687	13
14	0.9986	0.9984	0.9981	0.9978	0.9974	0.9970	0.9966	0.9961	0.9956	0.9950	0.9943	0.9935	0.9927	0.9918	0.9908	0.9897	0.9886	0.9873	0.9859	0.9844	14
15	0.9995	0.9994	0.9993	0.9992	0.9990	0.9988	0.9986	0.9984	0.9982	0.9979	0.9976	0.9972	0.9969	0.9964	0.9959	0.9954	0.9948	0.9941	0.9934	0.9926	15
16	0.9998	0.9998	0.9997	0.9997	0.9996	0.9996	0.9995	0.9994	0.9993	0.9992	0.9990	0.9989	0.9987	0.9985	0.9983	0.9980	0.9978	0.9974	0.9971	0.9967	16
17	0.9999	0.9999	0.9999	0.9999	0.9999	0.9998	0.9998	0.9998	0.9997	0.9997	0.9996	0.9996	0.9995	0.9994	0.9993	0.9992	0.9991	0.9989	0.9988	0.9986	17
18	1.0000	1.0000	1.0000	1.0000	1.0000	0.9999	0.9999	0.9999	0.9999	0.9999	0.9999	0.9998	0.9998	0.9998	0.9997	0.9997	0.9996	0.9996	0.9995	0.9994	18
19	1.0000	1.0000	1.0000	1.0000	1.0000	1.0000	1.0000	1.0000	1.0000	1.0000	1.0000	0.9999	0.9999	0.9999	0.9999	0.9999	0.9999	0.9998	0.9998	0.9998	19
20	1.0000	1.0000	1.0000	1.0000	1.0000	1.0000	1.0000	1.0000	1.0000	1.0000	1.0000	1.0000	1.0000	1.0000	1.0000	1.0000	1.0000	0.9999	0.9999	0.9999	20

r	μ 8.0	8.1	8.2	8.3	8.4	8.5	8.6	8.7	8.8	8.9	9	9.1	9.2	9.3	9.4	9.5	9.6	9.7	9.8	9.9	ρ
0	0.0003	0.0003	0.0003	0.0002	0.0002	0.0002	0.0002	0.0002	0.0002	0.0001	0.0001	0.0001	0.0001	0.0001	0.0001	0.0001	0.0001	0.0001	0.0001	0.0001	0
1	0.0030	0.0028	0.0025	0.0023	0.0021	0.0019	0.0018	0.0016	0.0015	0.0014	0.0012	0.0011	0.0010	0.0009	0.0009	0.0008	0.0007	0.0007	0.0006	0.0005	1
2	0.0138	0.0127	0.0118	0.0109	0.0100	0.0093	0.0086	0.0079	0.0073	0.0068	0.0062	0.0058	0.0053	0.0049	0.0045	0.0042	0.0038	0.0035	0.0033	0.0030	2
3	0.0424	0.0396	0.0370	0.0346	0.0323	0.0301	0.0281	0.0262	0.0244	0.0228	0.0212	0.0198	0.0184	0.0172	0.0160	0.0149	0.0138	0.0129	0.0120	0.0111	3
4	0.0996	0.0940	0.0887	0.0837	0.0789	0.0744	0.0701	0.0660	0.0621	0.0584	0.0550	0.0517	0.0486	0.0456	0.0429	0.0403	0.0378	0.0355	0.0333	0.0312	4
5	0.1912	0.1822	0.1736	0.1653	0.1573	0.1496	0.1422	0.1352	0.1284	0.1219	0.1157	0.1098	0.1041	0.0986	0.0935	0.0885	0.0838	0.0793	0.0750	0.0710	5
6	0.3134	0.3013	0.2896	0.2781	0.2670	0.2562	0.2457	0.2355	0.2256	0.2160	0.2068	0.1978	0.1892	0.1808	0.1727	0.1649	0.1574	0.1502	0.1433	0.1366	6
7	0.4530	0.4391	0.4254	0.4119	0.3987	0.3856	0.3728	0.3602	0.3478	0.3357	0.3239	0.3123	0.3010	0.2900	0.2792	0.2687	0.2584	0.2485	0.2388	0.2294	7
8	0.5925	0.5786	0.5647	0.5507	0.5369	0.5231	0.5094	0.4958	0.4823	0.4689	0.4557	0.4426	0.4296	0.4168	0.4042	0.3918	0.3796	0.3676	0.3558	0.3442	8
9	0.7166	0.7041	0.6915	0.6788	0.6659	0.6530	0.6400	0.6269	0.6137	0.6006	0.5874	0.5742	0.5611	0.5479	0.5349	0.5218	0.5089	0.4960	0.4832	0.4705	9
10	0.8159	0.8058	0.7955	0.7850	0.7743	0.7634	0.7522	0.7409	0.7294	0.7178	0.7060	0.6941	0.6820	0.6699	0.6576	0.6453	0.6329	0.6205	0.6080	0.5955	10
11	0.8881	0.8807	0.8731	0.8652	0.8571	0.8487	0.8400	0.8311	0.8220	0.8126	0.8030	0.7932	0.7832	0.7730	0.7626	0.7520	0.7412	0.7303	0.7193	0.7081	11
12	0.9362	0.9313	0.9261	0.9207	0.9150	0.9091	0.9029	0.8965	0.8898	0.8829	0.8758	0.8684	0.8607	0.8529	0.8448	0.8364	0.8279	0.8191	0.8101	0.8009	12
13	0.9658	0.9628	0.9595	0.9561	0.9524	0.9486	0.9445	0.9403	0.9358	0.9311	0.9261	0.9210	0.9156	0.9100	0.9042	0.8981	0.8919	0.8853	0.8786	0.8716	13
14	0.9827	0.9810	0.9791	0.9771	0.9749	0.9726	0.9701	0.9675	0.9647	0.9617	0.9585	0.9552	0.9517	0.9480	0.9441	0.9400	0.9357	0.9312	0.9265	0.9216	14
15	0.9918	0.9908	0.9898	0.9887	0.9875	0.9862	0.9848	0.9832	0.9816	0.9798	0.9780	0.9760	0.9738	0.9715	0.9691	0.9665	0.9638	0.9609	0.9579	0.9546	15
16	0.9963	0.9958	0.9953	0.9947	0.9941	0.9934	0.9926	0.9918	0.9909	0.9899	0.9889	0.9878	0.9865	0.9852	0.9838	0.9823	0.9806	0.9789	0.9770	0.9751	16
17	0.9984	0.9982	0.9979	0.9977	0.9973	0.9970	0.9966	0.9962	0.9957	0.9952	0.9947	0.9941	0.9934	0.9927	0.9919	0.9911	0.9902	0.9892	0.9881	0.9870	17
18	0.9993	0.9992	0.9991	0.9990	0.9989	0.9987	0.9985	0.9983	0.9981	0.9978	0.9976	0.9973	0.9969	0.9966	0.9962	0.9957	0.9952	0.9947	0.9941	0.9935	18
19	0.9997	0.9997	0.9997	0.9996	0.9995	0.9995	0.9994	0.9993	0.9992	0.9991	0.9989	0.9988	0.9986	0.9985	0.9983	0.9980	0.9978	0.9975	0.9972	0.9969	19
20	0.9999	0.9999	0.9999	0.9998	0.9998	0.9998	0.9998	0.9997	0.9997	0.9996	0.9996	0.9995	0.9994	0.9993	0.9992	0.9991	0.9990	0.9989	0.9987	0.9986	20
21	1.0000	1.0000	1.0000	0.9999	0.9999	0.9999	0.9999	0.9999	0.9999	0.9998	0.9998	0.9998	0.9998	0.9997	0.9997	0.9996	0.9996	0.9995	0.9995	0.9994	21
22	1.0000	1.0000	1.0000	1.0000	1.0000	1.0000	1.0000	1.0000	1.0000	0.9999	0.9999	0.9999	0.9999	0.9999	0.9999	0.9999	0.9998	0.9998	0.9998	0.9997	22
23	1.0000	1.0000	1.0000	1.0000	1.0000	1.0000	1.0000	1.0000	1.0000	1.0000	1.0000	1.0000	1.0000	1.0000	1.0000	0.9999	0.9999	0.9999	0.9999	0.9999	23

Binomial distribution function

r \ p	0.05	0.1	0.15	$\frac{1}{6}$	0.2	$\frac{1}{4}$	0.3	$\frac{1}{3}$	0.35	0.4	0.45	0.5	0.55	0.6	0.65	$\frac{2}{3}$	0.7	$\frac{3}{4}$	0.8	$\frac{5}{6}$	0.85	0.9	0.95	r
$n=7$																								
0	0.6983	0.4783	0.3206	0.2791	0.2097	0.1335	0.0824	0.0585	0.0490	0.0280	0.0152	0.0078	0.0037	0.0016	0.0006	0.0005	0.0002	0.0001	0.0000	0.0000	0.0000	0.0000	0.0000	0
1	0.9556	0.8503	0.7166	0.6698	0.5767	0.4449	0.3294	0.2634	0.2338	0.1586	0.1024	0.0625	0.0357	0.0188	0.0090	0.0069	0.0038	0.0013	0.0004	0.0001	0.0001	0.0000	0.0000	1
2	0.9962	0.9743	0.9262	0.9042	0.8520	0.7564	0.6471	0.5706	0.5323	0.4199	0.3164	0.2266	0.1529	0.0963	0.0556	0.0453	0.0288	0.0129	0.0047	0.0020	0.0012	0.0002	0.0000	2
3	0.9998	0.9973	0.9879	0.9824	0.9667	0.9294	0.8740	0.8267	0.8002	0.7102	0.6083	0.5000	0.3917	0.2898	0.1998	0.1733	0.1260	0.0706	0.0333	0.0176	0.0121	0.0027	0.0002	3
4	1.0000	0.9998	0.9988	0.9980	0.9953	0.9871	0.9712	0.9547	0.9444	0.9037	0.8471	0.7734	0.6836	0.5801	0.4677	0.4294	0.3529	0.2436	0.1480	0.0958	0.0738	0.0257	0.0038	4
5	1.0000	1.0000	0.9999	0.9999	0.9996	0.9987	0.9962	0.9931	0.9910	0.9812	0.9643	0.9375	0.8976	0.8414	0.7662	0.7366	0.6706	0.5551	0.4233	0.3302	0.2834	0.1497	0.0444	5
6	1.0000	1.0000	1.0000	1.0000	1.0000	0.9999	0.9998	0.9995	0.9994	0.9984	0.9963	0.9922	0.9848	0.9720	0.9510	0.9415	0.9176	0.8665	0.7903	0.7209	0.6794	0.5217	0.3017	6
7	1.0000	1.0000	1.0000	1.0000	1.0000	1.0000	1.0000	1.0000	1.0000	1.0000	1.0000	1.0000	1.0000	1.0000	1.0000	1.0000	1.0000	1.0000	1.0000	1.0000	1.0000	1.0000	1.0000	7
$n=8$																								
0	0.6634	0.4305	0.2725	0.2326	0.1678	0.1001	0.0576	0.0390	0.0319	0.0168	0.0084	0.0039	0.0017	0.0007	0.0002	0.0002	0.0001	0.0000	0.0000	0.0000	0.0000	0.0000	0.0000	0
1	0.9428	0.8131	0.6572	0.6047	0.5033	0.3671	0.2553	0.1951	0.1691	0.1064	0.0632	0.0352	0.0181	0.0085	0.0036	0.0026	0.0013	0.0004	0.0001	0.0000	0.0000	0.0000	0.0000	1
2	0.9942	0.9619	0.8948	0.8652	0.7969	0.6785	0.5518	0.4682	0.4278	0.3154	0.2201	0.1445	0.0885	0.0498	0.0253	0.0197	0.0113	0.0042	0.0012	0.0004	0.0002	0.0000	0.0000	2
3	0.9996	0.9950	0.9786	0.9693	0.9437	0.8862	0.8059	0.7414	0.7064	0.5941	0.4770	0.3633	0.2604	0.1737	0.1061	0.0879	0.0580	0.0273	0.0104	0.0046	0.0029	0.0004	0.0000	3
4	1.0000	0.9996	0.9971	0.9954	0.9896	0.9727	0.9420	0.9121	0.8939	0.8263	0.7396	0.6367	0.5230	0.4059	0.2936	0.2586	0.1941	0.1138	0.0563	0.0307	0.0214	0.0050	0.0004	4
5	1.0000	1.0000	0.9998	0.9996	0.9988	0.9958	0.9887	0.9803	0.9747	0.9502	0.9115	0.8555	0.7799	0.6846	0.5722	0.5318	0.4482	0.3215	0.2031	0.1348	0.1052	0.0381	0.0058	5
6	1.0000	1.0000	1.0000	1.0000	0.9999	0.9996	0.9987	0.9974	0.9964	0.9915	0.9819	0.9648	0.9368	0.8936	0.8309	0.8049	0.7447	0.6329	0.4967	0.3953	0.3428	0.1869	0.0572	6
7	1.0000	1.0000	1.0000	1.0000	1.0000	1.0000	0.9999	0.9998	0.9998	0.9993	0.9983	0.9961	0.9916	0.9832	0.9681	0.9610	0.9424	0.8999	0.8322	0.7674	0.7275	0.5695	0.3366	7
8	1.0000	1.0000	1.0000	1.0000	1.0000	1.0000	1.0000	1.0000	1.0000	1.0000	1.0000	1.0000	1.0000	1.0000	1.0000	1.0000	1.0000	1.0000	1.0000	1.0000	1.0000	1.0000	1.0000	8
$n=9$																								
0	0.6302	0.3874	0.2316	0.1938	0.1342	0.0751	0.0404	0.0260	0.0207	0.0101	0.0046	0.0020	0.0008	0.0003	0.0001	0.0001	0.0000	0.0000	0.0000	0.0000	0.0000	0.0000	0.0000	0
1	0.9288	0.7748	0.5995	0.5427	0.4362	0.3003	0.1960	0.1431	0.1211	0.0705	0.0385	0.0195	0.0091	0.0038	0.0014	0.0010	0.0004	0.0001	0.0000	0.0000	0.0000	0.0000	0.0000	1
2	0.9916	0.9470	0.8591	0.8217	0.7382	0.6007	0.4628	0.3772	0.3373	0.2318	0.1495	0.0898	0.0498	0.0250	0.0112	0.0083	0.0043	0.0013	0.0003	0.0001	0.0000	0.0000	0.0000	2
3	0.9994	0.9917	0.9661	0.9520	0.9144	0.8343	0.7297	0.6503	0.6089	0.4826	0.3614	0.2539	0.1658	0.0994	0.0536	0.0424	0.0253	0.0100	0.0031	0.0011	0.0006	0.0001	0.0000	3
4	1.0000	0.9991	0.9944	0.9910	0.9804	0.9511	0.9012	0.8552	0.8283	0.7334	0.6214	0.5000	0.3786	0.2666	0.1717	0.1448	0.0988	0.0489	0.0196	0.0090	0.0056	0.0009	0.0000	4
5	1.0000	0.9999	0.9994	0.9989	0.9969	0.9900	0.9747	0.9576	0.9464	0.9006	0.8342	0.7461	0.6386	0.5174	0.3911	0.3497	0.2703	0.1657	0.0856	0.0480	0.0339	0.0083	0.0006	5
6	1.0000	1.0000	1.0000	0.9999	0.9997	0.9987	0.9957	0.9917	0.9888	0.9750	0.9502	0.9102	0.8505	0.7682	0.6627	0.6228	0.5372	0.3993	0.2618	0.1783	0.1409	0.0530	0.0084	6
7	1.0000	1.0000	1.0000	1.0000	1.0000	0.9999	0.9996	0.9990	0.9986	0.9962	0.9909	0.9805	0.9615	0.9295	0.8789	0.8569	0.8040	0.6997	0.5638	0.4573	0.4005	0.2252	0.0712	7
8	1.0000	1.0000	1.0000	1.0000	1.0000	1.0000	1.0000	0.9999	0.9999	0.9997	0.9992	0.9980	0.9954	0.9899	0.9793	0.9740	0.9596	0.9249	0.8658	0.8062	0.7684	0.6126	0.3698	8
9	1.0000	1.0000	1.0000	1.0000	1.0000	1.0000	1.0000	1.0000	1.0000	1.0000	1.0000	1.0000	1.0000	1.0000	1.0000	1.0000	1.0000	1.0000	1.0000	1.0000	1.0000	1.0000	1.0000	9
$n=10$																								
0	0.5987	0.3487	0.1969	0.1615	0.1074	0.0563	0.0282	0.0173	0.0135	0.0060	0.0025	0.0010	0.0003	0.0001	0.0000	0.0000	0.0000	0.0000	0.0000	0.0000	0.0000	0.0000	0.0000	0
1	0.9139	0.7361	0.5443	0.4845	0.3758	0.2440	0.1493	0.1040	0.0860	0.0464	0.0233	0.0107	0.0045	0.0017	0.0005	0.0004	0.0001	0.0000	0.0000	0.0000	0.0000	0.0000	0.0000	1
2	0.9885	0.9298	0.8202	0.7752	0.6778	0.5256	0.3828	0.2991	0.2616	0.1673	0.0996	0.0547	0.0274	0.0123	0.0048	0.0034	0.0016	0.0004	0.0001	0.0000	0.0000	0.0000	0.0000	2
3	0.9990	0.9872	0.9500	0.9303	0.8791	0.7759	0.6496	0.5593	0.5138	0.3823	0.2660	0.1719	0.1020	0.0548	0.0260	0.0197	0.0106	0.0035	0.0009	0.0003	0.0001	0.0000	0.0000	3
4	0.9999	0.9984	0.9901	0.9845	0.9672	0.9219	0.8497	0.7869	0.7515	0.6331	0.5044	0.3770	0.2616	0.1662	0.0949	0.0766	0.0473	0.0197	0.0064	0.0024	0.0014	0.0001	0.0000	4
5	1.0000	0.9999	0.9986	0.9976	0.9936	0.9803	0.9527	0.9234	0.9051	0.8338	0.7384	0.6230	0.4956	0.3669	0.2485	0.2131	0.1503	0.0781	0.0328	0.0155	0.0099	0.0016	0.0001	5
6	1.0000	1.0000	0.9999	0.9997	0.9991	0.9965	0.9894	0.9803	0.9740	0.9452	0.8980	0.8281	0.7340	0.6177	0.4862	0.4407	0.3504	0.2241	0.1209	0.0697	0.0500	0.0128	0.0010	6
7	1.0000	1.0000	1.0000	1.0000	0.9999	0.9996	0.9984	0.9966	0.9952	0.9877	0.9726	0.9453	0.9004	0.8327	0.7384	0.7009	0.6172	0.4744	0.3222	0.2248	0.1798	0.0702	0.0115	7
8	1.0000	1.0000	1.0000	1.0000	1.0000	1.0000	0.9999	0.9996	0.9995	0.9983	0.9955	0.9893	0.9767	0.9536	0.9140	0.8960	0.8507	0.7560	0.6242	0.5155	0.4557	0.2639	0.0861	8
9	1.0000	1.0000	1.0000	1.0000	1.0000	1.0000	1.0000	1.0000	1.0000	0.9999	0.9997	0.9990	0.9975	0.9940	0.9865	0.9827	0.9718	0.9437	0.8926	0.8385	0.8031	0.6513	0.4013	9
10	1.0000	1.0000	1.0000	1.0000	1.0000	1.0000	1.0000	1.0000	1.0000	1.0000	1.0000	1.0000	1.0000	1.0000	1.0000	1.0000	1.0000	1.0000	1.0000	1.0000	1.0000	1.0000	1.0000	10

p	0.05	0.1	0.15	1/6	0.2	1/4	0.3	1/3	0.35	0.4	0.45	0.5	0.55	0.6	0.65	2/3	0.7	3/4	0.8	5/6	0.85	0.9	0.95	
r	$n = 11$										$n = 11$								$n = 11$					*r*
0	0.5688	0.3138	0.1673	0.1346	0.0859	0.0422	0.0198	0.0116	0.0088	0.0036	0.0014	0.0005	0.0002	0.0000	0.0000	0.0000	0.0000	0.0000	0.0000	0.0000	0.0000	0.0000	0.0000	0
1	0.8981	0.6974	0.4922	0.4307	0.3221	0.1971	0.1130	0.0751	0.0606	0.0302	0.0139	0.0059	0.0022	0.0007	0.0002	0.0001	0.0000	0.0000	0.0000	0.0000	0.0000	0.0000	0.0000	1
2	0.9848	0.9104	0.7788	0.7268	0.6174	0.4552	0.3127	0.2341	0.2001	0.1189	0.0652	0.0327	0.0148	0.0059	0.0020	0.0014	0.0006	0.0001	0.0000	0.0000	0.0000	0.0000	0.0000	2
3	0.9984	0.9815	0.9306	0.9044	0.8389	0.7133	0.5696	0.4726	0.4256	0.2963	0.1911	0.1133	0.0610	0.0293	0.0122	0.0088	0.0043	0.0012	0.0002	0.0001	0.0000	0.0000	0.0000	3
4	0.9999	0.9972	0.9841	0.9755	0.9496	0.8854	0.7897	0.7110	0.6683	0.5328	0.3971	0.2744	0.1738	0.0994	0.0501	0.0386	0.0216	0.0076	0.0020	0.0006	0.0003	0.0000	0.0000	4
5	1.0000	0.9997	0.9973	0.9954	0.9883	0.9657	0.9218	0.8779	0.8513	0.7535	0.6331	0.5000	0.3669	0.2465	0.1487	0.1221	0.0782	0.0343	0.0117	0.0046	0.0027	0.0003	0.0000	5
6	1.0000	1.0000	0.9997	0.9994	0.9980	0.9924	0.9784	0.9614	0.9499	0.9006	0.8262	0.7256	0.6029	0.4672	0.3317	0.2890	0.2103	0.1146	0.0504	0.0245	0.0159	0.0028	0.0001	6
7	1.0000	1.0000	1.0000	0.9999	0.9998	0.9988	0.9957	0.9912	0.9878	0.9707	0.9390	0.8867	0.8089	0.7037	0.5744	0.5274	0.4304	0.2867	0.1611	0.0956	0.0694	0.0185	0.0016	7
8	1.0000	1.0000	1.0000	1.0000	1.0000	0.9999	0.9994	0.9986	0.9980	0.9941	0.9852	0.9673	0.9348	0.8811	0.7999	0.7659	0.6873	0.5448	0.3826	0.2732	0.2212	0.0896	0.0152	8
9	1.0000	1.0000	1.0000	1.0000	1.0000	1.0000	1.0000	0.9999	0.9998	0.9993	0.9978	0.9941	0.9861	0.9698	0.9394	0.9249	0.8870	0.8029	0.6779	0.5693	0.5078	0.3026	0.1019	9
10	1.0000	1.0000	1.0000	1.0000	1.0000	1.0000	1.0000	1.0000	1.0000	1.0000	0.9998	0.9995	0.9986	0.9964	0.9912	0.9884	0.9802	0.9578	0.9141	0.8654	0.8327	0.6862	0.4312	10
11	1.0000	1.0000	1.0000	1.0000	1.0000	1.0000	1.0000	1.0000	1.0000	1.0000	1.0000	1.0000	1.0000	1.0000	1.0000	1.0000	1.0000	1.0000	1.0000	1.0000	1.0000	1.0000	1.0000	11
r	$n = 12$										$n = 12$								$n = 12$					*r*
0	0.5404	0.2824	0.1422	0.1122	0.0687	0.0317	0.0138	0.0077	0.0057	0.0022	0.0008	0.0002	0.0001	0.0000	0.0000	0.0000	0.0000	0.0000	0.0000	0.0000	0.0000	0.0000	0.0000	0
1	0.8816	0.6590	0.4435	0.3813	0.2749	0.1584	0.0850	0.0540	0.0424	0.0196	0.0083	0.0032	0.0011	0.0003	0.0001	0.0000	0.0000	0.0000	0.0000	0.0000	0.0000	0.0000	0.0000	1
2	0.9804	0.8891	0.7358	0.6774	0.5583	0.3907	0.2528	0.1811	0.1513	0.0834	0.0421	0.0193	0.0079	0.0028	0.0008	0.0005	0.0002	0.0000	0.0000	0.0000	0.0000	0.0000	0.0000	2
3	0.9978	0.9744	0.9078	0.8748	0.7946	0.6488	0.4925	0.3931	0.3467	0.2253	0.1345	0.0730	0.0356	0.0153	0.0056	0.0039	0.0017	0.0004	0.0001	0.0000	0.0000	0.0000	0.0000	3
4	0.9998	0.9957	0.9761	0.9636	0.9274	0.8424	0.7237	0.6315	0.5833	0.4382	0.3044	0.1938	0.1117	0.0573	0.0255	0.0188	0.0095	0.0028	0.0006	0.0002	0.0001	0.0000	0.0000	4
5	1.0000	0.9995	0.9954	0.9921	0.9806	0.9456	0.8822	0.8223	0.7873	0.6652	0.5269	0.3872	0.2607	0.1582	0.0846	0.0664	0.0386	0.0143	0.0039	0.0013	0.0007	0.0001	0.0000	5
6	1.0000	0.9999	0.9993	0.9987	0.9961	0.9857	0.9614	0.9336	0.9154	0.8418	0.7393	0.6128	0.4731	0.3348	0.2127	0.1777	0.1178	0.0544	0.0194	0.0079	0.0046	0.0005	0.0000	6
7	1.0000	1.0000	0.9999	0.9998	0.9994	0.9972	0.9905	0.9812	0.9745	0.9427	0.8883	0.8062	0.6956	0.5618	0.4167	0.3685	0.2763	0.1576	0.0726	0.0364	0.0239	0.0043	0.0002	7
8	1.0000	1.0000	1.0000	1.0000	0.9999	0.9996	0.9983	0.9961	0.9944	0.9847	0.9644	0.9270	0.8655	0.7747	0.6533	0.6069	0.5075	0.3512	0.2054	0.1252	0.0922	0.0256	0.0022	8
9	1.0000	1.0000	1.0000	1.0000	1.0000	1.0000	0.9998	0.9995	0.9992	0.9972	0.9921	0.9807	0.9579	0.9166	0.8487	0.8189	0.7472	0.6093	0.4417	0.3226	0.2642	0.1109	0.0196	9
10	1.0000	1.0000	1.0000	1.0000	1.0000	1.0000	1.0000	1.0000	0.9999	0.9997	0.9989	0.9968	0.9917	0.9804	0.9576	0.9460	0.9150	0.8416	0.7251	0.6187	0.5565	0.3410	0.1184	10
11	1.0000	1.0000	1.0000	1.0000	1.0000	1.0000	1.0000	1.0000	1.0000	1.0000	0.9999	0.9998	0.9992	0.9978	0.9943	0.9923	0.9862	0.9683	0.9313	0.8878	0.8578	0.7176	0.4596	11
12	1.0000	1.0000	1.0000	1.0000	1.0000	1.0000	1.0000	1.0000	1.0000	1.0000	1.0000	1.0000	1.0000	1.0000	1.0000	1.0000	1.0000	1.0000	1.0000	1.0000	1.0000	1.0000	1.0000	12
r	$n = 13$										$n = 13$								$n = 13$					*r*
0	0.5133	0.2542	0.1209	0.0935	0.0550	0.0238	0.0097	0.0051	0.0037	0.0013	0.0004	0.0001	0.0000	0.0000	0.0000	0.0000	0.0000	0.0000	0.0000	0.0000	0.0000	0.0000	0.0000	0
1	0.8646	0.6213	0.3983	0.3365	0.2336	0.1267	0.0637	0.0385	0.0296	0.0126	0.0049	0.0017	0.0005	0.0001	0.0000	0.0000	0.0000	0.0000	0.0000	0.0000	0.0000	0.0000	0.0000	1
2	0.9755	0.8661	0.6920	0.6281	0.5017	0.3326	0.2025	0.1387	0.1132	0.0579	0.0269	0.0112	0.0041	0.0013	0.0003	0.0002	0.0001	0.0000	0.0000	0.0000	0.0000	0.0000	0.0000	2
3	0.9969	0.9658	0.8820	0.8419	0.7473	0.5843	0.4206	0.3224	0.2783	0.1686	0.0929	0.0461	0.0203	0.0078	0.0025	0.0016	0.0007	0.0001	0.0000	0.0000	0.0000	0.0000	0.0000	3
4	0.9997	0.9935	0.9658	0.9488	0.9009	0.7940	0.6543	0.5520	0.5005	0.3530	0.2279	0.1334	0.0698	0.0321	0.0126	0.0088	0.0040	0.0010	0.0002	0.0000	0.0000	0.0000	0.0000	4
5	1.0000	0.9991	0.9925	0.9873	0.9700	0.9198	0.8346	0.7587	0.7159	0.5744	0.4268	0.2905	0.1788	0.0977	0.0462	0.0347	0.0182	0.0056	0.0012	0.0003	0.0002	0.0000	0.0000	5
6	1.0000	0.9999	0.9987	0.9976	0.9930	0.9757	0.9376	0.8965	0.8705	0.7712	0.6437	0.5000	0.3563	0.2288	0.1295	0.1035	0.0624	0.0243	0.0070	0.0024	0.0013	0.0001	0.0000	6
7	1.0000	1.0000	0.9998	0.9997	0.9988	0.9944	0.9818	0.9653	0.9538	0.9023	0.8212	0.7095	0.5732	0.4256	0.2841	0.2413	0.1654	0.0802	0.0300	0.0127	0.0075	0.0009	0.0000	7
8	1.0000	1.0000	1.0000	1.0000	0.9998	0.9990	0.9960	0.9912	0.9874	0.9679	0.9302	0.8666	0.7721	0.6470	0.4995	0.4480	0.3457	0.2060	0.0991	0.0512	0.0342	0.0065	0.0003	8
9	1.0000	1.0000	1.0000	1.0000	1.0000	0.9999	0.9993	0.9984	0.9975	0.9922	0.9797	0.9539	0.9071	0.8314	0.7217	0.6776	0.5794	0.4157	0.2527	0.1581	0.1180	0.0342	0.0031	9
10	1.0000	1.0000	1.0000	1.0000	1.0000	1.0000	0.9999	0.9998	0.9997	0.9987	0.9959	0.9888	0.9731	0.9421	0.8868	0.8613	0.7975	0.6674	0.4983	0.3719	0.3080	0.1339	0.0245	10
11	1.0000	1.0000	1.0000	1.0000	1.0000	1.0000	1.0000	1.0000	1.0000	0.9999	0.9995	0.9983	0.9951	0.9874	0.9704	0.9615	0.9363	0.8733	0.7664	0.6635	0.6017	0.3787	0.1354	11
12	1.0000	1.0000	1.0000	1.0000	1.0000	1.0000	1.0000	1.0000	1.0000	1.0000	1.0000	0.9999	0.9996	0.9987	0.9963	0.9949	0.9903	0.9762	0.9450	0.9065	0.8791	0.7458	0.4867	12
13	1.0000	1.0000	1.0000	1.0000	1.0000	1.0000	1.0000	1.0000	1.0000	1.0000	1.0000	1.0000	1.0000	1.0000	1.0000	1.0000	1.0000	1.0000	1.0000	1.0000	1.0000	1.0000	1.0000	13

n = 14

r \ *p*	0.05	0.1	0.15	1/6	0.2	1/4	0.3	1/3	0.35	0.4	0.45	0.5	0.55	0.6	0.65	2/3	0.7	3/4	0.8	5/6	0.85	0.9	0.95	*r*
0	0.4877	0.2288	0.1028	0.0779	0.0440	0.0178	0.0068	0.0034	0.0024	0.0008	0.0002	0.0001	0.0000	0.0000	0.0000	0.0000	0.0000	0.0000	0.0000	0.0000	0.0000	0.0000	0.0000	0
1	0.8470	0.5846	0.3567	0.2960	0.1979	0.1010	0.0475	0.0274	0.0205	0.0081	0.0029	0.0009	0.0003	0.0001	0.0000	0.0000	0.0000	0.0000	0.0000	0.0000	0.0000	0.0000	0.0000	1
2	0.9699	0.8416	0.6479	0.5795	0.4481	0.2811	0.1608	0.1053	0.0839	0.0398	0.0170	0.0065	0.0022	0.0006	0.0001	0.0001	0.0000	0.0000	0.0000	0.0000	0.0000	0.0000	0.0000	2
3	0.9958	0.9559	0.8535	0.8063	0.6982	0.5213	0.3552	0.2612	0.2205	0.1243	0.0632	0.0287	0.0114	0.0039	0.0011	0.0007	0.0002	0.0000	0.0000	0.0000	0.0000	0.0000	0.0000	3
4	0.9996	0.9908	0.9533	0.9310	0.8702	0.7415	0.5842	0.4755	0.4227	0.2793	0.1672	0.0898	0.0426	0.0175	0.0060	0.0040	0.0017	0.0003	0.0000	0.0000	0.0000	0.0000	0.0000	4
5	1.0000	0.9985	0.9885	0.9809	0.9561	0.8883	0.7805	0.6898	0.6405	0.4859	0.3373	0.2120	0.1189	0.0583	0.0243	0.0174	0.0083	0.0022	0.0004	0.0001	0.0000	0.0000	0.0000	5
6	1.0000	0.9998	0.9978	0.9959	0.9884	0.9617	0.9067	0.8505	0.8164	0.6925	0.5461	0.3953	0.2586	0.1501	0.0753	0.0576	0.0315	0.0103	0.0024	0.0007	0.0003	0.0000	0.0000	6
7	1.0000	1.0000	0.9997	0.9993	0.9976	0.9897	0.9685	0.9424	0.9247	0.8499	0.7414	0.6047	0.4539	0.3075	0.1836	0.1495	0.0933	0.0383	0.0116	0.0041	0.0022	0.0002	0.0000	7
8	1.0000	1.0000	1.0000	0.9999	0.9996	0.9978	0.9917	0.9826	0.9757	0.9417	0.8811	0.7880	0.6627	0.5141	0.3595	0.3102	0.2195	0.1117	0.0439	0.0191	0.0115	0.0015	0.0000	8
9	1.0000	1.0000	1.0000	1.0000	1.0000	0.9997	0.9983	0.9960	0.9940	0.9825	0.9574	0.9102	0.8328	0.7207	0.5773	0.5245	0.4158	0.2585	0.1298	0.0690	0.0467	0.0092	0.0004	9
10	1.0000	1.0000	1.0000	1.0000	1.0000	1.0000	0.9998	0.9993	0.9989	0.9961	0.9886	0.9713	0.9368	0.8757	0.7795	0.7388	0.6448	0.4787	0.3018	0.1937	0.1465	0.0441	0.0042	10
11	1.0000	1.0000	1.0000	1.0000	1.0000	1.0000	1.0000	0.9999	0.9999	0.9994	0.9978	0.9935	0.9830	0.9602	0.9161	0.8947	0.8392	0.7189	0.5519	0.4205	0.3521	0.1584	0.0301	11
12	1.0000	1.0000	1.0000	1.0000	1.0000	1.0000	1.0000	1.0000	1.0000	0.9999	0.9997	0.9991	0.9971	0.9919	0.9795	0.9726	0.9525	0.8990	0.8021	0.7040	0.6433	0.4154	0.1530	12
13	1.0000	1.0000	1.0000	1.0000	1.0000	1.0000	1.0000	1.0000	1.0000	1.0000	1.0000	0.9999	0.9998	0.9992	0.9976	0.9966	0.9932	0.9822	0.9560	0.9221	0.8972	0.7712	0.5123	13
14	1.0000	1.0000	1.0000	1.0000	1.0000	1.0000	1.0000	1.0000	1.0000	1.0000	1.0000	1.0000	1.0000	1.0000	1.0000	1.0000	1.0000	1.0000	1.0000	1.0000	1.0000	1.0000	1.0000	14

n = 15

r \ *p*	0.05	0.1	0.15	1/6	0.2	1/4	0.3	1/3	0.35	0.4	0.45	0.5	0.55	0.6	0.65	2/3	0.7	3/4	0.8	5/6	0.85	0.9	0.95	*r*
0	0.4633	0.2059	0.0874	0.0649	0.0352	0.0134	0.0047	0.0023	0.0016	0.0005	0.0001	0.0000	0.0000	0.0000	0.0000	0.0000	0.0000	0.0000	0.0000	0.0000	0.0000	0.0000	0.0000	0
1	0.8290	0.5490	0.3186	0.2596	0.1671	0.0802	0.0353	0.0194	0.0142	0.0052	0.0017	0.0005	0.0001	0.0000	0.0000	0.0000	0.0000	0.0000	0.0000	0.0000	0.0000	0.0000	0.0000	1
2	0.9638	0.8159	0.6042	0.5322	0.3980	0.2361	0.1268	0.0794	0.0617	0.0271	0.0107	0.0037	0.0011	0.0003	0.0001	0.0000	0.0000	0.0000	0.0000	0.0000	0.0000	0.0000	0.0000	2
3	0.9945	0.9444	0.8227	0.7685	0.6482	0.4613	0.2969	0.2092	0.1727	0.0905	0.0424	0.0176	0.0063	0.0019	0.0005	0.0003	0.0001	0.0000	0.0000	0.0000	0.0000	0.0000	0.0000	3
4	0.9994	0.9873	0.9383	0.9102	0.8358	0.6865	0.5155	0.4041	0.3519	0.2173	0.1204	0.0592	0.0255	0.0093	0.0028	0.0018	0.0007	0.0001	0.0000	0.0000	0.0000	0.0000	0.0000	4
5	0.9999	0.9978	0.9832	0.9726	0.9389	0.8516	0.7216	0.6184	0.5643	0.4032	0.2608	0.1509	0.0769	0.0338	0.0124	0.0085	0.0037	0.0008	0.0001	0.0000	0.0000	0.0000	0.0000	5
6	1.0000	0.9997	0.9964	0.9934	0.9819	0.9434	0.8689	0.7970	0.7548	0.6098	0.4522	0.3036	0.1818	0.0950	0.0422	0.0308	0.0152	0.0042	0.0008	0.0002	0.0001	0.0000	0.0000	6
7	1.0000	1.0000	0.9994	0.9987	0.9958	0.9827	0.9500	0.9118	0.8868	0.7869	0.6535	0.5000	0.3465	0.2131	0.1132	0.0882	0.0500	0.0173	0.0042	0.0013	0.0006	0.0000	0.0000	7
8	1.0000	1.0000	0.9999	0.9998	0.9992	0.9958	0.9848	0.9692	0.9578	0.9050	0.8182	0.6964	0.5478	0.3902	0.2452	0.2030	0.1311	0.0566	0.0181	0.0066	0.0036	0.0003	0.0000	8
9	1.0000	1.0000	1.0000	1.0000	0.9999	0.9992	0.9963	0.9915	0.9876	0.9662	0.9231	0.8491	0.7392	0.5968	0.4357	0.3816	0.2784	0.1484	0.0611	0.0274	0.0168	0.0022	0.0001	9
10	1.0000	1.0000	1.0000	1.0000	1.0000	0.9999	0.9993	0.9982	0.9972	0.9907	0.9745	0.9408	0.8796	0.7827	0.6481	0.5959	0.4845	0.3135	0.1642	0.0898	0.0617	0.0127	0.0006	10
11	1.0000	1.0000	1.0000	1.0000	1.0000	1.0000	0.9999	0.9997	0.9995	0.9981	0.9937	0.9824	0.9576	0.9095	0.8273	0.7908	0.7031	0.5387	0.3518	0.2315	0.1773	0.0556	0.0055	11
12	1.0000	1.0000	1.0000	1.0000	1.0000	1.0000	1.0000	1.0000	0.9999	0.9997	0.9989	0.9963	0.9893	0.9729	0.9383	0.9206	0.8732	0.7639	0.6020	0.4678	0.3958	0.1841	0.0362	12
13	1.0000	1.0000	1.0000	1.0000	1.0000	1.0000	1.0000	1.0000	1.0000	1.0000	0.9999	0.9995	0.9983	0.9948	0.9858	0.9806	0.9647	0.9198	0.8329	0.7404	0.6814	0.4510	0.1710	13
14	1.0000	1.0000	1.0000	1.0000	1.0000	1.0000	1.0000	1.0000	1.0000	1.0000	1.0000	1.0000	0.9999	0.9995	0.9984	0.9977	0.9953	0.9866	0.9648	0.9351	0.9126	0.7941	0.5367	14
15	1.0000	1.0000	1.0000	1.0000	1.0000	1.0000	1.0000	1.0000	1.0000	1.0000	1.0000	1.0000	1.0000	1.0000	1.0000	1.0000	1.0000	1.0000	1.0000	1.0000	1.0000	1.0000	1.0000	15

r $n = 16$

p	0.05	0.1	0.15	1/6	0.2	1/4	0.3	1/3	0.35	0.4	0.45	0.5	0.55	0.6	0.65	2/3	0.7	3/4	0.8	5/6	0.85	0.9	0.95	r
0	0.4401	0.1853	0.0743	0.0541	0.0281	0.0100	0.0033	0.0015	0.0010	0.0003	0.0001	0.0000	0.0000	0.0000	0.0000	0.0000	0.0000	0.0000	0.0000	0.0000	0.0000	0.0000	0.0000	0
1	0.8108	0.5147	0.2839	0.2272	0.1407	0.0635	0.0261	0.0137	0.0098	0.0033	0.0010	0.0003	0.0001	0.0000	0.0000	0.0000	0.0000	0.0000	0.0000	0.0000	0.0000	0.0000	0.0000	1
2	0.9571	0.7892	0.5614	0.4868	0.3518	0.1971	0.0994	0.0594	0.0451	0.0183	0.0066	0.0021	0.0006	0.0001	0.0000	0.0000	0.0000	0.0000	0.0000	0.0000	0.0000	0.0000	0.0000	2
3	0.9930	0.9316	0.7899	0.7291	0.5981	0.4050	0.2459	0.1659	0.1339	0.0651	0.0281	0.0106	0.0035	0.0009	0.0002	0.0001	0.0000	0.0000	0.0000	0.0000	0.0000	0.0000	0.0000	3
4	0.9991	0.9830	0.9209	0.8866	0.7982	0.6302	0.4499	0.3391	0.2892	0.1666	0.0853	0.0384	0.0149	0.0049	0.0013	0.0008	0.0003	0.0000	0.0000	0.0000	0.0000	0.0000	0.0000	4
5	0.9999	0.9967	0.9765	0.9622	0.9183	0.8103	0.6598	0.5469	0.4900	0.3288	0.1976	0.1051	0.0486	0.0191	0.0062	0.0040	0.0016	0.0003	0.0000	0.0000	0.0000	0.0000	0.0000	5
6	1.0000	0.9995	0.9944	0.9899	0.9733	0.9204	0.8247	0.7374	0.6881	0.5272	0.3660	0.2272	0.1241	0.0583	0.0229	0.0159	0.0071	0.0016	0.0002	0.0000	0.0000	0.0000	0.0000	6
7	1.0000	0.9999	0.9989	0.9979	0.9930	0.9729	0.9256	0.8735	0.8406	0.7161	0.5629	0.4018	0.2559	0.1423	0.0671	0.0500	0.0257	0.0075	0.0015	0.0004	0.0002	0.0000	0.0000	7
8	1.0000	1.0000	0.9998	0.9996	0.9985	0.9925	0.9743	0.9500	0.9329	0.8577	0.7441	0.5982	0.4371	0.2839	0.1594	0.1265	0.0744	0.0271	0.0070	0.0021	0.0011	0.0001	0.0000	8
9	1.0000	1.0000	1.0000	1.0000	0.9998	0.9984	0.9929	0.9841	0.9771	0.9417	0.8759	0.7728	0.6340	0.4728	0.3119	0.2626	0.1753	0.0796	0.0267	0.0101	0.0056	0.0005	0.0000	9
10	1.0000	1.0000	1.0000	1.0000	1.0000	0.9997	0.9984	0.9960	0.9938	0.9809	0.9514	0.8949	0.8024	0.6712	0.5100	0.4531	0.3402	0.1897	0.0817	0.0378	0.0235	0.0033	0.0001	10
11	1.0000	1.0000	1.0000	1.0000	1.0000	1.0000	0.9997	0.9992	0.9987	0.9951	0.9851	0.9616	0.9147	0.8334	0.7108	0.6609	0.5501	0.3698	0.2018	0.1134	0.0791	0.0170	0.0009	11
12	1.0000	1.0000	1.0000	1.0000	1.0000	1.0000	1.0000	0.9999	0.9998	0.9991	0.9965	0.9894	0.9719	0.9349	0.8661	0.8341	0.7541	0.5950	0.4019	0.2709	0.2101	0.0684	0.0070	12
13	1.0000	1.0000	1.0000	1.0000	1.0000	1.0000	1.0000	1.0000	1.0000	0.9999	0.9994	0.9979	0.9934	0.9817	0.9549	0.9406	0.9006	0.8029	0.6482	0.5132	0.4386	0.2108	0.0429	13
14	1.0000	1.0000	1.0000	1.0000	1.0000	1.0000	1.0000	1.0000	1.0000	1.0000	0.9999	0.9997	0.9990	0.9967	0.9902	0.9863	0.9739	0.9365	0.8593	0.7728	0.7161	0.4853	0.1892	14
15	1.0000	1.0000	1.0000	1.0000	1.0000	1.0000	1.0000	1.0000	1.0000	1.0000	1.0000	1.0000	0.9999	0.9997	0.9990	0.9985	0.9967	0.9900	0.9719	0.9459	0.9257	0.8147	0.5599	15
16	1.0000	1.0000	1.0000	1.0000	1.0000	1.0000	1.0000	1.0000	1.0000	1.0000	1.0000	1.0000	1.0000	1.0000	1.0000	1.0000	1.0000	1.0000	1.0000	1.0000	1.0000	1.0000	1.0000	16

r $n = 17$

r	0.05	0.1	0.15	1/6	0.2	1/4	0.3	1/3	0.35	0.4	0.45	0.5	0.55	0.6	0.65	2/3	0.7	3/4	0.8	5/6	0.85	0.9	0.95	r
0	0.4181	0.1668	0.0631	0.0451	0.0225	0.0075	0.0023	0.0010	0.0007	0.0002	0.0000	0.0000	0.0000	0.0000	0.0000	0.0000	0.0000	0.0000	0.0000	0.0000	0.0000	0.0000	0.0000	0
1	0.7922	0.4818	0.2525	0.1983	0.1182	0.0501	0.0193	0.0096	0.0067	0.0021	0.0006	0.0001	0.0000	0.0000	0.0000	0.0000	0.0000	0.0000	0.0000	0.0000	0.0000	0.0000	0.0000	1
2	0.9497	0.7618	0.5198	0.4435	0.3096	0.1637	0.0774	0.0442	0.0327	0.0123	0.0041	0.0012	0.0003	0.0001	0.0000	0.0000	0.0000	0.0000	0.0000	0.0000	0.0000	0.0000	0.0000	2
3	0.9912	0.9174	0.7556	0.6887	0.5489	0.3530	0.2019	0.1304	0.1028	0.0464	0.0184	0.0064	0.0019	0.0005	0.0001	0.0000	0.0000	0.0000	0.0000	0.0000	0.0000	0.0000	0.0000	3
4	0.9988	0.9779	0.9013	0.8604	0.7582	0.5739	0.3887	0.2814	0.2348	0.1260	0.0596	0.0245	0.0086	0.0025	0.0006	0.0003	0.0001	0.0000	0.0000	0.0000	0.0000	0.0000	0.0000	4
5	0.9999	0.9953	0.9681	0.9496	0.8943	0.7653	0.5968	0.4777	0.4197	0.2639	0.1471	0.0717	0.0301	0.0106	0.0030	0.0019	0.0007	0.0001	0.0000	0.0000	0.0000	0.0000	0.0000	5
6	1.0000	0.9992	0.9917	0.9853	0.9623	0.8929	0.7752	0.6739	0.6188	0.4478	0.2902	0.1662	0.0826	0.0348	0.0120	0.0080	0.0032	0.0006	0.0001	0.0000	0.0000	0.0000	0.0000	6
7	1.0000	0.9999	0.9983	0.9965	0.9891	0.9598	0.8954	0.8281	0.7872	0.6405	0.4743	0.3145	0.1834	0.0919	0.0383	0.0273	0.0127	0.0031	0.0005	0.0001	0.0000	0.0000	0.0000	7
8	1.0000	1.0000	0.9997	0.9993	0.9974	0.9876	0.9597	0.9245	0.9006	0.8011	0.6626	0.5000	0.3374	0.1989	0.0994	0.0755	0.0403	0.0124	0.0026	0.0007	0.0003	0.0000	0.0000	8
9	1.0000	1.0000	1.0000	0.9999	0.9995	0.9969	0.9873	0.9727	0.9617	0.9081	0.8166	0.6855	0.5257	0.3595	0.2128	0.1719	0.1046	0.0402	0.0109	0.0035	0.0017	0.0001	0.0000	9
10	1.0000	1.0000	1.0000	1.0000	0.9999	0.9994	0.9968	0.9920	0.9880	0.9652	0.9174	0.8338	0.7098	0.5522	0.3812	0.3261	0.2248	0.1071	0.0377	0.0147	0.0083	0.0008	0.0000	10
11	1.0000	1.0000	1.0000	1.0000	1.0000	0.9999	0.9993	0.9981	0.9970	0.9894	0.9699	0.9283	0.8529	0.7361	0.5803	0.5223	0.4032	0.2347	0.1057	0.0504	0.0319	0.0047	0.0001	11
12	1.0000	1.0000	1.0000	1.0000	1.0000	1.0000	0.9999	0.9997	0.9994	0.9975	0.9914	0.9755	0.9404	0.8740	0.7652	0.7186	0.6113	0.4261	0.2418	0.1396	0.0987	0.0221	0.0012	12
13	1.0000	1.0000	1.0000	1.0000	1.0000	1.0000	1.0000	1.0000	0.9999	0.9995	0.9981	0.9936	0.9816	0.9536	0.8972	0.8696	0.7981	0.6470	0.4511	0.3113	0.2444	0.0826	0.0088	13
14	1.0000	1.0000	1.0000	1.0000	1.0000	1.0000	1.0000	1.0000	1.0000	0.9999	0.9997	0.9988	0.9959	0.9877	0.9673	0.9558	0.9226	0.8363	0.6904	0.5565	0.4802	0.2382	0.0503	14
15	1.0000	1.0000	1.0000	1.0000	1.0000	1.0000	1.0000	1.0000	1.0000	1.0000	1.0000	0.9999	0.9994	0.9979	0.9933	0.9904	0.9807	0.9499	0.8818	0.8017	0.7475	0.5182	0.2078	15
16	1.0000	1.0000	1.0000	1.0000	1.0000	1.0000	1.0000	1.0000	1.0000	1.0000	1.0000	1.0000	1.0000	0.9998	0.9993	0.9990	0.9977	0.9925	0.9775	0.9549	0.9369	0.8332	0.5819	16
17	1.0000	1.0000	1.0000	1.0000	1.0000	1.0000	1.0000	1.0000	1.0000	1.0000	1.0000	1.0000	1.0000	1.0000	1.0000	1.0000	1.0000	1.0000	1.0000	1.0000	1.0000	1.0000	1.0000	17

$n = 18$

r	0.05	0.1	0.15	1/6	0.2	1/4	0.3	1/3	0.35	0.4	0.45	0.5	0.55	0.6	0.65	2/3	0.7	3/4	0.8	5/6	0.85	0.9	0.95	r
0	0.3972	0.1501	0.0536	0.0376	0.0180	0.0056	0.0016	0.0007	0.0004	0.0001	0.0000	0.0000	0.0000	0.0000	0.0000	0.0000	0.0000	0.0000	0.0000	0.0000	0.0000	0.0000	0.0000	0
1	0.7735	0.4503	0.2241	0.1728	0.0991	0.0395	0.0142	0.0068	0.0046	0.0013	0.0003	0.0001	0.0000	0.0000	0.0000	0.0000	0.0000	0.0000	0.0000	0.0000	0.0000	0.0000	0.0000	1
2	0.9419	0.7338	0.4797	0.4027	0.2713	0.1353	0.0600	0.0326	0.0236	0.0082	0.0025	0.0007	0.0001	0.0000	0.0000	0.0000	0.0000	0.0000	0.0000	0.0000	0.0000	0.0000	0.0000	2
3	0.9891	0.9018	0.7202	0.6479	0.5010	0.3057	0.1646	0.1017	0.0783	0.0328	0.0120	0.0038	0.0010	0.0002	0.0000	0.0000	0.0000	0.0000	0.0000	0.0000	0.0000	0.0000	0.0000	3
4	0.9985	0.9718	0.8794	0.8318	0.7164	0.5187	0.3327	0.2311	0.1886	0.0942	0.0411	0.0154	0.0049	0.0013	0.0003	0.0001	0.0000	0.0000	0.0000	0.0000	0.0000	0.0000	0.0000	4
5	0.9998	0.9936	0.9581	0.9347	0.8671	0.7175	0.5344	0.4122	0.3550	0.2088	0.1077	0.0481	0.0183	0.0058	0.0014	0.0009	0.0003	0.0000	0.0000	0.0000	0.0000	0.0000	0.0000	5
6	1.0000	0.9988	0.9882	0.9794	0.9487	0.8610	0.7217	0.6085	0.5491	0.3743	0.2258	0.1189	0.0537	0.0203	0.0062	0.0039	0.0014	0.0002	0.0000	0.0000	0.0000	0.0000	0.0000	6
7	1.0000	0.9998	0.9973	0.9947	0.9837	0.9431	0.8593	0.7767	0.7283	0.5634	0.3915	0.2403	0.1280	0.0576	0.0212	0.0144	0.0061	0.0012	0.0002	0.0000	0.0000	0.0000	0.0000	7
8	1.0000	1.0000	0.9995	0.9989	0.9957	0.9807	0.9404	0.8924	0.8609	0.7368	0.5778	0.4073	0.2527	0.1347	0.0597	0.0433	0.0210	0.0054	0.0009	0.0002	0.0001	0.0000	0.0000	8
9	1.0000	1.0000	0.9999	0.9998	0.9991	0.9946	0.9790	0.9567	0.9403	0.8653	0.7473	0.5927	0.4222	0.2632	0.1391	0.1076	0.0596	0.0193	0.0043	0.0011	0.0005	0.0000	0.0000	9
10	1.0000	1.0000	1.0000	1.0000	0.9998	0.9988	0.9939	0.9856	0.9788	0.9424	0.8720	0.7597	0.6085	0.4366	0.2717	0.2233	0.1407	0.0569	0.0163	0.0053	0.0027	0.0002	0.0000	10
11	1.0000	1.0000	1.0000	1.0000	1.0000	0.9998	0.9986	0.9961	0.9938	0.9797	0.9463	0.8811	0.7742	0.6257	0.4509	0.3915	0.2783	0.1390	0.0513	0.0206	0.0118	0.0012	0.0000	11
12	1.0000	1.0000	1.0000	1.0000	1.0000	1.0000	0.9997	0.9991	0.9986	0.9942	0.9817	0.9519	0.8923	0.7912	0.6450	0.5878	0.4656	0.2825	0.1329	0.0653	0.0419	0.0064	0.0002	12
13	1.0000	1.0000	1.0000	1.0000	1.0000	1.0000	1.0000	0.9999	0.9997	0.9987	0.9951	0.9846	0.9589	0.9058	0.8114	0.7689	0.6673	0.4813	0.2836	0.1682	0.1206	0.0282	0.0015	13
14	1.0000	1.0000	1.0000	1.0000	1.0000	1.0000	1.0000	1.0000	1.0000	0.9998	0.9990	0.9962	0.9880	0.9672	0.9217	0.8983	0.8354	0.6943	0.4990	0.3521	0.2798	0.0982	0.0109	14
15	1.0000	1.0000	1.0000	1.0000	1.0000	1.0000	1.0000	1.0000	1.0000	1.0000	0.9999	0.9993	0.9975	0.9918	0.9764	0.9674	0.9400	0.8647	0.7287	0.5973	0.5203	0.2662	0.0581	15
16	1.0000	1.0000	1.0000	1.0000	1.0000	1.0000	1.0000	1.0000	1.0000	1.0000	1.0000	0.9999	0.9997	0.9987	0.9954	0.9932	0.9858	0.9605	0.9009	0.8272	0.7759	0.5497	0.2265	16
17	1.0000	1.0000	1.0000	1.0000	1.0000	1.0000	1.0000	1.0000	1.0000	1.0000	1.0000	1.0000	1.0000	0.9999	0.9996	0.9993	0.9984	0.9944	0.9820	0.9624	0.9464	0.8499	0.6028	17
18	1.0000	1.0000	1.0000	1.0000	1.0000	1.0000	1.0000	1.0000	1.0000	1.0000	1.0000	1.0000	1.0000	1.0000	1.0000	1.0000	1.0000	1.0000	1.0000	1.0000	1.0000	1.0000	1.0000	18

$n = 19$

r	0.05	0.1	0.15	1/6	0.2	1/4	0.3	1/3	0.35	0.4	0.45	0.5	0.55	0.6	0.65	2/3	0.7	3/4	0.8	5/6	0.85	0.9	0.95	r
0	0.3774	0.1351	0.0456	0.0313	0.0144	0.0042	0.0011	0.0005	0.0003	0.0001	0.0000	0.0000	0.0000	0.0000	0.0000	0.0000	0.0000	0.0000	0.0000	0.0000	0.0000	0.0000	0.0000	0
1	0.7547	0.4203	0.1985	0.1502	0.0829	0.0310	0.0104	0.0047	0.0031	0.0008	0.0002	0.0000	0.0000	0.0000	0.0000	0.0000	0.0000	0.0000	0.0000	0.0000	0.0000	0.0000	0.0000	1
2	0.9335	0.7054	0.4413	0.3643	0.2369	0.1113	0.0462	0.0240	0.0170	0.0055	0.0015	0.0004	0.0001	0.0000	0.0000	0.0000	0.0000	0.0000	0.0000	0.0000	0.0000	0.0000	0.0000	2
3	0.9868	0.8850	0.6841	0.6070	0.4551	0.2631	0.1332	0.0787	0.0591	0.0230	0.0077	0.0022	0.0005	0.0001	0.0000	0.0000	0.0000	0.0000	0.0000	0.0000	0.0000	0.0000	0.0000	3
4	0.9980	0.9648	0.8556	0.8011	0.6733	0.4654	0.2822	0.1879	0.1500	0.0696	0.0280	0.0096	0.0028	0.0006	0.0001	0.0001	0.0000	0.0000	0.0000	0.0000	0.0000	0.0000	0.0000	4
5	0.9998	0.9914	0.9463	0.9176	0.8369	0.6678	0.4739	0.3519	0.2968	0.1629	0.0777	0.0318	0.0109	0.0031	0.0007	0.0004	0.0001	0.0000	0.0000	0.0000	0.0000	0.0000	0.0000	5
6	1.0000	0.9983	0.9837	0.9719	0.9324	0.8251	0.6655	0.5431	0.4812	0.3081	0.1727	0.0835	0.0342	0.0116	0.0031	0.0019	0.0006	0.0001	0.0000	0.0000	0.0000	0.0000	0.0000	6
7	1.0000	0.9997	0.9959	0.9921	0.9767	0.9225	0.8180	0.7207	0.6656	0.4878	0.3169	0.1796	0.0871	0.0352	0.0114	0.0074	0.0028	0.0005	0.0000	0.0000	0.0000	0.0000	0.0000	7
8	1.0000	1.0000	0.9992	0.9982	0.9933	0.9713	0.9161	0.8538	0.8145	0.6675	0.4940	0.3238	0.1841	0.0885	0.0347	0.0241	0.0105	0.0023	0.0003	0.0001	0.0000	0.0000	0.0000	8
9	1.0000	1.0000	0.9999	0.9996	0.9984	0.9911	0.9674	0.9352	0.9125	0.8139	0.6710	0.5000	0.3290	0.1861	0.0875	0.0648	0.0326	0.0089	0.0016	0.0004	0.0001	0.0000	0.0000	9
10	1.0000	1.0000	1.0000	0.9999	0.9997	0.9977	0.9895	0.9759	0.9653	0.9115	0.8159	0.6762	0.5060	0.3325	0.1855	0.1462	0.0839	0.0287	0.0067	0.0018	0.0008	0.0000	0.0000	10
11	1.0000	1.0000	1.0000	1.0000	1.0000	0.9995	0.9972	0.9926	0.9886	0.9648	0.9129	0.8204	0.6831	0.5122	0.3344	0.2793	0.1820	0.0775	0.0233	0.0079	0.0041	0.0003	0.0000	11
12	1.0000	1.0000	1.0000	1.0000	1.0000	0.9999	0.9994	0.9981	0.9969	0.9884	0.9658	0.9165	0.8273	0.6919	0.5188	0.4569	0.3345	0.1749	0.0676	0.0281	0.0163	0.0017	0.0000	12
13	1.0000	1.0000	1.0000	1.0000	1.0000	1.0000	0.9999	0.9996	0.9993	0.9969	0.9891	0.9682	0.9223	0.8371	0.7032	0.6481	0.5261	0.3322	0.1631	0.0824	0.0537	0.0086	0.0002	13
14	1.0000	1.0000	1.0000	1.0000	1.0000	1.0000	1.0000	0.9999	0.9999	0.9994	0.9972	0.9904	0.9720	0.9304	0.8500	0.8121	0.7178	0.5346	0.3267	0.1989	0.1444	0.0352	0.0020	14
15	1.0000	1.0000	1.0000	1.0000	1.0000	1.0000	1.0000	1.0000	1.0000	0.9999	0.9995	0.9978	0.9923	0.9770	0.9409	0.9213	0.8668	0.7369	0.5449	0.3930	0.3159	0.1150	0.0132	15
16	1.0000	1.0000	1.0000	1.0000	1.0000	1.0000	1.0000	1.0000	1.0000	1.0000	0.9999	0.9996	0.9985	0.9945	0.9830	0.9760	0.9538	0.8887	0.7631	0.6357	0.5587	0.2946	0.0665	16
17	1.0000	1.0000	1.0000	1.0000	1.0000	1.0000	1.0000	1.0000	1.0000	1.0000	1.0000	1.0000	0.9998	0.9992	0.9969	0.9953	0.9896	0.9690	0.9171	0.8498	0.8015	0.5797	0.2453	17
18	1.0000	1.0000	1.0000	1.0000	1.0000	1.0000	1.0000	1.0000	1.0000	1.0000	1.0000	1.0000	1.0000	0.9999	0.9997	0.9995	0.9989	0.9958	0.9856	0.9687	0.9544	0.8649	0.6226	18
19	1.0000	1.0000	1.0000	1.0000	1.0000	1.0000	1.0000	1.0000	1.0000	1.0000	1.0000	1.0000	1.0000	1.0000	1.0000	1.0000	1.0000	1.0000	1.0000	1.0000	1.0000	1.0000	1.0000	19

$n = 20$

r \ p	0.05	0.1	0.15	1/6	0.2	1/4	0.3	1/3	0.35	0.4	0.45	0.5	0.55	0.6	0.65	2/3	0.7	3/4	0.8	5/6	0.85	0.9	0.95	r
0	0.3585	0.1216	0.0388	0.0261	0.0115	0.0032	0.0008	0.0003	0.0002	0.0000	0.0000	0.0000	0.0000	0.0000	0.0000	0.0000	0.0000	0.0000	0.0000	0.0000	0.0000	0.0000	0.0000	0
1	0.7358	0.3917	0.1756	0.1304	0.0692	0.0243	0.0076	0.0033	0.0021	0.0005	0.0001	0.0000	0.0000	0.0000	0.0000	0.0000	0.0000	0.0000	0.0000	0.0000	0.0000	0.0000	0.0000	1
2	0.9245	0.6769	0.4049	0.3287	0.2061	0.0913	0.0355	0.0176	0.0121	0.0036	0.0009	0.0002	0.0000	0.0000	0.0000	0.0000	0.0000	0.0000	0.0000	0.0000	0.0000	0.0000	0.0000	2
3	0.9841	0.8670	0.6477	0.5665	0.4114	0.2252	0.1071	0.0604	0.0444	0.0160	0.0049	0.0013	0.0003	0.0000	0.0000	0.0000	0.0000	0.0000	0.0000	0.0000	0.0000	0.0000	0.0000	3
4	0.9974	0.9568	0.8298	0.7687	0.6296	0.4148	0.2375	0.1515	0.1182	0.0510	0.0189	0.0059	0.0015	0.0003	0.0000	0.0000	0.0000	0.0000	0.0000	0.0000	0.0000	0.0000	0.0000	4
5	0.9997	0.9887	0.9327	0.8982	0.8042	0.6172	0.4164	0.2972	0.2454	0.1256	0.0553	0.0207	0.0064	0.0016	0.0003	0.0002	0.0000	0.0000	0.0000	0.0000	0.0000	0.0000	0.0000	5
6	1.0000	0.9976	0.9781	0.9629	0.9133	0.7858	0.6080	0.4793	0.4166	0.2500	0.1299	0.0577	0.0214	0.0065	0.0015	0.0009	0.0003	0.0000	0.0000	0.0000	0.0000	0.0000	0.0000	6
7	1.0000	0.9996	0.9941	0.9887	0.9679	0.8982	0.7723	0.6615	0.6010	0.4159	0.2520	0.1316	0.0580	0.0210	0.0060	0.0037	0.0013	0.0002	0.0000	0.0000	0.0000	0.0000	0.0000	7
8	1.0000	0.9999	0.9987	0.9972	0.9900	0.9591	0.8867	0.8095	0.7624	0.5956	0.4143	0.2517	0.1308	0.0565	0.0196	0.0130	0.0051	0.0009	0.0001	0.0000	0.0000	0.0000	0.0000	8
9	1.0000	1.0000	0.9998	0.9994	0.9974	0.9861	0.9520	0.9081	0.8782	0.7553	0.5914	0.4119	0.2493	0.1275	0.0532	0.0376	0.0171	0.0039	0.0006	0.0001	0.0000	0.0000	0.0000	9
10	1.0000	1.0000	1.0000	0.9999	0.9994	0.9961	0.9829	0.9624	0.9468	0.8725	0.7507	0.5881	0.4086	0.2447	0.1218	0.0919	0.0480	0.0139	0.0026	0.0006	0.0002	0.0000	0.0000	10
11	1.0000	1.0000	1.0000	1.0000	0.9999	0.9991	0.9949	0.9870	0.9804	0.9435	0.8692	0.7483	0.5857	0.4044	0.2376	0.1905	0.1133	0.0409	0.0100	0.0028	0.0013	0.0001	0.0000	11
12	1.0000	1.0000	1.0000	1.0000	1.0000	0.9998	0.9987	0.9963	0.9940	0.9790	0.9420	0.8684	0.7480	0.5841	0.3990	0.3385	0.2277	0.1018	0.0321	0.0113	0.0059	0.0004	0.0000	12
13	1.0000	1.0000	1.0000	1.0000	1.0000	1.0000	0.9997	0.9991	0.9985	0.9935	0.9786	0.9423	0.8701	0.7500	0.5834	0.5207	0.3920	0.2142	0.0867	0.0371	0.0219	0.0024	0.0000	13
14	1.0000	1.0000	1.0000	1.0000	1.0000	1.0000	1.0000	0.9998	0.9997	0.9984	0.9936	0.9793	0.9447	0.8744	0.7546	0.7028	0.5836	0.3828	0.1958	0.1018	0.0673	0.0113	0.0003	14
15	1.0000	1.0000	1.0000	1.0000	1.0000	1.0000	1.0000	1.0000	1.0000	0.9997	0.9985	0.9941	0.9811	0.9490	0.8818	0.8485	0.7625	0.5852	0.3704	0.2313	0.1702	0.0432	0.0026	15
16	1.0000	1.0000	1.0000	1.0000	1.0000	1.0000	1.0000	1.0000	1.0000	1.0000	0.9997	0.9987	0.9951	0.9840	0.9556	0.9396	0.8929	0.7748	0.5886	0.4335	0.3523	0.1330	0.0159	16
17	1.0000	1.0000	1.0000	1.0000	1.0000	1.0000	1.0000	1.0000	1.0000	1.0000	1.0000	0.9998	0.9991	0.9964	0.9879	0.9824	0.9645	0.9087	0.7939	0.6713	0.5951	0.3231	0.0755	17
18	1.0000	1.0000	1.0000	1.0000	1.0000	1.0000	1.0000	1.0000	1.0000	1.0000	1.0000	1.0000	0.9999	0.9995	0.9979	0.9967	0.9924	0.9757	0.9308	0.8696	0.8244	0.6083	0.2642	18
19	1.0000	1.0000	1.0000	1.0000	1.0000	1.0000	1.0000	1.0000	1.0000	1.0000	1.0000	1.0000	1.0000	1.0000	0.9998	0.9997	0.9992	0.9968	0.9885	0.9739	0.9612	0.8784	0.6415	19
20	1.0000	1.0000	1.0000	1.0000	1.0000	1.0000	1.0000	1.0000	1.0000	1.0000	1.0000	1.0000	1.0000	1.0000	1.0000	1.0000	1.0000	1.0000	1.0000	1.0000	1.0000	1.0000	1.0000	20

Random number table

23505	32408	00972	75953	27778	02240	98776	76255	35351	35730	08152	72896
84744	91235	85344	22723	97935	84372	13292	28162	29463	44734	78899	85427
67135	02152	34778	47377	18502	50742	24514	06370	95004	36152	79402	45594
33591	15583	96977	60192	44159	59332	74309	01316	50806	75379	24918	87070
57787	57248	89698	41776	52546	33597	87944	63145	81875	39865	38955	22913
14766	54285	52523	20721	06003	72173	69945	97350	93210	03392	01552	63127
45926	78875	83938	97158	27463	85241	47795	07886	43518	64679	75598	16054
78938	89726	40968	70279	84951	22225	30919	51919	43655	72606	28982	12951
59369	01028	62588	03664	52707	06334	87055	11072	63671	21627	93451	62562
30962	04554	03732	30794	79584	97655	59797	67924	00586	78601	56219	03616
27874	57190	80930	57080	51382	37294	34005	28287	31632	05457	21800	64011
47651	74469	43485	56113	93057	93002	63100	51545	03706	27973	88920	89087
31431	83282	43730	43006	61880	86284	50366	26745	53003	64851	39700	40594
75415	03441	26142	78621	52412	60563	74920	29887	05783	86082	96278	15336
24897	35940	68834	59572	72705	29934	05583	93433	99119	21725	67146	16111
32969	70116	59106	34163	24965	50167	81174	37670	20332	69765	83978	71301
87833	57604	12336	50895	20521	68370	22046	23914	42885	76263	56131	38447
91167	03154	31650	54722	23301	36239	31521	97178	64194	18987	31871	70386
58277	79938	16566	24973	35469	45707	85744	86713	76028	00206	30121	44449
56471	15945	06962	32146	47734	70351	29692	24522	67548	32305	29239	56412
58151	67936	18313	92678	58406	21529	64561	55445	69088	16396	09472	20839
99436	19991	11724	74357	49339	12500	21952	87811	45155	39804	23392	22234
25808	78720	23661	13185	62144	14508	34347	06116	84674	63687	57372	43121
22522	69270	18011	79451	55023	83396	65459	19040	92354	12263	54542	83984
83960	82719	47408	41168	60103	39533	59263	15805	57077	56831	36050	85170
65484	72838	34333	78561	54851	09256	51502	86839	65282	42651	07545	10767
03625	20785	62258	63419	39187	97844	91034	99656	50821	97208	15212	20855
78131	16692	70347	00396	85685	97622	00500	60893	09574	44518	13208	20855
15673	22384	39905	13615	78544	00238	77761	60618	77041	24887	66032	36594
57485	97418	61853	31389	69163	63071	69175	60365	25143	86177	00083	03325
93761	46283	22399	57562	32438	93946	03023	92889	48511	87837	08272	12847
74412	45238	37078	37604	57874	81430	48212	60258	02791	33847	46434	99449
31770	53625	22111	64275	07936	57060	56774	15297	77697	20999	47345	28709
38744	77136	07184	57549	54621	97858	55852	48261	95110	75124	73032	29955
28359	44192	51026	64701	48047	28759	88462	29886	33070	79827	65324	50024

APPENDIX I

Geometric distribution

The following is an algebraic derivation of the results established in Chapter 6, *Probability distribution*, concerning the relationship between the parameter, p and the mean μ, and the variance σ^2, of a random variable which is geometrically distributed: Geo(p).

Assume $X \sim \text{Geo(p)}$,

i.e. $P(X = n) = pq^{(n-1)}$; $q = 1 - p$ and $n = 1, 2, 3, 4, \ldots$

The mean, μ, is:

$$\mu = E(X) = \sum_{n=1}^{\infty} npq^{(n-1)}$$

Written out term by term:

$$\mu = 1 \times p + 2 \times pq + 3 \times pq^2 + 4 \times pq^3 + \ldots$$

Multiplying this by q yields:

$$q\mu = 1 \times pq + 2 \times pq^2 + 3 \times pq^3 + 4pq^4 + \ldots$$

Subtracting:

$$\begin{aligned}\mu - q\mu &= 1 \times p + 2pq + 3pq^2 + 4pq^3 + \ldots - 1 \times pq - 2pq^2 - 3pq^3 - 4pq^4 - \ldots \\ &= 1 \times p + (2-1)pq + (3-2)pq^2 + (4-3)pq^3 + \ldots \\ &= p + pq + pq^2 + pq^3 + \ldots\end{aligned}$$

This series is just the sum of all the probabilities and so is 1. Hence:

$$\mu - q\mu = 1$$

Factorising yields:

$$\mu(1 - q) = 1 \Rightarrow \mu = \frac{1}{1-q} = \frac{1}{p}$$

This is the conclusion reached in Chapter 6, *Probability distribution*.

A similar approach can be adopted for the variance, σ^2.

$$\sigma^2 = E(X^2) - \mu^2 = \sum_{n=1}^{\infty} (n^2pq^{(n-1)}) - \mu^2$$

Writing out the expected value of the square, $E(X^2)$ term by term yields:

$$(\sigma^2 + \mu^2) = E(X^2) = 1^2p + 2^2pq + 3^2pq^2 + 4^2pq^3 + \ldots$$

Multiplying this by q gives:

$$q(\sigma^2 + \mu^2) = 1^2pq + 2^2pq^2 + 3^2pq^3 + 4^2pq^3 + \ldots$$

Subtracting:

$$(1 - q)(\sigma^2 + \mu^2) = 1^2p + 2^2pq + 3^2pq^2 + 4^2pq^3 + \ldots$$
$$- 1^2pq - 2^2pq^2 - 3^2pq^3 - 4^2pq^3 - \ldots$$

$$\therefore p(\sigma^2 + \mu^2) = 1^2p + (2^2 - 1^2)pq + (3^2 - 2^2)pq^2 + (4^2 - 3^2)pq^3 + \ldots$$
$$= p + 3pq + 5pq^2 + 7pq^3 + \ldots$$

This series can be split into two parts as:

$$p(\sigma^2 + \mu^2) = (p + 2pq + 3pq^2 + 4pq^3 + \ldots) + (pq + 2pq^2 + 3pq^3 + \ldots)$$
$$= (p + 2pq + 3pq^2 + 4pq^3 + \ldots) + q(p + 2pq + 3pq^2 + \ldots)$$

Each of the series in brackets is equal to the mean, thus:

$$p(\sigma^2 + \mu^2) - \mu + q\mu$$

But q is $(1 - p)$ and p is $\frac{1}{\mu}$.

$$\frac{1}{\mu}(\sigma^2 + \mu^2) = \mu + \left(1 - \frac{1}{\mu}\right)\mu$$

$$\sigma^2 + \mu^2 = \mu(2\mu - 1)$$

$$\sigma^2 = 2\mu^2 - \mu - \mu^2$$

Hence the result established in Chapter 6, *Probability distribution*, is seen.

$$\sigma^2 = \mu^2 - \mu = \mu(\mu - 1)$$

APPENDIX II

Least squares regression line

A linear model of the form:

$\hat{y}_i = a + bx_i$

is proposed as a fit for data (x_i, y_i), $i = 1, \ldots, n$. The model sought is one where the total of the squared deviations of the fit, $\hat{y}_i$, from the data value, y_i, is a minimum. This requires that the sum of squares, S, of the residuals:

$$S = \sum_{i=1}^{n} \left\{ \left(\hat{y}_i - y_i \right)^2 \right\}$$

is a minimum.

Replace $\hat{y}_i$ *by* $a + bx_i$ and this gives:

$$S = \sum \left\{ \left(a + bx_i - y_i \right)^2 \right\}$$

The intention is to find the values of a and b which make S a minimum. So it will be necessary to treat S as a function of a and as a function of b.

First: Expand S as a function of a:

$$S = \sum \left\{ \left(a^2 + 2\left(bx_i - y_i\right)a + \left(bx_i - y_i\right)^2 \right) \right\}$$

$$= \sum a^2 + 2a \sum \left(bx_i - y_i\right) + \sum \left(bx_i - y_i\right)^2$$

$$= na^2 + 2n\left(b\bar{x} - \bar{y}\right)a + \sum \left(bx_i - y_i\right)^2$$

Thus S is essentially a quadratic function of the parameter a. It is like comparing this to a general quadratic phynomial:

$$Q(z) = \alpha z^2 + \beta z + \delta$$

which has its minimum value at $z = -\dfrac{\beta}{2\alpha}$. So S will have a minimum when:

$$a = -\frac{2n\left(b\bar{x} - \bar{y}\right)}{2n}$$

i.e. $a = -\left(b\bar{x} - \bar{y}\right)$

This can be rearranged to give:

$$\bar{y} = a + b\bar{x}$$

which says that S is a minimum when the straight line model passes through the mean point $\left(\bar{x}, \bar{y}\right)$.

Second: Expand S as a function of b:

$$S = \sum \left\{ b^2 x_i + 2(a - y_i) b x_i + (a - y_i) \right\}$$

$$= b^2 \sum x_i^{\,2} + 2b \sum x_i (a - y_i) + \sum (a - y_i)^2$$

Thus S is essentially a quadratic function of the parameter b. This means that its minimum occurs when:

$$b = -\frac{2\sum x_i (a - y_i)}{2\sum x_i^{\,2}}$$

Cross multiply to get:

$$b \sum x_i^{\,2} = -\sum x_i (a - y_i)$$

$$= \sum x_i y_i - \sum a x_i$$

But $a = \bar{y} - b\bar{x}$, hence:

$$b \sum x_i^{\,2} = \sum x_i y_i - (\bar{y} - b\bar{x}) \sum x_i$$

It helps to divide by n to yield:

$$b \frac{\sum x_i^{\,2}}{n} = \frac{\sum x_i y_i}{n} - (\bar{y} - b\bar{x})\bar{x}$$

Then, collecting the terms in b gives:

$$b \left\{ \frac{\sum x_i^{\,2}}{n} - \bar{x}^2 \right\} = \frac{\sum x_i y_i}{n} - \bar{x}\,\bar{y}$$

which is the same as saying:

$$b s_x^2 = S_{xy}$$

as required.